Environmental and Hydrological Systems Modelling

Environmental and Hydrological Systems Modelling

A. W. Jayawardena

CRC Press
Taylor & Francis Group
Boca Raton   London   New York

CRC Press is an imprint of the
Taylor & Francis Group, an **informa** business

CRC Press
Taylor & Francis Group
6000 Broken Sound Parkway NW, Suite 300
Boca Raton, FL 33487-2742

© 2014 by Taylor & Francis Group, LLC
CRC Press is an imprint of Taylor & Francis Group, an Informa business

No claim to original U.S. Government works

Printed on acid-free paper
Version Date: 20131126

International Standard Book Number-13: 978-0-415-46532-8 (Paperback)

Visit the Taylor & Francis Web site at
http://www.taylorandfrancis.com

and the CRC Press Web site at
http://www.crcpress.com

Contents

Preface

Systems science, or systems theory, perhaps has its origin in the general systems theory postulated by biologist Ludwig von Bertalanffy in the 1930s as a modelling tool that accommodates the interrelationships and overlap between separate disciplines. In the early 20th century, systems modelling started with two approaches, an *a priori* approach in which some assumptions are made followed by proposing some mathematical equations that when solved lead to some deductions, and an *a posteriori* approach in which mathematical equations that potentially explain the phenomena are fitted to observations. The advantage of the systems theory approach in problem solving is that it does not require *a priori* understanding of the detailed mechanisms of the underlying processes that govern the system.

Systems science is an interdisciplinary field of studies that covers a broad range of areas, including nature, society, science, engineering, and medicine. It covers well-defined fields such as cybernetics, dynamical systems and chaos, control theory, operational research, ecology, and many others. In view of the diversity of fields covered, systems theory has developed from different fields and its applications extend from microscopic scale to very large scale.

Most environmental and hydrological problems are quite complex, involve the interaction of many variables, and are often little understood, at least at the present time. The databases available for model development and calibration are also of a coarse resolution. There are also unresolved issues of scale effect. Therefore, it is difficult, if not impossible, to model such systems starting from first principles. Systems approach offers an alternative. The earliest application of systems theory may have been the introduction of the Dirac delta function and the unit step function as input functions for linear systems. They enable the determination of responses to arbitrary input functions via the principles of superposition and proportionality. Linear approaches have stood the test of time owing to their simplicity, generality, and extrapolation capability. However, they are not close to reality. More recently, researchers, in addition to attempts made to work from first principles, are moving into data-driven, non-linear approaches, which of course also have limitations and lack generality but can be made closer to reality. Approaches such as artificial neural networks, support vector machines, fuzzy logic, fuzzy neural systems, genetic algorithms, and genetic programming are becoming popular particularly when other approaches are difficult to implement or infeasible. Many such approaches that require repeated calculations of simple recursive equations could not have been possible some years ago. With the advent of high-speed computers, such calculations have now become feasible. This book attempts to bridge the gap between the linear assumption and reality. It is targeted towards a readership of researchers, graduate students, and professional with general interest in the hydro-environment and specialist interest in data-driven techniques of environmental and hydrological systems modelling.

The contents of the book are built on graduate courses given by the author in the Department of Civil Engineering of the University of Hong Kong and the International Centre for Water Hazard and Risk Management (ICHARM) under the auspices of UNESCO, Public Works Research Institute, Tsukuba, Japan, and the related research work carried out by the author and his co-workers and graduate students.

The author would like to record his appreciation to his co-workers at different times, in particular to Pengcheng Xu of the Chinese Academy of Sciences, Beijing, China; Bellie Sivakumar, now of the University of New South Wales, Australia; I.W. Johnson, formerly of the Institute of Hydrology, Wallingford, UK; Dulakshi S.K. Karunasingha of the Faculty of Engineering, University of Peradeniya, Sri Lanka; and N. Muttil of Victoria University, Australia. The author would also like to pay tribute to his graduate students at different times, both at the Department of Civil Engineering of the University of Hong Kong and at the International Centre for Water Hazard and Risk Management (ICHARM). In particular, the contributions by P.H. Lui, Lai Feizhou, Achela D.K. Fernando, T.M.K.G Fernando, A.B. Gurung, Zhou Maichun, S.P.P. Mahanama, Tian Ying, Somchit Amnatsan, Robin Kumar Biswas, Manish Maharjan, A.K.M. Saifudeen, Prem Raj Ghimire, J.D. Amarasekara, and Zhu Bing are highly appreciated. Sincere appreciation also goes to K. Takeuchi, Director of the International Centre for Water Hazard and Risk Management, for providing an opportunity to be academically active after leaving the University of Hong Kong.

The author also benefitted from the published work of others. Appropriate acknowledgements have been made in citing such work in relevant parts of the text. Last but not least, the author expresses his gratitude to Tony Moore, senior editor of the publisher Taylor & Francis, who initially proposed the idea, and gave continuous encouragement. His patience is gratefully appreciated. Thanks are also due to two Taylor & Francis staff members: Simon Bates, who looked after the project for some time, and Stephanie Morkert, who did the production work from the author's manuscript, and Amor Nanas from Manila Typesetting Company, who skillfully did the copy editing.

The book contains some 1072 equations, some simple and some not so simple. Despite the care taken to ensure the correctness of these equations and other material presented, it is still possible that there may be typographical errors and/or omissions due to oversight. The author would be grateful if the readers would kindly bring to his attention if they find any such errors and/or omissions. After all, to err is human and to forgive is divine.

MATLAB® is a registered trademark of The MathWorks, Inc. For product information, please contact:

The MathWorks, Inc.
3 Apple Hill Drive
Natick, MA 01760-2098 USA
Tel: 508-647-7000
Fax: 508-647-7001
E-mail: info@mathworks.com
Web: www.mathworks.com

Author

A.W. Jayawardena obtained his undergraduate degree BSc (Eng) Hons from the University of Ceylon (now Sri Lanka) and postgraduate degrees MEng from the University of Tokyo, MS from the University of California at Berkeley, and PhD from the University of London. He is a Chartered Engineer, a Fellow of the UK Institution of Civil Engineers, a Fellow of the Hong Kong Institution of Engineers, and a Life Member of the American Society of Civil Engineers. His academic career includes many years of teaching in the Department of Civil Engineering of the University of Hong Kong; he was also a Research and Training Advisor to the International Centre for Water Hazard and Risk Management (ICHARM) under the auspices of UNESCO and hosted by the Public Works Research Institute of Japan, a Professor at the National Graduate Institute for Policy Studies, Japan, and an Honorary Professor in the Department of Statistics and Actuarial Sciences of the University of Hong Kong. He is currently an Adjunct Professor in the Department of Civil Engineering of the University of Hong Kong, Technical Advisor to the Research and Development Centre, Nippon Koei Co. Ltd. (Consulting Engineers), Japan, and a Guest Professor of Beijing Normal University, China. He has been a specialist consultant for UNESCO and for several engineering consulting companies in Hong Kong, including as an expert witness and provider of expert opinion in several legal cases in Hong Kong.

Chapter 1

Introduction

Systems science is an interdisciplinary field of studies that covers a broad range of areas that include nature, society, science, engineering, and medicine. It covers well-defined fields such as complex systems, cybernetics, dynamical systems and chaos, control theory, operational research, ecology, and many others. In view of the diversity of fields covered, systems theory has developed from different fields and its applications extend from microscopic scale to very large scale.

In reality, however, it is difficult, if not impossible, to compartmentalize problems in nature into any of the above tightly bounded disciplines. They invariably involve interactions across disciplines. Therefore, a holistic approach is necessary to understand and solve real-life problems. Systems theory provides an approach that can be considered as a field of inquiry rather than looking from a specific discipline.

The advantage of the systems theory approach in problem solving is that it does not require *a priori* understanding of the detailed mechanisms of the underlying processes in the system. What it needs is a set of inputs and a set of corresponding outputs from which the system parameters (or functions) are estimated. The relationship between the input variables and the corresponding output variables may be assumed to be linear, in which case the analysis and subsequent generalization is easy, piece-wise linear, or completely non-linear.

1.1 SOME DEFINITIONS

1.1.1 System

Several definitions of a system have come at different times. For example,

- A system is a group of interacting components that conserves some identifiable set of relations with the sum of the components plus their relations (i.e., the system itself) conserving some identifiable set of relations to other entities (including other systems).
- A system is a complex of interacting components together with the relationships among them that permit the identification of a boundary-maintaining entity or process.
- A system is a combination of interacting elements that performs a function not possible with any of the individual elements. The elements can include hardware, software, bioware, facilities, policies, and processes.
- A system is a set of social, biological, technological, or material partners cooperating on a common purpose.

A system accepts inputs, over which it has no direct control, and transforms them into outputs. A system should have a well-defined boundary. Fitting into this single definition are many types of systems, some with states and some without.

Systems can be categorized as memoryless or dynamic. In a memoryless system, the outputs depend only on the present values of its inputs. In a dynamic system, the outputs depend on the present and past values of its inputs. For dynamic systems, the concept of a *state* must be defined.

1.1.2 State of a system

The state of a system makes the system's history irrelevant. The state of the system contains all the information needed to calculate responses to present and future inputs without reference to the history of inputs and outputs. The state of the system, the present inputs, and the sequence of future inputs allow computation of all future states (and outputs).

Some dynamic systems are modelled best with state equations, while others are modelled best with state machines.[1] State-equation systems are modelled with equations. For example, a projectile's motion can be modelled with state equations for position and velocity, which are functions of time. State-machine systems focus less on physical variables and more on logical attributes. Therefore, such systems have memory and are modelled with finite state machines.[2] Most computer systems are modelled with finite state machines.

At each instant of time, a dynamic system is in a specific state. State-equation systems can have one or many state variables. At any time, the system's state is defined as the unique values for each of the state variables. State-machine systems can be modelled with one or many concurrent state machines. At any time, each of the concurrent state machines must be in one and only one state. A state is a unique snapshot that is specified by values for a set of variables, characterizes the system for a period of time, and is different from other states. Each state is different from other states in either the inputs it responds to, the outputs it produces, or the transitions it take. A transition is a response to an input that may cause a change of state.

A *closed system* is one where interactions occur only among the system components and not with the external environment. An *open system* is one that receives input from the external environment and/or releases output to the environment. The basic characteristic of an open system is the dynamic interaction of its components, while the basis of a cybernetic model is the feedback cycle. Open systems can tend towards higher levels of organization (negative entropy), while closed systems can only maintain increasing entropy.

Systems modelling in the early 20th century started using two basic approaches: First, used mainly by mathematicians, is the *a priori* approach in which some assumptions are made, a set of mathematical equations are proposed, some deductions are made from the solutions to the mathematical equations, and finally such results are tested against observations. This

---

[1] A state machine is a device that stores the status of something at a given time that changes its status upon receiving inputs. For example, a computer is a state machine. Each machine instruction is input that changes one or more states and may cause other actions to take place. A simple example of a state machine is a turnstile. It has three arms and it can be in a locked or an unlocked state. When a coin or token is inserted, the turnstile is unlocked and the arm can be moved one-third of a rotation. It will not rotate more than one-third to prevent more than one person passing through. After a one-third rotation, the arms are locked again until a second coin or token is inserted. It has two inputs, inserting a coin or token and rotating the arm. In the locked state, the arms do not rotate. In the unlocked state, inserting additional coins has no effect (does not allow the arm to rotate). When a person passes through the turnstile, the machine again reverts back to the locked position. Other examples of simple state machines include vending machines, elevators, traffic lights, etc.

[2] Finite state machines are controllers of machines very common in daily life. For example, a traffic light signal that has three states, red, amber, and green, for vehicles and two states for pedestrians, red and green.

approach seems to work in situations where there are well-defined patterns in the behaviour of the system variables. The second approach, mainly used by statisticians, is the *a posteriori* approach, which begins with observations, fitting mathematical equations to the observations, and attempting to explain the underlying phenomena. Some degree of randomness is inherent in phenomena fitting into this category.

1.2 GENERAL SYSTEMS THEORY (GST)

The general systems theory (GST) was originally proposed by the biologist Ludwig von Bertalanffy in the 1930s as a modelling tool that accommodates the interrelationships and overlap between separate disciplines. His idea of the concept of systems is exemplified in the following passage:

> The 19th century and the first half of the 20th century conceived of the *world as chaos*. Chaos was the oft-quoted blind play of atoms, which, in mechanistic and positivistic philosophy, appeared to represent ultimate reality, with life as an accidental product of physical processes, and mind as an epi-phenomenon. It was chaos when, in the current theory of evolution, the living world appeared as a product of chance, the outcome of random mutations and survival in the mill of natural selection. In the same sense, human personality, in the theories of behaviourism as well as of psychoanalysis, was considered a chance product of nature and nurture, of a mixture of genes and an accidental sequence of events from early childhood to maturity.
>
> Now we are looking for another basic outlook on the world – *the world as organization*. Such a conception – if it can be substantiated – would indeed change the basic categories upon which scientific thought rests, and profoundly influence practical attitudes. This trend is marked by the emergence of a bundle of new disciplines such as cybernetics, information theory, general system theory, theories of games, of decisions, of queuing and others; in practical applications, systems analysis, systems engineering, operations research, etc. They are different in basic assumptions, mathematical techniques and aims, and they are often unsatisfactory and sometimes contradictory. They agree, however, in being concerned, in one way or another, with "systems," "wholes" or "organizations"; and in their totality, they herald a new approach. (Lilienfeld, 1978, pp. 7–8)

As highlighted in the second paragraph above, the reality is that new disciplines begin to emerge as the knowledge base expands. When scientists and philosophers first tried to explain how things worked in the universe, there were no separate disciplines. There were questions to be answered and problems to be solved. With the progress in understanding of the universe, science branched out into disciplines such as chemistry, physics, biology, mathematics, etc., and the investigations of a component of a problem were confined to the boundaries of the relevant discipline. With exponential increase in the size of the knowledge base, well-established disciplines were further subdivided into specialized fields. This process is likely to continue with time.

Real-world phenomena are usually too complex to be understood by modelling all their parts and interactions. Traditionally, scientists have simplified natural complexity by viewing individual items of observation in isolation. Since the time of the French mathematician and philosopher Renè Descartes (1596–1650), scientific methods had progressed under two assumptions. One was that a system could be broken down into individual parts and analysed as independent entities, and the other was that the parts so broken down can be linearly connected together to describe the whole. The logic at the time has been that it is better to have a deeper understanding of the parts than a not-so-deep understanding of the whole. The interactions of the parts that together constituted the whole were not the main concern. Bertalanffy (1934, 1968), on the contrary, postulated that a system is characterized by the interactions of its parts and the non-linearity of their interactions. The systems approach attempts to view the whole as well as the interrelationships among the constituent parts.

Systems theory and cybernetics have often been used synonymously, although cybernetics refers to a subset of systems that involve feedback loops. Bertalanffy identified cybernetics as the theory of control mechanisms in technology. It is founded on the concepts of information and feedback, but as part of a GST. Bertalanffy reiterated that although it has wide applications, it should not be identified with 'systems theory' in general. Cybernetics arose from engineering, whereas GST arose from biology.

Systems theory is a philosophical doctrine of describing systems as abstract organizations independent of substance, type, time, and space. Systems theory has influenced many sciences, including engineering fields, management science, mathematics, political science, psychology, sociology, life sciences, and many more. In the chapters that follow, the description and analysis are confined to environmental and hydrological fields only.

1.3 ECOLOGICAL SYSTEMS (ECOSYSTEMS)

An ecosystem is a community of living and non-living things that work together. In nature, everything is connected. It can be as small as a microorganism or as large as the entire universe. The first attempt to make predictive generalization of the whole world as an ecosystem with large cyclic entities interacting with each other and with the environment at the natural level of integration was perhaps by Odum (1950) in his PhD dissertation, although the idea of ecosystem was conceived much earlier (Tansley, 1935). He sought to give a general statement of natural selection applicable to large entities in the same way as it was to small entities such as those traditionally studied in biology. His first published papers (Odum, 1960a,b) on passive analogues set out the theoretical proposition that Ohm's law from electrical engineering was analogous to the thermodynamic functioning of ecosystems. He further explored this idea by constructing and simulating an electrical circuit of the Silver Springs ecosystem in which resistances at locations were considered as analogous to the producing and consuming populations, batteries as analogous to the energy sources from the sun and organic matter from the external environment, and wires of electrical circuits as analogous to the flow of food energy to consumers. In the electric circuit, electron flow represented the material flow in the ecosystem, the charge in the capacitor represented the storage of a material, and the scale of the model was determined by adjusting the sizes of electrical components. This concept became the foundation of Odum's approach to the development of ecological systems. Passive analogues had been used since the 1930s for simulating water flows, neurons, and other systems where Ohm's law could act as a basis for modelling; however, Odum was the only ecologist to use them for simulating ecosystems (Kangas, 2004).

1.4 EQUI-FINALITY

A system can be represented by a simple single-parameter model or by a more complex multi-parameter model. A model with all the characteristics of the system lumped into a single parameter is easy to calibrate. As the number of parameters increases, the degree of difficulty in calibration also increases. In multiparameter systems, the calibration is usually done using some kind of optimization technique. A problem that arises in such situations is that the same results may be arrived at with different sets of parameters. This is attributed to the principle of equi-finality, which states that in open systems, a given state can be reached in many potential paths or trajectories. This principle was postulated by Ludwig von Bertanffy (1934, 1968) in his GST. In a closed system, a direct cause–effect relationship exists between the initial and final states of the system.

As a calibration strategy that has the ability to simultaneously incorporate several objective functions, multiobjective optimization has its roots in late 19th century welfare economics in the original work of Francis Ysidro Edgeworth (1881) and later generalized by Vilfredo Pareto (1906). The Pareto set of solutions represent trade-offs with the property that moving from one solution to another results in the improvement of one objective while causing deterioration in one or more others. The Pareto set represents the minimum uncertainty that can be achieved for the parameters via calibration.

Equi-finality for multiparameter optimization means that the same final result may be arrived at from different initial conditions and in different ways. Two models are said to be equi-final if they lead to equally acceptable results. It is a key concept to assess the uncertainty of real-world predictions. There is no unique set of parameter values, but rather a feasible parameter space from which a Pareto set of optimal solutions is sought. A Pareto set is a set of states of objective parameters satisfying the criterion of Pareto optimality for multiobjective optimization problems. A state A (a set of target parameters) is said to be Pareto optimal if there is no other state B dominating the state A with respect to a set of objective functions. A state A dominates a state B, if A is better than B in at least one objective function and not worse with respect to all other objective functions.

1.5 SCOPE AND LAYOUT

The contents of this book are arranged in 14 chapters. Chapter 2 gives a review of the historical development of hydrological modelling, outlining the concepts originated under the assumption of linearity. Time and frequency domain analysis and their applications to linear systems through standard input functions such as Dirac delta and unit step functions are described next, followed by a description of the concepts of unit hydrograph, linear reservoir, linear channel, linear cascade, and time–area diagram, all in the context of linear hydrological systems. A brief introduction to random processes and linear systems, nonlinear systems, and multilinear systems is given next. The chapter ends with a description of flood routing using the hydrologic approach as well as the hydraulic approach, reservoir routing, a review of some rainfall–runoff models, and highlighting the challenges that lie ahead in hydrological modelling.

Chapter 3 gives a description of population dynamics starting from the Malthusian exponential growth theory. Although it has been modified, refuted, or superseded by the works of subsequent researchers, it still stands as the backbone of many population dynamics models. The studies of Verhulst, Pearl and Reed; the Lotka-Volterra predator–prey model; and the many oscillations of the logistic map, including bifurcation and chaos, are briefly reviewed. The chapter ends with a description of exponential cell growth, cell division by binary fission, mitosis and meiosis, cell growth in a bioreactor, bacterial growth and binary fission, Monod kinetics, and radioactive decay and carbon dating. In summary, the chapter gives a review of the historical development of the study of population dynamics, including those of some microbiological species.

Chapter 4 describes reaction kinetics, which is the process of concentration variation in a chemical or biological reaction. The concept of half-life, the relationship between the reaction rate and the substrate concentration, the relationship between the growth and consumption rates of the biomass produced and the substrate consumed, as well as two of the widely used models for describing the kinetics of such reactions, the Michaelis Menten equation and the Monod equation, are introduced.

Chapter 5 describes water quality systems. Starting with dissolved oxygen systems, the chapter describes biochemical oxygen demand, nitrification and de-nitrification, oxygen sag curve, re-oxygenation and de-oxygenation coefficients, as well as the dynamics of a completely mixed body of water, including the governing equations and their solutions to specific input functions. The chapter ends with a description of the water quality variation in rivers and streams due to specific waste loads.

Chapter 6 describes longitudinal dispersion, an important topic in environmental engineering. Starting from Fick's law of diffusion, the chapter first describes turbulent diffusion, shear flow dispersion, Taylor's approximations, and turbulent mixing coefficients. Next, a detailed description of the longitudinal dispersion coefficient, including how it is estimated by various methods; the analytical solution of the dispersion equation for certain specific inputs, boundary, and initial conditions; and the numerical solution of the dispersion equation using the finite difference method, finite element method, and moving finite element method are given. The chapter ends with a brief description of dispersion through porous media and an introduction to some general-purpose water quality models.

Chapter 7 describes time series analysis and forecasting. It gives the basic properties of time series, homogeneity tests, decomposition of a time series, tests for identification of trends and periodicities, time and frequency domain analyses including the convolution integral and Fourier transforms, correlation and spectral analysis, representation of the dependent stochastic component by various stochastic models, generation of synthetic data using the probability distribution of the independent residuals, forecasting, and basic concepts of Kalman filtering. Some illustrative examples of time series analysis are also given.

Chapter 8 describes artificial neural networks as one of the widely used data-driven techniques in environmental and hydrological systems modelling. Starting from the biological neuron, the chapter describes the development of the artificial neuron, perceptron, multilayer perceptron, types of activation functions, types of artificial neural networks such as feed-forward, recurrent, Kohonen, product unit, different types of learning, backpropagation algorithm, back-propagation through time, data preprocessing methods, and application areas in hydrology and environmental systems.

Chapter 9 describes radial basis functions as another form of artificial neural networks. It is an approach that can be used to solve interpolation problems in multidimensional space. Different types of radial basis function networks are introduced together with hybrid learning methods in which the selection of centres is done in an unsupervised mode, whereas the optimization of the weights is done in a supervised mode. Selection of centres is a key component of radial basis function networks, and a review of both supervised and unsupervised method of centre selection is given. Finally, some example applications in the hydrological context are given.

Chapter 10 is about fractals and chaos. Fractals are geometric objects that can be subdivided into parts, each of which is a copy of the whole, a property known as self-similarity. They have fine structures at infinitely small scales and can be generated by simple and well-defined recursive equations. Examples in nature include clouds, mountains, coastlines, trees, ferns, river networks, cauliflowers, system of blood vessels, snowflakes, etc. In the chapter, basic concepts of fractals and chaos, definitions and estimation methods of fractal dimension, and invariant measures for chaotic systems, as well as some examples of well-known fractals such as Cantor set, Sierpinski (gasket) triangle, Koch curve, Koch snowflake (or Koch star), Mandelbrot set, Julia set, and the perimeter–area $(P-A)$ relationship of fractals, are given. In the context of chaos, an introduction, some basic definitions, the butterfly effect, the 'n'-body problem, and the invariants of chaotic systems such as Lyapunov exponent and various measures of entropy are described. Finally, some examples of chaotic maps, such as the logistic map, Hénon map, Lorenz map, Duffing, Rössler and Chua's equations, including some application areas, are given.

Chapter 11 is about dynamical systems approach to modelling. Topics covered include a comparison between random and chaotic deterministic systems, dynamical systems and their sensitivity to initial conditions, embedding, embedding dimension and methods of estimating embedding dimension, phase–space reconstruction and phase–space predictions, some descriptions about non-linearity and determinism, including tests for determinism, noise and methods of noise reduction, noise level estimation, and some application areas.

Chapter 12 is about support vector machines, which belong to the class of supervised learning and which can be used for optical character recognition; pattern recognition such as handwriting, speech, images, etc.; classification and regression analysis; and time series prediction. Issues such as binary classification, linear and non-linear soft margin classification, linear support vector regression, non-linear support vector regression, parameter selection, kernel tricks, and some application areas are presented.

Chapter 13 describes fuzzy logic systems. Basic concepts of fuzzy logic; types of membership functions; rule bases; fuzzy inference systems such as Mamdani, Takagi–Sugeno–Kang, Tsukamoto, and Larsen; fuzzification and de-fuzzification methods; neuro-fuzzy systems; and adaptive neuro-fuzzy systems, including some application areas, are presented in the chapter.

Chapter 14, which is the last chapter, gives a brief introduction to genetic algorithms and genetic programming. As two of the relatively recent variations of evolutionary programs, genetic algorithms and genetic programming have become popular search techniques for parameter optimization in complex non-linear spaces. The chapter gives the basic components of genetic algorithms, coding, genetic operators, and the basic differences between genetic algorithms and genetic programming. Some application areas for both types are also given.

REFERENCES

Bertalanffy, L. von (1934): Untersuchungen über die Gesetzlichkeit des Wachstums. I. Allgemeine Grundlagen der Theorie; mathematische und physiologische Gesetzlichkeiten des Wachstums bei Wassertieren. *Archives Entwicklungsmech*, 131, 613–652.

Bertalanffy, L. von (1968): *General System Theory: Essays on its Foundation and Development*, Revised Edition. George Braziller, New York.

Edgeworth, F.Y. (1881): *Mathematical Psychics*. P. Keagan, London.

Kangas, P. (2004): The role of passive electrical analogs in H.T. Odum's systems thinking. *Ecological Modelling*, 178, 101–106.

Lilienfeld, R. (1978): *The Rise of Systems Theory: An Ideological Analysis*. Wiley, New York.

Odum, H.T. (1950): The biogeochemistry of strontium with discussion on the ecological integration of elements. PhD dissertation, Yale University, New Haven. 373 pp.

Odum, H.T. (1960a): Ecological potential and analogue circuits for the ecosystem. *American Science*, 48, 1–8.

Odum, H.T. (1960b): Ten classroom sessions in ecology. *American Biology Teacher*, 22, 71–78.

Pareto, V. (1906): *Manuale di Economia Politica*. Societa Editrice Libraria, Milan, Italy. Translated into English by A.S. Schwier as *Manual of Political Economy*. Macmillan, New York, 1971.

Tansley, A.G. (1935): The use and abuse of vegetational terms and concepts. *Ecology*, 16(3), 284–307.

Chapter 2

Historical development of hydrological modelling

A system can be considered as deterministic or stochastic. A deterministic system defines a cause–effect relationship, and therefore predictions are possible if the cause is known. The cause is the driving function (input) and effect is the response function (output). A stochastic system, on the other hand, is non-deterministic and has one or more parts attributed to chance and therefore random. Unlike in a deterministic system, a stochastic system does not always produce the same output for a given input.

Deterministic systems can be considered as lumped parameter systems or as distributed parameter systems. In general, from a mathematical point of view, lumped parameter systems are represented by ordinary differential equations and distributed parameter systems are represented by partial differential equations. In a linear system, the transformation from input to output takes place via a linear operator.

2.1 BASIC CONCEPTS AND GOVERNING EQUATION OF LINEAR SYSTEMS

A linear system can be represented by an ordinary differential equation of the form

$$a_1 y(t) + a_2 y'(t) + a_3 y''(t) + \ldots = b_1 x(t) + b_2 x'(t) + b_3 x''(t) + \ldots \tag{2.1}$$

where $x(t)$ represents an input function, $y(t)$ represents an output function, a_i's and b_i's are system parameters, and the superscripts in x and y indicate the corresponding derivatives with respect to time.

All linear systems satisfy the principles of proportionality and superposition; that is, if $y_1(t)$ and $y_2(t)$ are output functions corresponding to the input functions $x_1(t)$ and $x_2(t)$, then $y_1(t) + y_2(t)$ will be the output function corresponding to the input function $x_1(t) + x_2(t)$ and $\alpha y_1(t)$ will be the output corresponding to the input function $\alpha x_1(t)$ (α is a constant). Linear systems are relatively easier to analyse as the methods of analysis are well established. Analysis can be carried out in the time domain or in the frequency domain.

2.1.1 Time domain analysis

To make use of the principles of proportionality and superposition, the input functions to any linear system can be described as a linear summation of standard input functions. Several types of standard input functions can be identified.

9

2.1.1.1 Types of input functions

a. Unit impulse (or Dirac delta) function

The unit impulse (Figure 2.1) is of the Dirac delta function (Dirac, 1958) type, which has the following properties:

$$\delta(t - a) = 0 \text{ for } t \neq a \tag{2.2a}$$

$$\int_0^\infty \delta(t - a)\,dt = 1 \tag{2.2b}$$

If it is applied at $t = 0$,

$$\delta(t) = 0 \text{ for } t \neq 0 \tag{2.3a}$$

$$\int_0^\infty \delta(t)\,dt = 1 \tag{2.3b}$$

The Dirac delta function is not a real function. It is an abstract function that only exists in mathematics. The closest it can be to a real function is when an input of very high intensity is applied for a very short duration. In an environmental systems context, it can be equivalent to a unit mass of pollutants instantaneously discharged into a water body. In the hydrological context, it is equivalent to the instantaneous unit rainfall excess. Despite its abstractness, the Dirac delta function is a useful mathematical concept that has many applications in linear systems analysis.

b. Unit step function

The unit step function, $u(t)$ (Figure 2.2), has the following properties:

$$u(t) = 0 \text{ for } t < 0 \tag{2.4a}$$

$$u(t) = 1 \text{ for } t > 0 \tag{2.4b}$$

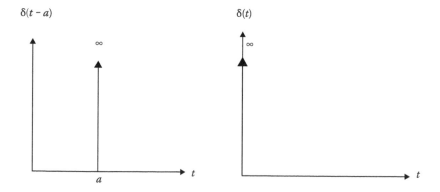

Figure 2.1 Dirac delta function.

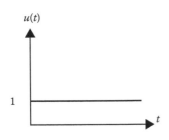
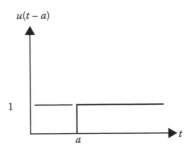

Figure 2.2 Unit step function.

If it is applied at $t = a$, then

$$u(t - a) = 0 \text{ for } t < a \qquad (2.5a)$$

$$u(t - a) = 1 \text{ for } t \geq a \qquad (2.5b)$$

The delta function and the unit step function are related to each other by

$$\int_{-\infty}^{t} \delta(\alpha - \tau)\,d\alpha = u(t - \tau) \text{ or } \frac{du(t)}{dt} = \delta(t) \qquad (2.6)$$

The case of a unit mass of pollutant discharged into a water body continuously or a continuous unit rainfall excess can be considered as unit step functions. The resulting output function corresponding to a step function in the context of the unit hydrograph is the S-curve. Two S-curves separated by a known time interval can be used to convert a unit hydrograph of one unit duration to that of another unit duration (see Section 2.2.2).

c. Arbitrary function

Any real input function can be expressed as a sum of functions of either the delta function or the step function, or both. Decomposing into delta functions, an input function $x(t)$ can be written as

$$x(t) = \int_{-\infty}^{\infty} x(\tau)\delta(t - \tau)\,d\tau \qquad (2.7)$$

Decomposing into unit step functions, an input function $x(t)$ can be written as

$$x(t) = \int_{0}^{\infty} \dot{x}(\tau)u(t - \tau)\,d\tau \qquad (2.8)$$

where the $\dot{x}(\tau)$ represents the derivative.

In discrete form for real systems, Equations 2.7 and 2.8 can be written as

$$x(j) = \sum_{k=0}^{j} x(k)\delta(j - k) \qquad (2.7a)$$

and

$$x(j) = \sum_{k=0}^{j} \dot{x}(k)u(j-k) \tag{2.8a}$$

2.1.1.2 System response function – convolution integral

The input and output functions of a linear system are related to each other by the convolution integral, which takes the form

$$y(t) = \int_{0}^{t} h(\tau)x(t-\tau)\,d\tau \tag{2.9a}$$

where $h(t)$ is the system response function. For real systems, the variables in the convolution integral are interchangeable; that is,

$$y(t) = \int_{0}^{t} x(\tau)h(t-\tau)\,d\tau \tag{2.9b}$$

The system response function $h(t)$ corresponds to the output function when the input function is of the impulse (delta) type. It is therefore called the impulse response function (IRF). A linear system is completely known if the IRF is known.

If the system is linear time variant, then the corresponding equation is of the form

$$y(t) = \int_{0}^{t} h(\tau,t)x(t-\tau)\,d\tau \tag{2.10}$$

implying $h(\tau,t)$ is a function of t.

For discrete systems, the convolution summation can be written as

$$y(j) = \sum_{k=0}^{j} x(j-k)h(k) \tag{2.11}$$

Any linear system can be analysed either in the time domain or in the frequency domain.

2.1.2 Frequency domain analysis

2.1.2.1 Fourier transform – frequency response function (FRF)

If a linear time invariant system is physically realizable and stable,[1] then it can be described by a frequency response function (FRF) in the frequency domain. The FRF $H(f)$ is the Fourier transform (after Jean Baptiste Joseph Fourier, 1768–1830) of the IRF and is a function of

---

[1] If the input function is bounded, then the output function is also bounded; that is, if $|x(t)| < N$, then $|y(t)| < M$; N, M are constants.

the frequency f in the same way as $h(t)$ is a function of t. Fourier transform enables decomposition of any wavy signal into sine and cosine waves, thus forming an alternative way of representing the behaviour of the function. This is possible because any time domain signal can be represented by a series of sines and cosines.

For a complex function, the Fourier transforms[2] are given by (Kreyszig, 1999, pp. 565–570)

$$H(f) = \sqrt{\frac{1}{2\pi}} \int_{-\infty}^{\infty} h(t)e^{-2\pi jft}\, dt \qquad (2.12a)$$

and the inverse Fourier transform by

$$h(t) = \sqrt{\frac{1}{2\pi}} \int_{-\infty}^{\infty} H(f)e^{2\pi jft}\, df \qquad (2.12b)$$

In Equation 2.12, $j^2 = -1$.

Since $e^{j\theta} = \cos\theta + j\sin\theta$ (Euler formula), Equations 2.12a and 2.12b can be expressed as

$$H(f) = \sqrt{\frac{1}{2\pi}} \int_{-\infty}^{\infty} h(t)e^{-2\pi jft}\, dt = \sqrt{\frac{1}{2\pi}} \int_{-\infty}^{\infty} h(t)\{\cos(2\pi ft) - j\sin(2\pi ft)\}\, dt \qquad (2.13a)$$

$$h(t) = \sqrt{\frac{1}{2\pi}} \int_{-\infty}^{\infty} H(f)e^{2\pi jft}\, df = \sqrt{\frac{1}{2\pi}} \int_{-\infty}^{\infty} H(f)\{\cos(2\pi ft) + j\sin(2\pi ft)\}\, df \qquad (2.13b)$$

In these equations, $H(f)$ represents the amplitude of the real and imaginary parts of the sinusoids at each frequency.

For an even function, the Fourier cosine transformation is given by

$$H(f) = \sqrt{\frac{2}{\pi}} \int_{0}^{\infty} h(t)\cos(2\pi ft)\, dt \qquad (2.14a)$$

and the inverse Fourier cosine transformation by

$$h(t) = \sqrt{\frac{2}{\pi}} \int_{0}^{\infty} H(f)\cos(2\pi ft)\, df \qquad (2.14b)$$

---

[2] The general form of a Fourier transform pair in terms of the angular frequency ω ($\omega = 2\pi f$) takes the form $H(\omega) = \int_{-\infty}^{\infty} h(t)e^{-j\omega t}\, dt$; $h(t) = \frac{1}{2\pi}\int_{-\infty}^{\infty} H(\omega)e^{j\omega t}\, d\omega$, which in discrete form can be written as $H(k) = \sum_{i=0}^{N-1} x(i)e^{-\frac{2\pi ijk}{N}}$, $x(i) = \frac{1}{N}\sum_{i=0}^{N-1} H(k)e^{\frac{2\pi ijk}{N}}$. Different forms of the constant normalizing term can be found in the literature.

For an odd function, the Fourier sine transformation is given by

$$H(f) = \sqrt{\frac{2}{\pi}} \int_0^\infty h(t) \sin(2\pi ft) \, dt \qquad (2.15a)$$

and the inverse Fourier sine transformation by

$$h(t) = \sqrt{\frac{2}{\pi}} \int_0^\infty H(f) \sin(2\pi ft) \, df \qquad (2.15b)$$

Similarly, the Fourier transforms of $x(t)$ and $y(t)$ can, respectively, be expressed as

$$X(f) = \int_{-\infty}^\infty x(t)\{\cos(2\pi ft) - j \sin(2\pi ft)\} \, dt \qquad (2.16a)$$

$$Y(f) = \int_{-\infty}^\infty y(t)\{\cos(2\pi ft) - j \sin(2\pi ft)\} \, dt \qquad (2.16b)$$

Through Fourier transformation, the convolution integral that relates the input to the output in the time domain can be represented by an algebraic equation in the frequency domain as follows:

$$Y(f) = X(f)H(f) \qquad (2.17)$$

(All real parts are even functions, whereas the imaginary parts are odd functions.)

2.1.2.2 Laplace transform

The Laplace (after Pierre Simon Marqueis De Laplace, 1749–1827) transforms convert the problem of solving differential equations to that of solving algebraic equations. The Laplace transforms of many functions are given in the form of tables in many mathematics books (e.g., Kreyszig, 1999, pp. 296–299). It is one of most important integral transforms used in engineering. The Laplace transform $H(s)$ of $h(t)$ is given by (Kreyszig, 1999, pp. 251)

$$H(s) = \int_0^\infty h(t)e^{-st} \, dt \qquad (2.18)$$

where s is the variable of the Laplace transform, which has the dimension $[T]^{-1}$. $H(s)$ is dimensionless. Similar to Fourier transform, the convolution integral transforms to an algebraic equation by Laplace transformation as well

$$Y(s) = X(s)H(s) \qquad (2.19)$$

where $Y(s)$ and $X(s)$ are the Laplace transforms of $y(t)$ and $x(t)$.

a. Properties of $H(s)$

Because $h(t)$ is a bounded function of exponential order, $H(s)$ is finite for all positive values of s. Since $h(t)$ and e^{-st} are both positive functions in the range of integration, $H(s)$ is non-negative. Since $h(t)$ is of exponential order, $H(s)$ must approach zero as s increases.

$$\text{For } s = 0, H(0) = \int_0^\infty h(t)\,dt = 1 \tag{2.20a}$$

$$\text{For } s = 1, \frac{dH(s)}{ds} = -t_L \tag{2.20b}$$

$$[\text{because } \int_0^\infty h(t)t\,dt = t_L \quad (\text{lag time})]$$

$$\text{Also, } \frac{dH(s)}{ds} = \left(\frac{dH(s)}{dt}\right) \text{ is negative.}$$

The problem with Laplace transforms is that it is not easy to find the inverse transform to obtain $h(t)$. Usually $H(s)$ is obtained in a discrete form, and an empirical equation is fitted to the discrete form before inverse transformation is carried out.

2.1.2.3 z-Transform

Originated as a discrete-time counterpart of Laplace transform, the concept of converting a discrete time domain signal to a complex frequency domain representation was introduced as z-transform by Ragazzini and Zadeh in 1952. The one-sided z-transform is given as

$$H(z) = \sum_{i=0}^\infty x_i z^{-i} \tag{2.21a}$$

and its inverse as

$$x_i = \frac{1}{2\pi j} \oint_{\text{unit circle}} H(z) z^{i-1}\,dz \tag{2.21b}$$

where z is a complex variable. The two-sided z-transform is given as

$$H(z) = \sum_{i=-\infty}^\infty x_i z^{-i} \tag{2.22}$$

Similar to Laplace transform,

$$Y(z) = H(z)X(z) \tag{2.23}$$

and $H(z)$ is called the pulsed transfer function.

2.2 LINEAR SYSTEMS IN HYDROLOGICAL MODELLING

2.2.1 Hydrological systems

Within the broad definitions given at the beginning, a hydrological system is a system concerned with water and its interconnections. It can be as large as the entire hydrological cycle, or as small as the trapping of methane in a water molecule to produce methane hydrate. The mechanism of producing an output from a hydrological system depends on the nature of the inputs it receives, the nature of the physical laws governing the system behaviour, and the nature of the system itself. In the systems approach, the complex details governed by physical laws are either simplified or not considered at all, but instead a relationship between the inputs and corresponding outputs is sought.

The hydrological cycle in a wider sense is a closed system as there is no net transfer of water from the earth. However, hydrologists are concerned about what happens within the confines of a certain spatial domain, such as, for example, a catchment. In this situation, the hydrological system is an open system because there is transfer of water across the boundaries of the domain. There are many processes that contribute to the overall hydrological cycle and each of such processes can be considered as a system in its own right or as components of a much larger system. However, the main concern of hydrologists is the catchment scale in which a relationship between the input rainfall and the corresponding output runoff is sought. Some of the important processes and how they are modelled is first briefly described next.

The principal processes in the hydrological cycle are precipitation, evaporation and evapotranspiration, interception, infiltration, runoff as overland flow, interflow and/or baseflow, and subsurface flow under saturated conditions or as moisture flow under partially saturated conditions.

At any instant, the atmosphere contains about 13×10^3 km^3 (Table 10.4, World Water Development Report 3 [2009]) of water as vapour, liquid, or solid. Precipitation is the process by which this water is deposited on the earth's surface. It can take place in one of several forms, such as rain, snow, mist, hail, sleet, dew, and frost, depending on the prevailing environmental conditions. The effect of liquid precipitation on the hydrological cycle is immediate, whereas that of solid precipitation is slow and attenuated. Evaporation, which is the process of converting liquid water to gaseous water, or water vapour, returns the water deposited on the earth's surface back to the atmosphere. These two are the processes of exchange of water between the earth's surface and the atmosphere and forms a bridge between hydrology and meteorology. Evaporation is an energy-absorbing process, whereas condensation, which is the process of changing phase from a gaseous phase back to liquid phase, is energy releasing. Many of the extreme events taking place in the atmosphere are fuelled by the energy from the sun and the latent heat of condensation released when water vapour becomes liquid water.

The next important process in the hydrological cycle is infiltration, which transfers the water on the surface of the earth to the subsurface. The infiltrated water reappears as base flow, percolate into deeper layers of the subsoil and form groundwater, or increase the water content stored in the soil.

Quantitative information about these three processes is usually obtained by measurement, and as such, they become input data to any hydrological modelling system. They can, however, be empirically estimated using measurements of the predominant factors that affect them. Predictions based on the laws of physics are too complicated, and it is doubtful whether such laws that are defined for an infinitesimally small spatial domain can be applied to a larger domain typically of interest to hydrologists. The scale issue is yet unresolved.

The most important catchment process of interest to hydrologists is the transformation of rainfall to runoff. Rainfall, which is measurable to a high degree of accuracy, becomes the input and the corresponding runoff, which can also be measured but not so easily and not with the same degree of accuracy and reliability, becomes the corresponding output. The problem, therefore, is to predict runoff as the output from a hydrological model that takes in rainfall as the input. This is the focus of the hydrological systems modelling part of this book.

2.2.2 Unit hydrograph

The unit hydrograph method is one of the basic tools in hydrological computations. It was originally presented by L.K. Sherman in 1932. It is founded upon the assumption that the rainfall excess–direct runoff process is linear, which implies that the principles of superposition and proportionality hold. It is defined as the discharge hydrograph resulting from a unit depth of rainfall excess lasting for a specified period called the unit duration and assumed to be uniformly distributed over the catchment. Rainfall excess is that portion of the rainfall that comes out as direct runoff. Direct runoff is the difference between the total runoff and base flow. It is that component directly resulting from rainfall.

Under the principles of the unit hydrograph, the hydrograph corresponding to any effective rainfall excess can be obtained via the convolution integral. The limitation for the use of the convolution integral is that the unit hydrograph duration should be the same as the unit interval of the rainfall excess hyetograph. Otherwise, methods of converting a unit hydrograph of one unit duration to that of a different unit duration should be followed. If the required unit duration is an integer multiple of the known unit duration, superposition of a number of unit hydrographs of the given unit duration, each displaced by the given unit duration, and dividing the cumulative hydrograph by the number of unit hydrographs superposed will give the unit hydrograph of the required unit duration. For example, a 4-h unit hydrograph can be obtained by the superposition of two unit hydrographs each of 2-h unit duration and dividing by two, or by the superposition of four unit hydrographs each of 1-h unit duration and dividing by four, etc. If the required unit duration is not an integer multiple of the known unit duration, the S-curve method, which is obtained by summing up an infinite number of unit hydrographs of the known unit duration, each displaced by the given unit duration, shifting it by the time interval of the required unit duration, taking the difference between the two S-curves so obtained and diving by the ratio of the two unit durations, should be used. The summation of a large number of unit hydrographs is equivalent to the hydrograph resulting from a continuous rainfall excess of intensity equal to the reciprocal of the given unit duration. For example, if the given unit duration is t_0 hours, the summation is equivalent to a continuous rainfall excess of intensity $\frac{1}{t_0}$. The resulting hydrograph obtained by subtracting one S-curve from another is not a unit hydrograph and therefore should be divided by $\frac{t_0'}{t_0}$, where t_0' is the required unit duration. A property of the S-curve is that it will level off after a finite number of superpositions.

The principles of proportionality and superposition lead to very useful practical applications of the unit hydrograph concept. For example, a hydrograph of discharge resulting from a series of rainfall excesses may be constructed by summing up the hydrographs due to each single unit of rainfall excess. They also imply that the time base of direct runoff hydrographs resulting from rainfall excesses of same unit duration is the same regardless of the intensity.

The unit duration corresponds to the rate at which the catchment is filled up to a rainfall excess of unit depth. Application of a given amount of rainfall excess at different rates gives different responses. For example, the consequence of a 100-mm rain falling in a duration of 1 h is quite different from the same amount of rain falling in 1 day. The rate of application of the rainfall excess therefore plays a significant role in the runoff generation characteristic of the catchment.

The catchment can be thought of as a tank or reservoir that receives the rainfall excess and releases it through an outlet. The catchment characteristics are reflected in the discharge coefficient of the tank or reservoir outlet, which affects the recession part of the hydrograph. The rising part is affected by the rate of application of the unit rainfall excess.

2.2.2.1 Unit hydrograph for a complex storm

Using the convolution method, which is a linear superposition involving multiplication, translation in time, and addition, the discharge corresponding to a multiple unit rainfall excess can be obtained. The convolution integral is given as follows:

$$y_t = \int_0^t x_\tau u_{t-\tau} \, d\tau = \int_0^t x_{t-\tau} u_\tau \, d\tau \tag{2.24}$$

In discrete form, it can be written as

$$y_t = \sum_{\tau=0}^t x_\tau u_{t-\tau} \tag{2.25}$$

For digital computations, the following forms are convenient:

$$y_t = \sum_{j=1}^t x_j u_{t-j+1} \text{ if } t \le r \tag{2.26a}$$

and

$$= \sum_{j=1}^r x_j u_{t-j+1} \text{ if } r \le t \le m \tag{2.26b}$$

because $u_t = 0$ for $t \le 0$ and $t \ge m$. Expanding Equations 2.26a and 2.26b, the following set of equations can be obtained:

$$
\left.
\begin{aligned}
y_1 &= u_1 x_1 \\
y_2 &= u_2 x_1 + u_1 x_2 \\
y_3 &= u_3 x_1 + u_2 x_2 + u_1 x_3 \\
y_4 &= u_4 x_1 + u_3 x_2 + u_2 x_3 + u_1 x_4 \\
&\cdots \\
y_r &= u_r x_1 + u_{r-1} x_2 + u_{r-2} x_3 + \ldots u_2 x_{r-1} + u_1 x_r \\
y_{r+1} &= u_{r+1} x_1 + u_r x_2 + u_{r-1} x_3 + \ldots u_3 x_{r-1} + u_2 x_r \\
&\cdots \\
&\cdots \\
y_{r+m-2} &= 0 + 0 + \ldots + x_r u_{m-1} + x_{r-1} u_m \\
y_{r+m-1} &= 0 + 0 + \ldots + x_r u_m
\end{aligned}
\right\} \tag{2.27}
$$

In Equation 2.27, x_1, x_2,..., x_r are the rainfall excess values, u_1, u_2,...,u_m are the unit hydrograph ordinates, y_1, y_2,...,y_n are the direct runoff hydrograph ordinates, and $n = r + m - 1$. With this system of equations, it is relatively easy to determine the direct runoff hydrograph if the input rainfalls and the unit hydrograph ordinates are known by direct substitution. In practice, however, the main issue is deriving the unit hydrograph ordinates from a given set of rainfall excess and corresponding direct runoff. The following procedure can be adopted for this purpose.

Equation 2.27, in matrix form, can be written as

$$
\begin{bmatrix}
x_1 0 \\
x_2 x_1 0 \\
x_3 x_2 x_1 0 \\
\cdots \\
\cdots \\
x_r x_{r-1} x_{r-2} \\
0 \quad x_r x_{r-1} \quad x_2 x_1 0 \\
0 \quad 0 \quad x_r \\
\qquad 0 \qquad\qquad x_r x_r \ 1 \\
\qquad\qquad x_r
\end{bmatrix}
\begin{bmatrix}
u_1 \\
u_2 \\
u_3 \\
\cdot \\
\cdot \\
\cdot \\
\cdot \\
u_m
\end{bmatrix}
=
\begin{bmatrix}
y_1 \\
y_2 \\
y_3 \\
\\
y_r \\
y_{r+1} \\
\cdot \\
\cdot \\
y_n
\end{bmatrix}
\tag{2.28}
$$

which is of the form

$$[x][u] = [y] \tag{2.28a}$$

where
- $[x]$ is a matrix of dimension $(m + r - 1) \times m$
- $[u]$ is a vector of dimension m
- $[y]$ is a vector of dimension $m + r - 1$

The solution of Equation 2.28 is not straightforward because $[x]$ is not a square matrix. There are more equations than the number of unknowns in the system. To get the optimal solution that will minimize the error as estimated by $\sum_{t=1}^{n} \left(y(t) - y_{expected}(t)\right)^2$ (least squares method), the following matrix manipulation is performed:

Let $[z] = [x]^T[x]$

It would be a square matrix of dimension $m \times m$ because, $[x]^T$ is of dimension $m \times (m + r - 1)$ and $[x]$ is of dimension $(m + r - 1) \times m$, and can therefore be inverted.

Then,

$$[z][u] = [x]^T[y] \tag{2.29}$$

from which

$$[u] = [z]^{-1}[x]^T[y] \tag{2.30}$$

Equation 2.30 does not lead to an exact solution to the problem but gives a unique solution. Solutions could also be obtained in a sequential manner as follows:

First row of Equation 2.27 gives $u_1 = \dfrac{y_1}{x_1}$; substituting for u_1 in the second row of Equation 2.27 gives u_2; substituting for u_1 and u_2 in the third row of Equation 2.27 gives u_3; and so on.

Therefore, all the equations can be solved. This method, however, does not lead to accurate solutions because the errors become accumulated with each equation solved, and the solutions will be different depending on whether the procedure is started from the beginning or from the end. There are more equations than unknowns. A unit hydrograph can be made dimensionless by dividing the discharge by a reference discharge and the time by a reference time.

A variety of unit hydrographs and dimensionless unit hydrographs can be found in the literature: the general dimensionless unit hydrograph based on the cascade of linear reservoirs and dependent only on the number of linear reservoirs' and the Courant number, which in the context of unit hydrograph is equal to the ratio of the duration of the unit hydrograph and the storage coefficient of the reservoir (Ponce, 1989a,b); a geomorphological unit hydrograph in which Horton's stream laws have been used to integrate the delay effects of streams and expresses the unit hydrograph as a probability density function of travel time to the catchment outlet (Rodríguez-Iturbe and Valdés, 1979); a unit hydrograph formulated by combining a time–area diagram with a linear reservoir to take into account both translation and storage effects (Clark, 1945); a spatially distributed unit hydrograph that is similar in concept to the geomorphological unit hydrograph but uses GIS to describe the connectivity of the links in the streamflow network, thereby eliminating the need to use a probability density function (Maidment, 1993); a synthetic unit hydrograph defined by the time to peak, time base, peak discharge, and the widths of the hydrograph at 50% and 75% of the peak discharge, which are all related to catchment characteristics (Snyder, 1938); and a dimensionless unit hydrograph obtained from a large number of unit hydrographs from a number of catchments of different sizes and locations in which the dimensionless discharge is expressed as the ratio of discharge to peak discharge and dimensionless time is expressed as the ratio of time to the time to peak of the hydrograph (US Soil Conservation Service, 1972, 1985).

2.2.2.2 Instantaneous unit hydrograph (IUH)

The instantaneous unit hydrograph (IUH) $u(0,t)$ for a catchment is the hydrograph of direct runoff resulting from a finite volume (or depth) of rainfall excess falling in an infinitesimally short time. In terms of the delta function, such a rainfall excess may be expressed as $V_0\delta(0)$, where V_0 is the volume of rainfall excess. A t_0 hour unit hydrograph $u(t_0, t)$ is the direct runoff hydrograph resulting from a finite volume (or depth) of rainfall excess falling in t_0 hours. The conversion of the IUH to one of finite duration can be done using superposition.

2.2.2.3 Empirical unit hydrograph

The actual unit hydrograph derived from a particular storm, also referred to as the empirical unit hydrograph, can be obtained by selecting a storm that is isolated, intense, and uniform over the catchment and time by scanning through rainfall and runoff records with the duration of the storm not greater than the period of rise (time of concentration), separating the base flow and determining the direct runoff hydrograph, finding the depth (or volume) of runoff as the area under the hydrograph, normalizing the direct runoff hydrograph by dividing the ordinates by the depth (or volume) of runoff, and subject to the requirement that the area under the unit hydrograph is unity (e.g., 1 mm or 1 cm). The area may be expressed as a volume of runoff (discharge × time = volume) or as a depth of runoff by dividing by the catchment area.

A catchment can have an infinite number of unit hydrographs because each duration of rainfall excess produces its own unit hydrograph. The unit duration t_0 is determined by assuming that the volume (or depth) of direct runoff is equal to the volume (or depth) of rainfall excess. The difference between the recorded rainfall and the direct runoff is taken as losses due to infiltration and other causes. The recorded rainfall is therefore adjusted using various methods of accounting for infiltration losses to give a rainfall depth equal to the depth of direct runoff. The effective duration of the adjusted rainfall is the unit duration.

2.2.2.4 Unit pulse response function

The unit pulse response function represents the runoff hydrograph from a constant rainfall excess of intensity $\dfrac{1}{\Delta t}$ and of duration Δt.

$$y(t) = \frac{1}{\Delta t}[S(t) - S(t - \Delta t)] = \frac{1}{\Delta t}\int_{t-\Delta t}^{t} h(\tau)\,d\tau \tag{2.31}$$

It can be obtained as the normalized difference between two S-curves lagged by a time interval of Δt.

2.2.3 Linear reservoir

In a linear reservoir (Figure 2.3), the output (discharge) is assumed to be proportional to the storage S as follows:

$$S(t) = Ky(t) \tag{2.32}$$

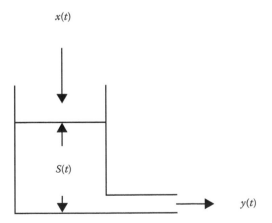

Figure 2.3 Linear reservoir.

From continuity considerations,

$$x(t) - y(t) = \frac{dS(t)}{dt} \qquad (2.33)$$

Combining

$$x(t) - y(t) = K \frac{dy(t)}{dt} \qquad (2.34)$$

or

$$\frac{dy(t)}{dt} + \frac{1}{K} y(t) = \frac{1}{K} x(t) \qquad (2.35)$$

which is the same as Equation 2.1 with $a_1 = \frac{1}{K}$, $a_2 = 1$, $b_1 = \frac{1}{K}$, and all other $a_i's$ and $b_i's$ zero. Equation 2.35, which is an ordinary differential equation, can also be written as

$$\frac{1}{K} dt = \frac{dy(t)}{x(t) - y(t)} \qquad (2.36)$$

which when integrated and after substituting the initial condition $y(t) = 0$ at $t = 0$ gives the solution

$$y(t) = x(t) \left\{ 1 - e^{-\frac{t}{K}} \right\} \qquad (2.37)$$

As $t \to \infty$, $y(t) \to x(t)$ in the above equation (steady state condition). If the inflow stops at time t_0 since outflow began, the outflow at time t in terms of the outflow $y(t_0)$ at time t_0 is obtained by substituting the initial condition $y(t) = y(t_0)$ at $t = t_0$ as

$$y(t) = y(t_0) e^{-\frac{(t-t_0)}{K}} \qquad (2.38)$$

Substituting the condition $S(t_0) = Ky(t_0)$ into Equation 2.38 and the condition that $t_0 = 0$ when the inflow is instantaneous gives

$$y(t) = \frac{S(t_0)}{K} e^{-\frac{t}{K}} \qquad (2.39)$$

For unit input, $S(t_0) = 1$, and therefore the IRF of a linear reservoir is

$$h(t) = \frac{1}{K} e^{-\frac{t}{K}} \qquad (2.40)$$

The shape of the response function of a linear reservoir to an instantaneous input (delta function type) consists of a suddenly increasing part at the time of inflow up to a finite outflow followed by an exponentially decreasing part that approaches zero asymptotically.

The storage coefficient K is approximated by the lag time t_L, which is the time difference between the centroids of rainfall excess hyetograph and the direct runoff hydrograph. One disadvantage of the linear reservoir is that the discharge takes place without any translation. There is no time delay from the time of application of input to the time of the appearance of the corresponding output. Linear reservoir is a one-parameter unit hydrograph.

2.2.4 Linear cascade

Linear cascade, originally proposed by Nash (1957, 1958, 1959, 1960), is a series of linear reservoirs arranged in series in such a way that the output from the upstream reservoir is used as the input to the downstream reservoir in the form of a cascade. The resulting response function takes the form

$$y_1(t) = h_1(t) = \frac{1}{K} e^{-\frac{t}{K}} \tag{2.41a}$$

$$y_2(t) = h_2(t) = \int_0^t h(\tau)x(t-\tau)\,d\tau = \int_0^t \frac{1}{K} e^{-\frac{\tau}{K}} \frac{1}{K} e^{-\frac{(t-\tau)}{K}}\,d\tau$$

$$= \frac{1}{K^2} \int_0^t e^{-\frac{\tau}{K}} e^{-\frac{(t-\tau)}{K}}\,d\tau = \frac{1}{K^2} \int_0^t e^{-\frac{\tau}{K} - \frac{(t-\tau)}{K}}\,d\tau$$

$$= \frac{1}{K^2} e^{-\frac{t}{K}} \int_0^t d\tau = \frac{1}{K^2} t e^{-\frac{t}{K}} \tag{2.41b}$$

$$y_3(t) = h_3(t) = \int_0^t h(\tau)x(t-\tau)\,d\tau = \int_0^t \frac{1}{K} e^{-\frac{\tau}{K}} \frac{1}{K^2} \tau e^{-\frac{(t-\tau)}{K}}\,d\tau$$

$$= \frac{1}{K^3} e^{-\frac{t}{K}} \int_0^t \tau\,d\tau = \frac{1}{K^3} e^{-\frac{t}{K}} \frac{t^2}{2} = \frac{1}{K^3} \frac{t^2}{2} e^{-\frac{t}{K}} \tag{2.41c}$$

$\cdots$

$\cdots$

$\cdots$

$$y_n(t) = h_n(t) = \frac{1}{(n-1)!K^n} t^{(n-1)} e^{-\frac{t}{K}} \tag{2.41d}$$

$$= \frac{1}{\Gamma(n)K^n} t^{(n-1)} e^{-\frac{t}{K}} \tag{2.41e}$$

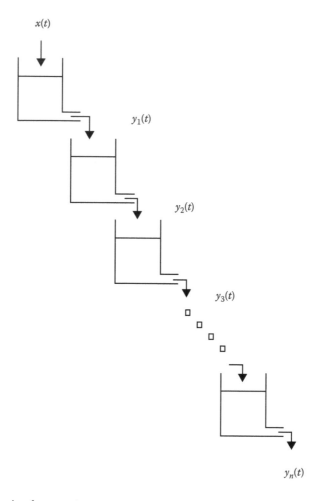

Figure 2.4 Linear cascade of reservoirs.

where $\Gamma(n) = (n-1)!$. This is the IUH of a linear cascade of reservoirs (Figure 2.4), which mathematically is representing a two-parameter (n is the shape parameter and K the scale parameter) Gamma distribution. It is a two-parameter unit hydrograph. By taking the first and second moments about the origin of the IUH defined by Equation 2.41e, it has been shown (Nash, 1957) that

$$m_1 = nK \tag{2.42a}$$

$$m_2 = n(n + 1)K^2 \tag{2.42b}$$

where m_1 and m_2 are the first and second moments.[3] Equations 2.42a and 2.42b can then be used to get the best-fit values of the parameters K and n for any empirically derived IUH. The IUH can also be determined by assuming suitable values for K and n if known, or by trial and error. The values assumed should then be verified by the method of moments.

When the direct runoff hydrograph and the rainfall excess hyetographs are available, the parameters K and n can be estimated as functions of the first and second moments about the origin of the direct runoff hydrograph and the rainfall excess hyetograph (Nash, 1957) as

---

[3] Zeroth moment, $m_0 = \int h(t)\,dt$; first moment, $m_1 = \int h(t)t\,dt$; second moment, $m_2 = \int h(t)t^2\,dt$.

$$m_{drh1} - m_{reh1} = nK \tag{2.42c}$$

$$m_{drh2} - m_{reh2} = n(n + 1)K^2 + 2nKm_{reh1} \tag{2.42d}$$

where the suffix 'drh' corresponds the direct runoff hydrograph, 'reh' to the rainfall excess hyetograph, and 1 and 2 refer to the first and second moments, respectively.

If K is variable, the cascade becomes non-linear. A non-linear model obtained by combining two linear cascades of different n and K in parallel and in which the input split into two parts by weighting parameters α and β takes the form (Diskin et al., 1978)

$$h(t) = \frac{\alpha}{\Gamma(n_1)K_1^{n_1}} t^{(n_1-1)} e^{-\frac{t}{K_1}} + \frac{\beta}{\Gamma(n_2)K_2^{n_2}} t^{(n_2-1)} e^{-\frac{t}{K_2}} \tag{2.43}$$

where $\beta = 1 - \alpha$.

2.2.5 Linear channel

The response to a linear system consists of storage effects as well as translation effects. Storage effects have been considered by the linear reservoirs and cascades. When an input function is routed through a linear channel, its shape is unchanged, but translated downstream. Analogous to the linear reservoir concept in which the discharge is assumed to have a linear relationship with storage, the linear channel concept assumes a similar linear relationship between discharge and area (Dooge, 1959). This implies that the velocity at any point is constant for all discharges but may vary from point to point. The time lag between the input and output functions is called the translation time (also, time of concentration). A linear channel is characterized by the continuity and a linear rating curve of the form (Ramiréz, 2000)

$$\frac{\partial Q}{\partial x} + \frac{\partial A}{\partial t} = 0 \tag{2.44}$$

$$A = C(x)Q \tag{2.45}$$

where $C(x)$ is some function describing the rating curve. For no lateral flow, the unit IRF of such a linear channel is given as

$$h(t) = \delta(t - T) \tag{2.46}$$

Equation 2.46 implies that the downstream outflow is the same as the upstream inflow but delayed by a lag time of T, which represents the travel time through the system. The combination of a linear reservoir and a linear channel involves the shifting of the time scale. The outflow from such a combination in series, regardless of the order in which they occur, due to an instantaneous inflow of $V_0\delta(0)$ is given by

$$y(t) = \frac{V_0}{K} e^{-\frac{(t-a)}{K}} \text{ with } t \geq a \tag{2.47}$$

where a is the translation time due to the linear channel. The principle of superposition can be used to obtain the effect of having a number of linear channels placed in series. The total translation time is equal to the sum of individual translation times.

The linear cascade model can be modified to account for translation effects by combining linear channels and linear reservoirs. Dooge (1959) developed a general unit hydrograph theory on the basis that the catchment can be represented by a cascade of linear channels and linear reservoirs in series. For this, the catchment is divided into subareas using isochrones with each subarea represented by a linear channel and a linear reservoir in series. For a system with n equal reservoirs with parameter K and n, and equal linear channels of lag-time parameter T arranged in series, the unit hydrograph takes the form (Ramiréz, 2000)

$$h_n(t) = \frac{1}{K\Gamma(n)} \left(\frac{t - nT}{K} \right)^{n-1} e^{-\frac{t-nT}{K}} \tag{2.48}$$

2.2.6 Time–area diagram

If only translation effects are considered, the system response function is given by the time–area diagram, which is a function that shows the percentage of the total catchment area contributing to runoff at different times. It is of the form

$$h(t) = A(t) \tag{2.49}$$

At different times, different areas of the catchment will be contributing to the runoff at the outlet. The lines that divide the catchment into contours of equal travel time are called isochrones. For example, from a 1-h isochrone, the rain falling between that isochrone and the catchment outlet will reach the outlet in 1 h; from a 2-h isochrone, the rain falling between that isochrone and the catchment outlet will reach the outlet in 2 h; and so on. The cumulative area contributing to the runoff at time $i\Delta t$ is given as

$$A(i\Delta t) = \sum_{k=1}^{i} A(k) \tag{2.50}$$

The vertical axis in the time–area histogram is the percentage of the catchment area that contributes to the runoff at the outlet at that particular time. The unit hydrograph in this case will be the time–area histogram itself. The outflow hydrograph is then given by the convolution integral as

$$y(t) = \sum_{k=0}^{t} x(k)h(t - k) \tag{2.51}$$

The time–area histogram method does not take into account the storage effects of the catchment. A unit hydrograph known as Clark unit hydrograph combines the storage effects and time–area effects, thereby eliminating this deficiency (Clark, 1945).

Example 2.1

Given the following input and time–area functions, determine the output function.

Time (h)	1	2	3
A(t)	0.25	0.5	0.25
x(t)	1	3	2

$$y(t) = \sum_{k=0}^{t} x(k)h(t-k)$$

$$
\begin{aligned}
y(0) &= 1 \times 0.25 & & & = 0.25 \\
y(1) &= 1 \times 0.50 & +3 \times 0.25 & & = 1.25 \\
y(2) &= 1 \times 0.25 & +3 \times 0.25 & +2 \times 0.25 & = 2.25 \\
y(3) &= & 3 \times 0.25 & +2 \times 0.50 & = 1.75 \\
y(4) &= & & +2 \times 0.25 & = 0.50
\end{aligned}
$$

A variation of the above concepts by considering the catchment response to reflect storage effects and translation effects by introducing two kernel functions, $h(t)$ and $w(t)$, consists of two linear reservoirs and a linear channel (Singh, 1964) as follows:

$$h(t) = \frac{1}{K_1 - K_2} \int_0^t \left\{ e^{-\frac{(t-\tau)}{K_1}} - e^{-\frac{(t-\tau)}{K_2}} \right\} w(\tau)\,d\tau \tag{2.52}$$

where $w(\tau) = \dfrac{dA}{d\tau}$.

2.3 RANDOM PROCESSES AND LINEAR SYSTEMS

In the systems theory approach, the usual practice is to have deterministic inputs and outputs. However, it is also possible to have stochastic inputs and outputs in a linear system and still estimate the system response function using their statistical properties. In such situations, the stochasticity can only be described statistically since the complete phenomenon is never known. Information about a sample(s) only is available. Inferences about the population are made on the basis of the information gathered from the sample(s). This places some restrictions. For example, it is assumed that the phenomenon is stationary and ergodic. Normality and linearity are also often implied.

Statistical properties of a function include the moments, correlation, probability density, etc. For the rainfall–runoff process with random variables $x(t)$ and $y(t)$,

$$y(t) = \int_0^t h(t)x(t-\tau)\,d\tau \tag{2.53}$$

where the system function $h(t)$ is deterministic.

The following statistical parameters of the input and output functions can be obtained (Lattermann, 1991, pp. 103–104):

- Mean value

$$E[y(t)] = E\left[\int_{-\infty}^{\infty} h(\tau)x(t-\tau)\,d\tau \right] = \left[\int_{-\infty}^{\infty} h(\tau)E\{x(t-\tau)\}\,d\tau \right] \tag{2.54}$$

- Autocorrelation

$$R_{yy}(u) = \int_{-\infty}^{\infty} h(\tau)R_{xy}(u-\tau)\,d\tau \tag{2.55}$$

- Cross correlation

$$R_{yx}(u) = \int_{-\infty}^{\infty} h(\tau)R_{xx}(u-\tau)\,d\tau \tag{2.56}$$

The cross-correlation functions can be obtained by minimizing the mean square error between the measured and calculated output functions:

$$\text{Min } E\left[\left| y(t) - \int_{-\infty}^{\infty} h(\tau)x(t-\tau)\,d\tau \right|^2 \right] \tag{2.57}$$

The result of the above minimization is (for derivation, see pp. 105–106 of Lattermann, 1991)

$$R_{yx}(u) = \int_{-\infty}^{\infty} h(\tau)R_{xx}(u-\tau)\,d\tau \tag{2.58}$$

where $R_{yx}(u)$ is the cross correlation between y and x; $R_{xx}(u)$ is the autocorrelation of x.

These results are applicable to time series of noise-free data. Equation 2.58 is known as the Wiener filter for non-causal systems. For causal systems, the limits of integration are 0 to ∞, and the resulting form of Equation 2.58 is known as Wiener–Hopf integral equation. In discrete form, it is

$$R_{yx}(m) = \sum_{k=0}^{m} h(k)R_{xx}(m-k) \tag{2.59a}$$

which can be written as

$$R_{yx}(0) = h(0)R_{xx}(0) + h(1)R_{xx}(-1) + (\ldots) + h(m)R_{xx}(0-m)$$

$$R_{yx}(1) = h(0)R_{xx}(1) + h(1)R_{xx}(0) + (\ldots) + h(m)R_{xx}(1-m)$$

$$R_{yx}(2) = h(0)R_{xx}(2) + h(1)R_{xx}(1) + (\ldots) + h(m)R_{xx}(2-m)$$

$$\ldots \tag{2.59b}$$

$$\ldots$$

$$\ldots$$

$$R_{yx}(m) = h(0)R_{xx}(m) + h(1)R_{xx}(m-1) + \ldots + h(m)R_{xx}(m-m)$$

or as

$$[R_{yx}] = [R_{xx}][h] \tag{2.59c}$$

It is also called the Yule–Walker equation. From Equation 2.59, $h(t)$ can be determined.

2.4 NON-LINEAR SYSTEMS

When the system considered is non-linear, the representation becomes very complex. For example, if the general linear representation is to be considered in a non-linear model, the representation will be of the form

$$y(t) = \int_0^\infty h_1(\tau)x(t-\tau)\,d\tau + \int_0^\infty\int_0^\infty h_2(\tau_1,\tau_2)x(t-\tau_1)x(t-\tau_2)\,d\tau_1\,d\tau_2 \tag{2.60}$$

$$+ \int_0^\infty\int_0^\infty\int_0^\infty h_3(\tau_1,\tau_2,\tau_3)x(t-\tau_1)x(t-\tau_2)x(t-\tau_3)\,d\tau_1\,d\tau_2\,d\tau_3 + \dots$$

For causal[4] systems, $\int_0^\infty = \int_0^t$ and therefore the limits of integration in Equation 2.60 can be changed to $\int_0^t$.

This is similar to the Volterra series and has been used to solve many non-linear systems, such as the unit hydrograph (Amorocho, 1963; Lattermann, 1991).

2.4.1 Determination of the kernel functions

For a second-order system, with an input function of the form

$$x(t) = a\delta(t) \tag{2.61}$$

the output function is

$$y(t) = \int_0^t h_1(\tau)a\delta(t-\tau)\,d\tau + \int_0^t\int_0^t h_2(\tau_1,\tau_2)a\delta(t-\tau_1)a\delta(t-\tau_2)\,d\tau_1\,d\tau_2 = ah_1(t) + a^2h_2(t,t) \tag{2.62}$$

(because $h(t) = y(t) = \int_0^t h(\tau)\delta(t-\tau)\,d\tau$)

---

[4] A system where the present value of the output depends only on the past and present values of the inputs and not on future values.

Therefore,

$$\frac{y(t)}{a} = h_1(t) + ah_2(t,t) = HU(t) \tag{2.63}$$

where $h_1(t)$ is the first-order unit response function and $h_2(t,t)$ is the second-order unit response function. The relationship

$$\frac{y(t)}{a} = HU(t) \tag{2.64}$$

represents a linear system.

The system functions of the second-order model should satisfy

$$\int_0^\infty h_1(t)\,dt = 1 \tag{2.65}$$

and

$$\int_0^\infty h_2(t,t)\,dt = 0 \tag{2.66}$$

Considering two independent input functions a_1 and a_2 (e.g., effective rainfall), producing different unit hydrographs $HU_1(t)$ and $HU_2(t)$ in a linear model, the first- and second-order kernel functions of the non-linear Volterra series can be obtained as follows (Lattermann, 1991, pp. 130–131):

$$HU_1(t) = h_1(t) + a_1 h_2(t,t) \tag{2.67a}$$

$$HU_2(t) = h_1(t) + a_2 h_2(t,t) \tag{2.67b}$$

Subtracting,

$$HU_1(t) - HU_2(t) = (a_1 - a_2)\, h_2(t,t)$$

$$h_2(t,t) = \frac{HU_1(t) - HU_2(t)}{(a_1 - a_2)} \tag{2.68a}$$

Eliminating $h_2(t,t)$ from Equation 2.67,

$$h_1(t) = \frac{a_2 HU_1(t) - a_1 HU_2(t)}{(a_2 - a_1)} \tag{2.68b}$$

It is implied that the linear system models $HU_1(t)$ and $HU_2(t)$ for the two events contain some non-linearity that is assumed to be relatively small. It is also possible to have three input functions a_1, a_2, a_3. Then h_1, h_2, and h_3 can be simultaneously determined. If there are three input functions and two kernel functions, then some optimization will be needed.

2.5 MULTILINEAR OR PARALLEL SYSTEMS

Linear systems can be combined to form multilinear models. Depending on the arrangement, non-linear effects of the system can be incorporated in such systems. In general (Lattermann, 1991, pp. 135–136),

$$y(t) = \sum_{i=1}^{n} y_i(t) = \sum_{i=1}^{n} \int_0^t h_i(\tau) x_i(t-\tau) d\tau \qquad (2.69)$$

with

$$\sum_{i=1}^{n} x_i(t) = x(t) \qquad (2.70)$$

The input and output functions are divided into parts, which are then used to determine their system parameters. The subdivision to a considerable extent influences the effect on non-linear modelling and therefore cannot be done arbitrarily. Once the individual response functions are determined, the output function for any given input function can be obtained by superposition.

2.6 FLOOD ROUTING

Channel storage can be significant in large rivers. When a flood wave propagates downstream, the effect of channel storage should be taken into consideration in the computation of the downstream hydrograph from an upstream hydrograph. This can be done using a hydrologic approach or a hydraulic approach. The hydrologic approach is a kind of systems theory approach in which the input is the inflow (upstream) hydrograph $I(t)$, and the output is the outflow (downstream) hydrograph $O(t)$, which is assumed to be related to the inflow hydrograph in a form

$$O(t) = f[c,I(t)] \qquad (2.71)$$

where c is a channel characteristic and f is a transfer function. Flood routing is a hydrological modelling approach in which the continuity equation is used with an outflow–storage relationship. The commonly used methods are the inventory method, the Muskingum method, the modified Puls method, and the Muskingum–Cunge method.

2.6.1 Inventory method

The continuity equation is

$$I - O = \frac{dS}{dt} \qquad (2.72)$$

where I is inflow, O is outflow, and S is storage. For a time interval Δt, it is possible to write

$$S_2 - S_1 = \bar{I}\Delta t - \bar{O}\Delta t \qquad (2.73a)$$

or

$$\Delta S = \frac{(I_1 + I_2)}{2}\Delta t - \frac{(O_1 + O_2)}{2}\Delta t \qquad (2.73b)$$

where the overbar indicates the mean value over the time interval. In this equation, I_1, I_2, O_1, and S_1 are known; only S_2 and O_2 are unknowns. Therefore, another relationship is needed to solve for these two unknowns. If the outflow–storage relationship is known, Equation 2.73 can be solved sequentially.

2.6.2 Muskingum method

The Muskingum method (McCarthy, 1938) considers a prism storage (KO) and a wedge storage $Kc(I - O)$ when added together gives the total storage as $S = KO + Kc(I - O)$. Prism storage refers to the volume between the stream bed and a line parallel to the bed and intersecting the water surface at the end of the reach considered, whereas wedge storage refers to the volume contained between this line and the water surface.

It assumes a single stage–discharge relationship that may not always be valid. Under certain flow conditions, the rising part and the recession part of the hydrograph may have different friction slopes, thereby producing a loop (hysteresis) rather than a single relationship. Manning's equation is sometimes used as the stage–discharge relationship. The combined storage can also be obtained as follows:

Assuming storage–discharge relationships of the form

$$S_I = KI^x \qquad (2.74)$$

$$S_O = KO^x \qquad (2.75)$$

where S_I is inflow storage (I is the inflow), S_O is outflow storage (O is the outflow), and K is a storage constant for the reach that has the units of time, it is possible to write the storage within a reach as a weighted sum of the input and output storages

$$S = cS_I + (1 - c)S_O \qquad (2.76)$$

where c is a weighting factor between 0 and 0.5 (usually about 0.2). Δt is chosen such that it is less than 20% of the time to peak. In Equation 2.76, $S = S_O$, when $c = 0$. This condition leads to the maximum attenuation and is called convex routing method, and is useful for reservoir routing. When $c = 0.5$, $S = 0.5(S_I + S_O)$ (no attenuation, translation only). Combining Equations 2.74 through 2.76,

$$S = K[cI^x + (1 - c)O^x] \qquad (2.77)$$

For rectangular channels, $x = 0.6$ (McCuen, 2004, p. 606). For natural channels, x is larger and Muskingum method assumes a value of $x = 1$. Then, Equation 2.77 becomes

$$S = K[cI + (1 - c)O] \qquad (2.78a)$$

Stream flow data generally show that K is approximately equal to the travel time of the reach.

Equation 2.78a can be written as

$$S_2 = K[cI_2 + (1 - c)O_2] \qquad (2.78b)$$

and

$$S_1 = K[cI_1 + (1 - c)O_1] \qquad (2.78c)$$

or

$$S_2 - S_1 = K[c(I_2 - I_1) + (1 - c)(O_2 - O_1)] \qquad (2.78d)$$

Combining with the inventory equation, Equation 2.73, $S_2 - S_1 = \dfrac{(I_1 + I_2)}{2}\Delta t - \dfrac{(O_1 + O_2)}{2}\Delta t$

$$\Rightarrow K = \frac{0.5\Delta t[(I_2 + I_1) - (O_2 + O_1)]}{[c(I_2 - I_1) + (1 - c)(O_2 - O_1)]} = \frac{Y}{X} \qquad (2.79a)$$

or,

$$\Rightarrow K = \frac{0.5\Delta t[(I_2 + I_1) - (O_2 + O_1)]}{[c(I_2 - I_1) + (1 - c)(O_2 - O_1)]} = \frac{Y}{X} \qquad (2.79b)$$

which can be written as

$$O_2 = C_0 I_2 + C_1 I_1 + C_2 O_1 \qquad (2.80)$$

where

$$C_0 = \frac{-Kc + 0.5\Delta t}{K(1 - c) + 0.5\Delta t} \qquad (2.81a)$$

$$C_1 = \frac{Kc + 0.5\Delta t}{K(1 - c) + 0.5\Delta t} \qquad (2.81b)$$

$$C_2 = \frac{K(1 - c) - 0.5\Delta t}{K(1 - c) + 0.5\Delta t} \qquad (2.81c)$$

It can be seen that $C_0 + C_1 + C_2 = 1$.

Therefore, if the input hydrograph is known, the output hydrograph can be determined for the reach. The initial value of the outflow hydrograph, the routing time step Δt, and routing parameters K and c must be specified. Implied in this method are the assumptions that the water surface profile within the reach is uniform and continuous and that K and c are constants for the reach.

2.6.2.1 Estimation of the routing parameters K and c

The parameters K and c can be estimated if the inflow and outflow hydrographs are known for the reach. It can be done graphically or statistically. The statistical method is more precise. They can also be related to the channel characteristics such as the slope, roughness as measured by Manning's 'n', hydraulic radius, or other equivalent, etc. This approach requires extensive field studies to establish a relationship between site characteristics and the routing parameters.

The value of K is estimated to be the travel time through the reach. An ambiguity occurs as to whether the travel time should be estimated using the average flow, peak flow, or some other reference flow. It may be estimated as the kinematic travel time or a travel time based on Manning's equation. The second routing parameter c must lie between 0.0 and 0.5. It is a weighting factor for inflow and outflow. A low value of c implies that the inflow has little or no effect on storage. A reservoir, for example, has a c value of zero as the dominant process there is attenuation. Typical values used in natural streams range from about 0.2 to 0.3, but 0.4 to 0.5 may be used for streams with no storage effects. A value of $c = 0.5$ would represent equal weighting between inflow and outflow and would produce translation with little or no attenuation.

a. Graphical method

By assuming different values for c, the numerator and denominator of Equation 2.79 are calculated for each time interval of the inflow and outflow hydrographs and they are summed and plotted against each other (denominator X on the x-axis and numerator Y on the y-axis). Note that the cumulative values of X and Y will increase first and then decrease to form a loop. The plot that generally has a loop and the one that is closest to a straight line would be the best choice for K and c. The slope of the line gives the value of K. The value of c that produced the closest deviation from a straight line is taken as its representative value.

b. Statistical method

In this approach, K and c are obtained by the least squares method. The error to be minimized is the difference between the actual downstream hydrograph and the computed hydrograph defined in Equation 2.80:

$$\text{Error, } E = \sum \left(O_2^{\text{computed}} - O_2^{\text{observed}} \right)^2 \tag{2.82}$$

Differentiating E with respect to the three parameters C_0, C_1, and C_2, the normal equations can be obtained. They are

$$C_0 \sum I_{t+1}^2 + C_1 \sum I_{t+1} I_t + C_2 \sum O_t I_{t+1} = \sum I_{t+1} O_{t+1} \tag{2.83a}$$

$$C_0 \sum I_{t+1} I_t + C_1 \sum I_t^2 + C_2 \sum O_t I_t = \sum I_t O_{t+1} \tag{2.83b}$$

$$C_0 \sum O_t I_{t+1} + C_1 \sum O_t I_t + C_2 \sum O_t^2 = \sum O_t O_{t+1} \tag{2.83c}$$

Once C_0, C_1, and C_2 have been determined, the Muskingum parameters K and c can be obtained from the relationships given in Equation 2.81. The additional condition $C_0 + C_1 + C_2 = 1$ may be used as a check. The estimated parameters can also be used to determine the downstream hydrograph, which can be compared with the known hydrograph for verification.

To avoid numerical instability (such as negative values for discharges), the time step should be chosen to satisfy the condition

i. $2Kc < \Delta t < K$.

Other criteria used include

ii. $2Kc < \Delta t < 2K(1 - c)$ (Hjelmfelt, 1985).

iii. $\dfrac{K}{3} < \Delta t < K$ (Maidment, 1993, p. 10.7).

iv. $\dfrac{1}{2(1-c)} \leq \dfrac{K}{N\Delta t} \leq \dfrac{1}{2c}$ (USACE, 1990); N is the number of subreaches so that the total travel time through the reach is K.

When $c = 0.5$, the conditions (i) and (ii) are equivalent.

2.6.2.2 Limitations of the Muskingum method

The Muskingum method is limited to slow-rising hydrographs when routed along mild sloping channels. The method ignores backwater effects such as when there is a dam downstream, constrictions, bridges, and tidal influences. It may also give negative flows in the initial portion of the hydrograph.

2.6.3 Modified Puls method

The modified Puls method of routing is often applied to reservoir routing, but may also be applied to river routing under certain channel situations. It is based on the finite difference form of the continuity equation. The method requires a stage–discharge–storage relationship to be known, assumed, or derived (see also Section 2.7).

2.6.4 Muskingum–Cunge method

Cunge (1969) developed a routing method that is equivalent to the Muskingum method under certain conditions. It is therefore referred to as the Muskingum–Cunge method. It is based on combining the continuity equation and the diffusion wave approximation form of the momentum equation followed by a linearization, and the formulation is considered an approximate solution of the convective diffusion equation. It uses physical characteristics to estimate the routing parameters. The physical characteristics consist of the reach length L, the slope S_0, the kinematic wave celerity v, and a characteristic unit discharge q_0. The wave celerity is defined as

$$v = \frac{dQ}{dA} = \frac{1}{B}\frac{dQ}{dh}, \text{which is also equal to } \frac{dx}{dt} \tag{2.84}$$

where Q is the discharge, h is the depth of flow, A is the cross-sectional area of flow, and B is the width of the water surface. The term $\dfrac{dQ}{dh}$ represents the slope of the discharge–stage relationship at the water surface elevation. The actual discharge q (discharge per unit width), or the mean or peak discharge may be taken as the characteristic discharge q_0.

Cunge (1969) gave the following expression for the weighting parameter c:

$$c = \frac{1}{2}\left(1 - \frac{Q_0}{BvS_0L}\right) \tag{2.85a}$$

which may also be written as

$$c = 0.5\left(1 - \frac{q_0}{v S_0 L}\right)$$
(2.85b)

The Muskingum–Cunge method uses the expression

$$K = \frac{L}{v}$$
(2.86)

for K, which uses Equation 2.84 for v. Combining the continuity equation

$$\bar{I} - \bar{O} = \frac{dS}{dt}$$
(2.87)

and the Muskingum equation (Equation 2.78a),[5] it is possible to write

$$K \frac{d}{dt}[cI + (1 - c)O] = \bar{I} - \bar{O}$$
(2.88)

In finite difference form, $(I = O_i ; O = O_{i+1})$

$$\frac{K}{\Delta t}\left\{cO_i^{t+1} + (1-c)O_{i+1}^{t+1} - cO_i^t - (1-c)O_{i+1}^t\right\} = 0.5\left\{O_i^{t+1} + O_i^t - O_{i+1}^{t+1} - O_{i+1}^t\right\}$$
(2.89)

If $K = \frac{\Delta x}{v}$, where v is the wave celerity, Equation 2.89 is a finite difference form of the continuity equation with no lateral inflow:

$$\frac{\partial O}{\partial t} + v \frac{\partial O}{\partial x} = 0$$
(2.90)

From Equation 2.89,

$$O_{i+1}^{t+1} = C_0 O_i^{t+1} + C_1 O_i^t + C_2 O_{i+1}^t$$
(2.91)

where the coefficients C_0, C_1, and C_2 have the same definitions as in the Muskingum method (i.e., Equation 2.81).

$$C_0 = \frac{-Kc + 0.5\Delta t}{K(1 - c) + 0.5\Delta t}$$

$$C_1 = \frac{Kc + 0.5\Delta t}{K(1 - c) + 0.5\Delta t}$$

$$C_2 = \frac{K(1 - c) - 0.5\Delta t}{K(1 - c) + 0.5\Delta t}$$

---

[5] $S = K[cI + (1 - c)O]$.

Since $K = \dfrac{\Delta x}{v}$ is the time of travel for each reach of length Δx and wave velocity v, Cunge showed that v is the celerity of the kinematic wave. When K and Δt are constants, the solution given by Equation 2.91 is an approximation of the kinematic wave. It has also been shown that when the wave celerity is defined by Equation 2.84 (and therefore $K = \dfrac{\Delta x}{v}$) and the weighting factor c is defined by Equation 2.85a, the solution given by Equation 2.91 is an approximation to a modified form of diffusion equation. When $c = 0.5$, $v\dfrac{\Delta t}{\Delta x} = 1.0$ (compare with Equation 2.84), implying only translation with no attenuation. Negative values of C_0 could be avoided if $\dfrac{\Delta t}{K} > 2c$.

2.6.5 Hydraulic approach

Hydraulic routing, which is also known as dynamic routing, is carried out by solving the St. Venant equations. It is a reference routing method for comparison of other routing methods. The solution of the St. Venant equations is generally accomplished by the method of characteristics, or by direct methods (explicit and implicit). The solutions, either way, are resource intensive. The most popular implicit method is to use a four-point weighted finite difference scheme. It is also possible to solve the problem using finite element methods, but the formulation is more complex. One-dimensional (1D) problems are usually solved by the finite difference methods, whereas 2D and 3D problems often use the finite element method.

Dynamic routing heavily relies on physically measured parameters rather than using calibration techniques, which require past flood data. Added advantages include solutions to the sediment transport and pollutant transport problems, which can be coupled to the water transport problem. On the negative side, computational instability may occur in situations where the flood wave travels through the subreach considered in a time less than the time interval used for computations. To get over this possibility, the Courant stability criterion must be satisfied; that is

$$\Delta t \le \frac{\Delta x}{v}$$

where v is the wave celerity and Δx is the reach length.

2.6.5.1 Solution of the St. Venant equations

The St. Venant equations are also referred to as the dynamic wave equations or the shallow water equations. The problem described by these equations is unsteady and non-uniform flow. In a 1D framework, the continuity and momentum equations can be written, respectively, as

$$\frac{\partial q}{\partial x} + \frac{\partial h}{\partial t} = (i - f) = Q \tag{2.92}$$

$$\frac{\partial v}{\partial t} + v\frac{\partial v}{\partial x} + g\frac{\partial h}{\partial x} + \frac{Qv}{h} = g(S_0 - S_f) \tag{2.93}$$

where
    q = discharge per unit width
    h = depth of flow
    Q = lateral inflow per unit length per unit width
    v = velocity of flow

S_0 = bed slope of the flow plane
S_f = friction slope of the flow plane
i = rainfall rate
f = infiltration rate
x, t = distance along the flow plane, and time

Writing $q = hv$, Equation 2.92 becomes

$$h\frac{\partial v}{\partial x} + v\frac{\partial h}{\partial x} + \frac{\partial h}{\partial t} = Q \tag{2.94}$$

Replacing the derivatives of Equations 2.94 and 2.93 by the finite differences,

$$h_{i,j}\frac{v_{i+1,j} - v_{i-1,j}}{2\Delta x} + v_{i,j}\frac{h_{i+1,j} - h_{i-1,j}}{2\Delta x} + \frac{h_{i,j+1} - h_{i,j}}{\Delta t} = Q_{i,j} \tag{2.95}$$

and

$$\frac{v_{i,j+1} - v_{i,j}}{\Delta t} + v_{i,j}\frac{v_{i+1,j} - v_{i-1,j}}{2\Delta x} + g\frac{h_{i+1,j} - h_{i-1,j}}{2\Delta x} + \frac{Q_{i,j}v_{i,j}}{h_{i,j}} = g(S_0 - S_f) \tag{2.96}$$

The problem has two degrees of freedom, h and v. It is also non-linear. Hence, an iterative simultaneous solution should be attempted. The boundary conditions for this problem are usually of the Dirichlet type. Prescribed values of variable must be given.

The frictional slope S_f is usually represented by the Manning or Chezy equations.

2.6.5.2 Diffusion wave approximation

The diffusion wave approximation considers the continuity equation (Equation 2.92) and simplified form of the momentum equation obtained by neglecting the acceleration terms and the lateral momentum term which then takes the form

$$g\frac{\partial h}{\partial x} = g(S_0 - S_f) \tag{2.97}$$

The flow is usually considered as quasi-steady non-uniform. Changes in water surface profile are caused only by inflows. Again, solutions are sought using numerical techniques.

2.6.5.3 Kinematic wave approximation

The momentum (or dynamic) equation under certain circumstances can be reduced to a more simple form by neglecting the inertia, inflow momentum, and free surface slope terms. Kinematics, as the name implies, refers to the description of flows without considering forces. By an order of magnitude analysis, it can be seen that these terms are small compared with the bed and friction slopes. Therefore, it is possible to simplify the dynamic equation to

$$g(S_0 - S_f) = 0, \text{ or, } (S_0 - S_f) = 0 \tag{2.98}$$

This condition implies uniform flow in which the weight of the fluid element is balanced by the frictional force due to boundary shear. Using boundary layer concepts to define the boundary shear stress the momentum equation can be reduced to the form

$$q = \alpha h^m \qquad (2.99)$$

for both laminar and turbulent flows.

This approximation is known as the kinematic wave approximation, and has been applied to model overland flow, flood flow, and traffic flow. The pair of Equations 2.92 and 2.99

$$\frac{\partial q}{\partial x} + \frac{\partial h}{\partial t} = Q$$

$$q = \alpha h^m$$

gives a quasi steady-state solution in which the changes in water surface profile are caused by changes in flow rate q only. The solutions are not valid at the two boundaries where the condition $S_0 = S_f$ is not satisfied; that is, the bed slope and the free surface slope are not the same. For turbulent flow, $\alpha = \frac{\sqrt{S_0}}{n}$ and $m = \frac{5}{3}$ have been used. For natural surfaces, $m = 2$ has been used.

An analytical solution of Equations 2.92 and 2.99 has been provided by Henderson and Wooding (1964), and later critically reviewed by Eagleson (1970) for constant intensity rainfall. For other input conditions, subject to boundary and initial conditions, they can be solved by a number of numerical methods such as the finite difference method or the finite element method. An important property of the kinematic wave approximation is that the solution is always upstream driven. A disturbance made at the downstream has no influence on the upstream. Dynamic waves, on the other hand, can travel both upstream and downstream. Solutions to kinematic waves can be obtained with only one boundary condition, whereas solutions to dynamic waves require both upstream and downstream boundary conditions.

In natural rivers, flood waves consist of kinematic waves as well as dynamic waves. It has been proven that the velocity of the main part of a flood wave approximates that of a kinematic wave (Lighthill and Whitham, 1955). Solutions to the kinematic wave approximation using finite elements in space and finite differences in time have also been obtained and applied to a real catchment, which is subdivided into strips that run from the top of the slope to the main drainage paths in directions normal to the topographical contours (Jayawardena and White, 1977, 1979).

The kinematic wave equations may also be written as

$$\frac{\partial Q}{\partial x} + \frac{\partial A}{\partial t} = Q'(x,t) \qquad (2.100)$$

$$S_0 = S_f$$

where Q' is the lateral inflow rate and S_f is the frictional slope.

The momentum equation may also be written as

$$A = \alpha Q^\beta \qquad (2.101)$$

which upon differentiating gives

$$\frac{\partial A}{\partial t} = \alpha \beta Q^{\beta-1} \frac{\partial Q}{\partial t} \qquad (2.102)$$

Eliminating A in the governing equation gives

$$\frac{\partial Q}{\partial x} + \alpha\beta Q^{\beta-1}\frac{\partial Q}{\partial t} = Q' \tag{2.103}$$

By writing the momentum equation as

$$Q = \alpha A^{\gamma} \tag{2.104}$$

which when differentiated gives

$$\frac{\partial Q}{\partial A} = \alpha\gamma A^{\gamma-1} \tag{2.105}$$

Substituting the relationship $\dfrac{\partial Q}{\partial x} = \dfrac{\partial Q}{\partial A}\dfrac{\partial A}{\partial x}$ in Equation 2.100 gives

$$\frac{\partial A}{\partial t} + \frac{\partial Q}{\partial A}\frac{\partial A}{\partial x} = Q' \tag{2.106}$$

which simplifies to

$$\frac{\partial A}{\partial t} + \alpha\gamma A^{\gamma-1}\frac{\partial A}{\partial x} = Q' \tag{2.107}$$

Equations 2.103 and 2.107 are two forms of the governing equation; in Equation 2.103, Q is the dependent variable and in Equation 2.107, A is the dependent variable. Taking logarithms of the kinematic wave approximation for A (Equation 2.101),

$$\ln A = \ln\alpha + \beta\ln Q \tag{2.108}$$

which when differentiated gives

$$\frac{dQ}{Q} = \frac{1}{\beta}\frac{dA}{A} \tag{2.109}$$

Similarly, taking logarithms of the kinematic approximation for Q (Equation 2.104),

$$\ln Q = \ln\alpha + \gamma\ln A \tag{2.110}$$

which when differentiated gives

$$\frac{dA}{A} = \frac{1}{\gamma}\frac{dQ}{Q}, \text{ or, } \frac{dQ}{Q} = \gamma\frac{dA}{A} \tag{2.111}$$

Equation 2.111 gives relative errors of Q and A.

Either forms of the above kinematic wave formulations can be used to solve the flood propagation problem using a chosen numerical method.

2.7 RESERVOIR ROUTING

The continuity equation can be written as

$$I - O = \frac{dS}{dt} \tag{2.112}$$

where I represents the inflow, O, the outflow and S, the storage. Over a finite interval of time between t and $t + \Delta t$,

$$\left(\frac{I_1 + I_2}{2}\right) - \left(\frac{O_1 + O_2}{2}\right) = \frac{S_2 - S_1}{\Delta t}$$

which can be written as

$$(I_1 + I_2) + \left(\frac{2S_1}{\Delta t} - O_1\right) = \left(\frac{2S_2}{\Delta t} + O_2\right) \tag{2.113}$$

In this equation, the left-hand side is known and the right-hand side contains two unknowns, S_2 and O_2. An additional condition is necessary to solve Equation 2.113. This is the functional relationship between $\frac{2S}{\Delta t} + O$ and O, which is obtained using the stage–storage relationship and the stage–discharge relationship. The procedure is illustrated in the following example:

Example 2.2

A stormwater detention basin is estimated to have the following storage characteristics:

Stage (m)	5.0	5.5	6.0	6.5	7.0	7.5	8.0
Storage (m³)	0	694	1525	2507	3652	4973	6484

The discharge weir from the detention basin has a crest elevation of 5.5 m, and the weir discharge, Q, is given by

$$Q = 1.83H^{3/2}$$

where H is the height of the water surface above the crest of the weir. The inflow hydrograph is given by

Time (min)	0	30	60	90	120	150	180	210	240	270	300	330	360	390
Runoff (m³/s)	0	2.4	5.6	3.4	2.8	2.4	2.2	1.8	1.5	1.2	1.0	0.56	0.34	0

If the prestorm stage in the detention basin is 5.0 m, estimate the outflow hydrograph from the detention basin.

Solution

This requires runoff routing through a reservoir. Using a time step of 30 min ($\Delta t = 30 \times 60 = 1800$ s), the calculations can be done as shown in Table 2.1.

Table 2.1 Stage–storage–discharge characteristics

Stage (m)	S (m³)	O (m³/s)	2S/Δt + O (m³/s)
5.0	0	0	0
5.5	694	0	0.771
6.0	1525	0.647	2.34
6.5	2507	1.83	4.62
7.0	3652	3.36	7.42
7.5	4973	5.18	10.7
8.0	6484	7.23	14.4

In Table 2.1, columns 1 and 2 are given. Column 3 is obtained from the weir formula. Column 4 is obtained using columns 2 and 3 (stage → storage S → discharge $O → \frac{2S}{\Delta t} + O$). Columns 3 and 4 then gives a relationship between discharge O and $\frac{2S}{\Delta t} + O$, which will be used in subsequent calculations.

At the beginning, $I_1 = 0$, $I_2 = 2.4$; therefore $I_1 + I_2 = 2.4$; $O_1 = 0$; $S_1 = 0$ (because the weir operates only when the head exceeds 5.5 m). Therefore, $\frac{2S}{\Delta t} - O = 0$. For the next time step (30 min), from continuity equation,

$$(I_1 + I_2) + \left(\frac{2S_1}{\Delta t} - O_1\right) = \left(\frac{2S_2}{\Delta t} + O_2\right)$$

$$\Rightarrow \frac{2S_2}{\Delta t} + O_2 = (I_1 + I_2) + \left(\frac{2S_1}{\Delta t} - O_1\right) = 0 + 2.4 + 0 = 2.4$$

Next, O_2 corresponding to $\frac{2S_2}{\Delta t} + O_2 = 2.4$ is obtained by linear interpolation using the relationship between O and $\frac{2S}{\Delta t} + O$ given in columns 3 and 4 of Table 2.1. This works out to 0.68. Then, $\frac{2S_1}{\Delta t} - O_1$ is obtained from the already calculated values of $\frac{2S_2}{\Delta t} + O_2$ and O_2 as

$$\left(\frac{2S_1}{\Delta t} - O_1\right) = \left(\frac{2S_2}{\Delta t} + O_2\right) - 2O_2 = 2.4 - 2 \times 0.68 = 1.04$$

The rest of the calculations are carried out recursively and given in Table 2.2. The computational sequence may be summarized as follows:

Stage → storage S → discharge $O → \frac{2S}{\Delta t} + O$. Initially S, O, and therefore $\frac{2S}{\Delta t} + O$ are all zero. In the next time step, I_1 and I_2 are known and therefore $\left(\frac{2S_2}{\Delta t} + O_2\right)$ can be calculated.

Find O corresponding to the calculated value of $\frac{2S}{\Delta t} + O$ by linear interpolation in the $\frac{2S}{\Delta t} + O$ versus O characteristic in columns 3 and 4 of Table 2.1.

Find $\frac{2S}{\Delta t} - O$ as $\frac{2S}{\Delta t} + O - 2O$, and repeat the procedure.

Table 2.2 Routing through the reservoirs

Time (min)	I_1 (m³/s)	I_1+I_2 (m³/s)	$\dfrac{2S}{\Delta t} - O$ (m³/s)	$\dfrac{2S}{\Delta t} + O$ (m³/s)	O (m³/s)
0	0	2.4	0	0	0
30	2.4	8.0	1.04	2.4	0.68
60	5.6	9.0	0.52	9.04	4.26
90	3.4	6.2	0.469	9.52	4.52
120	2.8	5.2	0.769	6.669	2.95
...	2.4	...	...	...	...
...	...	...	...	...	...

2.8 RAINFALL–RUNOFF MODELLING

Rainfall–runoff models of the deterministic type can be broadly classified into hydrologic models and hydraulic models. Hydrologic models are usually based on the continuity equation only, whereas the hydraulic models require an additional equation, which may be a (simplified) form of the dynamic equation. Rainfall–runoff models may also be classified as continuous-time simulation models or event simulation models. The former is usually of the hydrologic type, whereas the latter can be of the hydraulic type. Hydrologic-type models widely use the linear reservoir, the linear channel, and their combinations as the basic building blocks.

Rainfall–runoff models can also be classified as analog (now outdated), conceptual, physics-based, or data driven. They may also be classified as lumped or distributed, linear or non-linear, continuous-time simulation or event simulation, and/or, according to the process description (deterministic, conceptual, stochastic), domain representation (lumped, semidistributed, distributed), temporal scale (annual, monthly, daily, hourly, etc.), solution technique, and model application.

The first rainfall–runoff model was probably the rational method (Mulvany, 1845), which relates the peak runoff to the rainfall intensity. The unit hydrograph method (Sherman, 1932) is a conceptual model that linearly relates the rainfall excess to direct runoff and which has stood the test of time due to its simplicity. Non-linear versions of the unit hydrograph (Amorocho, 1963) as well as many other versions that employ the systems theory approach have been suggested and used over the years (e.g., Delleur and Rao, 1971 and Lattermann, 1991).

In addition to the rational method and the unit hydrograph method, other historical highlights of rainfall–runoff models include the Stanford watershed model (Crawford and Linsley, 1966), Tank model (Sugawara et al., 1984), Xinanjiang model (Zhao, 1977, 1984; Zhao and Liu, 1995; Zhao et al., 1980), HEC-1 model (USACE, 1998), Topmodel (Beven and Kirkby, 1979), Système Hydrologique Européen (Abbott et al., 1986a,b), and ARNO model (Todini, 1996).

Since the advent of the digital computer, there has been a proliferation of models and modelling techniques giving rise to a plethora of models. Among them, the more widely used ones include HEC-HMS, the standard model used in private sector in the United States; NWS, the standard flood forecasting model in the United States; HSPF/BASINS, the standard water quality model in the United States; WATFLOOD, a popular model used in Canada; RORB, a popular model used in Australia; HBV, the standard model used in Scandinavia; SHE, the standard distributed model in Europe; Tank, a popular model used in

Japan; Xinanjiang, a popular model used in China; ARNO, TOPIKAPI, and LCS, popular models used in Italy; and VIC, a popular model used in the United States.

There are also other less widely used models such as SWMM – Storm Water Management Model; SWAT – Soil and Water Assessment Tool; QUALHYMO – Storm Water Runoff Model; HSPF – Hydrological Simulation Program Fortran; AGNPS – Agricultural Non-Point Source Pollution Model; PRMS – Precipitation–Runoff Modelling System; SSARR – Streamflow Synthesis and Reservoir Regulation Model; HELP – Hydrological Evaluation of Landfill Performance; WATER BUDGET – Cumming Cockburn Limited Water Budget Model; SAC-SMA – Sacramento Soil Moisture Accounting model, used by the National Weather River Forecasting System for flood forecasting in the United States; and ANSWERS – Areal Non-Point Source Watershed Environment.

Despite the abundance, there is no perfect model that suits all purposes. Each model has its own pros and cons. Moreover, more and more models are being developed continuously. One should take note of the saying that 'all models are wrong, but some are useful', and exercise careful judgment in choosing a model or a modelling approach for a specific purpose.

2.8.1 Conceptual-type hydrologic models

2.8.1.1 Stanford watershed model (SWM)

Originated in Stanford University (Crawford and Linsley, 1966), the Stanford watershed model is suitable for hourly or daily simulation. It uses the lumped parameter approach.

Inputs consist of hourly or daily precipitation data, evaporation data, and catchment parameters. The model can carry out continuous-time simulation and is based on the continuity equation:

$$P = E + R + \Delta S \tag{2.114}$$

where P is precipitation, E is evapo-transpiration, R is runoff, and ΔS is total change in storage in each zone.

At each step, all hydrological activities (precipitation, interception, depression storage, infiltration, percolation, surface runoff, groundwater flow, etc.) are simulated and balanced before proceeding to the next time step. Interception, evapo-transpiration, infiltration, overland flow, interflow, which are all parameterized are modelled. Calibration of the model is iterative.

2.8.1.2 Tank model

The Tank model, proposed and developed by Sugawara and his colleagues in the late 1950s, is the most widely used rainfall–runoff model in Japan. The publication of the details of this model in the English language came much later (Sugawara et al., 1984) and its application has since then spread to regions outside Japan (e.g., Jayawardena, 1988). It is a deterministic, non-linear, lumped, continuous, and time-invariant model that transforms the measured rainfall into a corresponding runoff without having to estimate losses and base flows separately.

The model is composed of tanks (usually four) arranged vertically in series. Precipitation is fed into the top tank, and evaporation is subtracted sequentially from the top tank downwards. As each tank is emptied, the evaporation shortfall is taken from the next tank below until all tanks are empty. The outputs from the side outlets are the calculated runoffs. The output from the top tank is considered as surface runoff, output from the second tank as intermediate runoff, output from the third tank as subbase runoff, and output from the fourth tank as base flow.

The basic principle of the Tank model is that discharge is proportional to storage. The catchment is considered as a reservoir that temporarily stores the rainfall and subsequently releases as runoff. The discharge is considered as consisting of surface runoff, subsurface runoff, and groundwater flow, each component having its own response time. The inputs to the Tank model are precipitation and potential evaporation.

In the model, the tanks can have different outlets in the same tank, or different tanks arranged in series. There are threshold values in each tank to account for the initial rainfall that does not produce any runoff. Flooding can be modelled by having two outlets in the top tank. Highly porous surfaces can be modelled by having a bottom outlet in the top tank with a large discharge coefficient. In the latter case, the discharge will be mainly from the second tank. Large catchments can be modelled by having the series arrangement of tanks in parallel.

The soil moisture is modelled by having a primary soil moisture storage in the top tank with a saturation capacity. There will be no runoff or infiltration until the storage in the top tank attains this value. The difference between the storage in the top tank and the saturation capacity of the primary soil moisture storage is the free water available.

After the primary soil moisture storage is saturated, there will be transfer of water from the primary to secondary. When evaporation takes place, primary soil moisture is depleted first and then water is drawn laterally from the secondary storage. If there is no free water in the second tank, vertical transfer takes place from the free water in the third and fourth tanks.

Calibration can be done either by trial and error or by systematically optimizing an objective function. In the latter case, the objective function used is

$$CRHY = \frac{1}{2}(MSEQ + MSELQ) \tag{2.115}$$

where $MSEQ = \dfrac{\text{mean square error}}{\text{mean discharge}}$, and $MSELQ$ = mean square error of log Q.

The Tank model can be used for continuous-time simulation as well as for event simulation. In the latter case, losses due to evaporation can be ignored in flood analysis.

2.8.1.3 HEC series

The Hydrologic Engineering Centre (HEC) is the technical arm of the United States Army Corps of Engineers (USACE) dealing with hydrologic engineering. It was established in 1964 and started with hydrological software that included HEC-1 (Watershed Hydrology), HEC-2 (River Hydraulics), HEC-3 (Reservoir Analysis and Conservation), and HEC-4 (Stochastic Streamflow Generation). Subsequently, this series of software has been superseded with a more up-to-date family that incorporates the same basic concepts as the first series but with improved numerical procedures. The new family of software includes HEC-HMS (Hydrologic Modelling System), HEC-RAS (River Analysis System), HEC-FDA (Flood Damage Reduction Analysis), and HEC-ResSim (Reservoir System Simulation).

a. HEC-1

The HEC-1 model (USACE, 1990, 1998) has several components that are linked to each other. The catchment is divided into a number of subcatchments on the basis of the topography and geography. Each subcatchment is modelled to simulate the land surface runoff component, river routing component, reservoir component, and an abstractions and inflows component.

To model the land surface runoff, rainfall is the input variable with losses assumed according to some empirical relationship. In each subcatchment, rainfall and losses are assumed to be uniform. Infiltration losses are assumed as an initial and uniform loss rate or as an exponential loss rate. Routing the effective rainfall is by the unit hydrograph or by the kinematic wave method. If unit hydrograph method is used, synthetic unit hydrographs such as Clark's, Snyder's or SCS are used. If kinematic routing is adopted, the subcatchment is divided into flow planes, collector channels, and a main channel. Solution for different sections is by the finite difference method. For the river routing component, the inflow hydrograph from the land surface area is routed downstream along the river using standard flood routing procedures such as the Muskingum method. Reservoir component is routed in the same way as for rivers, except that the reservoir provides storage only. Abstractions and inflows, if present, are linked to the model nodes. The parameters of HEC-1 model are obtained mostly by optimization.

b. HEC-HMS

HEC-HMS (USACE, 2000) is a generalized modelling system, developed in 1992 to supersede HEC-1, and is the standard model used in private sector in the United States. It is, in some sense, distributed in that the catchment is subdivided into smaller subcatchments. Any mass or energy flux in the hydrological cycle can then be represented by a mathematical model. In most cases, several model choices are available for representing each flux. Each mathematical model included in the program is suitable in different environments and under different conditions. Making the correct choice requires knowledge of the catchment, the goals of the hydrologic study, and engineering judgment. HEC-HMS provides a number of options for simulating the rainfall–runoff process. These include precipitation modelled using either actual gauged events or hypothetical, frequency-based storms, rainfall losses represented empirically (SCS) or with physically based algorithms (Green and Ampt, 1911); runoff generated from unit hydrographs or kinematic wave method; stream routed by Muskingum, Muskingum–Cunge, and kinematic wave methods; and reservoir routing, base flow, and diversions modelling.

The inputs to HEC-HMS consist of three components:

- A basin component, which is a description of the different elements of the hydrologic system (subbasins, channels, junctions, sources, sinks, reservoirs, and diversions), including their hydrologic parameters and topology
- A precipitation component, which is a description–in space and time–of the precipitation event to be modelled, and consists of time series of precipitation at specific points or areas and their relation to the hydrologic elements (precipitation distribution can be by arithmetic mean, Thiessen polygon, or isohyetal method)
- A control component, which defines the time window for the precipitation event and for the calculated flow hydrograph

Parameters are estimated by optimization. The software can support GIS and is well documented. The source code is not freely available.

Evaporation, as modelled in HEC-HMS, includes evaporation from open water surfaces as well as transpiration from vegetation combined into evapo-transpiration (ET). In this input, monthly-varying ET values are specified, along with an ET coefficient. The potential ET rate for all time periods within the month is computed as the product of the monthly value and the coefficient.

c. HEC-RAS

HEC-RAS is a program released in 1995 that can carry out 1D steady flow, unsteady flow, sediment transport/mobile bed computations, and water temperature modelling. Steady flow is modelled by solving the 1D energy equation. The momentum equation is used in situations where the flow is rapidly varied, such as in the case of a hydraulic jump. Unsteady flow is modelled by solving the full, dynamic, St. Venant equations using an implicit finite difference method.

HEC-RAS can model a network of channels, a dendritic system, or a single river reach as well as subcritical, supercritical, and mixed flows with the effects of bridges, culverts, and other hydraulic structures. HEC-RAS is well tested, peer reviewed, and can be downloaded freely. On the negative side, numerical instability may occur in unsteady flow computations under steep flow conditions.

d. HEC-FDA

HEC-FDA provides the capability to perform an integrated hydrologic engineering and economic analysis during the formulation and evaluation of flood risk management plans. It is designed to assist in using risk analysis procedures for formulating and evaluating flood risk management measures.

e. HEC-ResSim

HEC-ResSim is designed to be used to model reservoir operations at one or more reservoirs whose operations are defined by multiple goals and constraints.

2.8.1.4 Xinanjiang model

The Xinanjiang model has been successfully and widely applied in humid and semihumid regions in China since its development in the 1970s (Zhao, 1977, 1984; POYB, 1979; Zhao et al., 1980; Zhang, 1990; Zhao and Liu, 1995). The Xinanjiang model provides an integral structure to statistically describe the non-uniform distribution of runoff producing areas, which features it as one of the conceptual, semidistributed hydrological models developed. The original Xinanjiang model uses a single parabolic curve to represent the spatial distribution of the soil moisture storage capacity over the catchment, where the exponent parameter b measures the non-uniformity of this distribution. The same function is also employed in the variable infiltration capacity (VIC) model (Wood et al., 1992) and in the ARNO model (Todini, 1996).

The Xinanjiang model consists of a runoff generating component and a runoff routing component. The runoff generation of the Xinanjiang model acts on three vertical layers identified as upper, lower, and deep. The upper layer refers to the vegetation, water surface, and the very thin top soil. The lower layer refers to the soil in which the vegetation roots dominate and the moisture transportation is mainly driven by the potential gradient. The deep layer refers to the soil beneath the lower layer where only the deep-rooted vegetation can absorb water, and the potential gradient is very small. Replenishment and depletion of soil moisture storage take place, respectively, via rainfall and evapo-transpiration. In the upper layer, evapo-transpiration takes place at the potential rate. On exhaustion of moisture content in the upper layer, evapo-transpiration proceeds to the lower layer at a decreased rate that is proportional to the moisture content in that layer. Only when the total evapo-transpiration in the upper and lower layers is less than a preset threshold, represented as a fraction of the potential evapo-transpiration, does it further proceed to the deep layer to keep this preset minimum value.

During rainfall, evaporation is first subtracted and the runoff generation is computed by considering the respective soil moisture states and the storage capacities on the three layers. The runoff is separated into three components: immediate runoff, surface runoff, and groundwater runoff, according to their generating levels in the vertical profile. Immediate runoff is the net rainfall directly falling on the impervious and saturated area, which is expressed as a percentage of the total watershed area. After the storage deficit is satisfied in the upper and lower layers, the remaining excess rainfall infiltrates into the deep layer at a constant rate through the bottom of the lower layer to form the groundwater runoff. Surface flow and interflow are generated to form the surface runoff only when the excess rainfall intensity is greater than this constant infiltration. Immediate runoff, surface flow, and interflow together contribute to the direct runoff.

In general, the Xinanjiang model is used with spatially uniform rainfall events. For non-uniform rainfalls, the basin can be divided into several subbasins. The Xinanjiang model is then applied to each subbasin with the average rainfall of each subbasin as the input. The complete Xinanjiang model has a total of 11 parameters of which 8 are for runoff generation and 3 for runoff routing.

The Xinanjiang model simulates the runoff generation on partial areas by considering the non-uniformity of the spatial distribution of soil moisture storage capacity over the catchment. In the original Xinanjiang model, it is represented by a soil moisture storage capacity curve (Figure 2.5a) defined as follows

$$\frac{f}{F} = 1 - \left(1 - \frac{W'_m}{W'_{mm}}\right)^b ; \quad 0 \le \frac{W'_m}{W'_{mm}} \le 1 \tag{2.116}$$

where f is the partial pervious area of the watershed whose soil moisture storage capacity is less than or equal to the ordinate W'_m, which varies from zero to its maximum value W'_{mm} (Figure 2.5a); F is the total pervious area of the watershed; and b is the exponent of the curve.

The original Xinanjiang model has subsequently been modified (Jayawardena and Zhou, 2000) by introducing a double parabolic curve in place of the original single parabolic curve. The double parabolic curve (Figure 2.5b) takes the form

$$\frac{f}{F} = (0.5 - c)^{1-b} \left(\frac{W'_m}{W'_{mm}}\right)^b ; \quad 0 \le \frac{W'_m}{W'_{mm}} \le 0.5 - c \tag{2.117a}$$

and

$$\frac{f}{F} = 1 - (0.5 - c)^{1-b} \left(1 - \frac{W'_m}{W'_{mm}}\right)^b ; \quad 0.5 - c \le \frac{W'_m}{W'_{mm}} \le 1 \tag{2.117b}$$

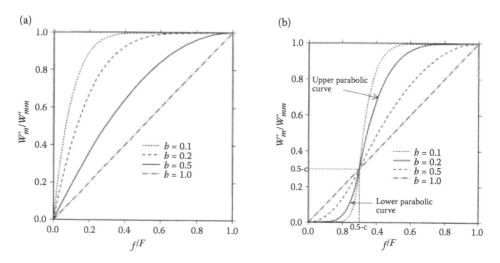

Figure 2.5 (a) Soil moisture storage capacity by a single parabolic curve. (b) Soil moisture storage capacity by a double parabolic curve.

where all notations have the same meanings as in Equation 2.116 except for the additional parameter c, a weighting factor. It is continuous and smooth at the point where the two branches meet and has its function and first derivative values of $(0.5 - c)$ and (b/W'_{mm}). It can be seen that this double parabolic curve is more flexible and easier to be calibrated than the original single parabolic curve.

The double parabolic curve consists of lower and upper branches, taking weights of $(0.5 - c)$ and $(0.5 + c)$, respectively, as shown in Figure 2.5b. In the interval $[0, 1]$ of f/F, the lower branch occupies the $[0, 0.5 - c]$ part, whereas the upper branch occupies the remaining part, $(0.5 - c, 1)$. The parameter c varies in an interval of $[-0.5, 0.5]$. When the weighting parameter c takes a value of (-0.5), the double parabolic curve reduces to the single lower parabolic curve:

$$\frac{f}{F} = \left(\frac{W'_m}{W'_{mm}} \right)^b ; 0 \le \frac{W'_m}{W'_{mm}} \le 1 \tag{2.118}$$

When the weighting parameter c takes a value of 0.5, the double parabolic curve reduces to the single upper parabolic curve that is used in the original Xinanjiang model as Equation 2.116. Thus, the single parabolic curve is a special case of the double parabolic curve that uses an additional weighting parameter c to adjust the shape of the curve, so that it is closer to the spatial distribution of soil moisture storage capacity.

The Xinanjiang model gives only the runoff generating rainfall (or rainfall excess). It then needs to be routed along the channel network to obtain the required discharge. Several approaches of routing such as the Muskingum method, unit hydrograph method, and kinematic wave method can be used.

2.8.1.5 Variable infiltration capacity (VIC) model

The VIC model is a semidistributed conceptual hydrological model that characterizes a watershed as multilayered grids and employs distributed information of the watershed. It was developed originally as a soil–vegetation–atmosphere transfer scheme (SVATs) for the purpose of representing the land surface in general circulation models (GCMs) used for climate simulation and numerical weather prediction (Wood et al., 1992). Distinguishing characteristics of the VIC model include the representation of subgrid variability in land surface vegetation classes, the representation of subgrid variability in soil moisture storage capacity as a spatial probability distribution, and the representation of drainage from the lower soil moisture zone (baseflow) as a non-linear recession (Liang et al., 1994).

Using the infiltration formulation used in the Xinanjiang model (Zhao et al., 1980), the VIC model has been developed for decades to form a family of models that includes model structures from a single layer to three layers sequenced as VIC-1L, VIC-2L, and VIC-3L. The original VIC model, with one layer to represent the variation in infiltration capacity within a GCM grid cell (Wood et al., 1992), was progressed to consist of a two-layer characterization of soil column (i.e., VIC-2L) that uses an aerodynamic representation of the latent and sensible heat fluxes at the land surface with different land surface and evapo-transpiration description (Liang et al., 1994). The VIC-2L model was further modified to add one thin surface layer (i.e., VIC-3L) to better capture the dynamic behaviour of soil moisture content. In addition, the diffusion process was included in the representation of the VIC soil column (Liang et al., 1996b). Additional progress includes improved snow representation (Storck and Lettenmaier, 1999), improved ground heat flux parameterization (Liang et al., 1999), efficient parameterization of subgrid precipitation (Liang et al., 1996a), improvements for model performance in cold regions (Cherkauer and Lettenmaier, 1999), efforts to incorporate a double parabolic curve for describing the infiltration capacity (Jayawardena and Mahanama, 2002; Jayawardena and Zhou, 2000), and improved runoff generation strategies and interactions between surface and groundwater (Liang and Xie, 2001). In the Asian region, VIC-3L has been applied to predict daily discharges in Mekong River, which is an international river running through China, Myanmar, Cambodia, Lao, Thailand, and Vietnam (Jayawardena and Tian, 2005; Jayawardena, 2006; Tian, 2007).

Among the VIC series models, the VIC-3L is the latest version (as of 2007) and it can combine most model characteristics. The components of the model can be described as follows (Liang et al., 1996b; Matheussen et al., 2000):

- The land surface is described horizontally by $N + 1$ land cover (vegetation) types, where $n = 1, 2, ..., N$ represents N different types of vegetation, and $n = N + 1$ represents bare soil.
- The subsurface is characterized vertically with three soil layers identified as the upper layer, which is partitioned into a top thin layer (layer 0) and a thicker layer (layer 1), and a lower layer (layer 2). The top thin layer allows a quick response due to changes in surface conditions. The infiltration algorithm of the upper layer represents the dynamic fast response to rainfall. The lower layer characterizes the slowly varying soil moisture behaviour.
- Associated with each land cover class is a single canopy layer and three soil layers. The land cover classes are specified by their leaf area index (LAI), canopy resistance, and relative fraction of roots in each of the three soil layers, which depend on the vegetation class and the soil type.
- Total evapo-transpiration over a grid cell is computed as the summation of the canopy, vegetation, and bare soil components, weighted by the respective surface cover area fractions. The evapo-transpiration from each vegetation type within the grid cell is characterized by its potential evapo-transpiration, aerodynamic resistance to the transfer of water, and architectural resistance.
- Surface runoff (direct flow) generates only from the upper layer. There is vertical soil moisture movement by drainage and diffusion. Drainage from the top thin layer and the upper thick layer to the corresponding next layer is based on the hydraulic conductivity of the current layer. The lower layer produces subsurface flow (baseflow) through the ARNO model (Franchini and Pacciani, 1991). In this way, the model separates subsurface runoff from quick storm response.
- The infiltration, the fluxes of moisture between soil layers, and runoff vary with vegetation cover class. Surface runoff and subsurface runoff are computed for each cover type and summed over all land cover types within a grid cell as model output.

As the core of the model, the VIC model assumes that the infiltration capacity (defined as the maximum depth of water that can be stored in the soil column), and therefore runoff generation and evapo-transpiration, vary within an area (grid cell or catchment) due to variations in topography, soil, and vegetation (Wood et al., 1992). The variation in infiltration capacity i over an area is expressed (Figure 2.6) as

$$i = i_m[1-(1 - A)1/^B]$$ (2.119)

where A is the fraction of a grid cell area for which the infiltration capacity is less than i ($0 \leq A \leq 1$), i_m is the maximum infiltration capacity within the grid cell [mm], and B is the infiltration shape parameter [dimensionless], which is a measure of the spatial variability of the infiltration capacity.

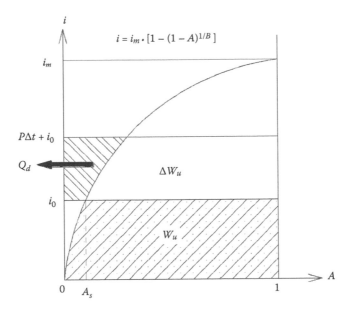

Figure 2.6 Schematic representation of the infiltration capacity curve. Note: P is the precipitation [mm/day], Q_d is the surface runoff [mm/day], W_u and ΔW_u are the soil moisture storage and the change in the soil moisture storage in the upper layer [mm], A_s is the fraction of an area that is saturated, i_0 is the corresponding point infiltration capacity [mm], and Δt is the time step. (Reproduced after Wood, E.F., Lettenmaier, D.P. and Zartarian, V.G., Journal of Geophysical Research, 97, 2717, 1992.)

Helpful information on summary of the VIC processes, model operation overview, and general guide for inputs and outputs processing as well as important references are available from the homepage of the VIC model: http://www.hydro.washington.edu/Lettenmaier/Models/VIC/VIChome.html.

The VIC model assumes that surface runoff (direct flow) is generated by those areas for which precipitation, when added to the soil moisture storage at the end of the previous time step, exceeds the storage capacity of the soil. When the soil is saturated, the drainage term follows the saturated hydraulic conductivity. Otherwise, the drainage term between soil layers are calculated based on the Clapp–Hornberger relationship (Clapp and Hornberger, 1978). Finally, the grid-based runoff is routed through the channel network in the same way as in the Xinanjiang model.

2.8.2 Physics-based hydrologic models

2.8.2.1 Système Hydrologique Europèen (SHE) model

Système Hydrologique Europèen (SHE) is a physically based, distributed, integrated hydrological modelling system produced jointly by the Danish Hydraulic Institute (DHI; http://www.dhisoftware.com), the British Institute of Hydrology (now the Centre for Ecology and Hydrology), and the French consulting company SOGREAH with the financial support of the Commission of the European Communities. Since 1987, SHE has been further developed independently by the University of Newcastle (UK), Laboratoire d'Hydraulique de France, and DHI. DHI's version of SHE, known as the MIKE SHE, represents significant new developments with respect to user interface, computational efficiency, and process descriptions. For the background to the development of SHE, as well as the descriptions of model structures and process equations in SHE, the reader is referred to the reviews by Abbott et al. (1986a,b), Refsgaard and Knudsen (1996), and Refsgaard (1997). The main modules consist of an evapo-transpiration component, an unsaturated zone flow component described by the 1D Richard's equation, a saturated zone flow component described by the 3D Boussinesq equation, and overland flow (2D) and channel flow (1D) components described by the diffusion wave approximation of the Saint Venant equations.

SHE enables the simulation of water, solutes, and sediments in the entire land phase of the hydrological cycle. It is a dynamic, user-friendly modelling tool for the analysis, planning, and management of a wide range of water resources and environmental problems related to surface water and groundwater, including surface water/groundwater interaction, conjunctive use of water, water resources management, irrigation management, changes in land use practices, contaminant transport in the subsurface, and farming practices including fertilizers and agrochemicals.

SHE is an integrated modelling system with a modular structure. Individual components can be used independently and customized to local needs depending on data availability and aims of the given study. Powerful preprocessing and results presentation tools are included in the SHE software package. SHE contains a number of process simulation modules, which, in combination, describes the entire land phase of the hydrological cycle.

MIKE SHE solves the partial differential equations governing overland and channel flow, unsaturated and saturated subsurface flow. In the complete model, processes such as snow melt, interception, and evapo-transpiration are also included. The flow equations are solved numerically using finite difference methods. In the horizontal plane, the catchment is discretized into a network of grid squares. The river system is assumed to run along their boundaries. Within each square, the soil profile is described at a number of nodes, which above the groundwater table may become partly saturated. Lateral subsurface flow is only considered in the saturated part of the profile. The commercially available MIKE-11 is a comprehensive, 1D modelling system for the simulation of flows, sediment transport, and water quality in estuaries, rivers, irrigation systems, and other water bodies. Its 2D counterpart MIKE-21 is a comprehensive modelling system for 2D free surface flows applicable to studies of lakes, reservoirs, estuaries, bays, coastal areas, and seas where stratification can be neglected.

Since the SHE model is a physically based one, the parameters should be physically identifiable and measurable, at least in theory. This implies that calibration is not necessary if sufficient data are available. In practice, however, it is difficult if not impossible to make such measurements in the field because of catchment heterogeneities. A compromise is often made in resorting to some calibration technique, thereby diluting the true meaning of a physically based distributed model. The goal then is to find an optimal set of parameters that are also realistic that will give the expected outcome.

2.8.3 Data-driven models

The type of hydrological modelling and prediction used for a particular situation depends on the richness of the theory as well as the richness of the data. For theory-rich situations, knowledge-based models that use the principles and laws of physics can be used. They generally lead to the problem of finding solutions to a set of partial differential equations. Even under the simplest assumptions, theoretical solutions to practical problems cannot be found, and therefore, resort is often made to numerical solutions in a simplified spatial domain. They are useful for understanding the system dynamics under often unrealistic assumptions and cannot be applied to real situations. The alternative then is to go for data-driven approaches that do not require a prior understanding of the dynamics of the system. Such approaches are particularly suited to data-rich, theory-weak situations.

The earliest data-driven approach may perhaps be the regression analysis in which a simple relationship is sought between dependent and independent variables of a system. A more traditional data-driven approach of modelling and predicting complex hydrological systems considers such systems as stochastic, which at least in theory can have an infinite number of degrees of freedom. Such approaches have been used in the past, but researchers are now turning towards the realization that certain systems that are seemingly stochastic are in fact driven by fully deterministic processes. Several new techniques of analysis and prediction have emerged in the last decade or so, and the frontier in this field is still moving forward. Emphasis has over the years shifted from global modelling in which attempts are made to find a function or a model that will fit into the entire data set to local modelling in which function approximations are done in a linear piecewise manner. Many of the new techniques have originated in mathematical and statistical domains, and subsequently made their way into engineering fields.

2.8.3.1 Why data-driven models?

Data-driven models have many advantages over other types of models. They are relatively simple to formulate and easy to implement. Data contain all the information about the system and their use is quite logical. Certain data-driven modelling techniques can learn from experience, generalize from examples, and there is less reliance on expertise. They are particularly suited for data-rich, theory-weak problems. They can also implement non-linear mapping. Data-driven models become the only choice when other approaches are not feasible.

Physics-based models consider the catchment processes from a physics point of view; however, their formulation, calibration, and implementation are quite resource and expertise demanding. Thus far, no fully physics-based model has been successfully applied to a catchment without making drastic assumptions and simplifications.

2.8.3.2 Types of data-driven models

There are many types of data-driven models. Regression models may perhaps be the earliest, followed by stochastic models. More recently, several types of artificial neural networks such as multilayer perceptrons, radial basis functions, recurrent neural networks, wavelet neural networks, product unit neural networks and fuzzy neural networks, fuzzy logic systems, support vector machines, genetic algorithms, genetic programming, and dynamical systems approach have found applications in hydrology. Details of some of these techniques will be given in the chapters that follow.

2.9 GUIDING PRINCIPLES AND CRITERIA FOR CHOOSING A MODEL

A model should be useful to solve or understand a particular problem under a given set conditions and constraints with a reasonable balance between the costs and benefits. Many models and modelling techniques add only a marginal value at an unjustifiable cost. Data-driven models are relatively easier to implement, but not without problems. Physics-based models are more difficult. Their formulation, calibration, and implementation are quite resource and expertise demanding. Their problems are also of a higher magnitude.

Opinions about whether a model should be resources driven or needs driven, whether it should be simple or complex, whether it is for a specific purpose to solve a problem or for an academic purpose for better understanding of the system, and whether it is the end result that matters or how it is obtained are divided. A model should be as simple as possible (parsimonius). What matters is not the physics or the mathematics or the degree of sophistication, but whether the model is reliable, whether it has the repeatability property, and whether it works or not. It should be dynamic and capable of updating as and when new data become available, robust, and well tested.

2.10 CHALLENGES IN HYDROLOGICAL MODELLING

The main challenges in the hydrological modelling include choosing between stochastic and deterministic approaches, lumped and distributed approaches, linear and non-linear approaches, and stationary and non-stationary assumptions. Most data-driven models are lumped. Linear assumption makes subsequent analysis and application simple; however, in many instances, it is far from reality. Non-linear assumption makes the problem more realistic, but at a cost and lacks generality. Similarly, stationarity assumption makes analysis and application simpler; however, with human influence in the hydrologic system, the stationarity assumption no longer holds in many situations.

For physics-based distributed models, it is necessary to define a set of governing equations. They are defined for a continuum, and whether such equations are valid in the scale of typical distributed models is an unresolved issue. Scale issue plays an important role. Governing equations include the St. Venant equations for overland flow; Richard's equation for soil water flow; diffusion type equation based on Darcy's law and continuity for groundwater flow; Horton, Philip, Green and Ampt type equations for infiltration; and mass transfer, aerodynamic, or combination type equations for evaporation. Inclusion of these equations and any other equations that represent processes such as interception, depression storage, etc., into the model requires some approximations and assumptions while their exclusion results in oversimplification. Despite the detailed description of the catchment processes and properties, measured quantitative data on land cover, canopy resistance, and fraction of roots are not available except in a few places. As a result, many hydrological modellers use default values set by the model developers and coarse data available in public domains. The validity of the use of such default values is certainly in question. Although this kind of distributed approach is expected to produce output variable values at each grid point in the mesh, they cannot be validated due to lack of data.

Identification of parameters is another issue. They may be physically identifiable and measurable but at a cost and may not be representative due to spatial and temporal heterogeneities. In many instances, the parameters are estimated using some optimization technique. When optimization option, which requires minimizing an error function, is adopted, it is also important to ensure that the objective function does not end at a local minimum. Popular global search methods include population-evolution-based search strategies, such as the Shuffled Complex Evolution (SCE) algorithm (Duan et al., 1993), simulated annealing (Corana et al., 1987; Goffe et al., 1994) and Genetic Algorithm (GA) (Wang, 1991). Sometimes it is desirable to use multiobjective optimization rather than single objective optimization. On the basis of the original SCE algorithm, recent studies have led to the development of the Shuffled Complex Evolution Metropolis (SCEM) and the Multi-objective Shuffled Complex Evolution Metropolis (MOSCEM) algorithms (Vrugt et al., 2003a,b). Direct comparison of these methods would be helpful in selecting the most suitable calibration algorithm from the extensively used SCE family of algorithms. For physics-based models, which are necessarily of a distributed nature, the use of optimization techniques for calibration defeats the purpose. As a result, most models that start with laws of physics end up as data-driven models, thereby defeating the very purpose of adopting such an approach.

The accuracy and reliability of the outcome of a model depends on the accuracy and reliability of the data used as inputs. For simple hydrological models, the basic input is rainfall, which varies spatially and temporally. Present-day rain gauges can measure rainfall to a very high degree of accuracy, but a reasonable spatial and temporal resolution is necessary to ensure that the data are representative. Averaging out data has the tendency to smooth out variations, thereby distorting the real situation. A compromise is often needed to strike a balance between the resources available and the accuracy of the expected result. The second most important hydrological variable for modelling is the discharge resulting from rainfall, which can be considered as an integrator of all catchment-scale processes. Direct measurements of discharges are rarely made under normal conditions. They are derived from stage measurements using rating curves. Stage measurements can be made quite precisely; however, the rating curves depend on many factors such as the techniques and instruments used to measure velocities and channel hydraulic parameters, and whether or not measurements cover the entire range of possible values. Very often, rating curves are established under normal flow conditions, and extrapolated to obtain discharges at high flow conditions, thereby introducing an uncertain error. Measurements at high flow conditions are usually not made because they are difficult, dangerous, and costly. For early warning of impending floods, it is desirable to use the stage as the dependent variable as it is more reliable than the discharge and much easier to comprehend.

In addition to hydrological data, geometrical and topographical data are needed for distributed type of models. On a local scale, such data can be found in limited situations. The resolutions vary and depend on the region and the catchment. On a global or regional scale, remotely sensed topographical data are available, particularly from satellite observations. Their resolutions also vary, but at the present time, the publicly available data sets do rarely have resolutions finer than 1 km × 1 km horizontally, and a few tens of metres vertically. The results of any distributed model that uses such coarse data will have inherent errors of the same order or higher, than those of the input topographical data.

Challenges in hydrological modelling arise as a result of inadequacy of resources for research, lack or insufficiency of relevant data, lack or insufficiency of expertise, and in many instances, a lack of a clear driving force for any modelling attempt. The choice should be needs driven rather than resources driven. When it is resources driven, the marginal potential benefit should be weighed against costs associated with uncertainties and inaccuracies of data, model formulation, and calibration issues. Hydrologic modelling based on the linear and stationary assumptions has reached saturation levels. Advances in analysis of non-linear systems in recent years have made it possible to explore the hydrologic system (which has always been non-linear but for simplicity has been assumed linear) within a non-linear framework. There is also evidence to believe that the stationarity assumption made in the analysis of hydrologic time series in the past is no longer valid due to human interference in the natural hydrologic cycle. Any future approach of hydrologic modelling should take into account this shift in paradigm and consider the vital role that non-linear non-stationary dynamics and scaling theories can play. It is also important to consider the costs and benefits associated with any modelling attempt.

The main hurdle for understanding the catchment-scale hydrologic cycle is the lack of data, and the coarse resolution even when available. There are more than adequate frameworks for physics-based models available in the literature, but none of them can be verified in the catchment scale due to lack of field data. To make any headway, modelling and data collection should go hand in hand. Remote sensing can play a vital role in this aspect. Satellite-based data collection can cover a large area, but their resolution and the calibration against ground-measured data are issues that are not yet resolved but receiving attention. Calibration of parameters and uncertainty analysis are other areas that need attention.

In environmental and hydrological modelling, two models are said to be equi-final if they lead to equally acceptable results. The concept of equi-finality originated in the general systems model of Bertalanffy (1934, 1968), meaning that the same final result may be arrived from different initial conditions and in different ways. In the context of multiparameter optimization, what this means is that there is no unique set of parameter values, but rather a feasible parameter space from which a Pareto set of solutions is sought.

2.11 CONCLUDING REMARKS

Hydrological systems modelling started with the rational formula and has since then advanced a long way. It is still developing and the frontiers are still moving forward. This chapter gives a glimpse of the developments made up to recent years. It is in no way comprehensive since new knowledge is added continuously.

Modelling is only an attempt to understand a complex system. In the context of hydrological modelling, developments in the past have taken place under two main assumptions. For deterministic modelling, the main assumption is linearity of the rainfall excess–direct runoff process, whereas for stochastic modelling it is the assumption of stationarity. Both these assumptions are far from reality but nevertheless have helped hydrologists to solve many practical problems.

At the next level of modelling, attempts should be made to get closer to reality. It is no longer necessary to assume that the hydrological system is linear or stationary. There are new techniques of analysis that can take care of non-linearities as well as non-stationarities. Developments of such techniques have come from disciplines as diverse as physics, mathematics, statistics, and control engineering, and are gradually finding their way to applications in hydrology. It is therefore important to keep abreast of developments in basic sciences fields to see where hydrologists can benefit. A few of the emerging fields in data-driven approaches of modelling will be presented in the remaining chapters.

REFERENCES

Abbott, M.B., Bathurst, J.C., Cunge, J.A., O'Connell, P.E. and Rasmussen, J. (1986a): An introduction to the European Hydrological System, Système hydrologique Europèen, SHE-1, History and philosophy of a physically based, distributed modelling system. *Journal of Hydrology*, 87, 45–59.

Abbott, M.B., Bathurst, J.C., Cunge, J.A., O'Connell, P.E. and Rasmussen, J. (1986b): An introduction to the European Hydrological System, Système hydrologique Europèen, SHE, 2, Structure of a physically based, distributed modelling system. *Journal of Hydrology*, 87, 61–77.

Amorocho, J. (1963): Measures of the linearity of hydrologic systems. *Journal of Geophysical Research*, 68(8), 2237–2249.

Bertalanffy, L. von (1934): Untersuchungen über die Gesetzlichkeit des Wachstums. I. Allgemeine Grundlagen der Theorie; mathematische und physiologische Gesetzlichkeiten des Wachstums bei Wassertieren. *Archives Entwicklungsmech*, 131, 613–652.

Bertalanffy, L. von (1968): *General System Theory: Essays on its Foundation and Development*, Revised Edition. George Braziller, New York.

Beven, K.J. and Kirkby, M.J. (1979): A physically based, variable source area model of basin hydrology. *Hydrological Sciences Bulletin*, 24, 43–69.

Cherkauer, K.A. and Lettenmaier, D.P. (1999): Hydrologic effects of frozen soils in the upper Mississippi River basin. *Journal of Geophysical Research-Atmospheres*, 104(D16), 19599–19610.

Clapp, R.B. and Hornberger, G.M. (1978): Empirical equations for some soil hydraulic properties. *Water Resources Research*, 14, 601–604.

Clark, C.O. (1945): Storage and the unit hydrograph. *Transactions, ASCE*, 110, 1419–1446.

Corana, A., Marchesi, M., Martini, C. and Ridella, S. (1987): Minimising multimodal functions of continuous variables with the simulated annealing algorithm. *ACM Transactions on Mathematical Software*, 13, 262–280.

Crawford, N.H. and Linsley, R.K. (1966): Digital simulation in hydrology: Stanford watershed model IV. Technical Report No. 39, Department of Civil Engineering, Stanford University, 210 pp.

Cunge, J.A. (1969): On the subject of a flood propagation computation method (Muskingum method). *Journal of Hydraulic Research*, 7(2), 205–230.

Delleur, J.W. and Rao, R.A. (1971): Linear systems analysis in hydrology – The transform approach, the kernel oscillations and the effect of noise. *Proc. US-Japan Bi-Lateral Seminar in Hydrology*, Honolulu, Jan., pp. 116–129.

Dirac, P. (1958): *Principles of Quantum Mechanics*, 4th edition. Oxford at the Clarendon Press, Oxford, UK.

Diskin, M.H., Ince, S. and Oben-Nyarko, K. (1978): Parallel cascades model for urban watersheds. *Journal of the Hydraulics Division, ASCE*, 104(HY2), 261–276.

Dooge, J.C.I. (1959): A general theory of the unit hydrograph. *Journal of Geophysical Research*, 64(2), 241–256, doi:10.1029/JZ064i002p00241.

Duan, Q., Gupta, V.K. and Sorooshian, S. (1993): A shuffled complex evolution approach for effective and efficient global minimization. *Journal of Optimization Theory and Applications*, 76(3), 501–521.

Eagleson, P.S. (1970): *Dynamic Hydrology*. McGraw Hill, New York.

Franchini, M. and Pacciani, M. (1991): Comparative analysis of several conceptual rainfall-runoff models. *Journal of Hydrology*, 122, 161–219.

Green, W.H. and Ampt, G. (1911): Studies of soil physics. Part I: The flow of air and water through soils. *J. Agricultural Sci.*, 4, 1–24.

Goffe, W.L., Ferrier, G.D. and Rogers, J. (1994): Global optimisation of statistical functions with simulated annealing. *Journal of Econometrics*, 60, 65–99.

Henderson, F.M. and Wooding, R.A. (1964): Overland flow and groundwater flow from a steady rainfall of finite duration. *Journal of Geophysical Research*, 69(8), 1531–1540.

Hjelmfelt, A.T., Jr. (1985): Negative outflows from Muskingum flood routing. *Journal of Hydraulic Engineering, ASCE*, 111(6), 1010–1014.

Jayawardena, A.W. (1988): Stream flow simulation using tank model: Application to Hong Kong catchments, Hong Kong Engineer. *Journal of the Hong Kong Institution of Engineers*, 16(7), 33–36.

Jayawardena, A.W. (2006): Calibration of VIC model for daily discharge prediction of Mekong River using MOSCEM algorithm. *Proceedings of the 3rd APHW Conference*, Bangkok, Thailand, Oct. 16–18, 2006 (abstract in CD ROM, p 256).

Jayawardena, A.W. and Mahanama, S.P.P. (2002): Meso-scale hydrological modelling: Application to Mekong and Chao Phraya Basins. *Journal of Hydrologic Engineering, ASCE*, 7(1), 12–26.

Jayawardena, A.W. and White, J.K. (1977): A finite element distributed catchment model, I – Analytical basis. *Journal of Hydrology*, 34(3–4), 269–286.

Jayawardena, A.W. and White, J.K. (1979): A finite element distributed catchment model, II – Application to real catchments. *Journal of Hydrology*, 42(3–4), 231–249.

Jayawardena, A.W. and Zhou, M.C. (2000): A modified spatial soil moisture storage capacity distribution curve for the Xinanjiang model. *Journal of Hydrology*, 227(1–4), 93–113.

Jayawardena, A.W. and Tian, Y. (2005): Flow modelling of Mekong River with variance in spatial scale. In: Herath, S., Dutta, D., Weesakul, U. and Gupta, A.D. (Eds.), *Proceedings of the International Symposium on "Role of Water Sciences in Transboundary River Basin Management"*, Ubon Ratchathani, Thailand, March 10–12, 2005, pp. 147–154.

Kreyszig, E. (1999): *Advanced Engineering Mathematics*, 8th edition. John Wiley & Sons, Inc., New York.

Lattermann, A. (1991): *System-Theoretical Modelling in Surface Water Hydrology*. Springer-Verlag, Berlin, 200 p.

Liang, X., Lettenmaier, D.P. and Wood, E.F. (1996a): One-dimensional statistical dynamic representation of subgrid spatial variability of precipitation in the two-layer Variable Infiltration Capacity model. *Journal of Geophysical Research*, 101(D16), 21403–21422.

Liang, X., Lettenmaier, D.P., Wood, E.F. and Burges, S.J. (1994): A simple hydrologically based model of land surface water and energy fluxes for general circulation models. *Journal of Geophysical Research*, 99(D7), 14415–14428.

Liang, X., Wood, E.F. and Lettenmaier, D.P. (1996b): Surface soil moisture parameterization of the VIC-2L model: Evaluation and modification. *Global and Planetary Change*, 13(1–4), 195–206.

Liang, X., Wood, E.F. and Lettenmaier, D.P. (1999): Modelling ground heat flux in land surface parameterization schemes. *Journal of Geophysical Research*, 104(D8), 9581–9600.

Liang, X. and Xie, Z. (2001): A new surface runoff parameterization with subgrid-scale soil heterogeneity for land surface models. *Advances in Water Resources*, 24(9–10), 1173–1193.

Lighthill, M.J. and Whitham, G.B. (1955): On kinematic waves I: flood movement in long rivers. *Proceedings, Royal Society of London, Series A*, 229(1178), 281–316.

Maidment, D.R. (1993): Application of geographic information systems in hydrology and water resources. *Proceedings of the Vienna Conference (HydroGIS 93)*, April 1993, IAHS Publication No. 211, pp. 181–192.

Matheussen, B., Kirschbaum, R.L., Goodman, I.A., O'Donnell, G.M. and Lettenmaier, D.P. (2000): Effects of land cover change on streamflow in the interior Columbia River Basin (USA and Canada). *Hydrological Processes*, 14(5), 867–885.

McCarthy, G.T. (1938): The unit hydrograph and flood routing. *Conference North Atlantic Division, US Army Corps of Engineers*, New London, CT.

McCuen, R.M. (2004): *Hydrologic Analysis and Design*. Prentice Hall, NJ, 859 pp.

Mulvany, W.T. (1845): Observations on regulating weirs. *Transactions of the Institution of Civil Engineers Ireland*, I, 83–93.

Nash, J.E. (1957): The form of the instantaneous unit hydrograph. *International Association of Scientific Hydrology Publication*, 45(3), 114–121.

Nash, J.E. (1958): Determining runoff from rainfall. *Proceedings of the Institution of Civil Engineers*, 10, 163–184.

Nash, J.E. (1959): Systematic determination of unit hydrograph parameters. *Journal of Geophysical Research*, 64(1), 111–115.

Nash, J.E. (1960): A unit hydrograph study, with particular reference to British catchments. *Proceedings of the Institution of Civil Engineers*, 17, 249–282.

Ponce, V.M. (1989a): A general dimensionless unit hydrograph. (Ponce.tv/general_dimensionless_unit_hydrograph.html).

Ponce, V.M. (1989b): *Engineering Hydrology: Principles and Practices*. Prentice Hall, Englewood Cliffs, NJ.

POYB (Planning Office of Yangtze Basin) (1979): *Methods of Hydrological Forecasting*. Water Resource and Electric Press, Beijing, pp. 89–112 (in Chinese).

Ragazzini, J.R. and Zadeh, L.A. (1952): The analysis of sampled-data systems. *Transactions of the American Insttitute of Electrical Engineers*, 71(II), 225–234.

Ramiréz, J.A. (2000): Prediction and modelling of flood hydrology and hydraulics. In: Wohl, E. (Ed.), *Chapter 11 of Inland Flood Hazards: Human, Riparian and Aquatic Communities*. Cambridge University Press, Cambridge.

Refsgaard, J.C. (1997): Parameterization, calibration and validation of distributed hydrological models. *Journal of Hydrology*, 198, 69–97.

Refsgaard, J.C. and Knudsen, J. (1996): Operational validation and inter-comparison of different types of hydrological models. *Water Resources Research*, 32(7), 2189–2202.

Rodríguez-Iturbe, I. and Valdés, J.B. (1979): The geomorphologic structure of hydrologic response. *Water Resources Research*, 15(6), 1409–1420.

Sherman, L.K. (1932): Streamflow from rainfall by the unit hydrograph method. *Engineering News Record*, 108, 501–505.

Singh, K.P. (1964): Non-linear unit hydrograph theory. *Journal of Hydraulics Division, Proceedings, ASCE*, 90(HY2), 313–347.

Snyder, F.F. (1938): Synthetic unit-graphs. *Transactions of the American Geophysical Union*, 19, 447–454.

Storck, P. and Lettenmaier, D.P. (1999): Predicting the effect of a forest canopy on ground snow accumulation and ablation in maritime climates. In: Troendle, C. (Eds.). *Proceedings of the 67th Western Snow Conference*, Lake Tahoe, California, April 19–22, pp. 1–12.

Sugawara, M., Watanabe, I., Ozaki, E. and Katsuyama, Y. (1984): Tank model with snow component. Research Notes No. 65, National Research Centre for Disaster Prevention, Japan.

Tian Y. (2007): Macro-scale flow modelling of the Mekong River with spatial variance. A thesis submitted in partial fulfillment of the requirements for the degree of Doctor of Philosophy at The University of Hong Kong, June 2007.

Todini, E. (1996): The ARNO rainfall-runoff model. *Journal of Hydrology*, 175(1–4), 339–382.

USACE (1990): *River Routing with HEC-1 and HEC-2*, September 1990, TD-30, US Army Corps of Engineers, Hydrologic Engineering Center, Davis, CA.

USACE (1998): *HEC1 Flood Hydrograph Package User's Manual*, June 1998, CPD-1A, Version 4.1, US Army Corps of Engineers, Hydrologic Engineering Center, Davis, CA.

USACE (2000): *HEC-HMS Hydrologic Modeling System User's Manual*. Hydrologic Engineering Center, US Army Corps of Engineers, Davis, CA.

US Soil Conservation Service (1972, 1985): *National Engineering Handbook*, Sec. 4, Hydrology. US Department of Agriculture, Washington, D.C.

Vrugt, J.A., Gupta, H.V., Bastidas, L.A., Bouten, W. and Sorooshian, S. (2003a): Effective and efficient algorithm for multi-objective optimization of hydrologic models. *Water Resources Research*, 39(8), SWC51–SWC519.

Vrugt, J.A., Gupta, H.V., Bouten, W. and Sorooshian, S. (2003b): A shuffled complex evolution metropolis algorithm for optimization and uncertainty assessment of hydrologic model parameters. *Water Resources Research*, 39(8), SWC11–SWC116.

Wang, Q.J. (1991): The genetic algorithm and its application to calibrating conceptual rainfall-runoff models. *Water Resources Research*, 27(9), 2467–2471.

Wood, E.F., Lettenmaier, D.P. and Zartarian, V.G. (1992): A land-surface hydrology parameterization with subgrid variability for general circulation models. *Journal of Geophysical Research*, 97(D3), 2717–2728.

World Water Development Report 3 (2009): Water in a changing world. World Water Assessment Programme, UNESCO.

Zhang, W.H. (1990): *Theory and Practices of Flood Forecasting of Storms*. Water Resource and Electric Press, Beijing, pp. 161–176 (in Chinese).

Zhao, R.J. (1977): *Flood Forecasting Method for Humid Regions of China*. East China College of Hydraulic Engineering, Nanjing, pp. 19–51 (in Chinese); see also 135–170 and 206–224.

Zhao, R.J. (1984): *Hydrological Simulation of Watersheds*. Water Resource and Electric Press, Beijing, pp. 32–47 (in Chinese); see also pp. 71–82 and 106–130.

Zhao, R.J. and Liu, X.R. (1995): The Xinanjiang model. In: Singh, V.P. (Ed.), *Computer Models of Watershed Hydrology*. Water Resources Publication, Littleton, CO, pp. 215–232.

Zhao, R.J., Zhuang, Y.L., Fang, L.R., Liu, X.R. and Zhang, Q.S. (1980): The Xinanjiang model. *Hydrological Forecasting Proceedings Oxford Symposium*, 129, IAHS Publication, pp. 351–356.

Chapter 3

Population dynamics

3.1 INTRODUCTION

The human population worldwide has always been increasing except for short-term falls in the 14th and 17th centuries. These falls were mainly due to pandemics caused by 'black death' and plague. It is also predicted that the population will continue to increase until about 2050. The positive trends can be attributed to a number of reasons, such as improved medical facilities, low infant mortality rates, increased life expectancy, and increased food production. The population, which in 1750 was 791 million, has exploded to over 7 billion (Figure 3.1) in October 2011 (Aubuchon, 2011), with the highest rate of growth of 2.2% per year recorded in 1963. Despite the past trends, it is also well known that continued increase is not sustainable as the competition for resources will begin to dominate the growth rate sooner or later.

Attempts to model population dynamics can be traced back to the late 18th century when the Malthusian exponential growth theory was introduced. Although it has been modified, refuted, or superseded by the works of subsequent researchers, it still stands as the backbone of many population dynamics models. In this chapter, a review of the historical development of the study of population dynamics, including those of some other biospecies, is presented.

3.2 MALTHUSIAN GROWTH MODEL

The first attempt to model population was done by the British economist Thomas Malthus (1766–1834) by linking economic growth to population. Historically, economic growth in the world has had periods of stagnation followed by periods of rapid growth, except in a few countries. Before the 1800s, the GDP per capita was low in many countries. With the emergence of the industrial revolution, Britain was the first country to take off, followed by Western Europe and the United States. This pattern can also be seen in the modern context in countries such as China and India.

The explanation as to why economic stagnation occurred, according to Malthus, is based on the hypothesis that economic development depended on agricultural production, which depended on land availability. He hypothesized that population increased exponentially while the resources (mainly food production) increased linearly. Depending on the growth rates, it is certain that the population would surpass the resource production, and this state is referred to as the Malthusian catastrophe or stagnation. He also hypothesized that the growth rate was positively related to the per capita income. Given the fact that the quantity of land is fixed, more population would lead to lower per capita income. Then, the growth becomes an increasing function of productivity and a decreasing function of population. If the negative effects of population growth are strong, the economy leads to stagnation. On

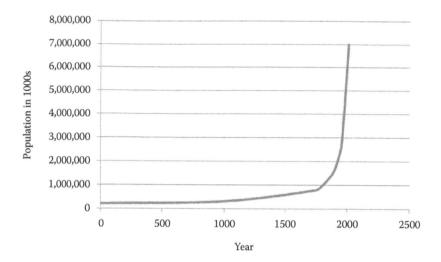

Figure 3.1 World population.

the other hand, if the per capita income is rising, the population growth will accelerate until it is offset by the productivity growth.

The Malthusian model was generally accepted until the emergence of the industrial revolution. It has enabled modern economies to avoid Malthusian catastrophes. Unlike in an agriculture-dominated economy, the amount of land was not so important in an industry-dominated economy. Productivity and technological advancements were more the key drivers of economy. In the preindustrial revolution time, population growth was a necessary ingredient because more children meant more help in the field to increase the productivity. The negative effect has been that the survival rates of children were not very high because of diseases, less resources per child, and more burdens on the parents to support. In the modern context, parents prefer to have fewer 'quality' children than quantity. Such phenomena can be seen in the modern world.

Exponential growth is the growth of a physical/biological entity in the absence of predators or limiting conditions. It takes the form

$$\frac{dP}{dt} = rP \tag{3.1}$$

which, when integrated gives

$$P_t = P_0 e^{rt} \tag{3.2}$$

(P_t is the population at time t; P_0 is the population at time $t = 0$; r is the growth rate), which is the model hypothesized by Malthus (1798). It can also be written as

$$x_{t+1} = (1 + r)x_t \tag{3.3}$$

According to this model, a 1% constant growth rate results in doubling the population every 70 years; 2% in every 35 years.

(Note: 70 is coincidentally related to 2 by $\ln(2) \approx 0.7 \rightarrow \times 100 = 70$.)

3.3 VERHULST GROWTH MODEL

Subsequently, Pierre-Francois Verhulst (1838) refined the Malthusian model by assuming that the rate of growth (reproduction) is proportional to the existing population and the amount of resources available, given all other conditions being equal. In equation form

$$\frac{dP}{dt} = rP\left(1 - \frac{P}{K}\right)$$

(3.4)

where the two parameters r and K are the growth rate and the carrying capacity, respectively. Equation 3.4 has a solution of the form

$$P_t = \frac{KP_0 e^{rt}}{K + P_0(e^{rt} - 1)}$$

(3.5)

At $t = 0$, $P_t = P_0$, and as $t \to \infty$, $P_t \to K$ (carrying capacity)

Substituting $x = \frac{P}{K}$, Equation 3.4 can be written as

$$\frac{dx}{dt} = rx(1 - x)$$

(3.6)

which has a solution of the form[1]

$$x_t = \frac{1}{1 + \left(\frac{1}{x_0} - 1\right)e^{-rt}}$$

(3.7)

where x_0 is the value of x at $t = 0$. Equation 3.7 is also referred to as the sigmoid function that is widely used as a non-linear mapping function in artificial neural networks. When $x_0 = \frac{1}{2}$, Equation 3.7 simplifies to

$$x_t = \frac{1}{1 + e^{-rt}}$$

(3.8)

---

[1] $\dfrac{1}{x(1-x)}\dfrac{dx}{dt} = r \Rightarrow \displaystyle\int_0^T \dfrac{1}{x(1-x)}\dfrac{dx}{dt}dt = \int_0^T r\,dt \Rightarrow \int_{x_0}^{x_T}\left\{\dfrac{1}{x} + \dfrac{1}{1-x}\right\}dx = rT \Rightarrow \left[\ln(x) - \ln(1-x)\right]_{x_0}^{x_T} = rT$

$\Rightarrow \left[\ln\left(\dfrac{x}{1-x}\right)\right]_{x_0}^{x_T} = rT \Rightarrow \ln\dfrac{(x_T)(1-x_0)}{x_0(1-x_T)} = rT \Rightarrow \dfrac{(x_T)(1-x_0)}{x_0(1-x_T)} = e^{rT} \Rightarrow x_T = \dfrac{\dfrac{x_0}{1-x_0}e^{rT}}{1 + \dfrac{x_0}{1-x_0}e^{rT}}$

$\Rightarrow \dfrac{x_0}{(1-x_0)e^{-rT} + x_0} \Rightarrow x_t = \dfrac{1}{1 + \left(\dfrac{1}{x_0} - 1\right)e^{-rt}}$

Pearl and Reed (1920), independently, or perhaps not being aware of Verhulst's work proposed an S-curve using US population data, which later became known as the logistic curve (Verhulst used the term 'logistique'). It has the same mathematical form as Verhulst's equation. In early 20th century, Robertson (1908a,b) applied the same equation to describe the growth in animals, plants, and humans; called the curve 'autocatalytic'; and later applied it to protozoa and bacteria populations.

Pearl and Reed (1920), recognizing that there are cycles of logistic curves for different spans of time, modified the original logistic equation to include more parameters (Kingsland, 1995):

$$P_t = \frac{K}{1 + me^{a_1t + a_2t^2 + a_3t^3 + \ldots a_nt^n}} \tag{3.9}$$

Although the above form of the equation can be fitted more closely to data, it lacks generality to represent the pattern.

3.4 PREDATOR–PREY (LOTKA–VOLTERRA) MODEL

US ecologist Alfred J. Lotka (1880–1949) and the Italian mathematician Vito Volterra (1860–1940) independently studied the predator–prey population interaction and concluded in 1925 and 1926, respectively, that there are periodic oscillations in the two populations. Their mathematical form, which later came to be known as the Lotka–Volterra equations, takes the forms

$$\frac{dP_1}{dt} = r_1 P_1 - k_1 P_1 P_2 \tag{3.10a}$$

$$\frac{dP_2}{dt} = k_2 P_1 P_2 - r_2 P_2 \tag{3.10b}$$

where P_1 and P_2 represent the prey and predator populations, respectively; r_1 is the growth rate for P_1; r_2 is the mortality rate for P_2; k_1 is the rate at which the prey is destroyed by the predator (proportional to the product of the populations of prey and predator); and k_2 is the rate at which the predator population increases (proportional to the product of the populations of prey and predator). If the prey has unlimited food supply, then it will grow exponentially in the absence of the predator. However, because of the predator, P_1 would decrease. Similarly, if there is no prey, the predator population would become extinct. However, it is increased because of the encounter between the predator and prey at a rate proportional to $P_1 P_2$. The first-order non-linear ordinary differential equations, Equation 3.10, can also be written as

$$\frac{dP_1}{dP_2} = \frac{P_1}{P_2} \frac{(r_1 - k_1 P_2)}{(k_2 P_1 - r_2)} \tag{3.11}$$

The equilibrium condition for these two equations occurs when $P_1 = 0$, $P_2 = 0$, or when $P_1 = \frac{r_2}{k_2}, P_2 = \frac{r_1}{k_1}$.

3.5 GOMPERTZ CURVE

The Gompertz curve (after Benjamin Gompertz [1779–1865]) is used for time series modelling and the equation is of the form (Gompertz, 1832)

$$x_t = ae^{be^{ct}} \qquad (3.12)$$

where a is the upper asymptote, c is the growth rate, and b, c are negative. In this curve, which is of the double exponential type, growth is slowest at the beginning and end of the time series. It is a refinement of Malthus' model and has been applied to model mortality rates and in the life insurance sector.

There are many forms of the logistic growth curve. These include, for example, a model that has been used to describe the incidence of infectious diseases such as AIDS, which takes the form

$$x_t = \frac{K}{1 + e^{a + bt}} \qquad (3.13)$$

where a, b are parameters and b is negative, an asymmetric five-parameter model that has been used to fit into immune assay and bioassay data by minimizing the weighted squared residuals, and which takes the form (also called Richard's curve, or generalized logistic curve; Figure 3.2) (Richards, 1959)

$$x_t = a + \frac{K}{1 + ce^{-b(t - t_m)^{\frac{1}{p}}}} \qquad (3.14)$$

where x is the dependent variable (e.g., weight, height, etc.), t is the time, a is the lower asymptote, K is the difference between the upper asymptote and the lower asymptote (if the lower asymptote is zero, then K is the carrying capacity), t_m is the time of maximum growth, b is the growth rate, c is a parameter that depends on x_0, and p is a parameter that affects

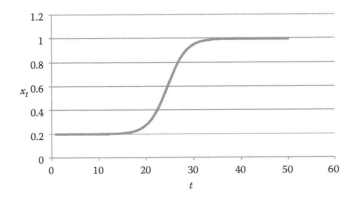

Figure 3.2 Generalized logistic curve (Equation 3.14 with a = 0.2, K = 0.8, b = 0.5, c = 0.8, t_m = 25, p = 1.0).

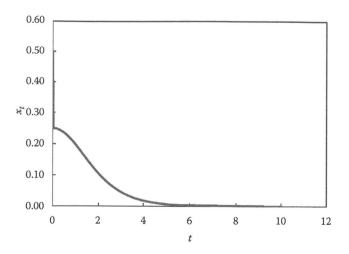

Figure 3.3 Hubbert's curve (Equation 3.15).

where the maximum growth occurs (near which asymptote), and a model used to describe the rate of oil extraction over time that takes the form (Hubbert, 1956)

$$x_t = \frac{e^{-t}}{(1+e^{-t})^2} = \frac{1}{(2+2\cosh t)} \tag{3.15}$$

and that resembles a Gaussian distribution function but not a Gaussian distribution function (Figure 3.3).

Sigmoid is also a special case of one form of the logistic function given by

$$x(t,a,m,n,r) = a\frac{1+me^{-rt}}{1+ne^{-rt}} \tag{3.16}$$

which simplifies to

$$x(t) = \frac{1}{1+e^{-rt}} \tag{3.17}$$

(when $a = 1$, $m = 0$, $n = 1$)

which is the same as Equation 3.8. It can also be expanded as an infinite Maclaurin series of the form

$$f(x) = f(0) + xf'(0) + \frac{x^2 f''(0)}{2!} + \frac{x^3 f'''(0)}{3!} + \dots \tag{3.18}$$

3.6 LOGISTIC MAP

A discrete version of the logistic equation first proposed by Verhulst (1838) has also used to identify chaotic behaviour of non-linear systems (May, 1976). It is referred to as the logistic map and takes the form

$$x_{n+1} = r x_n (1 - x_n) \tag{3.19}$$

where $x_n : (0,1)$; $r > 0$ represent the population and the rate of growth (reproduction) and decline (starvation), respectively. In this equation, which is in non-dimensional form, the behaviour will depend on the value of r as well as the initial condition. For example,

- $r(0,1)$: Population will eventually reduce to zero regardless of initial condition.
- $r(1,2)$: $x \to \dfrac{r-1}{r}$ independent of x_0.
- $r(2,3)$: $x \to \dfrac{r-1}{r}$, with initial oscillations around the limiting value, rate of convergence is linear except for $r = 3$ when it is less than linear convergence.
- $r\left(3, 1+\sqrt{6}\,(\approx 3.45)\right)$: x oscillates between two values indefinitely.
- $r\left(1+\sqrt{6}, 3.54\right)$: x oscillates between four values indefinitely.
- $r > 3.54$: x oscillates between 8, 16, 32, etc., values indefinitely.
- $r = 3.57$: Chaotic behaviour begins; no oscillations; slight change in initial conditions leads to completely different results. Within the chaotic region ($r \geq 3.57$), there exist some isolated values of $r\left(1+\sqrt{8} \approx 3.83\right)$ where there are some islands of stability, for example, oscillations between 3 values, 6 values, and 12 values, etc.
- $r > 4$: x goes outside $(0,1)$ and diverges for all initial values.

Period doubling bifurcation leads to a new behaviour with twice the period of the original system. Period halving bifurcation leads to a new behaviour with half the period of the original system. Period halving leads from chaos to order, whereas period doubling leads from order to chaos.

An inherent problem with the logistic map is that some initial conditions and some parameter values lead to negative population values. The Ricker model (Ricker, 1954), which also has chaotic behaviour, does not have this problem. The Ricker model, which was developed in the context of fish population, is given as

$$x_{n+1} = x_n e^{\left[r\left(1-\frac{x_t}{K}\right)\right]} \tag{3.20}$$

where r is the growth rate and K is the carrying capacity.

3.6.1 Specific points in the logistic map

$\dfrac{dx}{dt} = 0$ when $x = 0$, or $x = 1$. These are the steady-state values, or fixed points.

$\dfrac{dx}{dt} > 0$ when $rx(1-x) > 0$, or, $0 < x < 1$; x increases in this range

$\dfrac{dx}{dt} < 0$ when $rx(1-x) < 0$, or, $x > 1$; x decreases

x_t versus t is a trajectory.

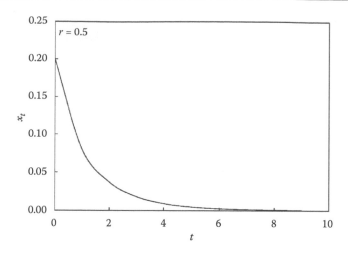

Figure 3.4 Logistic map with $r = 0.5$ ($x_t \to 0$ regardless of initial value).

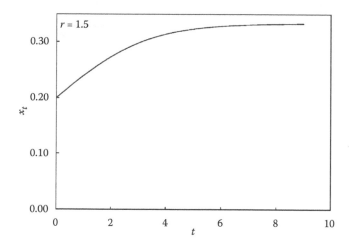

Figure 3.5 Logistic map with $r = 1.5$ ($x_t \to 0.333$ regardless of initial value).

For $r = 3.57$ and 4, the map is chaotic; however, for $r = 4$, there is a solution for $x_0 = 0.75$. Then, $x_{n+1} = 4 \times 0.75(1 - 0.75) = 0.75$, for all n. Therefore, x remains constant. Figures 3.4 and 3.5 illustrate the logistic map for some specific values of r.

3.7 CELL GROWTH

Cell is the smallest unit of any living organism and also its structural and functional unit. Some organisms are unicellular. Humans have about 10^{14} cells with a typical size of 10 μm (1 μm = 10^{-6} m) and an approximate mass of 1 ng (10^{-9} g). Cells have membranes to separate from the surrounding.

Each cell is self-contained, meaning that it takes in nutrients, converts them into energy, and carries out various cell functions, including reproduction. Basically, there are two types of cells, the prokaryote and the eukaryote. Examples of prokaryotes include bacteria and archaea, which are independent and have no nuclei or nuclear membrane. Eukaryotes are multicellular organisms and are approximately 10 times larger than prokaryotes. They have nuclei and membrane-bound compartments in which metabolic activities take place. Examples include fungi, plants, and animals.

All cells contain DNA, the hereditary material of genes. Prokaryotes are characterized by a circular DNA structure, whereas eukaryotes contain linear molecules called chromosomes. All cells of an individual have the same DNA that is found in the chromosomes. Humans have a chromosome number of 46, while mice and corn have 40 and 20, respectively. When a mother cell splits up into two daughter cells, they will have the same DNA as the original.

All cells in an organism are initially unspecialized. Such cells are called 'stem cells', which can turn into other specialized cells (such as muscle cells, blood cells, bone cells, etc.), or produce more stem cells. Stem cells divide without limit to repair and replenish other cells throughout the life of a living being. The main characteristic features of stem cells are that they are capable of dividing and remaining for long periods of time, that they are unspecialized, and that they can give rise to specialized cells. The latter process is called differentiation. The specialized cells will reproduce, some throughout life, while others stop early. The process of cell reproduction is called mitosis, which allows the repair and replacement of old cells. Stem cell research is a hot area of research because of the potential use in regeneration therapies.

3.7.1 Cell division

Division of cells is by binary fission, mitosis, and meiosis. Prokaryotic cells that lack a nucleus divide by binary fission. Eukaryotic cells divide by mitosis or meiosis. Mitosis is nuclear division followed by division of the cell called kinesis. The cell duplicates the chromosomes in its nucleus to generate two identical daughter cells. Both these processes are asexual. Meiosis is the process of reduction division in which the number of chromosomes per cell is cut in half. Meiosis is essential for sexual reproduction and therefore occurs in all eukaryotes, including single-celled organisms that reproduce sexually. In diploid cells, specialized cells called gametes (egg and sperm) are created through meiosis. Then, the sperm and egg join together to form a single cell. Cell growth cycle is a hot area of research because of its medical applications. Cancer cells grow exponentially, and an understanding of the growth cycle is needed to control their proliferation through drugs.

In prokaryotic organisms, the division is quite straightforward; cells grow to their critical size by adding materials to their cell membranes, and cell walls if present, then they divide and the process is repeated. The growth rate is exponential but is typically limited by nutrient availability and temperature. In eukaryotic organisms, which contain multicellular chromosomes within a nucleus, on the other hand, the cell mitotically divides into two genetically identical daughter cells.

The growth rate of cells during the exponential growth phase is expressed by the term 'generation time', which is the time taken to double the population under standard nutritional conditions. For *Escherichia coli* under laboratory conditions, the generation time is in the range of 15–20 min, whereas in the intestinal tract it can be in the range of 12–24 h. For most known bacteria that can be cultured, the generation time is in the range of 15 min to 1 h (Todar, 2007).

Cell growth can be modelled under various assumptions. Unstructured models assume fixed cell composition and do not consider stoichiometry and are often empirical in formulation. On the other hand, structured models consider cells microscopically, consider cells' phenotype stoichiometry, and are formulated based on experiments. Non-segregated models assume a homogenous culture, and treat the system as a black box, whereas segregated models treat the cells as discrete units and are therefore distributed in formulation. The combination of structured and segregated assumption is realistic but difficult. Unstructured and non-segregated combinations are simple and practical but there are limitations. They are normally applicable to the exponential phase of the growth curve under single-stage, steady-state assumption, but fail during the transient phase.

3.7.2 Exponential growth

Cell growth generally assumes exponential proliferation with equations such as

$$N_t = N_0 2^{\frac{t}{\Delta t}} \tag{3.21a}$$

or

$$N_t = N_0 2^n \tag{3.21b}$$

where N_t and N_0 denote the number of cells at time t and at an initial time, respectively; Δt is the cell division time; and $n = \dfrac{t}{\Delta t}$. This equation assumes that each cell divides into two daughter cells. To account for the presence of non-dividing cells, Sherley et al. (1995) proposed a model (Equation 3.22) with an additional parameter α, which is referred to as the mitotic fraction, which becomes Equation 3.21 when $\alpha = 1$. Equation 3.22 is used for modelling stem cell growth.

$$N_t = N_0 \left\{ 0.5 + \frac{1 - (2\alpha)^{n+1}}{2(1 - 2\alpha)} \right\} \tag{3.22}$$

Assuming that Δt is constant, dividing cells are capable of producing both dividing and non-dividing cells, and that non-dividing cells do not enter the cell cycle during the period of observation. Allowing for cell loss due to cell mortality and migration, Deasy et al. (2003) proposed the following:

$$N_t = N_0 \left\{ 0.5 + \frac{1 - (2\alpha)^{n+1}}{2(1 - 2\alpha)} \right\} - M_t \tag{3.23}$$

where M_t is the cumulative number of dead cells at time t.

3.7.3 Cell growth models in a batch (closed system) bioreactor

The governing equation for an unstructured unsegregated model for cell growth is

$$r_x = \frac{dx}{dt} = \mu x \tag{3.24}$$

where r_x is the rate of cell generation ($ML^{-3} T^{-1}$), x is the cell concentration (ML^{-3}), and μ is the specific growth rate (T^{-1}).

For a continuous feed reactor, the cell balance can be written as

$$Fx - Fx_f + V\frac{dx}{dt} = r_x V \tag{3.25}$$

where F is the flow rate ($L^3\ T^{-1}$), x is the concentration inside the reactor and in the outlet stream (ML^{-3}), x_f is the concentration of feed flow (ML^{-3}), V is the volume of the reactor (L^3), and r_x is the rate of cell generation reaction rate ($ML^{-3}\ T^{-1}$).

If the feed flow is sterile, $x_f = 0$; therefore,

$$\frac{dx}{dt} = (\mu - D)x \tag{3.26}$$

where D is the dilution rate $\left(= \dfrac{F}{V} \right)$.

At steady state, $\dfrac{dx}{dt} = 0$, and therefore,

$$\mu = D \tag{3.27}$$

The substrate balance can be written as

$$Fs - Fs_f + V\frac{ds}{dt} = r_s V \tag{3.28}$$

where s is the substrate concentration (ML^{-3}), s_f is the substrate concentration in the feed flow (ML^{-3}), and r_s is the rate of substrate consumption ($ML^{-3}\ T^{-1}$).

Introducing the yield parameter y, which is defined as

$$y = \frac{\text{mass of cells produced per unit volume per unit time}}{\text{mass of substrate consumed per unit volume per unit time}} = \frac{r_x}{-r_s} \tag{3.29}$$

and combining Equations 3.24, 3.28, and 3.29,

$$\frac{ds}{dt} = D(s_f - s) - \frac{\mu x}{y} \tag{3.30}$$

Equations 3.26 and 3.30 form a pair of coupled ordinary differential equations that describes the rates of changes of the cell and substrate concentrations.

At steady state, Equation 3.30 yields the cell concentration x as

$$x = y(s_f - s)$$

The two-parameter Monod's equation (Monod, 1949) for substrate limiting growth, which is empirical and does not take into account the lag and accelerating growth phases but capable of simulating exponential growth followed by a reduced growth rate, is of the form

$$\mu = \frac{\mu_{max} s}{k_s + s} \tag{3.31}$$

where μ is the specific growth rate (T^{-1}), s is the substrate (nutrient) concentration (ML^{-3}), k_s is the Monod coefficient or half-saturation coefficient for substrate (ML^{-3}) and corresponds to the concentration at which the specific growth rate μ is half its maximum value (k_s is also referred to as the saturation constant or substrate affinity constant in some literature), and μ_{max} is the maximum specific growth rate (T^{-1}). At steady state (when $\mu = D$), the substrate concentration s from Monod's equation is

$$s = \frac{Dk_s}{\mu_{max} - D} \tag{3.32}$$

Monod's equation is not valid for some bioprocesses with low nutrient concentrations. It is valid when the concentration of limiting nutrients is much greater than the half-saturation coefficient (or Monod coefficient) k_s.

In a batch reactor when there are two substrates, each one is capable of supporting cell growth. The microbial growth rate, however, is not additive, but rather sequential; that is

$$\mu_{mixed} \neq \mu_{max,1} + \mu_{max,2}, \text{ but } \mu_{mixed} = \mu_{max,1} \text{ or } \mu_{max,2}$$

In other words, the preferred substrate is consumed first. This phenomenon has been termed 'diauxie' by Monod, meaning two growth phases, in Greek, and a detailed discussion is given by Kompala et al. (1986).

When the average cellular size is small, such as in bacteria, the cell is considered as a well-mixed bioreactor. The Gillespie (1977) algorithm, which is considered as the 'golden standard' for modelling, stochastically generates an ensemble of trajectories with correct statistics for a set of biochemical reactions.

3.8 BACTERIAL GROWTH

Bacteria are unicellular organisms that existed on Earth for billions of years. They are prokaryotes, range in size from about 0.5 to 5 μm, and have three types of shapes: bacilli (or rod shape), cocci (or spherical), and spirilla (or curved). There are about 10 times more bacterial cells than human cells in a human body, mainly in the skin and in the digestive tract. Some bacteria are harmless, and sometimes useful and necessary, while some are disease causing or pathogenic.

Bacteria that form a parasitic association with other organisms are called pathogens. Infections such as tetanus, typhoid fever, diphtheria, syphilis, cholera, food-borne diseases, leprosy, tuberculosis in humans, leaf spot, fire blight and wilts in plants, and salmonella and anthrax in farm animals are caused by pathogenic bacteria.

Escherichia coli, or *E. coli*, is perhaps the most widely studied bacteria. They live in the human colon and are usually harmless. However, water and food contaminated with serotype (O157:H7) strain of *E. coli* cause infections that can at times be fatal.

Bacteria grow and live in three different environments: aerobic – where oxygen is necessary for growth and existence; anaerobic – where gaseous oxygen is not necessary and sometimes toxic and live in deep underwater sediments; and facultative anaerobic – where the presence of oxygen is preferred, but growth can continue in its absence.

3.8.1 Binary fission

The most common mechanism of bacterial growth is by binary fission. If the Malthusian exponential growth model is applied, the population will increase indefinitely; however, under normal conditions, this does not happen. Even under optimum conditions of energy and nutrient supply, pH, temperature, etc., the growth of bacteria in a controlled culture solution undergoes four phases: a slow lag phase during which the bacteria are getting acclimatized to the new environment, an exponential phase during which the bacteria multiply exponentially, a stationary phase during which there is no growth due to nutrient limitations, or due to the accumulation of inhibiting end products, or due to the lack of biological space, and a death phase during which the cells die at a faster rate than they are replaced. In the exponential growth phase, the cell count can be expressed as

$$N_t = N_0 2^n \tag{3.33}$$

where N_t and N_0 are the cell counts at time t and at some initial time; n is the number of generations in the time interval. Generation time, or the doubling time, is the rate of exponential growth.

This relationship can also be expressed as

$$N_t = N_0 e^{rt} \tag{3.34a}$$

From their equivalence,

$$e^{rt} = 2^{t/\Delta t}, \text{ which leads to } rt = \frac{t}{\Delta t}\ln(2), \text{ and hence } \Delta t = \frac{\ln(2)}{r} \tag{3.34b}$$

where r is the growth rate [T^{-1}]. Therefore, given the growth rate, the generation time can be obtained or vice versa. For *E. coli*, the generation time is about 15–20 min under laboratory conditions. In nature, however, the growth rate is not the same. For example, the generation time in the intestinal tract is estimated to be 12–24 h.

3.8.2 Monod kinetics

Bacterial growth generally follows the Monod kinetics, which takes into account the effect of nutrient limitation. The relationship can be expressed as

$$\frac{dx}{xdt} = \frac{rs}{k_s + s} \tag{3.35a}$$

or

$$\frac{dx}{dt} = x\frac{rs}{k_s + s} \tag{3.35b}$$

in which x represents the bacteria concentration [ML^{-3}], s the nutrient concentration [ML^{-3}], r the maximum growth rate [T^{-1}], and k_s the half-saturation constant [ML^{-3}]. The term on the left-hand side of Equation 3.35a represents the specific growth rate, while the term on the right-hand side is referred to as the Monod function. Monod function increases monotonically with increasing nutrient concentration and reaches the maximum growth rate asymptotically. When the Monod function attains half its maximum value, the corresponding nutrient concentration is referred to as the half-saturation constant or Monod coefficient. Monod also observed that

$$y\frac{ds}{dt} = -\frac{dx}{dt}$$

(3.36)

implying that the bacterial growth rate is proportional to the rate of nutrient consumption but opposite in sign. In this equation, y refers to the growth yield, which is the ratio of the rate of biomass production to the rate of nutrient consumption $\left(=-\dfrac{dx}{ds}\right)$. Combining Equations 3.35 and 3.36, the nutrient concentration gradient can be obtained as

$$\frac{ds}{dt} = -\frac{1}{y}x\frac{rs}{k_s + s}$$

(3.37)

Equations 3.35 and 3.37 form a pair of coupled ordinary differential equations. They can be solved subject to given initial conditions. Qualitatively, s decreases monotonically to zero whereas x increases to reach a steady-state value asymptotically. It should, however, be noted that bacteria need energy to maintain cellular functions and stay alive even without reproduction. This condition is modelled by assuming that the 'maintenance energy' per unit mass (m) is constant and modifying Equation 3.35 as

$$\frac{dx}{dt} = x\frac{rs}{k_s + s} - mx$$

(3.38)

Qualitatively, Equation 3.38 gives a solution that consists of a rising segment, a peak segment, and a recession segment that approaches zero asymptotically for x, while for s it consists of a monotonically decreasing segment that quickly approaches zero.

The Monod coefficient k_s typically ranges from 0.1 to 10 mg/l. The parameter values reported for *E. coli* in a glucose substrate at 30°C, are $k_s = 1$ mg/l, $r = 1.35/h$, and $y = 0.23$ (Smith and Waltman, 1995) from which the generation time works out to be approximately half an hour.

3.9 RADIOACTIVE DECAY AND CARBON DATING

Carbon dating is used as a method to estimate the age of material as old as 50,000 years. The method, developed by Libby in the 1950s (Libby, 1955), is based on the ratio of carbon 14 (which is a radioactive isotope of the more common carbon 12) to carbon 12 ($^{14}C/^{12}C$). ^{14}C is produced in the atmosphere when nitrogen atoms are bombarded with neutrons from cosmic rays.

$$^{14}_{7}N + ^{1}_{0}n \rightarrow ^{14}_{6}C + ^{1}_{1}p$$

This cosmic reaction equation means that when a nitrogen atom (which contains seven protons and seven neutrons and add up to its atomic mass number 14) is bombarded by a neutron from cosmic radiation produces one atom of carbon (which contains six protons and eight neutrons and add up to its atomic mass number 14) and one proton.

Free carbon, including ^{14}C, reacts with oxygen from the atmosphere to produce CO_2, which has a steady-state concentration of one atom of ^{14}C for every 10^{12} atoms of ^{12}C.

$$^{14}C + O \rightarrow CO_2$$

The CO_2 produced is absorbed by plants during photosynthesis, or building up food matter, which in turn is consumed by animals. Living things that consume CO_2 and air have the same $^{14}C/^{12}C$ ratio at any given instant of time. ^{12}C is stable, whereas ^{14}C is radioactive and always decaying but replaced continuously by cosmic action. When a living thing dies, there is no more carbon intake. At the time of death, the ratio $^{14}C/^{12}C$ is the same for all living things. ^{14}C decay is not replenished while ^{12}C does not decay.

After the 1940s, nuclear bombs and other nuclear devices released large amounts of neutrons into the atmosphere, thereby disturbing the natural cosmic balance. As a result, radioactive dating of any carbonaceous matter that died during the postnuclear era is not reliable.

The decaying of radioactive matter follows a first-order reaction:

$$\frac{dN}{dt} = -\lambda N \tag{3.39}$$

which has a solution of the form

$$N_t = N_0 e^{-\lambda t} \tag{3.40}$$

where N_t is the concentration of the matter at time t, N_0 is the concentration at some initial time, and λ is the decay constant. In radioactive decaying terminology, $t_{1/2}$, which is called the half-life, is used to define the rate of decay; that is,

$$\frac{N_0}{2} = N_0 e^{-\lambda t_{1/2}} \Rightarrow t_{1/2} = \frac{\ln(2)}{\lambda} = \frac{0.693}{\lambda} \text{ or } \lambda = \frac{\ln(2)}{t_{1/2}} \tag{3.41}$$

This property has been used to trace the history of the ruins of any substance of biological (carbonaceous) origin. It is called carbon dating and has been widely used after the work of Libby (1955). The widely used half life of ^{14}C as estimated by Libby is 5568 ± 30 years, but has been revised as Cambridge half life as 5730 ± 40 years. The origin of dead biological matter is estimated in radiocarbon years as the time before present (BP), and in this context, the year 1950 is taken as the present. Radiocarbon year is not the same as calendar year and a calibration, which is normally done using tree rings, is necessary for conversion. For the post 1950 era, there is no standard calibration curve.

3.10 CONCLUDING REMARKS

In this chapter, an attempt has been made to review population dynamics from a historical perspective. Various attempts to model the growth and decay of biospecies from the early exponential growth model hypothesized by Malthus to subsequent refinements made by Verhulst, Lotka–Volterra, Gompertz, among several others, are briefly described. It is followed by brief descriptions of cell growth, including various types of cell division and their growth under different assumptions, the two-parameter Monod's equation for substrate limiting growth, and bacterial growth by binary fission as well as according to Monod kinetics. The chapter ends with a summary description of radioactive decay and carbon dating.

REFERENCES

Aubuchon, V. (2011): World population growth history, Vaughn's summaries. Available at http://www.vaughns-1-pagers.com/history/world-population-growth.htm. Vaughn's summaries, Vaughn Aubuchon, 2003, 2004, 2011.

Deasy, B.M., Jankowski, R.J., Payne, T.R., Cao, B., Goff, J.P., Greenberger, J.S. and Huard, J. (2003): Modeling stem cell population growth: Incorporating terms for proliferative heterogeneity. *Stem Cells*, 21, 536–545.

Gillespie, D.T. (1977): Exact stochastic simulation of coupled chemical reactions. *Journal of Physical Chemistry*, 81(25), 2340–2361.

Gompertz, B. (1832): On the nature of the function expressive of the law of human mortality, and on a new mode of determining the value of life contingencies. *Philosophical Transactions of the Royal Society London*, 123, 513–585.

Hubbert, M.K. (1956): Nuclear energy and the fossil fuels. Drilling and Production Practice, American Petroleum Institute & Shell Development Co. Publication No. 95. See pp. 9–11, 21–22. http://www.hubbertpeak.com/hubbert/1956/1956.pdf.

Kingsland, S.E. (1995): *Modelling Nature: Episodes in the History of Population Ecology*. Chicago University Press, Chicago, p. 267.

Kompala, D.S., Ramkrishna, D., Jansen, N.B. and Tsao, G.T. (1986): Investigation of bacterial growth on mixed substrates. Experimental evaluation of cybernetic models. *Biotechnology and Bioengineering*, 28, 1044–1055.

Libby, W.F. (1955): *Radiocarbon Dating*, 2nd Edition. Chicago University Press, Chicago.

Lotka, A.J. (1925): *Elements of Physical Biology*. Williams and Wilkins, Baltimore.

Malthus, T. (1798): *Population: The First Essay*. Reprinted by University of Michigan Press, Ann Arbor, 1959.

May, R.M. (1976): Simple mathematical models with very complicated dynamics. *Nature*, 261, 459, doi:10.1038/261459a0.

Monod, J. (1949): The growth of bacterial cultures. *Annual Review of Microbiology*, 3, 371–394.

Pearl, R. and Reed, L.J. (1920): On the rate of growth of the population of the United States since 1790 and its mathematical representation. *Proceedings of the National Academy of Sciences of the United States of America*, 6, 275–288.

Richards, F.J. (1959): A flexible growth function for empirical use. *Journal of Experimental Botany*, 10, 290–300.

Ricker, W.E. (1954): Stock and recruitment. *Journal of Fisheries Board of Canada*, 11, 559–623.

Robertson, T.B. (1908a): On the normal rate of growth of an individual, and its biochemical significance. *Archiv für Entwicklungsmechanik der Organismen*, 25, 581.

Robertson, T.B. (1908b): Further remarks on the normal rate of growth of an individual, and its biochemical significance. *Archiv für Entwicklungsmechanik der Organismen*, 26, 108–118.

Sherley, J.L., Stadler, P.B. and Stadler, J.S. (1995): A quantitative method for the analysis of mammalian cell proliferation in culture in terms of dividing and non-dividing cells. *Cell Proliferation*, 28(3), 137–144.

Smith, H.L. and Waltman, P. (1995): *The Theory of the Chemostat*. Cambridge University Press, Cambridge, UK. math.la.asu.edu/~halsmith/bacteriagrow.pdf.

Todar, K. (2007): *Growth of Bacterial Populations*. University of Wisconsin-Madison, Department of Bacteriology. http://bioinfo.bact.wisc.edu/themicrobialworld/growth.html.

Verhulst, P.F. (1838): Notice sur la loi que la population poursuit dans son accroissement. *Correspondance Mathématique et Physique*, 10, 113–121.

Volterra, V. (1926): Variations and fluctuations of the numbers of individuls in animal species living together Reprinted in R.N. Chapman (1931), *Animal Ecology*. McGraw Hill, New York.

Chapter 4

Reaction kinetics

4.1 INTRODUCTION

Reaction kinetics relates the rate of concentration variation in a chemical or biological reaction. The mechanism of the reaction is dependent on the order of the reaction, which may be zero-, first-, or second-order. Higher orders are rarely used to describe chemical reactions.

A zero-order reaction can be described in differential form by

$$\frac{dc}{dt} = -k \tag{4.1}$$

which, when integrated takes the form

$$c = -kt + c_0 \tag{4.2}$$

where c_0 is the concentration at time $t = 0$ and k is the rate constant $[T^{-1}]$, which is an indicator of the speed of reaction.

A first-order reaction can be described in differential form by

$$\frac{dc}{dt} = -kc \tag{4.3}$$

which, when integrated takes the form

$$c = c_0 e^{-kt} \tag{4.4}$$

A second-order reaction can be described in differential form by

$$\frac{dc}{dt} = -kc^2 \tag{4.5}$$

which, when integrated takes the form

$$c = \frac{c_0}{1 + ktc_0} \tag{4.6}$$

The rate constant k in each case can be determined by plotting c versus t, $\ln(c)$ versus t, or $\frac{1}{c}$ versus t for the zero-, first-, and second-order reactions, respectively.

An important concept in reaction kinetics is the time required to halve the concentration from an initial state, which is referred to as the half-life. In each of the above three cases, the half-lives $(t_{1/2})$ are $\frac{c_0}{2k}$, $\frac{\ln(2)}{k}$, and $\frac{1}{kc_0}$. It can be seen that the second half-life is twice as long as the first. Half-lives for some typical radioactive elements are as follows: ^{238}uranium = 4.5 billion years; ^{234}uranium = 240,000 years; ^{14}carbon = 5730 years; ^{220}lead = 22 years; ^{222}radon = 3.8 days; ^{214}polonium = 160 ms.

In biological kinetics, the biological half-life of water in humans is about 7–10 days. Alcohol consumption, which causes dehydration, can drastically reduce the half-life. The concept of half-life is used in administering drugs for decontaminating the human body that has been contaminated with toxic material. Typical half-lives of some elements in the human body are as follows: cesium, 1–4 months; lead in bones, 10 years; cadmium in bones, 30 years; and plutonium in bones, 100 years.

In biochemical reactions, it is often necessary to determine the relationship between the reaction rate and the substrate concentration, or the relationship between the growth and consumption rates of the biomass produced and the substrate consumed. Two of the widely used models for describing the kinetics of such reactions are the Michaelis–Menten equation and the Monod equation. They are briefly described below.

4.2 MICHAELIS–MENTEN EQUATION

The relationship between the concentration of enzymes (proteins that act as catalysts in biological reactions) and the concentration of substrates (based on which the enzyme lives or acts) was first proposed by Victor Henri (1903), and subsequently examined by Leonor Michaelis and Maud Menten in 1913, and published in German. The Michaelis–Menten equation, which describes the kinetics of many enzymes, provides a relationship between the rate of an enzyme-catalysed reaction and the concentration of the enzyme's substrate. It is based on laws of mass action, which assumes Fickian diffusion and thermodynamically driven random collision, although most biochemical reactions do not satisfy these conditions. The present form of the equation is after Briggs and Haldane (1925). It starts as follows:

If an enzyme (E) combines with a substrate (S) to form a substrate complex (ES), which then forms a product (P) while the enzyme is recycled to be used again, then the reactions can be described by the following reaction equations

$$S+E \quad \underset{\leftarrow (k_{-1})}{\overset{(k_1) \rightarrow}{}} \quad ES \quad (k_2) \rightarrow \quad E+P$$

where k_1 is the rate constant for the association of substrate and enzyme, k_{-1} is the rate constant for the dissociation of unaltered substrate from the enzyme, and k_2 is the rate constant for the dissociation of the product (altered substrate) from the enzyme, sometimes also referred to as the catalytic rate constant. In the above reaction, E first binds S to form the enzyme–substrate complex ES, which then generates the product P. The rate of the reaction is limited by the conversion of ES to E and P, and depends on the reaction constant k_2 and concentration of ES, which is the concentration of enzyme bound to the substrate. Therefore, it is possible to write

$$r = k_2[ES] \; ; \; r = \frac{dP}{dt} \quad \text{(reaction rate)} \tag{4.7}$$

where [] indicates the concentration.

The reaction rate is defined as the number of reactions per unit time catalysed per mole of the enzyme, and sometimes referred to as the reaction velocity, or the rate of production of the product.

Assuming that $[S] \gg [E]$, and steady-state conditions, that is, $\frac{d[ES]}{dt} = 0$, implying that the formation and breakdown of ES are at the same rate, consideration of mass balance gives (formation of ES depends on k_1, $[S]$, and $[E]$; dissociation of ES can be by conversion of substrate to product and non-reactive dissociation of substrate from complex)

$$k_1[E][S] = k_{-1}[ES] + k_2[ES]$$

which, upon simplification gives

$$\frac{[E][S]}{k_M} = [ES] \tag{4.8}$$

where

$$k_M = \frac{k_{-1} + k_2}{k_1} \tag{4.9}$$

is the Michaelis–Menten constant, which is the substrate concentration when the reaction rate is half the maximum reaction rate. Since the reaction rate increases with increasing substrate concentration and approaches the maximum value asymptotically, it is more meaningful to use the substrate concentration at which the reaction rate attains half the maximum value.

The total amount of enzyme in the system $[E_0]$ remains constant either in free (unbound) state $[E]$, or in bound state with substrate $[ES]$; that is,

$$[E_0] = [E] + [ES] \Rightarrow [E] = [E_0] - [ES]$$

Substituting in Equation 4.8,

$$\frac{\{[E_0] - [ES]\}[S]}{k_M} = [ES] \Rightarrow \frac{[E_0][S]}{k_M + [S]} = [ES]$$

Substituting back in Equation 4.7 gives

$$r = \frac{dP}{dt} = k_2[ES] = k_2 \frac{[E_0][S]}{k_M + [S]} \tag{4.10}$$

The maximum rate would be achieved when $[E_0] = [ES]$; that is, when all $[E]$ will be in the form $[ES]$. Substituting in Equation 4.7 under this condition gives

$$r_{max} = k_2[E_0]$$

which when substituted into Equation 4.10 yields the Michaelis–Menten equation:

$$r = \frac{r_{max}[S]}{k_M + [S]} \tag{4.11}$$

This is Briggs–Haldane (1925) analysis of the reaction. Michaelis–Menten analysis further assumes that $k_2 \ll k_{-1}$, thus making the pre-equilibrium approximation leading to a simpler result

$$k_M = \frac{k_{-1}}{k_1} \tag{4.12}$$

In this case, k_M is the true dissociation constant. The parameters of the Michaelis–Menten equation, k_M and r_{max}, can be obtained experimentally by observing the variation of reaction rate with increasing substrate concentration. With a linear scale, Equation 4.11 plots as a rectangular hyperbolic curve (Figure 4.1). However, by transforming the Michaelis–Menten equation to the following reciprocal form (Lineweaver and Burk, 1934)

$$\frac{1}{r} = \frac{k_M}{r_{max}[S]} + \frac{1}{r_{max}} \tag{4.13}$$

the Lineweaver–Burk (double reciprocal, also known as Hanes–Woolf plot) plot of $\frac{1}{r}$ versus $\frac{1}{[S]}$ gives a straight line with the intercept on the vertical axis $\left(\frac{1}{r}\text{axis}\right)$ as $\frac{1}{r_{max}}$, the intercept on the horizontal axis $\left(\frac{1}{[S]}\text{axis}\right)$ as $-\frac{1}{k_M}$, and the slope as $\frac{k_M}{r_{max}}$. Equation 4.11 can also be written as

$$k_M \frac{r}{[S]} + r = r_{max} \tag{4.14}$$

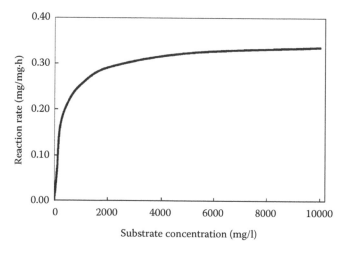

Figure 4.1 Michaelis–Menten curve of an enzyme reaction with r_{max} = 0.35 mg/mg·h and k_M = 400 mg/l.

which also gives a straight line when r is plotted against $\dfrac{r}{[S]}$. This is known as the Eadie–Hofstee plot. The two parameters can be estimated by either of the two plots.

The two parameters, which depend on the enzyme–substrate system, have a high degree of variation. A list of typical values of k_M and k_2 for several enzyme–substrate systems can be found in several websites, including the following: en.wikipedia.org/wiki/Michaelis%E2%80%93Menten_kinetics.

4.3 MONOD EQUATION

Monod equation was proposed by Jacques Monod (1942, 1949), introducing the concept of growth-controlling (-limiting) substrates. It relates the growth and consumption rates to the concentration of a single growth-controlling substrate via two parameters, the maximum specific growth rate (μ_{max}) or the maximum specific rate of substrate consumption (q_{max}), and the half saturation constant sometimes referred to as the substrate affinity constant (k_s). The Monod equation is a non-structured, non-segregated type of model. In a non-structured model

Substrates + cells → by-products + more cells

For a single growth-controlling substrate, Monod postulated the following equations:

$$\mu = \mu_{max}\frac{s}{k_s + s} \tag{4.15}$$

$$q = q_{max}\frac{s}{k_s + s} \tag{4.16}$$

where μ is the specific rate $\left(= \dfrac{1}{x}\dfrac{dx}{dt} \right)$ of biomass growth (T^{-1}), q is the specific rate $\left(= \dfrac{1}{x}\dfrac{ds}{dt} \right)$ of substrate consumption (T^{-1}), μ_{max} is the maximum specific rate of biomass growth (T^{-1}), q_{max} is the maximum specific rate of substrate consumption (T^{-1}), k_s is the half-saturation coefficient (ML^{-3}), s is the substrate concentration (ML^{-3}), and x is the biomass concentration (ML^{-3}). Equations 4.15 and 4.16 represent the biomass growth and substrate utilization kinetics, respectively, and they are sometimes referred to as the Monod functions (Figure 4.2). It can be seen that, when $s >>> k_s$, $\mu \to \mu_{max}$ and $q \to q_{max}$.

Equations 4.15 and 4.16 can explain and simulate the exponential and decelerating growth phases in a batch growth system when coupled with the mass balance equation for the substrate concentration. It does not, however, take into account the initial lag phase and the accelerating growth phase. It is empirical unlike the Michaelis–Menten equation (Michaelis and Menten, 1913), which is mechanistic. The Monod equation can be used to model the kinetics of decay of organic substances and growth of associated biomass. The Monod equations can also be written in a form that describes the time dependence of a growth-limiting substrate and biomass concentrations as follows:

$$\frac{ds}{dt} = -q_{max}\frac{s}{k_s + s}x \tag{4.17}$$

$$\frac{dx}{dt} = q_{max}y\frac{s}{k_s + s}x - bx \tag{4.18}$$

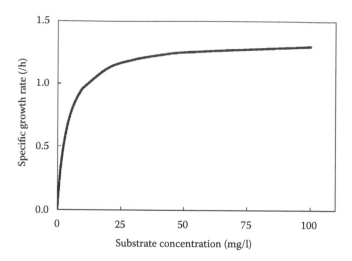

Figure 4.2 Monod function with μ_{max} = 1.35/h and k_s = 4 mg/l.

where y is the stoichiometric yield coefficient of the mass of biomass produced from a unit mass of substrate consumed $\left(y = -\dfrac{dx}{ds} \right)$, and b is the endogenous decay rate (T^{-1}). The yield coefficient gives a measure of the efficiency of converting a substrate to a cell material. The endogenous decay rate, which is very small (≈ 0.0042/h), accounts for the energy loss for cell maintenance and is given by

$$\frac{dx}{dt} = -bx \tag{4.19}$$

Equations 4.17 and 4.18 constitute a coupled set of non-linear ordinary differential equations that has asymptotic behaviour.

When $s >>> k_s$, Equations 4.17 and 4.18 simplify to

$$\frac{ds}{dt} = -q_{max}x \tag{4.20}$$

$$\frac{dx}{dt} = q_{max}yx - bx \tag{4.21}$$

and, when $s <<< k_s$, to

$$\frac{ds}{dt} = -q_{max}\frac{s}{k_x}x \tag{4.22}$$

$$\frac{dx}{dt} = q_{max}\frac{s}{k_s}yx - bx \tag{4.23}$$

Monod (1942) also postulated a linear link between the specific rate of biomass growth (μ) and the specific rate of substrate utilization (q) via the yield coefficient (y):

$$\mu = yq \tag{4.24}$$

The estimation of the parameters (q_{max}, μ_{max}, k_s, and y) of these equations is not trivial. They cannot be estimated uniquely by simple linear regression analysis and even non-linear methods have not yielded unique values of the parameters for certain substrates under certain conditions (Knightes and Peters, 2000). Linearization of Equations 4.17 and 4.18 enables the estimation of the parameters by linear regression, although the assumption of normally distributed errors in the method could be violated. A summary of the difficulties associated with different non-linear methods of parameter estimation is given by Knightes and Peters (2000). Several alternatives to the Monod equation are also available. They include

$$\mu = \mu_{max} \text{ for } s \geq 2k_s; \ \mu = \frac{\mu_{max}s}{2k_s} \text{ for } s \leq 2k_s \text{ (Blackman, 1905)} \tag{4.25}$$

$$\mu = \mu_{max}\left(1 - e^{-k_s}\right) \text{ (Tessier, 1942)} \tag{4.26}$$

Since the original Monod equation does not take into account the situation where the cells may need a substrate even after they stopped growing, it has been modified by introducing a threshold substrate concentration as a maintenance factor. The modified equations take the form (e.g., Alexander, 1994; Herbert, 1958; Kovárová et al., 1996)

$$s = k_s \frac{D}{\mu_{max} - D} + s_{min} \tag{4.27}$$

$$\mu = (\mu_{max} + m)\frac{s}{k_s + s} - m \tag{4.28}$$

where s_{min} is the threshold substrate concentration, D is the dilution rate, and m is the maintenance rate.

Mathematically, the Monod equation is similar to the Michaelis–Menten equation, but there is no relationship between k_s in the former and k_M in the latter except in special cases. It should also be noted that s_{min} is not applicable to situations such as batch cultures. The Monod equation can also be interpreted as a kind of predator–prey model (Hoppensteadt, 2006), in which the biomass can be considered as the predator that preys on the substrate. It is often referred to as the Jacob–Monod model (Jacob and Monod, 1961), and has been applied to model organisms such as bacteria that feed on nutrients.

A typical Monod parameter set for the aerobic biodegradation of polycyclic aromatic hydrocarbons with naphthalene and 2-methylnaphthalene as growth-limiting substrates, obtained by the maximum likelihood method of parameter estimation, have reported values of q_{max} = 0.636 mg/mg h^{-1}, k_s = 0.572 mg/l, and y = 0.413 mg/mg (Knightes and Peters, 2000). An experimentally determined set of these parameters for the growth of E. coli in a glucose substrate at 30°C has reported values of μ_{max} = 1.35/h, k_s = 4 mg/l, and y = 0.23

(Smith, 2006; Smith and Waltman, 1995). A more comprehensive list of kinetic parameter values (q_{max}, μ_{max}, k_s, and y) for different substrate–biomass systems has been compiled by Okpokwasili and Nweke (2005).

4.4 CONCLUDING REMARKS

This chapter describes the basic concepts of reaction kinetics and introduces two of the widely used equations that relate the growth and consumption rates of the biomass produced and the substrate consumed in biochemical reactions: the Michaelis–Menten equation and the Monod equation.

REFERENCES

Alexander, M. (1994): *Biodegradation and Bioremediation.* Academic Press Inc., San Diego, CA.

Blackman, F.F. (1905): Optima and limiting factors. *Annals of Botany*, 19, 281–295.

Briggs, G.E. and Haldane, J.B.S. (1925): A note on the kinetics of enzyme action. *Biochemical Journal*, 19, 339–339.

Henri, V. (1903): *Lois Générales de l'Action des Diastases.* Hermann, Paris.

Herbert, D. (1958): Continuous culture of microorganisms; some theoretical aspects. In: Málek, I. (Ed.), *Continuous Cultivation of Microorganisms*. Czech Academy of Sciences, Prague, Czech Republic, pp. 45–52.

Hoppensteadt, F. (2006): Predator–prey model. *Scholarpedia*, 1(10), 1563.

Jacob, F. and Monod, J. (1961): Genetic regulatory mechanisms in the synthesis of proteins. *Journal of Molecular Biology*, 3, 318–356.

Knightes, C.D. and Peters, C.A. (2000): Statistical analysis of nonlinear parameter estimation for Monod biodegradation kinetics using bivariate data. *Biotechnology and Bioengineering*, 69(2), 160–170.

Kovárová, K., Zehnder, A.J.B. and Egli, T. (1996): Temperature-dependent growth kinetics of *Escherichia coli* ML 30 in glucose-limited continuous culture. *Journal of Bacteriology*, 178, 4530–4539.

Lineweaver, H. and Burk, D. (1934): The determination of enzyme dissociation constants. *Journal of the American Chemical Society*, 56, 658–666, doi:10.1021/ja01318a036.

Michaelis, L. and Menten, M.L. (1913): Die Kinetik der Invertinwerkung. *Biochemische Zeitschrift*, 49, 333–369.

Monod, J. (1942): *Recherches sur la croissance des cultures bactériennes.* Hermann et Cie, Paris, France.

Monod, J. (1949): The growth of bacterial cultures. *Annual Review of Microbiology*, 3, 371–394.

Okpokwasili, G.C. and Nweke, C.O. (2005): Microbial growth and substrate utilization kinetics. *African Journal of Biotechnology*, 5(4), 305–317.

Smith, H.L. (2006): Bacterial growth. Department of Mathematics and Statistics, Arizona State University, Tempe, AZ, 85287, supported in part by NSF grant DMS 0107160. Available at http://math.la.asu.edu/~halsmith/bacteriagrow.pdf.

Smith, H.L. and Waltman, P. (1995): *The Theory of the Chemostat.* Cambridge University Press, Cambridge, UK.

Teissier, G. (1942): Croissance des populations bactériennes et quantité d'aliment disponible. *Review Science*, 80, 209.

Chapter 5

Water quality systems

5.1 DISSOLVED OXYGEN SYSTEMS

The health of a water body can be measured by the amount of dissolved oxygen (DO), which depends on its temperature, elevation, and salt content. Under pristine conditions, the concentration of DO would be at saturation level, which at 0°C is about 14.6 mg/l, at 30°C about 7.56 mg/l, and at 40°C about 6.41 mg/l. The DO concentration decreases with decreasing pressure and increasing salt content. In a heavily polluted water body, the concentration of DO may become zero. Under such conditions which give rise to a 'dead' water body, no living species can survive. For fish to survive in a water body, the DO concentration must be at least 4–6 mg/l.

The fluctuation of DO in a water body takes place as a result of oxygenation, which is the process of adding oxygen, and deoxygenation, which is the process by which oxygen is depleted. Oxygenation takes place via re-aeration, which is the process of oxygen transfer from the atmosphere to the water body through the air/water interface, from tributaries carrying water with higher DO concentration, and by photosynthesis. Deoxygenation takes place via the oxidation of carbonaceous biochemical oxygen demand (CBOD), nitrogenous biochemical oxygen demand (NBOD), sediment oxygen demand (SOD), and, algal respiration. CBOD refers to the reduction of organic carbon to CO_2 in the presence of micro-organisms such as bacteria; NBOD refers to the biological oxidation of ammonia (NH_3) to nitrates (NO_3^-); and SOD refers to aerobic decay of organic benthic material, which is negligible in flowing water.

5.1.1 Biochemical oxygen demand (BOD)

Biochemical oxygen demand, or BOD, in general consists of three components: CBOD, NBOD, and SOD. CBOD refers to the oxygen consumed in converting waste into CO_2 according to the reactions

$$\text{Oxidizable material} + \text{bacteria} + \text{nutrients} + O_2 \rightarrow CO_2 + H_2O + NO_3 + SO_4$$

and for reducing chemicals such as sulphides and nitrites

$$S^- + 2O_2 \rightarrow SO_4$$

$$NO_2^- + 0.5O_2 \rightarrow NO_3^-$$

CBOD can be further subdivided into particulate CBOD and dissolved CBOD. The former settles down and becomes part of SOD. CBOD kinetics can be written as follows assuming a first-order reaction:

$$\frac{dL}{dt} = -K_1 L \tag{5.1}$$

which, when integrated gives

$$L = L_0 e^{-K_1 t} \tag{5.2}$$

where L is the concentration of oxidizable carbonaceous material $[ML^{-3}]$, L_0 is the initial concentration of oxidizable carbonaceous material $[ML^{-3}]$, and K_1 is the rate of oxidation of the carbonaceous material $[T^{-1}]$, which has a reported average value of 0.35/day for untreated municipal waste at 20°C (Thomann and Mueller, 1987, Table 6.5). If the particulate CBOD and dissolved CBOD are considered separately, the combined effect can be obtained additively. Normally, these two effects are lumped together to obtain the combined effect. The oxygen consumed during this process is given as

$$y = L_0 - L = L_0(1 - e^{-K_1 t}) \tag{5.3}$$

Both y and L are measured as oxygen concentration equivalents and therefore have the same units. It is also to be noted that the ultimate CBOD is the same as the initial CBOD, L_0. The 5-day CBOD would then be

$$CBOD_5 = L_0(1 - e^{-5K_1}) \tag{5.4}$$

and the ratio of the ultimate CBOD to the 5-day CBOD is $\dfrac{1}{(1 - e^{-5K_1})}$.

Nitrogenous BOD (NBOD) refers to the oxygen required for nitrification, which is the biochemical oxidation of ammonia to nitrates according to the following two-stage chemical reaction:

$$NH_4 + 1.5O_2 \rightarrow 2H^+ + H_2O + NO_2^-$$

$$NO_2^- + 0.5O_2 \rightarrow NO_3^-$$

The corresponding reaction kinetics can be written as

$$\frac{dL^N}{dt} = -K_N L^N \tag{5.5}$$

where L^N refers to the overall concentration of oxidizable nitrogenous material, and K_N the overall NBOD oxidation rate. As before,

$$L^N = L_0^N e^{-K_N t} \tag{5.6}$$

and the oxygen consumed is

$$y^N = L_0^N - L^N = L_0^N(1 - e^{-K_N t}) \tag{5.7}$$

Figure 5.1 shows the functions given by Equations 5.2, 5.3, 5.6, and 5.7, and the summations of Equation 5.3 with Equation 5.7, and Equation 5.2 with Equation 5.6, for a water quality system with the following parameters: $L_0 = 300$ mg/l; $L_0^N = 200$ mg/l; $K_1 = 0.35$/day; $K_N = 0.3$/day.

Chemical oxygen demand, or COD, on the other hand, is the oxygen demand for biodegradable pollutants as well as non-biodegradable pollutants that are oxidizable. Ultimate BOD (or COD or SOD) is the total amount of oxygen consumed for complete oxidation of all organic wastes. In a waste material that contains both CBOD and NBOD, CBOD is exerted first, almost immediately. Since the process becomes slower with increasing time, the 30-day BOD is usually taken as the ultimate BOD (BOD_u). There is no general correlation between BOD_5 and BOD_u, or between BOD and COD. They depend on the type of biodegradable material.

Pollutants undergo decay with time resulting in decrease of BOD. Hence, BOD is a function of time. BOD_5, measured by a chemical procedure to determine how fast biological organisms consume oxygen in a water body, usually performed over a 5-day period at 20°C in the absence of sunlight, is considered as the reference level for comparison. The 5-day time originated in the United Kingdom in the early 20th century on the basis that it took less than 5 days for any pollutant within the country to reach the sea. It is not an absolute value but is now considered as a standard in many countries. Some countries, however, use other values – for example, France uses both 5-day and 21-day BOD as their reference values; Nordic countries use 7-day BOD as their reference. It should be noted that the values of BOD determined in the laboratory could be different from those under natural conditions.

Typical value of BOD_5 in pristine rivers is usually less than 1 mg/l. For moderately polluted rivers, it ranges from about 2 to 8 mg/l. Sewage that has undergone treatment up to the tertiary level has an approximate value of 20 mg/l, while raw sewage has a value of approximately 600 mg/l. It should, however, be noted that the BOD value of raw sewage depends on the dilution. In some countries such as the United States, the flushing volume is much higher than in some other countries, thereby reducing the BOD concentration.

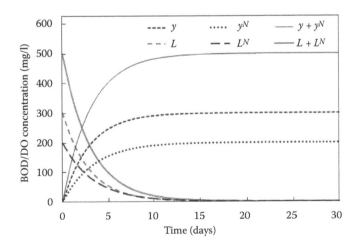

Figure 5.1 Typical BOD/DO concentration variations according to Equations 5.2 through 5.7.

Typical values of BOD concentrations in untreated US municipal wastes range from about 100 to 450 mg/l for $CBOD_5$, 120–580 mg/l for $CBOD_u$, 220 mg/l for NBOD, 5–35 mg/l for organic nitrogen, and 10–60 mg/l for ammonia nitrogen. For combined sewer overflows, the ranges are 40–503 mg/l for $CBOD_5$, 220 mg/l for $CBOD_u$, 0.08–25 mg/l for organic nitrogen, 0–11.5 mg/l for ammonia nitrogen, and for separate urban runoff 2–84 mg/l for $CBOD_5$, 0.2–4.8 for organic nitrogen, and 0.1–1.9 for ammonia nitrogen (Table 6.4, Thomann and Mueller, 1987). Approximate reaction rates for CBOD at 20°C for untreated municipal wastes in the United States range from about 0.3 to 0.4, and for wastes that have gone through primary and secondary treatment range from about 0.1 to 0.3, with corresponding ultimate CBOD to BOD_5 ratios of 1.2 and 1.6 (Table 6.5, Thomann and Mueller, 1987). Typical values of K_N range from about 0.1 to 0.5 per day. Both K_1 and K_N are temperature dependent. Their dependence on temperature is generally expressed as

$$K_{1(T)} = K_{1(20)}\theta^{T-20}$$
(5.8a)

$$K_{N(T)} = K_{N(20)}\theta^{T-20}$$
(5.8b)

where $K_{1(T)}$ and $K_{N(T)}$ are the CBOD and NBOD reoxygenation coefficients at a temperature $T°C$, respectively; $K_{1(20)}$ and $K_{N(20)}$ are their respective values at 20°C; and θ is a temperature correction factor (≈ 1.024).

5.1.2 Nitrification

Nitrification is a microbial process by which nitrogen compounds, especially ammonia (NH_3), are oxidized to nitrites (NO_2), and then to nitrates (NO_3). It can also be considered as the conversion of ammonia nitrogen to nitrate nitrogen. Effectively, it is the process by which an electron is taken from the nitrogen atom and added to the oxygen atom. It is an important component of the nitrogen cycle. The two-stage chemical reaction, which takes place in the presence of two types of bacteria, can be expressed as

$$NH_4^+ + O_2 + \text{bacteria } (Nitrosomonas) \rightarrow NO_2^- + 3H^+ + 2e^-$$

$$NO_2^- + H_2O + \text{bacteria } (Nitrobacter) \rightarrow NO_3^- + 2H^+ + 2e^-$$

Nitrification is an important process in wastewater treatment, and particularly relevant to fish culture zones. The toxic ammonia produced by fish excretions, food waste, and other organic matter such as algae are converted to less toxic nitrates. Fish survival depends on how efficiently the conversion takes place.

5.1.3 Denitrification

Denitrification is the process by which nitrates are converted into molecular nitrogen in the presence of denitrifying bacteria (*Paracoccus denitrificans*). It is an anaerobic process with the intermediate stages of converting nitrates (NO_3) to nitrites (NO_2), nitric oxide (NO), and nitrous oxide (N_2O).

$$NO_3^- \rightarrow NO_2^- \rightarrow NO \rightarrow N_2O \rightarrow N_2$$

Molecular nitrogen (N_2) is not accessible to living things, but is abundant in the atmosphere. Nitrogen fixation converts the gaseous N_2 to more biologically available forms using nitrogen-fixing bacteria (e.g., cyanobacteria). In nature, it takes place in soils and in the roots of leguminous plants.

Nitrification and denitrification are the two key processes in the nitrogen cycle (sometimes called the nitrogen fixation cycle). Nitrogen-fixing bacteria in leguminous plants and soils convert the nitrogen to ammonium $\left(NH_4^+\right)$, which becomes converted to nitrites $\left(NO_2^-\right)$ and then to nitrates $\left(NO_3^-\right)$ in the presence of nitrifying bacteria, which finally becomes converted to nitrogen (N_2) in the presence of denitrifying bacteria. The free N_2 is released into the atmosphere, which will return back to the earth system through rainfall. Plant and animal wastes also contribute to the build-up of ammonium. NH_3, which is highly toxic, is easily converted to NH_4^+, which is relatively less toxic. This conversion can take place both ways.

Nitrogen molecule is inert and therefore breaking it up to its atoms and combining with other atoms require substantial amounts of energy. There are three methods of achieving it:

i. Atmospheric fixation by lightning producing NH_3, oxides of nitrogen, and nitrates, which finally become deposited on Earth. This method accounts for about 10% of all nitrogen fixation.
ii. Biological fixation by microbes, for example, by cyanobacteria. N_2 is converted to NH_4^+, which is used up by leguminous plants or converted by bacteria to NO_3^-. Leguminous plants can take up the nitrates directly and other plants can take up after the death of leguminous plants.
iii. Industrial fixation via fertilizers.

5.1.4 Oxygen depletion equation in a river due to a single point source of BOD

When an oxygen-demanding pollutant is released into a water body, the DO in the water body is depleted. At the same time, a certain amount of reoxygenation also takes place since the water surface is in contact with the atmosphere. A mass balance for this deoxygenation and reoxygenation processes can be written as follows:

$$V \frac{dc}{dt} = O_{in} - O_{out} + (\text{rate of reoxygenation})V - (\text{rate of deoxygenation})V \qquad (5.9)$$

where V is the volume of the water body [L^3]; c, the DO concentration [ML^{-3}] (usually expressed as mg/l); O_{in} and O_{out}, the rates of external oxygen inflow and outflow [MT^{-1}]. Assuming that there are no external inflows and outflows contributing to the oxygen mass balance, Equation 5.9 can be written as

$$V \frac{dc}{dt} = \left(\frac{dc}{dt}\right)_{decay} V + \left(\frac{dc}{dt}\right)_{re\text{-}aeration} V \qquad (5.10a)$$

$$\frac{dc}{dt} = \left(\frac{dc}{dt}\right)_{decay} + \left(\frac{dc}{dt}\right)_{re\text{-}aeration} \qquad (5.10b)$$

Assuming a first-order decay, the rate of decay (or deoxygenation) and the rate of re-aeration (or reoxygenation) can be expressed respectively as

$$\left(\frac{dc}{dt}\right)_{decay} = -k_d L \tag{5.11}$$

and

$$\left(\frac{dc}{dt}\right)_{re\text{-}aeration} = k_r(c_s - c) \tag{5.12}$$

where k_d and k_r are the deoxygenation and reoxygenation coefficients; L, the BOD remaining in the water at time t; and c_s, the saturation value of DO concentration, which depends on the temperature. Equation 5.11 assumes that k_d is the overall deoxygenation rate that includes both oxidation of settled and soluble BOD. The Committee on Sanitary Engineering Research (1960) of the American Society of Civil Engineers (ASCE) has proposed an empirical equation to relate the saturation DO concentration to temperature, which takes the form

$$c_s = 14.652 - 0.41022T + 0.0079910T^2 - 0.000077774T^3 \tag{5.13}$$

where c_s is in mg/l and T is the temperature in °C. Figure 5.2 shows the decreasing trend of the saturation value with increasing temperature, according to Equation 5.13.

Substituting Equations 5.11 and 5.12 in Equation 5.10,

$$\frac{dc}{dt} = -k_d L + k_r(c_s - c) \tag{5.14}$$

This is the differential equation that describes the DO concentration variation in a water body subjected to BOD loading. In many situations, it is convenient to convert this equation to represent the oxygen deficit D, which is defined as

$$D = c_s - c \tag{5.15}$$

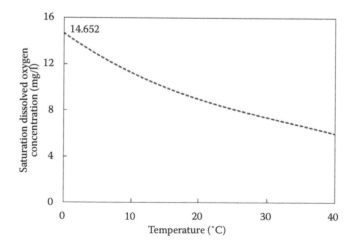

Figure 5.2 Variation of saturation DO concentration with temperature.

Then Equation 5.14 changes to

$$\frac{dD}{dt} = k_d L - k_r D \tag{5.16}$$

Assuming first-order decay for the BOD, Equation 5.16 can be written as

$$\frac{dD}{dt} + k_r D = k_d L_0 e^{-k_d t} \tag{5.17}$$

where L_0 is the ultimate BOD remaining in the water at time t. The solution to Equation 5.17, which is obtained by using an integrating factor[1] is of the form

$$D = \frac{k_d L_0}{k_r - k_d} (e^{-k_d t} - e^{-k_r t}) + D_0 e^{-k_r t} \tag{5.18}$$

In terms of the DO concentration,

$$c = c_s - \frac{k_d L_0}{k_r - k_d} \left(e^{-k_d t} - e^{-k_r t} \right) - D_0 e^{-k_r t} \tag{5.19}$$

In river systems, it is often desired to estimate the DO concentration in the downstream direction. This can easily be achieved by converting the time variable to a space variable $(t = \frac{x}{u}$; x is the distance from the outfall, u is the average velocity in the river). Then,

$$c = c_s - \frac{k_d L_0}{k_r - k_d} \left(e^{-k_d \frac{x}{u}} - e^{-k_r \frac{x}{u}} \right) - D_0 e^{-k_r \frac{x}{u}} \tag{5.20}$$

This solution to the mass balance differential equation has been first obtained by Streeter and Phelps (1925), and is referred to as the Streeter–Phelps equation, or the oxygen sag curve. It was first applied to study the water quality in Ohio River in the United States, and

[1] Multiplying Equation 5.16 by the integrating factor $e^{\int k_r dt} = e^{k_r t}$

$$\frac{dD}{dt}(e^{k_r t}) + k_r D(e^{k_r t}) = k_d L_0 e^{-k_d t}(e^{k_r t})$$

$$\frac{d}{dt}(D e^{k_r t}) = k_d L_0 e^{-k_d t}(e^{k_r t})$$

$$\frac{dD}{dt}(e^{k_r t}) + k_r D(e^{k_r t}) = k_d L_0 e^{-k_d t}(e^{k_r t})$$

$$\int d(D e^{k_r t}) = \int k_d L_0 e^{(k_r - k_d)t} dt$$

which, upon integration gives

$$D(e^{k_r t}) = \frac{k_d L_0}{k_r - k_d} e^{(k_r - k_d)t} + \text{constant}.$$

Using the initial condition $D = D_0$ at $t = 0$, constant $= D_0 - \frac{k_d L_0}{k_r - k_d}$.
Substituting and simplifying,

$$D = \frac{k_d L_0}{k_r - k_d}(e^{-k_d t} - e^{-k_r t}) + D_0 e^{-k_r t}$$

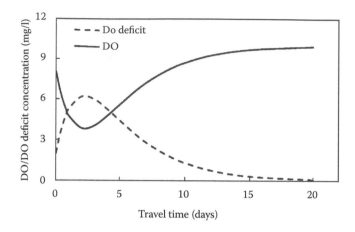

Figure 5.3 Dissolved oxygen sag and DO deficit curves (Equation 5.19 with k_r = 0.5/day, k_d = 0.3/day, L_0 = 20 mg/l, u = 0.35 m/s, D_0 = 2 mg/l, and c_s = 10 mg/l).

has since then become the basis of many applications of environmental modelling. Implicit in the Streeter–Phelps equation are the assumptions that the flow in the river is non-dispersive, steady-state flow BOD and DO reaction conditions, and the only reactions are deoxygenation by decay and reoxygenation by aeration. It should also be noted that the Streeter–Phelps equation is not valid when $k_r = k_d$. The critical (minimum) oxygen deficit can be estimated by setting $\dfrac{dD}{dt} = 0$. This occurs when

$$t_c = \frac{1}{k_r - k_d} \ln \left\{ \frac{k_r}{k_d} \left(1 - \frac{D_0 (k_r - k_d)}{k_d L_0} \right) \right\} \tag{5.21}$$

Figure 5.3 shows a DO deficit and DO variation with time for a typical set of parameters for which the time at which the minimum DO deficit occurs is 2.209 days (from Equation 5.21). The corresponding DO deficit and DO concentrations are 6.182 and 3.818 mg/l, respectively.

5.1.5 Reoxygenation coefficient

The re-aeration coefficient, k_r, depends on several physical factors such as the temperature, the velocity of flow, the level of turbulence, and the depth of flow. Its dependence on temperature is generally expressed as

$$k_{r(T)} = k_{r(20)} \theta^{T-20} \tag{5.22}$$

where $k_{r(T)}$ is the re-aeration coefficient at a temperature $T°C$, $k_{r(20)}$ is its value at 20°C, and θ is a temperature correction factor ($\approx$1.024). Empirically, the dependence on velocity and depth of flow is expressed as

$$k_{r(20)} = \alpha u^n h^m \tag{5.23}$$

where u is the average stream velocity; h the average depth; $k_{r(20)}$ is in units of day^{-1}; α, n, and m are empirical coefficients. Several investigators have proposed different versions of the above relationship:

$$k_{r(20)} = \frac{11.6u}{h^{1.67}} \quad \text{(Churchill et al., 1962)} \tag{5.24a}$$

$$k_{r(20)} = \frac{(D_L u)^{0.5}}{h^{1.5}} \quad \text{(O'Connor and Dobbins, 1956)} \tag{5.24b}$$

In both these equations, u is in feet per second, h in feet, and k_r in day^{-1}. D_L in Equation 5.24b is the oxygen diffusivity [L^{2}T^{-1}] at 20°C (≈ 0.000081 ft^2/s). When converted into SI units, Equation 5.24a and 5.24b, respectively, take the forms

$$k_{r(20)} = \frac{26.7u^{0.969}}{h^{1.673}} \tag{5.25a}$$

and

$$k_{r(20)} = \frac{3.28u^{0.5}}{h^{1.5}} \tag{5.25b}$$

Baecheler and Lazo (1999), summarizing the results of previous studies and using South American data, proposed the following equations for mild and medium slopes:

$$k_r = \frac{10.046u^{2.696}}{h^{3.902}} \quad \text{for mild slopes (up to 0.05\%)} \tag{5.26a}$$

$$k_r = \frac{1.923u^{1.325}}{h^{2.006}} \quad \text{for medium slopes (0.05–0.5\%)} \tag{5.26b}$$

Using Equation 5.26, they obtained values ranging from about 0.39 to 0.98 for mild slopes and 0.84 to 3.22 for medium slopes. The units of Equation 5.26 are as follows: k_r in day^{-1}, u in m/s, and h in m. They also give a comparison of the results obtained by using empirical equations developed by other investigators. As the depth increases, k_r tends to zero, implying there is very little reoxygenation in deep waters.

Another factor that is not often considered is the effect of photosynthesis. Dissolved oxygen generation also takes place during day time by aquatic plants in the build-up of their cell material by photosynthesis, which can be expressed in the following chemical equation:

$$6CO_2 + 6H_2O \rightarrow C_6H_{12}O_6 + 6O_2$$

It is expressed as a half sinusoid as follows (Chapra and Di Toro, 1991):

$$P = \frac{\pi T}{2f} P_{ave} \sin\left(\frac{\pi t}{f}\right) \quad 0 \leq t \leq f \tag{5.27}$$
$$= 0 \text{ otherwise}$$

Table 5.1 Mean and standard deviation of first-order BOD decay coefficients at 20°C

| | Decay coefficient at 20°C | | | | | | | | |
| | Mean | | | Standard deviation | | | Ultimate BOD/BOD$_5$ | | |
Source	BOD	CBOD	NBOD	BOD	CBOD	NBOD	BOD	CBOD	NBOD
San Joaquin River	0.087	0.11	0.057	0.019	0.022	0.017	2.8	1.7	4.0
Stockton Deep Water Ship Channel	0.094	0.11	0.076	0.034	0.023	0.038	2.7	1.7	3.2

Source: Litton, G., *Deposition Rates and Oxygen Demands in the Stockton Deep Water Ship Channel of the San Joaquin River.* San Joaquin River Dissolved Oxygen TMDL Steering Committee, Stockton, CA, USA, p. 185, 2003.

where P is the gross production of photosynthetic DO ([ML^{-3}T^{-1}], usually expressed as mg/l·h), P_{ave} is the daily average production of photosynthetic DO ([ML^{-3}T^{-1}], usually expressed as mg/l·h), T is the period ([T], 24 h), f is the photoperiod length (duration of sunlight) ([T], h), and, t is the time ([T], h), measured from sunrise. Photosynthesis can lead to oxygen supersaturation of the water body during the daytime.

In the field, the re-aeration coefficient is measured using tracer studies. A conservative tracer (usually rhodamine water tracing, or WT, dye) and a non-conservative tracer (usually propane gas) are concurrently injected into the stream, either as a slug or as a continuous injection, and the oxygen concentration profile is measured at two or more downstream stations. On the basis of the desorption of the gas tracer relative to the non-desorbing dye tracer, the re-aeration coefficient can be estimated.

5.1.6 Deoxygenation coefficient

The decay coefficient depends on the type of pollutant. Raw domestic sewage has a value of about 0.5/day, and this value tends to decrease as the sewage undergoes treatment. Table 5.1 shows some values of the decay coefficient for two rivers in California as compiled by Litton (2003) in a report prepared for San Joaquin River Dissolved Oxygen TMDL Steering Committee during the period June–November 2001. Its dependence on temperature generally follows Equation 5.8.

5.2 WATER QUALITY IN A COMPLETELY MIXED WATER BODY

Water quality systems can be considered under various assumptions. The well-mixed assumption implies that there are no concentration gradients in the horizontal and vertical directions. Concentration is assumed to vary only in the time domain. It is an idealized situation. In practice, well-mixed conditions do not exist in real water bodies. Nevertheless, the assumption enables an understanding of the gross effects of how a pollutant attains steady-state conditions from an initial state. The well-mixed assumption alone is not sufficient to obtain solutions to the governing equations. Several other assumptions are necessary. For example, the input waste load can be a constant or a time-variable type. Similarly, the output can also be either a constant or a time-variable type. The hydraulic parameters such as the flow rate and velocity of the water body as well as the reaction rates may be considered as time invariant or time varying. Finally, the system can be assumed to be either linear or non-linear. Non-linear approaches are rarely used for modelling water quality systems because of the difficulties associated with the modelling techniques as well as calibration. They also do not have general applicability. Under linear assumption, the parameters and inputs may also be considered as constant or time varying. Different combinations of these assumptions and their variations can lead to a large number of possible modelling systems.

5.2.1 Governing equations for a completely mixed system

The concentration gradient of a pollutant discharged into a completely mixed water body can be obtained from the law of conservation of mass as follows:

$$V\frac{dc}{dt} = W(t) - Qc - kVc \tag{5.28}$$

where V [L³], is the volume of the water body, c [ML⁻³] is the concentration of the pollutant, W [MT⁻¹] is the rate of application of the waste load, Q [L³T⁻¹] is the outflow from the water body, and k is the decay constant (T⁻¹). The ratio $\dfrac{V}{Q}$ can be considered as the detention time of the water body. Equation 5.28 can be rewritten as

$$V\frac{dc}{dt} + k'c = W(t) \tag{5.29}$$

where

$$k' = Q + kV \tag{5.30}$$

or

$$\frac{dc}{dt} + \frac{k'}{V}c = \frac{W(t)}{V} \tag{5.31}$$

In Equation 5.31, $W(t)$ represents the input; $c(t)$ represents the output; and k, V, and Q represent the system parameters. It is a linear first-order ordinary differential equation and therefore the principles of proportionality and superposition hold. Given the initial condition $c = c_0$ at time $t = 0$, the solution to Equation 5.31 is of the form

$$c(t) = \frac{1}{V}e^{-\frac{k'}{V}t}\int_0^t W(t)e^{\frac{k'}{V}t}dt + c_0 e^{-\frac{k'}{V}t} \tag{5.32}$$

When $W(t) = 0$, Equation 5.32 simplifies to

$$c(t) = c_0 e^{-\frac{k'}{V}t} \tag{5.33}$$

which is the effect of the initial condition. When $W(t) \neq 0$, it can take several forms: step (constant input for a period of time), periodic, impulse, arbitrary, or stochastic. It can be written as a constant and a variable part as

$$W(t) = \bar{W} + W'(t) \tag{5.34}$$

The solutions to each of these input functions can be obtained as follows.

5.2.2 Step function input

The unit step function is defined as follows:

$$
\begin{aligned}
u(t - t_0) &= 0, \text{ for } t < t_0 \\
&= 1, \text{ for } t > t_0
\end{aligned}
\tag{5.35}
$$

If a step input of magnitude $\bar{W}$ is imposed at $t_0 = 0$, the solution, by the principle of proportionality, called the step response,[2] is

$$c_u(t) = \frac{\bar{W}}{k'}\left[\left(1 - e^{-\frac{k'}{V}t}\right)\right] \tag{5.36}$$

In Equation 5.36, as $t \to \infty$,

$$c_u \to \frac{\bar{W}}{k'} = \frac{\bar{W}}{Q + kV} \tag{5.37}$$

If $k = 0$ (for a conservative substance), then

$$c_u = \frac{\bar{W}}{Q} \tag{5.38}$$

The total response (Equation 5.33 + Equation 5.36) is given by

$$c_u(t) = \frac{\bar{W}}{k'}\left[\left(1 - e^{-\frac{k'}{V}t}\right)\right] + c_0 e^{-\frac{k'}{V}t} \tag{5.39}$$

---

[2]
$$
\begin{aligned}
c_u(t) &= \left(\frac{1}{V}\right)e^{-\left(\frac{k'}{V}\right)t}\int \bar{W}(t)e^{\left(\frac{k'}{V}\right)t}\,dt \\
&= \left(\frac{1}{V}\right)e^{-\left(\frac{k'}{V}\right)t}\bar{W}(t)\int_0^t e^{\left(\frac{k'}{V}\right)t}\,dt \\
&= \left(\frac{1}{V}\right)e^{-\left(\frac{k'}{V}\right)t}\bar{W}(t)\left[\left(e^{\left(\frac{k'}{V}\right)t}\right)/\left(\frac{k'}{V}\right)\right]_0^t \\
&= \left(\frac{1}{V}\right)e^{-\left(\frac{k'}{V}\right)t}\bar{W}(t)\left(\frac{V}{k'}\right)\left[\left(e^{\left(\frac{k'}{V}\right)t} - 1\right)\right] \\
&= \left(\frac{\bar{W}}{k'}\right)\left[\left(1 - e^{\left(\frac{k'}{V}\right)t}\right)\right]
\end{aligned}
$$

The variation of the total response depends on the relative magnitude of the initial condition with respect to the ultimate steady-state concentration; that is, whether $c_0 > c_\infty$, or $c_0 < c_\infty$. In the former case, the two concentrations are additive, whereas as in the latter case, the total concentration will decrease and attain a new steady state. A typical example of a step function input is when a certain amount of waste loading enters a water body for a fixed period. The response to a combination of several step function type waste loads can be easily determined using the principle of superposition. The steady-state and maximum concentrations and the time to attain a specified concentration would be of particular interest.

Because of the linear assumption, it is possible to obtain solutions to other forms of $W(t)$. For example, if $W(t) = \bar{W} \pm bt$, the solution would take the form

$$c(t) = \frac{\bar{W}}{k'}\left[\left(1 - e^{-\frac{k'}{V}t}\right)\right] \pm \frac{b}{\left(\frac{k'}{V}\right)^2}\left[\frac{Q}{V}\right]\left[\left(1 - e^{-\frac{k'}{V}t} - \frac{k'}{V}t\right)\right] \qquad (5.40)$$

and for the exponential form $W(t) = \bar{W}e^{\pm bt}$, the solution is

$$c(t) = \frac{\bar{W}}{k' + bV}\left[e^{\pm bt} - e^{-\frac{k'}{V}t}\right] \qquad (5.41)$$

5.2.3 Periodic input function

A periodic input function can be written in the form

$$W(t) = \bar{W} + W_0 \sin(\omega t - \alpha) \qquad (5.42)$$

where W_0 is the amplitude [MT^{-1}] of the waste load, α is the phase shift angle in radians measured from $t = 0$ to the beginning of the positive portion of the sine curve, and ω is the angular frequency ($= 2\pi f = \frac{2\pi}{T}$, T is the period). Because of the linearity assumption, the principle of superposition can be made use of, and therefore only the time-dependent part of the load needs to be considered. The solution for the time-dependent input function $W_0 \sin(\omega t - \alpha)$ is

$$c(t) = W_0 A_m(\omega)\sin(\omega t - \alpha - \theta(\omega)) \qquad (5.43)$$

where

$$A_m(\omega) = \frac{\frac{1}{V}}{\left[\left(\frac{k'}{V}\right)^2 + \omega^2\right]^{\frac{1}{2}}} [L^{-3}T] \qquad (5.44)$$

$$\theta(\omega) = \arctan\left(\frac{\omega}{\frac{k'}{V}}\right) \text{ (radians)} \qquad (5.45)$$

The solution is therefore a function of $A_m(\omega)$ and $\theta(\omega)$, which are in turn functions of ω. The limiting cases are as follows:

When $\omega = 0$, $\quad A_m(\omega) = \dfrac{1}{k'}$; $\quad \theta(\omega) = \arctan(0) = 0$ radians

When $\omega \to \infty$, $\quad A_m(\omega) \to 0$; $\quad \theta(\omega) \to \dfrac{\pi}{2}$ radians

($\omega = 0$ when $T \to \infty$; $\omega \to \infty$ when $T \to 0$)

5.2.4 Fourier series input

Any periodic function can be expressed in the form of a Fourier series. For example, if $W(t)$ is periodic and has a fundamental period $T = \dfrac{2\pi}{\omega}$, then, it is possible to write

$$W(t) = \frac{a_0}{2} + \sum_{i=1}^{\infty}[a_i \cos(i\omega t) + b_i \sin(i\omega t)] \tag{5.46}$$

where the Fourier coefficients a_i and b_i are given by

$$a_i = \frac{2}{T}\int_0^T W(t)\cos(i\omega t)\,dt \tag{5.47a}$$

$$b_i = \frac{2}{T}\int_0^T W(t)\sin(i\omega t)\,dt \tag{5.47b}$$

An example of the use of Fourier series input is when $W(t)$ is given as the half sine wave. This is used in the study of DO variation (oxygen production due to photosynthesis takes place only during the daytime). The half sine wave can be written as

$$\begin{aligned} W(t) &= W_0 \sin(\omega t) \quad 0 \leq t \leq \frac{T}{2} \\ &= 0 \quad \frac{T}{2} \leq t \leq T \end{aligned} \tag{5.48}$$

When it is represented by a Fourier series,

$a_1 = 0$; $\quad b_1 = 1/2$

$$a_i = \frac{1}{\pi}\left[\frac{\cos(i\pi)+1}{1-i^2}\right] \quad \text{for } i > 1$$

$b_i = 0$ for $i > 1$

$$\frac{a_0}{2} = \mu = \frac{W_0}{\pi}$$

and

$$W(t) = W_0 \left\{ \frac{1}{\pi} + \frac{1}{2} \sin(\omega t) - \frac{2}{\pi} \left[\frac{\cos(2\omega t)}{3} + \frac{\cos(4\omega t)}{15} + \frac{\cos(6\omega t)}{35} + \ldots \right] \right\} \tag{5.49}$$

In a series system such as above, the total output will be the sum of the individual outputs for each term (for a linear time invariant system). The summation can be terminated after two or three terms.

To make use of the already computed response for the periodic sine input, the cosine terms of $W(t)$ can be changed to sine terms by shifting the phase angle by $\frac{\pi}{2}$ radians. Then,

$$W(t) \approx W_0 \left\{ \frac{1}{\pi} + \frac{1}{2} \sin(\omega t) - \frac{2}{3\pi} \cos(2\omega t) \right\} \tag{5.50}$$

$$= W_0 \left\{ \frac{1}{\pi} + \frac{1}{2} \sin(\omega t) - \frac{2}{3\pi} \sin\left(2\omega t + \frac{\pi}{2} \right) \right\} \tag{5.51}$$

The first term in this equation corresponds to the response due to steady term, the second term corresponds to the periodic response due to $\sin(\omega t)$, and the third term the periodic and the third term corresponds to the periodic response due to $\left(2\omega t + \frac{\pi}{2} \right)$.

5.2.5 General harmonic response

There are many aperiodic transient type inputs that can be transformed into the frequency domain. The frequency transfer function can then be used on the transformed input to obtain the transformed output. The inverse transformation of the output gives the output in the time domain.

The frequency transfer function $\phi(j\omega)$ is a complex function defined as

$$\phi(j\omega) = A_m(\omega) e^{j\theta(\omega)} \tag{5.52a}$$

$$= A_m(\omega)[\cos\theta(\omega) + j \sin\theta(\omega)] \tag{5.52b}$$

$$= P(\omega) + jQ(\omega) \tag{5.52c}$$

where

$$P(\omega) = A_m \cos\theta = \frac{1}{V} \left[\frac{\dfrac{k'}{V}}{\left(\dfrac{k'}{V} \right)^2 + \omega^2} \right] \tag{5.53a}$$

$$Q(\omega) = A_m \sin\theta = \frac{1}{V} \left[\frac{-\omega}{\left(\dfrac{k'}{V} \right)^2 + \omega^2} \right] \tag{5.53b}$$

$$A_m(\omega) = (P^2 + Q^2)^{1/2} \tag{5.53c}$$

$$\theta(\omega) = \arctan\frac{Q(\omega)}{P(\omega)} \tag{5.53d}$$

(The negative sign in $Q(\omega)$ is due to the fact that the phase angle is negative because the output lags behind the input. P and Q are fixed for a particular value of ω.)

A plot of P versus Q for different frequencies is called a Nyquist diagram. For a well-mixed body of water, the Nyquist diagram is a semicircle.

Returning to the general harmonic function, the Fourier transform pair for the input function, given in complex notation, is

$$W(t) = \frac{1}{2\pi} \int_{-\infty}^{\infty} W_F(j\omega)e^{j\omega t}\,d\omega \tag{5.54a}$$

$$W_F(j\omega) = \int_{-\infty}^{\infty} W(t)e^{-j\omega t}\,dt \tag{5.54b}$$

If the analysis is done in the frequency domain, the input and output are related to each other by an algebraic equation as follows:

$$C_F(j\omega) = \phi(j\omega) \cdot W_F(j\omega) \tag{5.55a}$$

The inverse transform of $C_F(j\omega)$ gives the output $c(t)$:

$$c(t) = \frac{1}{2\pi} \int_{-\infty}^{\infty} C_F(j\omega)e^{j\omega t}\,d\omega \tag{5.55b}$$

In the real domain, the transformations take the form

$$W(t) = \frac{2}{\pi} \int_{0}^{\infty} W_F(\omega)\cos(\omega t)\,d\omega \tag{5.56a}$$

$$W_F(\omega) = \int_{0}^{\infty} W(t)\cos(\omega t)\,dt \tag{5.56b}$$

$$c(t) = \frac{2}{\pi} \int_{0}^{\infty} C_F(\omega)\cos(\omega t)\,d\omega \tag{5.57a}$$

$$C_F(\omega) = \int_{0}^{\infty} c(t)\cos(\omega t)\,dt \tag{5.57b}$$

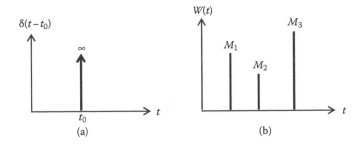

Figure 5.4 (a) Unit impulse input. (b) Delta functions.

5.2.6 Impulse input

A waste load can be applied at different rates. When the rate of application is very high (implying time of application is very short), it can be approximated by an impulse input, which is mathematically equivalent to the Dirac delta type function. Physically it is represented by the discharge of an amount M (kg) of waste in a very short time. Input function $W(t)$ is then given by (delta function has units of T^{-1})

$$W(t) = M\delta\,(t - t_0) \tag{5.58a}$$

In Equation 5.58a, $\delta\,(t - t_0)$ has the units of $[T^{-1}]$ because $\int_{-\infty}^{\infty} \delta(t - t_0)\mathrm{d}t = 1$. If a sequence of impulses are released at different times, then

$$W(t) = \sum_{r=1}^{n} M_r\delta(t - t_r) \tag{5.58b}$$

These two input functions are illustrated in Figure 5.4.
  If $t_0 = 0$, then $W(t) = M\delta(t)$. The solution to this case can be shown to be

$$c(t) = \frac{M}{V}e^{\left[-\left(\frac{k'}{V}\right)t\right]} \tag{5.59}$$

When $M = 1$, the response is referred to as the impulse response function, $I(t)$, the response due to an instantaneous unit load.

$$I(t) = \frac{1}{V}e^{-\frac{k'}{V}t} \tag{5.60}$$

5.2.7 Arbitrary input

The approach for an arbitrary input consists of approximating the input by a series of finite impulse inputs. The concept and the procedure is the same as that for the unit hydrograph theory.

$$c(t) = \int_{-\infty}^{t} W(\tau)I(t-\tau)\,d\tau \tag{5.61}$$

which is the well-known convolution integral.

Example 5.1

Calculate the response functions for a well-mixed water body with a volume of 300 million m³, an outflow of 0.6 million m³/day and a decay coefficient of 0.2/day given the following input conditions:

   i. Step waste input of 50,000 kg/day
   ii. An impulse waste input of 100 tons
   iii. A periodic waste input given by the function $W(t) = \bar{W} + 25{,}000\sin\left(\dfrac{2\pi}{7}t\right)$ kg/day
   iv. A periodic waste input given by a Fourier series function

Assume an initial concentration of 0.005 kg/m³. Plot the Nyquist diagram and verify the well-mixed assumption.
Data given are as follows:

$Q = 0.6 \times 10^6$ m³/day $\approx$ (6.9 m³/s)
$V = 300 \times 10^6$ m³
$K = 0.2$/day
$\bar{W} = 50{,}000$ kg/day
$W(t) = \bar{W} + 25{,}000\sin\left(\dfrac{2\pi}{7}t\right)$ kg/day
$C_0 = 0.005$ kg/m³ (5 mg/l)
$M = 100$ tons $= (100 \times 10^3$ kg)

   i. Step input function

   $k' = Q + kV = 60.6 \times 10^6$ m³/day

   $$c_u(t) = \frac{\bar{W}}{k'}\left[(1 - e^{-\frac{k'}{V}t})\right]$$

   $c(\infty) = 0.825 \times 10^{-3}$ kg/m³(0.825 mg/l)

   Step function response is given in Figure 5.5.
   ii. Impulse input
        The response to the impulse input function shown in Figure 5.6 is given by

   $$I(t) = \frac{M}{V}e^{-\frac{k'}{V}t}$$

   iii. Periodic function

        Angular frequency $\omega = \dfrac{2\pi}{7}$; therefore, period $T = 7$ days. The periodic input is equivalent to a steady input of 50,000 kg/day plus a sinusoidal periodic input with

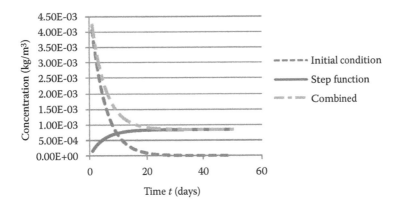

Figure 5.5 Effects of initial condition, step function, and their combination.

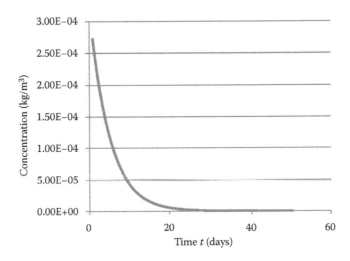

Figure 5.6 Response to impulse waste input.

a period of 7 days and amplitude of 25,000 kg/day. The phase shift angle is zero. The steady response has already been calculated (corresponding to $\omega = 0$). The periodic output is given by (Equation 5.43)

$$c(t) = W_0 A_m \sin(\omega t - \alpha - \theta(\omega))$$

where
$W_0 = 25,000$
$A_m(\omega) =$ already calculated for different ω
$\alpha = 0$
$\theta(\omega) =$ already calculated for different ω
    Therefore, for a period of 7 days

$$c_P(t) = 25,000 \times 0.00362 \times 10^{-6} \sin\left(\frac{2\pi t}{7} - 77°\right)$$

Total response = response due to step input + response due to periodic input.

The output function is delayed by 77°. The periodic input and output functions are shown in Figures 5.7 and 5.8.

iv. Fourier series input

Fourier series input is assumed to be of the form given in Equation 5.51 with only three terms considered:

$$W(t) \approx W_0 \left\{ \frac{1}{\pi} + \frac{1}{2}\sin(\omega t) - \frac{2}{3\pi}\sin\left(2\omega t + \frac{\pi}{2}\right) \right\}$$

This is shown in Figure 5.9.

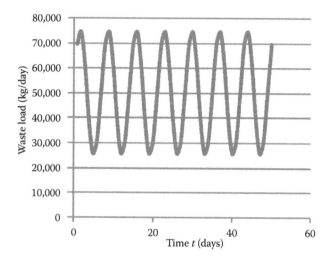

Figure 5.7 Periodic waste load.

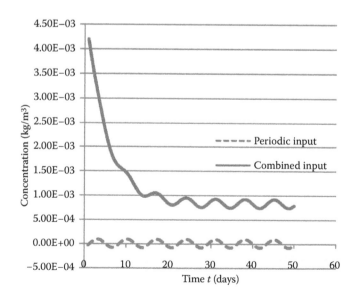

Figure 5.8 Response due to periodic waste input combined with step function input.

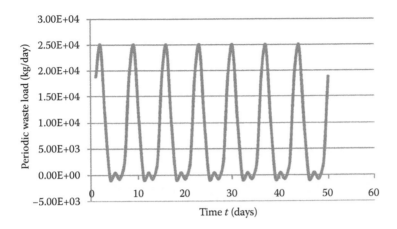

Figure 5.9 Fourier series–type periodic waste input.

The corresponding output is obtained by the principle of superposition by adding the steady input and the two components of the periodic part. The resulting output is shown in Figure 5.10.

Amplitude and phase characteristics are given by Equations 5.53a through 5.53d.

$\theta(\omega)$ in days is converted to θ (degrees) as follows:

$$\theta(\omega) \text{ days} = \theta(\omega) \text{ degrees} \times [\text{period (days)}/360 \text{ (degrees)}]$$

The Nyquist diagram is shown in Figure 5.11. The amplitude is measured from the origin while the phase angle is measured downward indicating a lag.

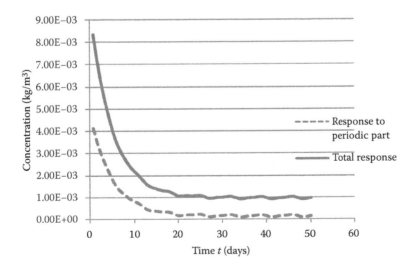

Figure 5.10 Response due to Fourier series–type waste input.

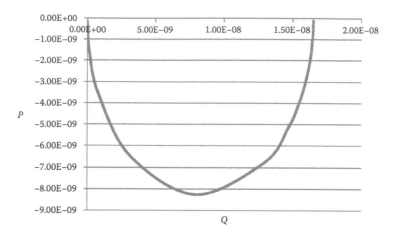

Figure 5.11 Nyquist diagram for Example 5.1.

5.3 WATER QUALITY IN RIVERS AND STREAMS

Water quality variation in a river system depends on many factors such as the hydraulic parameters, presence of tributaries and abstraction points, outfalls of waste material at fixed discharge points, non-point sources of pollution, and whether the system is considered to be at steady state or unsteady state. Different conditions lead to different formulations and solutions. The system should therefore be considered under specific assumptions and specific waste input conditions. The simplest is when there is a point source of waste loading in a river, which is assumed as a one-dimensional water body.

5.3.1 Point sources

The governing equation, or the mass balance equation, is formulated under three major assumptions. First, it is assumed that there is no concentration gradient across the width and depth of the river. This is an idealized condition that is justified only after some time (or distance) called the initial period (or mixing length) has lapsed. After the initial period, complete mixing is assumed to be achieved, at least in theory. Yotsukura (1968) and Kilpatrick et al. (1970) defined the mixing length as

$$L_m = 2.6u\frac{B^2}{h} \text{ for a side bank discharge} \tag{5.62a}$$

and

$$L_m = 1.3u\frac{B^2}{h} \text{ for a mid stream discharge} \tag{5.62b}$$

where L_m is the mixing length in feet, B is the average stream width in feet, h is the average stream depth in feet, and u is the average stream velocity in ft/s. The second assumption is that there is no dispersion in the longitudinal direction. This condition is also called plug flow system, or maximum gradient system or advective system. There is no mixing of one control volume of water with another control volume. The third assumption is that steady-state conditions prevail.

Three mass balance equations under these assumptions can be written for the flow and the waste material, respectively, as

$$Q_u c_u + Q_s c_s = Q_d c_d \tag{5.63}$$

$$W = Q_s c_s \tag{5.64}$$

$$Q_d = Q_u + Q_s \tag{5.65}$$

where Q_u, Q_d, and Q_s (all in $[L^3 T^{-1}]$) are the upstream, downstream, and point source flow discharges, respectively; c_u, c_d, and c_s (all in $[ML^{-3}]$) are the concentrations of the waste material upstream, downstream, and at the source, respectively; and W is waste load $[MT^{-1}]$. From Equations 5.63 and 5.65,

$$c_d = \frac{Q_u c_u + Q_s c_s}{Q_d} = \frac{Q_u c_u + Q_s c_s}{Q_u + Q_s} \tag{5.66}$$

If the upstream waste concentration is zero ($c_u = 0$), then

$$c_d = \frac{Q_s c_s}{Q_d} = \frac{W}{Q_d} \tag{5.67}$$

which gives the effect of dilution only. This condition can be applied to tributary inflows, which bring in waste concentrations. Because of the non-dispersive well-mixed assumptions, the concentrations downstream of the outfall will remain unaltered for a conservative pollutant until an external input of flow or waste material is added or taken away. For a conservative material, such as, for example, total dissolved solids, chlorides, and certain metals, there is no change in concentration between tributaries or waste inputs. The concentration changes only at a discharge point. It is also assumed that there is no leakage due to seepage. At a discharge point, the concentration will undergo a sharp increase or decrease depending on tributary inflows, outflows, and waste inputs. For non-conservative (biodegradable) materials, such as BOD, nutrients, bacteria, volatile chemicals, etc., the mass balance equation, assuming a first-order decay is

$$\frac{1}{A}\frac{d}{dx}(Qc) = -kc \tag{5.68a}$$

If Q is constant, then,

$$\frac{Q}{A}\frac{dc}{dx} + kc = 0 \tag{5.68b}$$

$$u\frac{dc}{dx} + kc = 0 \tag{5.68c}$$

In Equation 5.68, Q is the flow rate, A is the average cross sectional area of flow, u is the average velocity of flow, and k is a decay rate $[T^{-1}]$.

Equation 5.68c, with the boundary condition $c = c_0$ at $x = 0$, is a linear first-order ordinary differential equation that has a solution of the form

$$c = c_0 e^{-\frac{kx}{u}} \tag{5.69a}$$

which may also be written as

$$c = c_0 e^{-kt^*} \tag{5.69b}$$

where $t^* = \dfrac{x}{u}$ = time to travel a distance x at velocity u. In logarithmic form,

$$\ln(c) = -kt^* + \ln(c_0) \tag{5.69c}$$

which plots as a straight line from which the system parameters can be estimated for a known concentration–time profile. For multiple-point sources, the principle of superposition can be used. The total effect is the sum of individual effects plus the effects due to boundary conditions.

5.3.2 Distributed sources

Rivers and streams have sources and/or sinks along their lengths (e.g., wastes carried by runoff). The mass balance equation is

$$u\frac{dc}{dx} + kc = s_D \tag{5.70}$$

where s_D is the distributed source $(ML^{-3}T^{-1})$, which may be expressed as

$$S_D = \frac{w}{A} = \frac{wdx}{Adx}; \; A = (\text{width of river section}, B) \times (\text{depth of flow}, h)$$

and w is the external source of waste $(ML^{-1}T^{-1})$ (Figure 5.12).

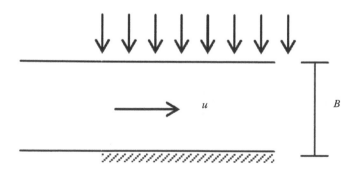

Figure 5.12 Distributed waste load along a river.

Boundary condition, for example, can be $c = 0$ at $x = 0$.
The solution of Equation 5.70 is then of the form

$$c = \frac{s_D}{k}(1 - e^{-kt*}) = \frac{s_D}{k}(1 - e^{-kx/u})$$ (5.71)

In Equation 5.71, as $x \to \infty$, $c \to \frac{s_D}{k}$. Thus, the concentration profile starts with zero, increases non-linearly, and reaches the steady-state value asymptotically.

It is also possible to have distributed sources from within the stream, for example, from benthal sources. In this case (e.g., sediment oxygen demand),

$$s_D = \frac{s_B B dx}{B h dx} = \frac{s_B}{h} (ML^{-3}T^{-1})$$ (5.72)

where h is the depth of flow. s_B has units of $ML^{-2}T^{-1}$; that is, mass per unit bed area per unit time.

5.3.3 Effect of spatial flow variation

Flow can vary spatially due to infiltration (sink) or ground water exfiltration (source). The mass balance equation is (Thomann, 1972)

$$\frac{Q}{A}\frac{dc}{dx} + (v + k)c = vc_I \pm s_D$$ (5.73)

where Q, A, c, v (spatial rate of change of flow), k, and s_D are spatially variable but temporally constant; $v = \frac{1}{A}\frac{dQ}{dx}$ (T^{-1}); c_I is the concentration of infiltrating/exfiltration flow (ML^{-3}); s_D is the distributed source/sink ($ML^{-3}T^{-1}$).

The solution of Equation 5.73, given by Li (1962), is

$$c = \{c_0 + I(x)\}e^{-\phi_I(x)}$$ (5.74)

where

$$\phi_1(x) = \int_0^x \frac{k+v}{u} dx$$ (5.75a)

$$I(x) = \int_0^x \frac{vc_I \pm s_D}{u} e^{\phi_I(x)} dx$$ (5.75b)

When

$$v = 0, \quad \phi_{\mathrm{I}}(x) = \frac{kx}{u}$$

5.3.3.1 Exponential spatial flow variation

If Q is assumed to follow an exponential variation, assuming constant area A and decay constant k, of the form

$$Q = Q_0 e^{qx} \tag{5.76}$$

then

$$v = \frac{1}{A}\frac{dQ}{dx} = \frac{1}{A}\{qQ_0 e^{qx}\} = qu_0 e^{qx}$$

where

$$u_0 = \frac{Q_0}{A}$$

Then, substituting in Equation 5.75a,

$$
\begin{aligned}
\phi_1(x) &= \int_0^x \left\{ \frac{k + qu_0 e^{qx}}{u_0 e^{qx}} \right\} dx \\
&= \int_0^x \left\{ \frac{k}{u_0 e^{qx}} \right\} dx + \int_0^x q\, dx \\
&= \frac{k}{u_0}\left[\frac{e^{-qx}}{-q} \right]_0^x + [qx]_0^x \\
&= \frac{k}{u_0}\left[\frac{e^{-qx} - 1}{-q} \right] + qx
\end{aligned}
\tag{5.77}
$$

If it is further assumed that there is no distributed source ($s_{\mathrm{D}} = 0$), and that the concentration of infiltrating/exfiltrating flow is zero ($c_{\mathrm{I}} = 0$), then $I(x) = 0$. Equation 5.74 then gives,

$$
\begin{aligned}
c &= c_0 e^{\frac{k}{qu_0}\left[e^{-qx}-1\right]-qx} \\
&= \frac{W}{Q_0} e^{\frac{k}{qu_0}\left[e^{-qx}-1\right]-qx}
\end{aligned}
\tag{5.78}
$$

Since $c_0 = \dfrac{W}{Q_0}$

5.3.4 Unsteady state

5.3.4.1 Non-dispersive systems

Unsteady state can be considered either as a non-dispersive system or as a dispersive system. In a non-dispersive system, there is no mixing in the longitudinal direction. This means that each parcel of water does not interfere with other parcels in front or behind. The condition is called plug flow. The governing equation is

$$\frac{\partial c}{\partial t} + u\frac{\partial c}{\partial x} + kc = 0 \tag{5.79}$$

The boundary condition is $c = c_0(t)$ at $x = 0$, which may be written as

$$c_0(t) = \frac{W(t)}{Q} \tag{5.80}$$

The solution of Equation 5.79, assuming that the waste load is a function of time, is

$$c(x,t) = \frac{W(t - t^*)}{Q} e^{-\frac{kx}{u}} \tag{5.81}$$

where t^* is the travel time. The concentration reduction is due to dilution $\left(\dfrac{W(t)}{Q}\right)$ and decay $(e^{-kx/u})$.

5.3.4.2 Dispersive systems

In real world, plug flow rarely exists. Instead, mixing of the waste load takes place along the longitudinal direction as well as in the vertical and lateral directions, primarily due to the respective velocity gradients. In addition, variations of geometrical parameters of the river channel also contribute to mixing. Longitudinal dispersion refers to the process of mixing in the longitudinal direction due mainly to velocity gradients. The governing equation has the form

$$\frac{\partial c}{\partial t} + u\frac{\partial c}{\partial x} + kc = D\frac{\partial^2 c}{\partial x^2} \tag{5.82}$$

where the dispersion coefficient D is of dimension L^2T^{-1}.

The solution to this equation depends on the input type.

a. Impulse input of mass M

This input condition is equivalent to a slug-type release, or sudden spill of a toxic material where the time of application is very short. The solution of Equation 5.82 then is given as

$$c(x,t) = \frac{M}{A\sqrt{4\pi Dt}} e^{\left[-\frac{(x-ut)^2}{4Dt} - kt\right]} \tag{5.83}$$

This is mathematically equivalent to the Gaussian probability density function with mean ut and variance $4Dt$. Thus, if 't' is fixed, the concentration–distance profile is symmetric around its peak value. On the other hand, if the concentration–time profile is considered at different locations, they will not be symmetric. A measure of spread around its peak value is the variance σ^2 (or standard deviation, σ), which, in this case, is $4Dt$. As t increases, σ also increases. The concentration profiles flatten out as the waste material is carried downstream resulting in reduction of the peak concentration. If $D = 0$, then, $\sigma = 0$ and the result will be plug flow.

Note: The impulse response function can be used to determine the response to other inputs using the principles of superposition and proportionality.

If the waste material is conservative, then $k = 0$. Equation 5.83 simplifies to

$$c(x,t) = \frac{M}{A\sqrt{4\pi Dt}} e^{\left[-\frac{(x-ut)^2}{4Dt}\right]}$$ (5.84a)

Where there is no advection, $u = 0$. Then, Equation 5.84a simplifies to

$$c(x,t) = \frac{M}{A\sqrt{4\pi Dt}} e^{\left[-\frac{x^2}{4Dt}\right]}$$ (5.84b)

b. Step input

Step input refers to an input over a fixed interval of time starting from zero, suddenly increasing to a fixed value, remaining at that value for a fixed interval of time and suddenly dropping back to zero. It has the shape of a rectangle. The general solution of the governing equation for a steady input with constant coefficients given by Thomann (1987, p. 73) is

$$c(x,t) = \frac{c_0}{2} e^{\frac{kx}{u}} \left\{ \mathrm{erf}\left[\frac{x - u(t-\tau)(1+\eta)}{\sqrt{4D(t-\tau)}}\right] - \mathrm{erf}\left[\frac{x - ut(1+\eta)}{\sqrt{4Dt}}\right] \right\}$$ (5.85)

where c_0 is the concentration of the input after mixing over the cross section, τ is the time interval of the input, $\eta = \dfrac{kD}{u^2}$ (dimensionless),[3] and

$$\mathrm{erf}(\phi) = \frac{2}{\sqrt{\pi}} \int_0^\phi e^{-z^2} dz \quad \{\text{Note}: \mathrm{erf}(-\phi) = -\mathrm{erf}(\phi)\}$$ (5.86)

If $\tau <<<< t^*$ (i.e., the time of application of the input is short in comparison to travel time), then, according to O'Loughlin and Bowner (1975) and Rose (1977), the time of travel of the peak concentration a distance x is given by

$$t_p = \frac{x + u\tau(1+\eta)}{u(1+\eta)}$$ (5.87)

---

[3] For upland streams, η (in Equations 5.85, 5.87, and 5.88) < 0.01 (Thomann, 1973); for main drainage rivers, η = 0.01–0.5; for large rivers, η = 0.5–1.0; that is, that longitudinal dispersion is not significant in upland streams

and the peak concentration c_p at distance x and time t_p is

$$c_p = \frac{c_0}{2} e^{-\frac{kx}{u}} \left\{ \text{erf}\left[\frac{x - u(t_p - \tau)(1 + \eta)}{\sqrt{4D(t_p - \tau)}} \right] - \text{erf}\left[\frac{x - ut_p(1 + \eta)}{\sqrt{4Dt_p}} \right] \right\} \qquad (5.88)$$

5.3.5 Tidal reaches

In tidal reaches such as estuaries, the mixing is caused by flow reversals as well as due to density differences. The dynamics of the water quality system therefore becomes more complicated. However, under certain assumptions, the system can still be modelled. The assumptions normally made include steady-state conditions; one-dimensional flow; the conditions described are average over a number of tidal cycles; and that the geometrical, flow, and reaction kinetic parameters are constant. The governing equation then takes the form (Thomann, 1972)

$$u\frac{\partial c}{\partial x} + kc = D\frac{\partial^2 c}{\partial x^2} \qquad (5.89)$$

When there are flow reversals, the spreading of the pollutant can be in both the downstream and upstream directions. The applicable dispersion coefficient is referred to the tidal dispersion coefficient, which is different from the longitudinal dispersion coefficient defined for unidirectional flow in rivers. The solution of Equation 5.89 for a non-conservative point source input of mass M at $x = 0$ with the boundary conditions

$c = c_0$ at $x = 0$

$c = 0$ at $x \rightarrow \pm \infty$

is

$$c(x) = \frac{M}{Aum} e^{\left[\frac{u}{2D}(1 \pm m)x \right]} \qquad (5.90)$$

where $u = \frac{Q}{A}$ (Q is the fresh water flow rate), and

$$m = \sqrt{1 + (4kD / u^2)} \qquad (5.91)$$

The negative sign in Equation 5.90 applies to the upstream of the discharge point.

5.3.5.1 Special case of no decay

If $k = 0$ (no decay), then $m = 1$. The governing equation becomes

$$\frac{udc}{dx} = D\frac{d^2 c}{dx^2} \qquad (5.92)$$

with the boundary conditions $c = 0$ at $x = -\infty$, and $c = c_0$ at $x = 0$. The solution to Equation 5.92 takes the form

$$c(x) = \frac{M}{Au} e^{\left[\frac{u}{2D}(1\pm1)x\right]}$$

(5.93a)

which for downstream is

$$c(x) = \frac{M}{Au} e^{\left[\frac{2ux}{2D}\right]} = \frac{M}{Au} e^{\frac{ux}{D}}$$

(5.93b)

and for upstream is

$$c(x) = \frac{M}{Au} e^0 = \frac{M}{Au}$$

(5.93c)

5.3.5.2 Special case of no dispersion

If there is no dispersion ($D = 0$). The governing equation becomes

$$u\frac{\partial c}{\partial x} + kc = 0$$

(5.94)

which has a solution given by

$$c(x) = \frac{M}{Au} e^{-\frac{kx}{u}}$$

(5.95)

There is no transport in the upstream direction. The upstream dispersion is small in comparison to the downstream advection.

5.4 CONCLUDING REMARKS

This chapter gives a description of water quality analysis and modelling using the systems theory approach. It should, however, be emphasized that it is only one of the approaches of water quality modelling and that there are many assumptions, in particular the linear assumption, made or implied leading to limitations in application. In situations where the system dynamics are not known *a priori*, the approach nevertheless enables an understanding of the gross effects of how the water quality in a water body attains a steady state from an initial state.

REFERENCES

Baecheler, J.V. and Lazo, O.L. (1999): Evaluation of water quality modeling parameters: Reaeration coefficient. *Proceedings, IAHR Congress*, Graz. http://www.iahr.org/membersonly/grazproceedings99/doc/000/000/267.htm.

Chapra, S.C. and Di Toro, D.M. (1991): Delta method for estimating primary production, respiration, and reaeration in streams. *Journal of Environmental Engineering*, 117(5), 640–655.

Churchill, M.A., Elmore, R.L. and Buckingham, R.A. (1962): The prediction of stream re-aeration rates. *Journal of the Sanitary Engineering Division, ASCE*, 88(7), 1–46.

Committee on Sanitary Engineering Research (1960): Solubility of atmospheric oxygen in water. *Journal of the Sanitary Engineering Division, ASCE*, 86(7), 41–53.

Kilpatrick, F.A., Martens, L.A. and Wilson, J.F. (1970): Measurement of time of travel and dispersion by dye tracing, Chapter A9, *US Geological Survey Techniques of Water Resources Investigations*, Book 3 (Applications of Hydraulics), US Department of the Interior, Washington, D.C.

Li, W.H. (1962): Unsteady dissolved oxygen sag in a stream. *Journal of the Sanitary Engineering Division, ASCE*, 88(SA3), 75–85.

Litton, G. (2003): *Deposition Rates and Oxygen Demands in the Stockton Deep Water Ship Channel of the San Joaquin River*. San Joaquin River Dissolved Oxygen TMDL Steering Committee, Stockton, CA, p. 185.

O'Connor, D.J. and Dobbins, W.E. (1956): The mechanisms of re-aeration in natural streams. *Journal of the Sanitary Engineering Division, ASCE*, 82(12), 1115–1130.

O'Loughlin and Bowner, K.H. (1975): Dilution and decay of aquatic herbicides in flowing channels. *Journal of Hydrology*, 26, 217–235.

Rose, D.A. (1977): Dilution and decay of aquatic herbicides in flowing channels—Comments. *Journal of Hydrology*, 32, 399–400.

Streeter, H.W. and Phelps, E.B. (1925): A study of the pollution and natural purification of the Ohio River. US Public Health Service Bulletin No. 146.

Thomann, R.V. (1972): *Systems Analysis and Water Quality Management*. McGraw Hill New York, p. 286.

Thomann, R.V. (1973): Effect of longitudinal dispersion on dynamic water quality of streams and rivers. *Water Resources Research*, 9(2), 355–366.

Thomann, R.V. and Mueller, J.A. (1987): *Principles of Surface Water Quality Modelling and Control*. Harper & Row, Publishers, New York. p. 644.

Yotsukura, N. (1968): As referenced in Kilpatrick, F.A., Martens, L.A. and Wilson, J.F. (1970): Measurement of time of travel and dispersion by dye tracing, Book 3, Chapter A9, US Geological Survey Techniques of Water Resources Investigations, Book 3 (Applications of Hydraulics), US Department of the Interior, Washington, D.C.

Chapter 6

Longitudinal dispersion

6.1 INTRODUCTION

Longitudinal dispersion is the process by which a solute introduced into a solvent dilutes by the combined action of molecular diffusion and cross-sectional velocity variation. It is different from molecular diffusion, which takes place due to molecular activity and is considered on a microscopic scale such as, for instance, in chemical and biological reactions. It is an important topic in environmental engineering. Knowledge of the dispersion characteristics is necessary for estimating the waste-receiving capacity and the distribution of pollutants in a water body.

6.2 GOVERNING EQUATIONS

Longitudinal dispersion is treated in accordance with Fick's (1855) law of diffusion, which is an extension to Fourier's law of heat conduction. For a one-dimensional process, Fick's law can be represented mathematically as

$$q = -D \frac{\partial c}{\partial x} \tag{6.1}$$

where q is the solute mass flux [MT^{-1}], D is the diffusion coefficient [L^2T^{-1}], and c is the mass concentration of solute [ML^{-3}].

By combining Fick's law with the continuity equation, which is dependent on the type of process by which the solute is transported, it is possible to obtain the partial differential equation that describes the process. In one-dimensional form, for a quiescent fluid, it is

$$\frac{\partial c}{\partial t} = D \frac{\partial^2 c}{\partial x^2} \tag{6.2}$$

The fundamental solution to the diffusion equation is the Gaussian distribution given by

$$c(x,t) = \frac{M}{A\sqrt{4\pi Dt}} e^{-\frac{x^2}{4Dt}} \tag{6.3}$$

for an initial slug of mass M introduced at $x = 0$, and $t = 0$, and A is the cross-sectional area of flow. This is equivalent to a Dirac[1] delta function type input. A property of the diffusion

---

[1] Dirac delta function is defined as $\delta(t) \neq 0$ at $t = 0$, $\delta(t) = 0$ at $t \neq 0$, and $\int_{-\infty}^{\infty} \delta(t) \, dt = 1$.

equation is that the variance of a concentration distribution always grows linearly with time. When the solvent medium is moving with a velocity u, the process is called advective[2] diffusion and is assumed to be a combination of transport by diffusion and advection. It is equivalent to assuming that the diffusion of solute in the solvent takes place in the moving solvent in the same way as in the stationary fluid; that is, transport by the mean motion of the fluid. The corresponding solute mass flux and the governing equation, respectively, are

$$q = uc + \left(-D\frac{\partial c}{\partial x} \right) \text{ (advective flux + diffusive flux)} \qquad (6.4)$$

$$\frac{\partial c}{\partial t} + u\frac{\partial c}{\partial x} = D\frac{\partial^2 c}{\partial x^2} \qquad (6.5)$$

In this type of diffusion, the flow is assumed to be laminar and therefore the molecular diffusion coefficients have constant values in all directions. This is called the advective diffusion equation, or simply the diffusion equation. Laminar flow in the environment is uncommon. Most flows in nature are turbulent. The Fickian diffusion coefficient for laminar flow is then replaced by an analogous coefficient called turbulent mixing coefficient ε_x. The analogous equation for turbulent flow can then be written in terms of ε_x as

$$\frac{\partial c}{\partial t} = \varepsilon_x \frac{\partial^2 c}{\partial x^2} \qquad (6.6)$$

It is possible to show that the coefficient ε_x can be compared with D since the mass flux can be expressed as

$$\overline{uc} = -\varepsilon_x \frac{\partial \overline{c}}{\partial x} \text{ (analogous to Equation 6.1)} \qquad (6.7)$$

where $\overline{u}$ and $\overline{c}$ are the time-averaged values of u and c. Thus, the flux is proportional to the concentration gradient. The coefficient ε_x is called the Fickian turbulent diffusion coefficient or eddy diffusivity.

6.2.1 Some characteristics of turbulent diffusion

Since the scale of distances for turbulent diffusion is very much greater than that for molecular diffusion, the former is more effective in dispersing on a macro scale. Turbulence is associated with eddies, large and small, with scales varying with the scale of the diffusion process.

Turbulent flow is something that cannot be expressed in exact terms. It has a random component in its behaviour, and therefore the characteristics of turbulent motion can only be described in statistical terms. This applies to the solute that mixes with the solvent in turbulent motion. By assuming further that the turbulence process is stationary and homogeneous, which implies that the variance of the velocity remains constant in time and space, it is possible to define statistical parameters of concentration such as mean concentration, variance of concentration–time curve, variance of concentration–distance curve, etc. These parameters define the overall character of turbulent diffusion.

---

[2] Advection strictly means transport in the horizontal direction, whereas convection refers to transport in the vertical direction. In the context of longitudinal dispersion, both terms are used to mean the same phenomenon.

6.2.2 Shear flow dispersion

Shear flow dispersion refers to the dispersion due to changes in velocity across a cross section. If the separation of particles by advection is greater than that caused by molecular diffusion, then it is called laminar shear flow.

In this type of flow also, the mass transport in the streamwise direction is proportional to the concentration gradient in the streamwise direction, which is identical to the result obtained for molecular diffusion. The difference being that in shear flow dispersion, both advection and diffusion are considered together, whereas in ordinary molecular diffusion, only molecular diffusion is considered. The diffusion coefficient is therefore replaced by a bulk transport coefficient, or dispersion coefficient. It expresses the diffusive property of the velocity distribution and is generally known as the longitudinal dispersion coefficient. The corresponding governing equation is

$$\frac{\partial \bar{c}}{\partial t} + \bar{u} \frac{\partial \bar{c}}{\partial x} = D \frac{\partial^2 \bar{c}}{\partial x^2} \tag{6.8}$$

in which the velocity and concentration values are cross sectionally averaged; that is,

$$\bar{u} = \frac{1}{A} \int_A u \, dA \tag{6.9}$$

$$\bar{c} = \frac{1}{A} \int_A c \, dA \tag{6.10}$$

where A is the cross-sectional area of flow.

Extending to three dimensions, the governing equation based on the law of conservation of mass can be written as the following partial differential equation:

$$\frac{\partial c}{\partial t} + u \frac{\partial c}{\partial x} + v \frac{\partial c}{\partial y} + w \frac{\partial c}{\partial z} = \frac{\partial}{\partial x}\left(\varepsilon_x \frac{\partial c}{\partial x}\right) + \frac{\partial}{\partial y}\left(\varepsilon_y \frac{\partial c}{\partial y}\right) + \frac{\partial}{\partial z}\left(\varepsilon_z \frac{\partial c}{\partial z}\right) \tag{6.11}$$

where c is the time-averaged concentration [ML^{-3}]; u, v, and w are the time-averaged velocities in the x, y, and z directions, respectively [LT^{-1}]; ε_x, ε_y, and ε_z are the turbulent mixing coefficients in the x, y, and z directions, respectively [L^2T^{-1}]; and t is the time [T]. The terms on the left-hand side of the equation correspond to advection along streamlines, while the terms on the right-hand side correspond to turbulent diffusion between streamlines. Molecular diffusion, which takes place at a microscopic scale, is small compared with turbulent diffusion and therefore ignored in the above formulation.

In this equation, a solute introduced into a turbulent flow field is considered to be transported by two mechanisms: advection by the time-averaged velocities and diffusion by turbulent fluctuations. In deriving the governing equation, it is assumed that the instantaneous variations are averaged over a period long enough to average out short-time fluctuations, but short enough to consider long-term changes in the time-averaged values.

There is no general analytical solution of Equation 6.11. Very often, assumptions are made in order to make the equation tractable. The lateral and transverse velocities, v and w, are very much small in comparison with the longitudinal velocity, and can be neglected without causing too much error. The turbulent mixing coefficients, ε_x, ε_y, and ε_z, are also assumed to be constants in which case the equation becomes linear.

6.2.3 Taylor's approximation

Taylor (1953, 1954), by studying uniform flow through long straight circular pipes, argued that in a coordinate system that moves with the average velocity of flow, the spreading of a solute should follow Fick's law of diffusion (Equation 6.1), which when combined with the continuity equation gives

$$\frac{\partial \bar{c}}{\partial t} = D \frac{\partial^2 \bar{c}}{\partial \xi^2} \tag{6.12}$$

where $\xi = x - \bar{u}t$, $\bar{u}$ and $\bar{c}$ are the cross-sectional means as defined by Equations 6.9 and 6.10. Taylor made further assumptions and approximations to Equation 6.11 to obtain what is generally known as the longitudinal dispersion equation. Among them, he assumed

$$c = \bar{c} + c' \tag{6.13a}$$

and

$$u = \bar{u} + u' \tag{6.13b}$$

where c' and u' are the deviations of the concentration and velocity from their respective mean values. He further assumed that $c' = c'(y,z)$, the longitudinal transport due to turbulence, $\frac{\partial}{\partial x}\left(\varepsilon_x \frac{\partial c}{\partial x}\right)$, is negligible; the rate of change of concentration at a point moving with the mean velocity $\frac{\partial c}{\partial t} + \bar{u}\frac{\partial c}{\partial x}$ is also negligible; and that the lateral and transverse velocities, v, w, are zero. With these drastic assumptions, Equation 6.11 becomes

$$u'\frac{\partial \bar{c}}{\partial \xi} = \frac{\partial}{\partial y}\left(\varepsilon_y \frac{\partial c'}{\partial y}\right) + \frac{\partial}{\partial z}\left(\varepsilon_z \frac{\partial c'}{\partial z}\right) \tag{6.14}$$

with $\frac{\partial c'}{\partial y} = \frac{\partial c'}{\partial z} = 0$ at the boundary. This is Taylor's simplified form of the governing equation. He obtained an expression for the dispersion coefficient (given below by Equation 6.17) using a solution for the above equation that assumes universal velocity distribution in pipes.

6.2.4 Turbulent mixing coefficients

Of the three turbulent mixing coefficients, ε_x in the longitudinal direction is assumed to be zero. In the y direction for the vertical transport of momentum (and assuming Reynold's analogy), ε_y can be derived theoretically for turbulent flow in an infinitely wide open channel since the distribution of velocity and shear are known. Assuming a logarithmic velocity distribution, it has been shown (Fischer, 1966) that ε_y takes the form

$$\varepsilon_y = \kappa \left(1 - \frac{y}{h}\right)\left(\frac{y}{h}\right)hu^* \tag{6.15}$$

where κ is the von Karman constant, h is the depth of flow, and u^* is the shear velocity $\left(= \sqrt{\dfrac{\tau_0}{\rho}}\right.$, τ_0 is the boundary shear stress and ρ is the density of fluid). Elder (1959), also using the logarithmic velocity distribution, has obtained the following expressions for the lateral mixing coefficient, ε_z, as

$$\varepsilon_z = 0.23hu^* \tag{6.16}$$

There are also several other expressions for the turbulent mixing coefficients derived under different assumptions.

6.3 DISPERSION COEFFICIENT

The dispersion coefficient is a dominant parameter in the dispersion equation. Earliest studies on the dispersion coefficient were after Taylor (1953, 1954), who carried out laboratory experiments in long straight circular pipes under laminar and turbulent flow conditions. His studies have become the basis of analysis by many investigators subsequently. Taylor, by assuming universal velocity distribution within a circular section, obtained the following relationship:

$$D = 10.1ru^* \tag{6.17}$$

where r is the radius of the pipe. Elder (1959), by assuming logarithmic velocity profile and Reynold's analogy[3] in the depth-wise direction in an infinitely wide channel, obtained the expression

$$D = 5.93hu^* \tag{6.18}$$

Both these expressions have been verified experimentally.

Using Taylor's analysis, Fischer (1966, 1968) derived the following expression for the dispersion coefficient, assuming that dispersion in natural streams is caused primarily by the transverse velocity variations (Elder considered the vertical velocity variations):

$$D = -\frac{1}{A}\int_0^W q'(z)\,dz\int_0^z \frac{1}{e_z h(z)}\,dz\int_0^z q'(z)\,dz \tag{6.19a}$$

in which

$$q'(z) = \int_0^{h(z)} u'(y,z)\,dy \tag{6.19b}$$

---

[3] Heat and momentum transfer are analogous.

where $u'(y,z)$ is the fluctuating component of the velocity at any point in the cross section, $h(z)$ is the depth of flow at any point in the cross section, e_z is the turbulent mixing coefficient in the transverse (z) direction and assumed to be equal to $0.23hu^*$, W is the channel width and y is the Cartesian coordinate in the depth-wise direction. Velocity is measured as a function of the cross-sectional position, and $u(z)$ and $u'(z)$ are depth averaged.

With extensive field data, the dispersion coefficient can be estimated by approximating Equation 6.19a to a summation as

$$D = -\frac{1}{A} \sum_{k=2}^{n} q'(k) \Delta z \left[\sum_{j=2}^{k} \frac{\Delta z}{\varepsilon_{z_j} h_j} \left(\sum_{i=1}^{j-1} q'_i \Delta z \right) \right] \tag{6.20}$$

in which $q'_i = \frac{1}{2}(h_i + h_{i+1})u'_i$, u_i is the mean velocity in the i-th vertical slice, $u'_i = u_i - \bar{u}$, $\bar{u}$ is the mean velocity of the cross section, h_i is the depth at the beginning of the i-th slice, Δz is the width of the vertical slice, u^* is the shear velocity $\left(= \sqrt{grS_f} \right)$, r is the hydraulic radius, and S_f is the slope of the energy grade line. An example application illustrating the procedure for computation of some of the above methods is given by Fischer (1968).

This method requires cross-sectional velocity profile as well as cross-sectional geometry, and therefore is data intensive. Subsequently, Fischer (1975) gave a simplified form of the above equation, which takes the form

$$D = I \frac{u'^2 l^2}{\varepsilon_y} \tag{6.21}$$

where I is a non-dimensional integral, whose value depends on the velocity distribution, and l is the distance from the point of maximum velocity to the most distant bank. Using the values that Fischer used ($I = 0.07$; $u'^2 = 0.2\bar{u}^2$; $l = 0.7W$; and the mean transverse turbulent diffusion coefficient $\varepsilon_y = 0.6hu^*$), Equation 6.21 becomes

$$D = 0.011 \frac{\bar{u}^2 W^2}{hu^*} \tag{6.22}$$

Using the same approach, Liu (1977) developed the following equation in which the parameter ψ relates to the channel cross-sectional profile and transverse velocity distribution:

$$D = \psi \frac{u^2 W^2}{hu^*}; \psi = 0.18 \left(\frac{u^*}{u} \right)^{3/2} \tag{6.23}$$

Seo and Cheong (1998) also developed an equation using hydraulic data that can be more easily obtained from natural rivers. Their equation is of the form

$$D = 5.915 hu^* \left(\frac{W}{h} \right)^{0.620} \left(\frac{u}{u^*} \right)^{1.428} \tag{6.24}$$

Other equations that have been derived include that by Deng et al. (2001) who used direct integration of the triple integral of Equation 6.19a, and that by Kashefipour and Falconer (2002) who used dimensional and regression analysis. The equations, respectively, are as follows:

$$D = \frac{0.15 h u^*}{8 \varepsilon_{t_0}} \left(\frac{W}{h} \right)^{5/3} \left(\frac{u}{u^*} \right)^2 \tag{6.25}$$

$$D = h u \frac{u}{u^*} \left[7.428 + 1.775 \left(\frac{W}{h} \right)^{0.620} \left(\frac{u}{u^*} \right)^{0.572} \right] \tag{6.26}$$

A review and a comparison of their application is given by Ahsan (2008). Under field conditions, the dispersion coefficient is reported to have values ranging from about 4.6 to 670 m²/s, with an exceptionally high value of 1500 m²/s for Missouri River (Fischer et al., 1979, Table 5.3). A theoretically exact method of determining the dispersion coefficient is the change of moment method, which is based on the properties of the diffusion equation. Under field conditions, however, the concentration profile has long tails and therefore it is difficult to calculate the moments. Dispersion coefficient by the change of moment method is given as

$$D = \frac{1}{2} \frac{d\sigma_\xi^2}{dt} \tag{6.27}$$

where σ_ξ^2 is the variance of the concentration–distance profile. This equation is valid during the diffusive period regardless of the shape of the concentration profile during the convective period. Because it is difficult to measure concentration–distance profiles, the spatial variance can be converted into a time variance for uniform steady flow by the transformation

$$\sigma_\xi^2 = \bar{u}^2 \sigma_t^2 \tag{6.28}$$

where σ_t^2 is the variance of the concentration–time profile at a fixed point. Therefore, by taking measurements at two fixed points, the dispersion coefficient can be estimated as

$$D = \frac{1}{2} \bar{u}^2 \frac{\sigma_{t_2}^2 - \sigma_{t_1}^2}{\bar{t}_2 - \bar{t}_1} \tag{6.29}$$

where the subscripts 1 and 2 refer to the two measuring points, and $\bar{t}$ refers to the time of passage of the centroid of the tracer cloud at the station.

6.3.1 Routing method

The routing method (Fischer, 1966, 1968) of estimating the dispersion coefficient consists of matching the measured concentration–time profile with a predicted concentration–time profile for an assumed value of the dispersion coefficient until the mismatch between the two profiles as measured by the sum of squared differences is a minimum. Using the properties of the underlying linear theory in the bulk diffusion process, the principle of superposition is invoked to obtain the concentration distribution profile at any given time if the profile at some initial time is known. If at $t = t_0$, the initial concentration profile $c = c_0(\xi, t_0)$, then at any subsequent time, the concentration profile is given by

$$c(\xi, t) = \int_{-\infty}^{\infty} c_0(\xi', t_0) \frac{e^{-\frac{(\xi - \xi')^2}{4D(t - t_0)}}}{\sqrt{4\pi D(t - t_0)}} d\xi' \qquad (6.30)$$

However, in practice, it is difficult to measure the concentration–distance profile in the field. Therefore, the concentration–distance profile is obtained from the concentration–time profile by the following transformation, which is approximate:

$$c(\xi, \bar{t}_0) = c(x_0, t) \qquad (6.31a)$$

where

$$\xi = \bar{u}(\bar{t}_0 - t) \qquad (6.31b)$$

Then the distance integration of Equation 6.30 becomes the following time integration:

$$c(x_1, t) = \int_{-\infty}^{\infty} c(x_0, \tau) \frac{\exp\left[\frac{-(\bar{u}(\bar{t}_1 - t) + \bar{u}(\tau - \bar{t}_0))^2}{4D(\bar{t}_1 - \bar{t}_0)} \right]}{\sqrt{4\pi D(\bar{t}_1 - \bar{t}_0)}} \bar{u} d\tau \qquad (6.32)$$

In Equation 6.32, $c(x_1, t)$ is the concentration–time profile at station x_1; $c(x_0, t)$ is the concentration–time profile at station x_0; $\bar{t}_0$ and $\bar{t}_1$ are the mean times of passage of the tracer cloud past stations x_0 and x_1, respectively; and $\bar{u}$ is the mean velocity of flow. In deriving Equation 6.32, $c(\xi, t)$ is replaced by $c(x_1, t)$, ξ by $\bar{u}(\bar{t}_1 - t)$, ξ' by $\bar{u}(\bar{t}_0 - \tau)$, $c_0(\xi', t_0)$ by $c(x_0, \tau)$, $(t - t_0)$ by $(\bar{t}_1 - \bar{t}_0)$, and $d\xi$ by $\bar{u} d\tau$.

The routing method assumes that no dispersion takes place while the tracer cloud passes through the measuring station. In actual practice, the concentration profiles at two stations are measured, and the upstream concentration profile is used as an input and the corresponding concentration profile at a downstream station is determined for an assumed value of D. It is then compared with the measured concentration profile at the downstream station, and the dispersion coefficient is adjusted iteratively until the two concentrations match within a certain specified tolerance.

6.3.2 Time scale – dimensionless time

There is one common limitation for all the methods described above. They are applicable after an initial period of time commonly referred to as the convective period. It can also be identified as the mixing period, or when converted into an equivalent length, the mixing length. Two types of time scales have been defined for this purpose: the Eulerian time scale and the Lagrangian time scale. They are, respectively, given as

$$T' = \frac{l^2}{0.23ru^*} \qquad (6.33a)$$

and

$$T = 0.30 \frac{l^2}{ru^*} \qquad (6.33b)$$

where r is the hydraulic radius, u^* is the shear velocity, and l is a characteristic length (normally taken as half the width for symmetrical channels, and the distance between the thread of maximum velocity and the furthest bank for non-symmetrical channels). In the context of Taylor's analysis, the dispersion process can be divided into three stages according to a dimensionless time $t' = \frac{t}{T}$ as convective period ($t' \sim 0\text{--}3$), transition period ($t' \sim 3\text{--}6$), and a diffusive period ($t' > 6$). Taylor's dispersion theory is applicable only during the dispersive period when the normally skewed concentration profile approaches a Gaussian distribution.

The dispersion coefficient based on the diffusive transport method can be estimated using the Lagrangian time scale as

$$D = \overline{u'^2}T \qquad (6.34)$$

where $\overline{u'^2}$ is the squared spatial velocity deviation from the cross-sectional mean.

All the methods proposed hitherto for estimating the dispersion coefficient are valid only after the convective period has elapsed. An expression derived by Fischer (1967) for the convective period has been reported to underestimate when applied to natural streams (Beltaos, 1980; Liu and Cheng, 1980). Attempts to explain this discrepancy has been made by proposing non-Fickian models in which the analogy between the one-dimensional dispersion and diffusion equations is not used (McQuivey and Keefer, 1976; Sabol and Nordin, 1978), and modified Fickian models using the concept of dead zone storage effects (Thackston and Schnelle, 1970; Valentine and Wood, 1977). It is argued that the presence of dead zones, channel irregularities, and non-uniformities merely prolong the time required to attain Fickian behaviour. In the latter type of models, the analogy between the one-dimensional dispersion equation and the diffusion equation is used with an extra term added to the dispersion equation to account for dead zone storage effects. Liu and Cheng (1980), using the same modified Fickian concept, dispensed with this extra term by introducing a time-dependent dispersion coefficient. However, the practical application of these methods to natural streams is limited.

Following the approach proposed by Liu and Cheng (1980) and embodying the arguments put forward by Sullivan (1971), Jayawardena and Lui (1983) proposed a time-dependent dispersion model that has been founded on a large number of laboratory tests. Their findings in which the dispersion coefficient is considered as a function of time is given by

$$D = D_F \left\{ 1 - \frac{T}{t}\left(1 - e^{-\frac{t}{T}}\right) \right\} = D_F f(t') \qquad (6.35)$$

In this equation, D is the time-dependent dispersion coefficient; D_F is the Fickian dispersion coefficient; T is the Lagrangian time scale, which is usually considered as a measure of the longest time during which on average a particle persists in motion; and $t' = \frac{t}{T}$ is the dimensionless time.

Through a large number of laboratory experiments, the function $f(t')$ has been empirically obtained and consists of three parts: a linear part for small t, a transitional part

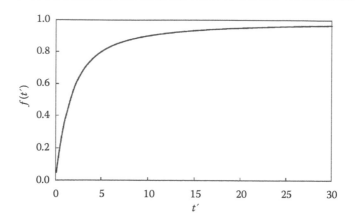

Figure 6.1 Function $f(t')$ of Equation 6.35.

for medium t, and an asymptotic part for large t. In this context, small t for practical purposes correspond to $\frac{t}{T} \leq 0.1$, intermediate t correspond to $0.1 \leq \frac{t}{T} \leq 20$, and large t correspond to $\frac{t}{T} > 20$. When $\frac{t}{T} > 20$, D lies between $0.95D_F$ and D_F. The function $f(t')$ is shown in Figure 6.1.

A field application of the above approach has been carried out for four narrow streams in Hong Kong using sodium chloride and rhodamine B as tracers (Lui and Jayawardena, 1983). The results, when compared with the dispersion coefficient estimated by the routing method, agree to within a factor of about 2. Dispersion prediction is considered as satisfactory if they differ by a factor of less than 4 (Fischer et al., 1979).

6.4 NUMERICAL SOLUTION

To understand the consequences of accidental discharges of pollutants into a water body as well as to understand possible environmental hazards under various scenarios, it is necessary to predict how the pollutant is dispersed as it is transported downstream. Such an attempt must start with the governing equations, including the boundary and initial conditions, simplifications and assumptions, estimation of the parameters of the processes involved, a geometrical framework upon which a solution is sought, and a numerical scheme to carry out the solution as there are no analytical solutions for most real-world problems. The one-dimensional dispersion equation that governs the spreading of a non-conservative pollutant introduced into a water course can be written as

$$\frac{\partial}{\partial t}(A\bar{c}) + \frac{\partial}{\partial x}(\bar{u}A\bar{c}) = \frac{\partial}{\partial x}\left(DA\frac{\partial \bar{c}}{\partial x}\right) - kA\bar{c} + Q_s \tag{6.36}$$

where A is the cross-sectional area of flow, k is a decay constant (assuming first-order decay), and Q_s is a source term. Other variables in this equation have the same meanings as those in Equation 6.8. The relevant boundary conditions can be written as

$$\bar{c} = \bar{c}_0, \text{ at } x = 0 \text{ for } t > 0 \tag{6.37a}$$

and

$$\frac{\partial \bar{c}}{\partial x} = 0, \text{ at } x = L \text{ for } t > 0 \tag{6.37b}$$

in which $\bar{c}_0$ may be either constant or time varying, and L is the length of the reach. Some initial condition is also necessary. The problem involves the solution of a parabolic-type partial differential equation with two independent variables, x and t. It can be done using the finite difference method or the finite element method or their combinations.

6.4.1 Finite difference method

The finite difference method is well established and has been used to solve many different types of field problems. All finite difference formulations are based on Taylor series expansion, which may be written as follows:

$$F(x + \Delta x) = F(x) + \Delta x F'(x) + \frac{(\Delta x)^2}{2!} F''(x) + \frac{(\Delta x)^3}{3!} F'''(x) + \frac{(\Delta x)^4}{4!} F''''(x)... \tag{6.38a}$$

$$F(x - \Delta x) = F(x) - \Delta x F'(x) + \frac{(\Delta x)^2}{2!} F''(x) - \frac{(\Delta x)^3}{3!} F'''(x) + \frac{(\Delta x)^4}{4!} F''''(x)... \tag{6.38b}$$

Ignoring terms containing $O(\Delta x)$ and higher, three types of first derivative representations can be obtained:

$$F'(x) = \frac{F(x + \Delta x) - F(x)}{\Delta x} \text{ (forward differencing)} \tag{6.39a}$$

$$F'(x) = \frac{F(x) - F(x - \Delta x)}{\Delta x} \text{ (backward differencing)} \tag{6.39b}$$

$$F'(x) = \frac{F(x + \Delta x) - F(x - \Delta x)}{2\Delta x} \text{ (central differencing)} \tag{6.39c}$$

Second-order derivatives can be obtained by addition of the two equations and ignoring $(\Delta x)^2$ and higher-order terms as

$$F''(x) = \frac{F(x + \Delta x) - 2F(x) + F(x - \Delta x)}{(\Delta x)^2} \tag{6.40}$$

Thus, the accuracy of all finite difference schemes depends on the step size, Δx. The forward and backward differencing schemes have trailing first-order terms and are therefore called first-order approximations. The central differencing scheme has trailing second-order terms and is therefore called second-order approximation. In a first-order scheme, the error would decrease by a factor of 2 if the mesh size (Δx) is halved. In a second-order scheme, the error would be reduced by a factor of 4 when the mesh size is halved. It does not, however, say anything about the magnitude of the error. A second-order scheme may not necessarily be capable of modelling a process more accurately than a first-order approximation. Formulation can be done to obtain the unknown value expressed explicitly in terms of a combination of known values, in which case solutions can be obtained in a recursive manner. However, explicit methods have the inherent problem resulting from the accumulation of round-off errors that leads to numerical instability. Small time steps, governed by certain stability conditions, are required to avoid this problem. In implicit methods, the unknown value is related to other unknown values as well as known values. Implicit methods are unconditionally stable. Solutions cannot be obtained in a recursive manner. Simultaneous solutions are therefore needed.

6.4.2 Finite element methods

The finite element method has the advantage of being more flexible than the finite difference method, although the formulation is more difficult. Widely used approaches of finite element formulation include the variational method, the method of weighted residuals of which Galerkin's method is a special case, boundary element methods, and adaptive methods. Each type has its pros and cons, but Galerkin's method by far has the widest application. General details of the finite element method can be found in several textbooks (e.g., Zienkiewicz, 1971, 1977; Oden and Carey, 1983). Specific details relevant to the solution of the dispersion equation can be found in Price et al. (1968), Smith et al. (1973), Guymon (1970), Jayawardena and Lui (1984), among others. Since the dispersion equation is of a parabolic type, the time domain is open, and a marching solution using finite differences in time and finite elements in space is normally carried out.

Before any such numerical model can be applied to field conditions, they should be verified for accuracy, consistency, convergence, and stability. Accuracy implies that the numerical solution does not deviate too much from the true exact solution. It depends on the mesh size, finite difference approximation (whether backward, forward, or central differencing), and the type of problem (e.g., presence of areas of sharp variations of the function to be evaluated). In general, the error is proportional to the square of the mesh size. Consistency implies that the numerical equation approaches the continuum equation as the mesh size decreases to zero. If the numerical solution approaches the exact solution as the mesh size decreases to zero, then the scheme is said to be convergent. It means that the difference equation approaches the differential equation as the mesh size tends to zero. It is difficult to prove this condition except for problems that have closed-form solutions. Stability arises from the accumulation of round-off errors. In any numerical solution, computations can only be done only up to a finite number of significant digits resulting in round-off errors. If these round-off errors do not build up as the number of computations increases, then the scheme is said to be stable. Stability depends on the type of scheme used. Explicit schemes are particularly vulnerable to stability problems.

An application of a numerical procedure using finite elements in space with Galerkin's formulation, and finite differences in time with the Crank–Nicolson scheme, in which the results have been compared with analytical solutions for specific assumptions, and subsequently applied to laboratory data as well as field data is given by Jayawardena and Lui (1984). In the finite element formulation in this application, linear and quadratic interpolation functions have been used. Their comparisons have been made against the analytical solutions for the following cases:

For the case with constant parameters without the decay and source terms, the governing equation simplifies to (same as Equation 6.8)

$$\frac{\partial \bar{c}}{\partial t} + \bar{u}\frac{\partial \bar{c}}{\partial x} = D\frac{\partial^2 \bar{c}}{\partial x^2}$$

subject to the boundary and initial conditions

$$\bar{c}(0,t) = \bar{c}_0, \quad \text{for } t > 0$$
$$\bar{c}(\infty,t) = 0, \quad \text{for } t > 0$$
$$\bar{c}(x,0) = 0, \quad \text{for all } x$$

An analytical solution for this case (Ogata and Banks, 1961) is given as

$$\bar{c}(x,t) = \frac{1}{2}\bar{c}_0 \left\{ \text{erfc}\left[\frac{x - \bar{u}t}{2\sqrt{Dt}}\right] + \exp\left[\left(\frac{\bar{u}x}{D}\right)\right]\text{erfc}\left[\frac{x + \bar{u}t}{2\sqrt{Dt}}\right]\right\} \tag{6.41}$$

where erfc is the complementary error function ($-1 - \text{erf}$), and the error function erf is defined as

$$\text{erf}(z) = \frac{2}{\sqrt{\pi}}\int_0^z e^{-t^2}dt \tag{6.42}$$

For the case with the decay term, the governing equation becomes

$$\frac{\partial \bar{c}}{\partial t} + \bar{u}\frac{\partial \bar{c}}{\partial x} = D\frac{\partial^2 \bar{c}}{\partial x^2} - k\bar{c} \tag{6.43}$$

subject to the boundary and initial conditions

$$\bar{c}(0,t) = \bar{c}_0, \quad \text{for } 0 \leq t \leq \delta_t$$
$$\bar{c}(0,t) = 0, \quad \text{for } t > \delta_t$$
$$\bar{c}(x,0) = 0, \quad \text{for all } x$$

where δ_t is a certain time interval.

An analytical solution for this case (Kinzelbach and Yang, 1981) is given as

$$\bar{c}(x,t) = \frac{1}{2}\bar{c}_0 \left\{ \begin{array}{l} \exp\left[\sqrt{\frac{\bar{u}^2}{4D}+k}\,\frac{x}{\sqrt{D}}\right] \mathrm{erfc}\left[\frac{x}{2\sqrt{Dt}}+\sqrt{\frac{\bar{u}^2}{4D}+k}\sqrt{t}\right] \\ +\exp\left[-\sqrt{\frac{\bar{u}^2}{4D}+k}\,\frac{x}{\sqrt{D}}\right] \mathrm{erfc}\left[\frac{x}{2\sqrt{Dt}}-\sqrt{\frac{\bar{u}^2}{4D}+k}\sqrt{t}\right] \end{array} \right\} \cdot \exp\left(\frac{\bar{u}x}{2D}\right)$$

$$-\frac{1}{2}\left\{ \begin{array}{l} \exp\left[\sqrt{\frac{\bar{u}^2}{4D}+k}\,\frac{x}{\sqrt{D}}\right]\mathrm{erfc}\left[\frac{x}{2\sqrt{D(t-\delta_t)}}+\sqrt{\frac{\bar{u}^2}{4D}+k}\sqrt{t-\delta_t}\,+\exp\left[-\sqrt{\frac{\bar{u}^2}{4D}+k}\,\frac{x}{\sqrt{D}}\right]\right] \\ \mathrm{erfc}\,\frac{x}{2\sqrt{D(t-\delta_t)}}-\sqrt{\frac{\bar{u}^2}{4D}+k}\sqrt{t-\delta_t} \end{array} \right\} \quad (6.44)$$

$$\cdot\exp\left(\frac{\bar{u}x}{2D}\right)\theta(t-\delta_t)$$

where $\theta(t - \delta_t) = 0$, for $t \le \delta_t$, and unity for $t > \delta_t$.

Comparisons of the finite element–finite difference solutions made with these analytical solutions were found to give very good agreements (Jayawardena and Lui, 1984). Stability criteria have been examined with reference to the Peclet number, p (sometimes referred to as the Reynolds cell number), defined as $p = \frac{\bar{u}\Delta x}{D}$, and the Courant number, c, defined as $c = \frac{\bar{u}\Delta t}{\Delta x}$ where Δx and Δt are the space and time increments. Successful application of the numerical procedure for laboratory and field data for four natural streams in Hong Kong is reported by Lui and Jayawardena (1983).

6.4.3 Moving finite elements

The numerical solution of the convective dispersion equation encounters difficulties when the problem is convection dominated. The normally parabolic-type partial differential equation becomes almost hyperbolic, causing the numerical solution with a fixed grid to oscillate and overshoot (Price et al., 1968; Ehlig, 1977; Lam, 1977; Jensen and Finlayson, 1978, 1980). Criteria for non-oscillatory solutions are normally expressed in terms of the Peclet number p and the Courant number c. To guarantee non-oscillatory solutions using finite differences centred in space or linear finite elements, the restriction on the Peclet number is given by Jensen and Finlayson (1978) as $p < 2$, and hence the mesh spacing Δx must satisfy the condition

$$\Delta x < \frac{2D}{\bar{u}} \qquad (6.45)$$

Using finite elements for discretization in space and centred finite differences for discretization in the time domain, Ehlig (1977) gives the restriction on the Courant number to prevent oscillations as $c < 0.1$. Using linear finite elements in space and a Crank–Nicholson finite difference approximation in time (a Crank–Nicholson Galerkin method), Jayawardena and Lui (1984) found that the condition $c < 0.1$ to be over-restrictive and that in practice the condition $c < 1$ is sufficient. The latter criterion gives the restriction on the size of the time step Δt to yield non-oscillatory numerical solutions as

$$\Delta t < \frac{\Delta x}{\bar{u}} \tag{6.46}$$

Combining these two criteria (Equations 6.45 and 6.46), the time increment Δt for non-oscillatory solutions of Equation 6.8 using linear finite elements in space and Crank–Nicholson scheme in time should satisfy the condition

$$\Delta t < \frac{2D}{\bar{u}^2} \tag{6.47}$$

When the dispersion process is convection dominated, it is well known that numerical solution with standard fixed mesh finite elements or finite differences requires large numbers of elements and small time steps in order to satisfy criteria for stability, to give oscillation-free accurate solutions. For such problems, numerical solution with fixed mesh methods is consequently inefficient and computationally expensive. For example, in the hypothetical case of $\bar{u} = 1$ m/s and $D = 0.001$ m²/s with $0 < x < 90$ m, $t = 60$ s, the above criteria for stability will require $\Delta x < 0.002$ m and $\Delta t < 0.002$ s, implying 4.5×10^4 finite elements in space and 3×10^4 time steps (Johnson and Jayawardena, 1998). This is certainly computationally expensive in terms of computational time and storage requirements. Failing to meet these criteria will make the solution oscillate.

A method to overcome this problem has been proposed by Johnson and Jayawardena (1998). They introduce the moving finite element (MFE) method, a moving mesh method, which has several advantages. In the MFE method, the finite element approximation is based on a mesh of nodes that move in time. A comprehensive account of MFE theory and applications is given by Baines (1994), and the particular application to the dispersion process by Johnson and Jayawardena (1998). Numerical solutions with highly accurate resolution of steep moving fronts can be generated using the MFE method using only a small number of elements and time steps greater than those afforded by fixed mesh methods. In the special case of linear dispersion, a highly efficient, explicit, direct solution procedure, requiring no matrix inversions has been presented.

6.5 DISPERSION THROUGH POROUS MEDIA

The equation for transport of solutes in a saturated porous medium is derived using the law of conservation of mass under the assumptions that the porous medium is homogeneous and isotropic, the medium is saturated and the flow is at steady state, and that Darcy's law is applicable. According to Darcy's law, the solute will be advected by the average linear velocity. If this is the only transport mechanism, the solute will move as a plug. However, owing to hydrodynamic dispersion, which is caused by variations in the microscopic velocity within each pore channel and from one channel to another, mixing of the solute will also take place.

Therefore, to describe the transport mechanism using macroscopic parameters, yet taking into account microscopic mixing, it is necessary to introduce a second mechanism of transport in addition to advection. The total transport of the solute per unit area per unit time, $q[ML^{-2}T^{-1}]$ therefore is

$$q = \bar{u}nc - nK\frac{\partial c}{\partial x} \tag{6.48}$$

where $\bar{u}$ is the average linear velocity $\left(=\frac{u}{n}\right)$, u is the Darcy velocity, n is the porosity of the medium, c is the mass concentration of solute $[ML^{-3}]$, and K is the dispersion coefficient for dispersion through the porous medium. The dispersion coefficient is related to the dispersivity α and the diffusion coefficient D by the relationship

$$K = \alpha\bar{u} + D \tag{6.49}$$

The dispersivity [L] is a characteristic property of the porous medium, whereas the coefficient of molecular diffusion is a property of diffusive transport process for the solute in the porous medium $[L^2T^{-1}]$.

The conservation of mass for a conservative solute together with Darcy's law gives

$$-\left\{\frac{\partial}{\partial x}(\bar{u}nc) - \frac{\partial}{\partial x}\left(nK\frac{\partial c}{\partial x}\right)\right\} = n\frac{\partial c}{\partial t} \tag{6.50}$$

which, when n is assumed to be a constant as for a homogeneous medium, simplifies to

$$-\bar{u}\frac{\partial c}{\partial x} + \frac{\partial}{\partial x}\left(K\frac{\partial c}{\partial x}\right) = \frac{\partial c}{\partial t} \tag{6.51}$$

In a homogeneous medium, K does not vary in space. Therefore,

$$\frac{\partial c}{\partial t} + \bar{u}\frac{\partial c}{\partial x} = K\frac{\partial^2 c}{\partial x^2} \tag{6.52}$$

This equation is mathematically the same as the longitudinal dispersion Equation 6.8. If u and K vary spatially, then

$$\frac{\partial c}{\partial t} + \frac{\partial}{\partial x}(uc) = \frac{\partial}{\partial x}\left(K\frac{\partial c}{\partial x}\right) \tag{6.53}$$

The resultant concentration profile for a tracer introduced continuously at the upstream end of a sample of porous medium maintaining a constant concentration c_0 at the upstream end, and subject to zero initial conditions and after steady state has been attained, can be derived analytically. Mathematically, the above problem has the following boundary and initial conditions:

$$c(x,0) = 0 \quad \text{for } x \geq 0$$
$$c(0,t) = c_0 \quad \text{for } t \geq 0$$
$$c(\infty,t) = 0 \quad \text{for } t \geq 0$$

Owing to molecular diffusion, the tracer appears at the outlet at a time before the water appears at the same point. The solution for the above boundary condition is given by

$$\frac{c}{c_0} = \frac{1}{2}\left\{ \text{erfc}\left[\frac{x - \bar{u}t}{2\sqrt{Kt}}\right] + \exp\left[\left(\frac{\bar{u}x}{K}\right)\right]\text{erfc}\left[\frac{x + \bar{u}t}{2\sqrt{Kt}}\right]\right\}$$

which is of the same form as Equation 6.41.

When the dispersivity of the porous medium is large or when x or t is large, the second term of the above equation becomes negligible; that is, the spreading of the tracer cloud is caused by diffusion and mechanical (or turbulent) mixing. At low velocities, $K \cong D$, and at high velocities, $K \cong \alpha\bar{u}$. The larger the value of α, the greater is the amount of mixing.

Marino (1974a,b) obtained an analytical solution to the dispersion equation in a semi-infinite medium for the following boundary and initial conditions:

$$c(0,t) = c_0 e^{\gamma t} \qquad \gamma \text{ is a constant } (\pm), t > 0$$
$$c(x,0) = 0 \qquad x \geq 0$$
$$c(\infty,t) = 0 \qquad t \geq 0$$

The solution is of the form

$$c(x,t) = c_0 e^{\gamma t}\left\{\exp\left[\frac{x(\bar{u} - \xi)}{2D}\right]\text{erfc}\left[\frac{x - \xi t}{2\sqrt{Dt}}\right] + \exp\left[\frac{x(\bar{u} + \xi)}{2D}\right]\text{erfc}\left[\frac{x + \xi t}{2\sqrt{Dt}}\right]\right\} \qquad (6.54)$$

where $\xi = (u^2 + 4D\gamma)^{1/2}$.

If $\gamma = 0$, the solution simplifies to

$$c(x,t) = \frac{c_0}{2}\left\{\text{erfc}\left[\frac{x - \bar{u}t}{2\sqrt{Dt}}\right] + \exp\left[\frac{\bar{u}x}{D}\right]\text{erfc}\left[\frac{x + \bar{u}t}{2\sqrt{Dt}}\right]\right\}$$

which is the same as Equation 6.41.

If the boundary and initial conditions are

$$c(0,t) = c_0(1 - e^{-\gamma t}) \qquad \text{for } t > 0$$
$$c(x,0) = 0 \qquad \text{for } x \geq 0$$
$$c(\infty,t) = 0 \qquad \text{for } t \geq 0$$

the corresponding solution is

$$c(x,t) = \frac{c_0}{2}\left\{\begin{array}{l}\text{erfc}\left[\dfrac{x - \bar{u}t}{2\sqrt{Dt}}\right] + \exp\left[\dfrac{\bar{u}x}{D}\right]\text{erfc}\left[\dfrac{x + \bar{u}t}{2\sqrt{Dt}}\right] - \\[2ex] \exp(-\gamma t)\left[\exp\left[\dfrac{x(\bar{u} - \phi)}{2D}\right]\text{erfc}\left[\dfrac{x - \phi t}{2\sqrt{Dt}}\right]\right] + \exp\left[\dfrac{x(\bar{u} + \phi)}{2D}\right]\text{erfc}\left[\dfrac{x + \phi t}{2\sqrt{Dt}}\right]\end{array}\right\} \qquad (6.55)$$

where $\phi = (u^2 - 4D\alpha)^{1/2}$.

In both solutions, the second term in the first solution and the second and fourth terms in the second solution are small. Therefore, approximate solutions may be obtained accordingly.

6.6 GENERAL-PURPOSE WATER QUALITY MODELS

Several water quality models that can simulate many different constituents are now available for general use. Among them are the Enhanced Stream Water Quality Model (QUAL2E) and its more recent version QUAL2K, both developed by the United States Environmental Protection Agency (USEPA), Water Quality Analysis Simulation Programme (WASP), and the One Dimensional Riverine Hydrodynamic and Water Quality Model (EPD-RIV1).

6.6.1 Enhanced Stream Water Quality Model (QUAL2E)

The Enhanced Stream Water Quality Model (QUAL2E) is applicable to well-mixed, dendritic streams. It simulates the major reactions of nutrient cycles, algal production, benthic and carbonaceous demand, atmospheric re-aeration, and their effects on the dissolved oxygen balance. The model assumes that the major transport mechanisms, advection and dispersion, are significant only along the longitudinal direction of flow. It can predict the following 15 water quality constituent concentrations:

- Dissolved oxygen
- Biochemical oxygen demand
- Temperature
- Algae as chlorophyll *a*
- Organic nitrogen as 'N'
- Ammonia as 'N'
- Nitrite as 'N'
- Nitrate as 'N'
- Organic phosphorus as 'P'
- Dissolved phosphorus as 'P'
- Coliform bacteria
- Arbitrary non-conservative constituent
- Three conservative constituents

It is intended as a water quality planning tool for developing total maximum daily loads, and can also be used in conjunction with field sampling for identifying the magnitude and quality characteristics of non-point sources. By operating the model dynamically, the user can study diurnal dissolved oxygen variations and algal growth. However, the effects of dynamic forcing functions, such as headwater flows or point source loads, cannot be modelled with QUAL2E. The model assumes that the stream flow and waste inputs are constant during the simulation time periods. QUAL2EU is an enhancement that allows users to perform three types of uncertainty analyses: sensitivity analysis, first-order error analysis, and Monte Carlo simulation. QUAL2K is an enhanced version of QUAL2E that takes into account the following:

- Unequally spaced river reaches and multiple loadings and abstractions in any reach
- Two forms of carbonaceous biological oxygen demand (slowly oxidizing and rapidly oxidizing) to represent organic carbon as well as non-living particulate organic matter
- Anoxia by reducing oxidation reactions to zero at low oxygen levels

- Sediment–water interactions
- Bottom algae
- Light extinction
- pH
- Pathogens

The Windows interface provides input screens to facilitate preparing model inputs and executing the model. It also has help screens and provides graphical viewing of input data and model results. More details of the software can be found at http://www.epa.gov/OST/QUAL2E_WINDOWS, and in The Enhanced Stream Water Quality Models QUAL2E and QUAL2E-UNCAS: Documentation and User's Manual (EPA 600/3-87-007), NTIS accession no. PB87 202 156.

6.6.2 Water Quality Analysis Simulation Programme (WASP)

This programme, which is based on the work of several researchers (Di Toro et al., 1983; Connolly and Winfield, 1984; Ambrose et al., 1988), can carry out dynamic compartment modelling of aquatic systems, including the water column as well as the benthos. It can analyse a number of pollutant types in one, two, or three dimensions. The programme can also be linked to hydrodynamic and sediment transport models. The pollutants it can handle include

- Nitrogen
- Phosphorus
- Dissolved oxygen
- Biochemical oxygen demand
- Sediment oxygen demand
- Algae
- Periphyton
- Organic chemicals
- Metals
- Mercury
- Pathogens
- Temperature

More information about WASP can be found at http://www.epa.gov/athens/wwqtsc/html/wasp.html.

6.6.3 One Dimensional Riverine Hydrodynamic and Water Quality Model (EPD-RIV1)

This is a system of programmes that performs one-dimensional (cross-sectionally averaged) hydraulic and water quality simulations. The hydrodynamic model is first applied and the results are then used as inputs to the water quality model. The model can simulate the following state variables:

- Dissolved oxygen
- Temperature
- Nitrogenous biochemical oxygen demand
- Carbonaceous oxygen demand

- Phosphorus
- Algae
- Iron
- Manganese
- Coliform bacteria
- Two arbitrary constituents

More information about EPD-RIV1 can be found at http://www.epa.gov/athens/wwqtsc/html/epd-riv1.html.

6.7 CONCLUDING REMARKS

Longitudinal dispersion is an important topic in environmental engineering because of its role in determining how a pollutant discharged into a water body spreads owing to the combined action of molecular diffusion and velocity gradients. In this chapter, governing equations starting from Fick's law of diffusion to turbulent diffusion to shear flow dispersion including Taylor's approximations are introduced. Methods of estimating the dispersion coefficient as well as solution of the dispersion equation using different numerical methods including comparison with known analytical solutions for specific cases of the governing equation are given. The chapter ends with a brief introduction of dispersion through porous media and some general-purpose water quality models.

REFERENCES

Ahsan, N. (2008): Estimating the coefficient of dispersion for a natural stream. *Proceedings of World Academy of Science, Engineering and Technology*, 34, October, 131–134. ISSN 2070–3740.

Ambrose, R.B. Jr., Wool, T.A., Connolly, J.P. and Schanz, R.W. (1988): WASP4, A hydrodynamic and Water Quality model - Model theory, user's manual and programmer's guide, Environmental Research Laboratory, Office of Research and Development, US Environmental Protection Agency, Athens, Georgia 30613 (EPA/600/3-87/039, January 1988).

Baines, M.J. (1994): *Moving Finite Elements*. Oxford University Press, Oxford.

Beltaos, S. (1980): Longitudinal dispersion in rivers. *Journal of the Hydraulic Division, ASCE*, 106(HY1), 151–172.

Connolly, J.P. and Winfield, R.P. (1984): A user's guide for WASTOX, a framework for modeling the fate of toxic chemicals in aquatic environments, Part 1. Exposure concentration, US Environmental Protection Agency, Gulf Breeze, FL. EPA-600/3/84/077.

Deng, Z.-Q., Singh, V.P. and Bengtsson, L. (2001): Longitudinal dispersion coefficient in straight rivers. *Journal of Hydraulic Engineering*, 127(11), 919–927.

Di Toro, D.M., Fitzpatrick, J.J. and Thomann, R.V. 1981, rev. (1983): Water Quality Analysis Simulation Program (WASP) and Model Verification Program (MVP) - Documentation. Hydroscience, Inc., Westwood, NJ. for US Environmental Protection Agency, Duluth, MN, Contract No. 68-01-3872.

Ehlig, C. (1977): Comparison of numerical methods for solutions of the one-dimensional diffusion-convection equation in one and two-dimensions. In: *Finite Elements in Water Resources*. Pentech Press, London, pp. 1.91–1.102.

Elder, J.W. (1959): The dispersion of marked fluid in turbulent shear flow. *Journal of Fluid Mechanics*, 5, 544–560.

Fick, A. (1855): *Poggendorf's Annalen*. 94, 59–86. In English: *Philosophical Magazine S4*, 10, 30–39 (1855).

Fischer, H.B. (1966): Longitudinal dispersion in laboratory and natural streams. W.M. Keck Laboratory of Hydraulics and Water Resources Division of Engineering and Applied Science, California Institute of Technology, Pasadena, CA, Report No. KH-R-12, June 1966, p. 250.

Fischer, H.B. (1967): The mechanics of dispersion in natural streams. *Journal of the Hydraulics Division, ASCE*, 93(HY6), 187–216.

Fischer, H.B. (1968): Dispersion prediction in natural streams. *Journal of the Sanitary Engineering Division, ASCE*, SA5, 927–943.

Fischer, H.B. (1975): Simple method for predicting dispersion in streams, Discussion by R.S. McQuivey and T.N. Keefer. *Journal of Environmental Engineering Division, ASCE*, 101(3), 453–455.

Fischer, H.B., List, E.T., Imberger, J. and Brooks, N.H. (1979): *Mixing in Inland and Coastal Waters*. Academic Press, New York, p. 136.

Guymon, G.L. (1970): A finite element solution of the one-dimensional diffusion-convection equation. *Water Resources Research*, 6(1), 204–210.

Jayawardena, A.W. and Lui, P.H. (1983): A time dependent dispersion model based on Lagrangian correlation. *Hydrological Sciences Journal*, 28(4), 455–473.

Jayawardena, A.W. and Lui, P.H. (1984): Numerical solution of the dispersion equation using a variable dispersion coefficient: method and applications. *Hydrological Sciences Journal*, 29(3), 293–309.

Jensen, O.K. and Finlayson, B.A. (1978): Solution of the convection–diffusion equation using a moving co-ordinate system. *Proc. 2nd Int. Congress Finite Elements in Water Resources*, 4, 21–32.

Jensen, O.K. and Finlayson, B.A. (1980): A moving co-ordinate solution of the transport equation. *Water Resources Research*, 30, 9–18.

Johnson, I.W. and Jayawardena, A.W. (1998): Efficient numerical solution of the dispersion equation using moving finite elements. *Finite Elements in Analysis and Design*, 28(3), 241–253.

Kashefipour, S.M. and Falconer, R.A. (2002): Longitudinal dispersion coefficients in natural channels. *Water Resources Research*, 36, 1596–1608.

Kinzelbach, W.K.H. and Yang, R.J. (1981): On finite difference methods for solving the one dimensional pollutant transport equation of a river. *Chinese Journal of Environmental Science*, 2(1), 25–30 (in Chinese).

Lam, D.C.L. (1977): Comparison of finite element and finite difference methods for nearshore advection-diffusion transport models. In: *Finite Elements in Water Resources*. Pentech Press, London, pp. 1.115–1.129.

Liu, H. (1977): Predicting dispersion coefficient of stream. *Journal of Environmental Engineering Division, ASCE*, 103(1), 59–69.

Liu, H. and Cheng, H.D.A. (1980): Modified Fickian model for predicting dispersion. *Journal of Hydraulics Division, ASCE*, 106(HY6), 1021–1040.

Lui, P.H. and Jayawardena, A.W. (1983): Application of a time dependent dispersion model for dispersion prediction in natural streams. *Hydrological Sciences Journal*, 28(4), 475–483.

Marino, M.A. (1974a): Longitudinal dispersion in saturated porous media. *Journal of Hydraulics Division, ASCE*, 101(HY1), 151–157.

Marino, M.A. (1974b): Numerical and analytical solutions of dispersion in a finite absorbing medium. *Water Resources Bulletin, AWRA*, 10(1), 81–90.

McQuivey, R.S. and Keefer, T.N. (1976): Convective model of longitudinal dispersion. *Journal of Hydraulics Division, ASCE*, 102(HY10), 1409–1437.

Oden, J.T. and Carey, G.F. (1983): *Finite Elements—Mathematical Aspects*, Vol. IV. Prentice Hall, Englewood Cliffs, p. 195.

Ogata, A. and Banks, R.B. (1961): A solution of the differential equation of longitudinal dispersion in porous media. US Geol. Survey Prof. Pap. 411-A. US Government Printing Office, Washington, DC.

Price, H.S., Cavendish, J.C. and Varga, R.S. (1968): Numerical methods of higher order accuracy for diffusion-convection equations. *Journal of the Society of Petroleum Engineers*, S3, 293–303.

Sabol, G.V. and Nordin, C.F. (1978): Dispersion in rivers as related to storage zones. *Journal of Hydraulics Division, ASCE*, 104(HY5), 693–708.

Seo, I.W. and Cheong, T.S. (1998): Predicting longitudinal dispersion coefficient in natural streams. *Journal of Hydraulic Engineering*, 124(1), 25–32.

Smith, I.M., Farraday, R.V. and O'Connor, B.A. (1973): Rayleigh-Ritz and Galerkin finite elements for diffusion–convection problems. *Water Resources Research*, 9(3), 593–600.

Sullivan, P.J. (1971): Longitudinal dispersion within a two-dimensional turbulent shear flow. *Journal of Fluid Mechanics*, 49, 551–576.

Taylor, G.I. (1953): Dispersion of soluble matter in solvent flowing through a tube. *Proceedings of the Royal Society of London*, A219, 186–203.

Taylor, G.I. (1954): The dispersion of matter in turbulent flow through a pipe. *Proceedings of the Royal Society of London*, 223, 446–468.

Thackston, E.L. and Schnelle, K.B. (1970): Predicting effects of dead zones on stream mixing. *Journal of Sanitary Engineering Division, ASCE*, 96(SA2), 319–331.

Valentine, E.M. and Wood, I.R. (1977): Longitudinal dispersion with dead zones. *Journal of Hydraulics Division, ASCE*, 103(HY9), 975–990.

Zienkiewicz, O.C. (1971, 1977): *Finite Elements Method in Engineering Science*. McGraw Hill, London, p. 570.

Chapter 7

Time series analysis and forecasting

7.1 INTRODUCTION

Most hydro-meteorological and environmental data are measured at regular intervals of time and can therefore be represented as functions of time or time series. These, when observed over a long period, exhibit certain patterns that, if identifiable, can be used for forecasting purposes. Time series analysis involves the identification of such patterns or properties by a process of decomposition and subsequent extrapolation by synthesizing the decomposed components such that the statistical character of the generated series remains the same as that of the historical series.

The objective of this chapter is to introduce basic properties of time series, homogeneity tests, decomposition of a time series, tests for identification of trends and periodicities, time and frequency domain analyses, representation of the dependent stochastic component by various stochastic models, generation of synthetic data using the probability distribution of the independent residuals, forecasting, and basic concepts of Kalman filtering. Illustrative examples will be given where possible.

7.2 BASIC PROPERTIES OF A TIME SERIES

7.2.1 Stationarity

In a stationary time series, the statistical properties of the series when computed from different samples do not change except due to sampling variations. This means that the probability density functions estimated from any sample of the time series is the same. To make the definition of stationarity more practical, this condition is restricted to the mean (expected value) and the variance only. A series that is stationary in the mean is called a first-order stationary series. A series that is stationary in the covariance and the mean is called second-order stationary or 'weakly stationary' (stationarity in the covariance[1] implies stationarity in the variance). If other statistical properties also are time invariant, the series is said to be 'strongly stationary'. A process can be stationary in one property and not in others.

A series whose statistical properties vary with time is non-stationary. A non-stationary series contains deterministic components such as trends and periodicities.

---

[1] $\text{cov}[x(t),\ x(t+\tau)] = \underset{T\to\infty}{\text{Limit}}\ \dfrac{1}{T}\int_0^T [x(t)-\mu][x(t+\tau)-\mu]\,dt = \text{cov}(\tau)$

$\text{Autocorrelation} = \dfrac{\text{Cov}(\tau)}{\text{Cov}(0)}.$

7.2.2 Ergodicity

A stationary time series is said to be ergodic if the time-averaged statistical properties are the same as the ensemble or spatial averages over the entire population. An ergodic series is necessarily stationary, but a stationary series may not be ergodic. For practical purposes, all stationary series are assumed to be ergodic.

7.2.3 Homogeneity

A series is said to be non-homogeneous when there is a jump (or drop) in the mean value of magnitude Δ after a certain number of observations. Homogeneity of a data series in statistical terms implies that the data belong to one population and therefore has a mean value, which is time invariant.

Homogeneity of data is affected by several factors, such as changes in the circumstances under which the data have been collected, changes in the data collection site and procedure, etc. This is particularly applicable to rainfall data. For example, non-homogeneity can occur due to changes in the height of exposure of the rain gauge, changes in site and environment, change of personnel, and change of instrument. One way of testing the homogeneity is to compare the cumulative values of the suspect data with the cumulative values of the mean of a number of nearby gauging stations, which are affected by similar meteorological conditions. The method, which leads to the double mass curve, is simple and provides an indication of the relative shift in the mean value, if any. It does not distinguish real changes in mean level from changes due to random fluctuations. Therefore, it is necessary to examine the significance of the non-homogeneity by using statistical tests.

7.3 STATISTICAL PARAMETERS OF A TIME SERIES

7.3.1 Sample moments

Sample moments around zero are defined as

$$m_r = \frac{1}{N} \sum x_t^r \tag{7.1}$$

where m_r is the r-th moment.

Sample moments around the mean $\bar{x}$ are defined as

$$m_r = \frac{1}{N} \sum (x_t - \bar{x})^r \tag{7.2}$$

For example, the first moment ($r = 1$) gives the mean, the second moment ($r = 2$) gives the variance, the third moment ($r = 3$) gives a measure of the skewness, and the fourth moment

($r = 4$) gives a measure of the kurtosis. Variance when multiplied by $\dfrac{N}{N-1}$ gives an unbiased[2] estimate.

When the third moment is calculated, positive and negative values can cancel out if the distribution is symmetrical. The degree to which the third moment deviates from zero is a measure of the skewness of the distribution and the sign of the third moment gives its direction. If the skewness coefficient is negative, the distribution is skewed to the right (mode > median and mean), and if it is positive it is skewed to the left (mode < median and mean). Skewness coefficient is defined as

$$\frac{1}{N} \sum \frac{(x_t - \bar{x})^3}{s^3} \tag{7.3}$$

where s is the sample standard deviation.

The coefficient of kurtosis, κ, which is a measure of peakedness, is defined as $\dfrac{m_4}{s^4}$, and kurtosis excess ε is defined as

$$\varepsilon = \kappa - 3 \tag{7.4}$$

A low value of the kurtosis indicates relatively few extreme values. It is an indicator of the peakedness or flatness. The coefficient of kurtosis for a normal distribution is 3. If it is greater than 3, then there are more high values concentrated near the mean. If it is less than 3, then there are more low values concentrated near the mean. When $\varepsilon > 0$ (i.e., high peak), the distribution is called leptokurtic. When $\varepsilon = 0$ (i.e., normal peak), the distribution is called mesokurtic. When $\varepsilon < 0$ (i.e., low peak), the distribution is called platykurtic.

7.3.2 Moving averages – low-pass filtering

Moving averages, also referred to as running averages or rolling averages, which has the effect of a low-pass filter,[3] will reduce high-frequency oscillations and are useful for eliminating short-term fluctuations. For example, a five-term moving average model of a series $x_1, x_2, x_3, x_4, x_5, x_6, x_7, x_8, x_9, x_{10},\dots$ will be of the form

$$y_1 = \frac{1}{5}(x_1 + x_2 + x_3 + x_4 + x_5)$$

$$y_2 = \frac{1}{5}(x_2 + x_3 + x_4 + x_5 + x_6)$$

$$y_3 = \frac{1}{5}(x_3 + x_4 + x_5 + x_6 + x_7) \tag{7.5}$$

$$\dots$$

$$\dots$$

As the number of terms in the moving average model increases, the number of terms of the new series is reduced.

---

[2] When the expected value of a statistic is equal to the corresponding population parameter, the statistic is called an unbiased estimator. For example, if the mean of the sample means is equal to the population mean, then the mean of samples is an unbiased estimator (i.e., mean of $\bar{x} = \mu$). The unbiased estimate is obtained by multiplying the biased estimate by $\dfrac{N}{N - \mathrm{DOF}}$, where DOF refers to the number of degrees of freedom. For variance, DOF = 1.

[3] A low-pass filter allows low-frequency (long period) signals to pass and smoothes high frequencies (short period). A high-pass filter does the opposite.

7.3.3 Differencing – high-pass filtering

Low-frequency oscillations, on the other hand, can be removed by using high-pass filtering methods such as

$$y'_t = x_t - x_{t-1} \quad \text{for } t = 2, 3, \dots \text{ (first order)} \tag{7.6a}$$

$$\begin{aligned} y_t &= y'_t - y'_{t-1} \\ &= x_t - 2x_{t-1} + x_{t-2} \end{aligned} \quad \text{for } t = 3, 4, \dots \text{ (second order)} \tag{7.6b}$$

In practice, however, most of these filtering methods may not be necessary. A visual observation of the data plot will often indicate whether a trend exists. If one exists, it may be described by a polynomial of the form

$$x_t = x_0 + a_1 t + a_2 t^2 + \dots a_n t^n + \zeta_1 \text{ (error)} \tag{7.7}$$

which is a linear trend if only the first two terms are applicable. Differencing is a 'whitening filter' and has the effect of transforming the distribution of the original series closer to a normal distribution.

7.3.4 Recursive means and variances

Recursive mean $\bar{x}_k$ is defined as

$$\bar{x}_k = \frac{1}{k} \sum_{i=1}^{k} x_i = \frac{1}{k} \left\{ \sum_{i=1}^{k-i} x_i + x_k \right\} \tag{7.8a}$$

$$= \frac{1}{k} \{(k-1)\bar{x}_{k-1} + x_k\} = \frac{(k-1)}{k} \bar{x}_{k-1} + \frac{1}{k} x_k \tag{7.8b}$$

$$= \bar{x}_{k-1} + \frac{1}{k} \{x_k - \bar{x}_{k-1}\} = \bar{x}_{k-1} - \frac{1}{k} \{\bar{x}_{k-1} - x_k\} \tag{7.8c}$$

Similarly, recursive variances σ_k^2 can be defined as

$$\sigma_k^2 = \frac{1}{k} \sum_{i=1}^{k} (x_i - \bar{x}_k)^2 \tag{7.9a}$$

which can be shown to be equivalent to

$$\sigma_k^2 = \frac{k-1}{k} \sigma_{k-1}^2 - \frac{1}{k-1} \{x_k - \bar{x}_k\}^2 \tag{7.9b}$$

Equations 7.8c and 7.9b are more efficient computationally.

7.4 TESTS FOR STATIONARITY

A test for stationarity can be done by dividing the time series into two or more subseries and checking whether the statistical character of each subseries is significantly different from one another. The standard normal variate z or the statistic Student's t can be used in this test. Student's t instead of z is used when the sample size is small (<30), or when the population standard deviation is replaced by the sample standard deviation.

Example 7.1: Testing stationarity

Stationarity test is illustrated using the annual rainfall data measured at the Observatory of Hong Kong (Hong Kong Observatory, Meteorological Results Part I) for the period 1884–1941 ($N = 58$ years), listed below and shown in Figure 7.1

1915.8	2766.7	1756.9	1684.3	3656.5	2040.7	1801.8	2974.7
2310.6	2538.9	2648.1	1164.3	1848.8	2540.7	1448.6	1846.7
1872.7	1417.0	2476.4	2378.7	2042.5	1802.1	1975.9	2376.0
2333.7	1923.6	1781.0	2300.0	1624.1	2126.8	2545.3	1931.1
2028.3	2069.7	2580.8	1934.0	2740.0	2472.2	1763.6	2711.1
2503.2	2224.4	2559.6	2740.0	1807.2	1773.5	2440.3	2041.8
2323.3	1583.5	2480.6	1811.6	1772.2	2095.5	1406.0	2202.3
2989.2	2433.1						

The series is split into three subseries of approximately equal length. The statistical parameters for the entire series together with those for the three subseries are computed. The mean value is used as the parameter for statistical tests. The results are summarized in Table 7.1.

In this example, the complete data series constitutes the population and the subseries constitute the samples. Since the population parameters (mean, μ; variance, σ^2) are known, the sample mean $\bar{x}$ is approximately normally distributed with mean μ and standard

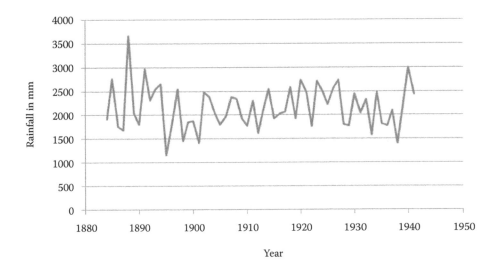

Figure 7.1 Annual rainfall in Hong Kong.

Table 7.1 Statistical parameters of the subseries

	Whole series, N = 58	First one-third, N = 19	Second one-third, N = 19	Last one-third, N = 20
Mean (mm)	2161.0	2143.0	2156.0	2183.0
Standard deviation (mm)	459.8	616.1	307.0	429.5
Skewness coefficient	0.4817	0.7048	0.2228	−0.00956
Student's t		−0.124	−0.069	0.223

deviation $\dfrac{\sigma}{\sqrt{N}}$, where N is the sample size. The approximation improves as the sample size N increases. For the above statement to be valid, there is no need to have any assumptions about the distribution of the population.

If the random sample comes from a normal population, then the above statement is exact regardless of the sample size N.

Therefore,

$$z = \frac{\bar{x} - \mu}{\left(\dfrac{\sigma}{\sqrt{N}}\right)}$$

follows a standard normal distribution. (This is approximate if the population is not normal.)

The null hypothesis H_0 and the alternative hypothesis H_a, respectively, are

H_0: $\bar{x} = 2161.0$, and the series is stationary

H_a: $\bar{x} \neq 2161.0$, and the series is not stationary

A two-tailed test is required. Under the null hypothesis H_0, for the first one-third

$$z = \frac{\bar{x} - \mu}{\left(\dfrac{\sigma}{\sqrt{19}}\right)} = \frac{2143 - 2161}{459.8}\sqrt{19} = -0.171$$

For a two-tailed test at 5% level of significance, adopt the following decision rule: Accept H_0 if 'z' lies within the interval $-z_{0.975}$ to $z_{0.975}$, which is the interval ± 1.96. Reject H_0 otherwise.

Since $z = -0.171$, H_0 is accepted at the 5% significance level.

For the second and last one-third, the corresponding z values are -0.0474 and 0.214, respectively, which lie well within the interval ± 1.96. Therefore, the hypothesis H_0 (that the series is stationary) is accepted.

7.5 TESTS FOR HOMOGENEITY

Homogeneity of a data series in statistical terms implies that the data belong to one population and therefore has a mean value that is time invariant. The tests are carried out by assuming that there is a jump (or drop) in the mean value of magnitude Δ after a certain number of observations. The null hypothesis is that the magnitude of Δ is not significantly different from zero. The alternative hypothesis is that there is a discontinuity in the mean value after a certain number of observations. It is rather vague because of the lack of any prior information about possible discontinuities. Certain statistics are defined, and their magnitudes are compared with the corresponding critical values at a chosen level of significance to accept or reject the null hypothesis.

Usually the tests assume that the data are independent, and have normal distribution. The first assumption is easily satisfied for annual data. The second assumption may not always be satisfied. However, with slight deviations from a normal distribution, the tests can still be used. Rather than applying the tests to a single series, it would be more conclusive if they are applied in a relative sense; that is, if two series are sufficiently correlated, relative tests are more powerful than absolute tests. Some test statistics used for testing homogeneity are defined below:

7.5.1   von Neumann ratio

The null hypothesis in this test is that the time series variable is independently and identically distributed (random). The alternate hypothesis is that the series is not random. The von Neumann ratio NR (von Neumann, 1941) is defined as

$$NR = \frac{\sum_{t=1}^{N-1}(x_t - x_{t+1})^2}{\sum_{t=1}^{N}(x_t - \bar{x})^2} \tag{7.10}$$

where x_t is the series of sample size N and mean value $\bar{x}$. Under the null hypothesis, $E(NR) = 2$. The mean of NR tends to be smaller than 2 for a non-homogeneous series. Critical values of NR for normally distributed samples, as given by Owen (1962) and Wijngaard et al. (2003), are shown in Table 7.2.

In this test, if a break occurs, implying non-homogeneity, the value of NR tends to be lower than 2; however, it does not give any indication about the location of the break.

7.5.2 Cumulative deviations

Also in this test, the null hypothesis is that the time series variable is independently and identically distributed. The alternate hypothesis is that there is a shift in the mean value after a certain time.

In the cumulative deviations test (Buishand, 1982), the departure from homogeneity is tested using the statistics Q and R, which are defined as

Table 7.2 Critical values of NR (1% and 5% significance levels) for von Neumann ratio test

N	20	30	40	50	70	100
1%	1.04	1.20	1.29	1.36	1.45	1.54
5%	1.30	1.42	1.49	1.54	1.61	1.67

Source: Owen, D.B., *Handbook of Statistical Tables*, Addison Wesley, Reading, MA, 1962; Wijngaard, J.B., Klein Tank, A.M.G. and Können, G.P., *International Journal of Climatology* 2003, 23, 679.

Table 7.3 Critical values of $\dfrac{Q}{\sqrt{N}}$ and $\dfrac{R}{\sqrt{N}}$ (1% and 5% significance levels) for cumulative deviations test

N	10	20	30	40	50	100	∞
1%	1.29 (1.38)	1.42 (1.60)	1.46 (1.70)	1.50 (1.74)	1.52 (1.78)	1.55 (1.86)	1.63 (2.00)
5%	1.14 (1.28)	1.22 (1.43)	1.24 (1.50)	1.26 (1.53)	1.27 (1.55)	1.29 (1.62)	1.36 (1.75)

Source: Buishand, T.A., *Journal of Hydrology* 1982, 58, 11.

Note: $\dfrac{R}{\sqrt{N}}$ is in bold face inside parenthesis.

$$Q = \max_{0 \le k \le N} \left| S_k^{**} \right| \tag{7.11}$$

and

$$R = \max_{0 \le k \le N} S_k^{**} - \min_{0 \le k \le N} S_k^{**} \tag{7.12}$$

in which the rescaled adjusted partial sums S_k^{**} are given by

$$S_k^{**} = \frac{S_k^*}{s_x} \quad k = 0, 1, 2, \ldots, N \tag{7.13}$$

where s_x is the sample standard deviation.

The adjusted partial sums (or cumulative deviations from the mean), S_k^*, are given by

$$S_0^* = 0 \tag{7.14a}$$

$$S_k^* = \sum_{t=1}^{k} (x_t - \bar{x}), \quad k = 1, 2, \ldots, N \tag{7.14b}$$

where x_t and $\bar{x}$ have the same meanings as before. High values of Q and R indicate departure from homogeneity. Critical values of $\dfrac{Q}{\sqrt{N}}$ and $\dfrac{R}{\sqrt{N}}$ based on 19,999 synthetic Gaussian random numbers, obtained by Buishand (1982), are given in Table 7.3.

In this test, if a break occurs in year k^*, then the statistic S_k^* attains its maximum (negative shift) or minimum (positive shift) near the year $k = k^*$.

Example 7.2: Homogeneity and trend tests

Homogeneity and trend tests are illustrated using the annual maximum temperature (°C) in Dhaka City, Bangladesh[4] for the period 1980–2009 (30 years), which is listed below and shown in Figure 7.2.

---

[4] The example illustrated using this data set is taken from the thesis submitted by A.K.M. Saifuddin (2010) as partial fulfillment for the degree of Master of Disaster Management of the National Graduate Institute for Policy Studies in Japan, under the supervision of the author.

30.29	30.01	30.34	30.13	30.48	31.00	31.29	31.48	31.49	31.38
30.03	30.11	30.68	30.20	30.93	31.48	31.68	30.62	30.87	31.53
30.13	30.47	30.30	30.23	30.52	30.83	31.28	30.54	30.52	31.61

The mean value of the data set is 30.75°C and the standard deviation is 0.55°C.

For the data given in Example 7.2, the minimum value of the test statistic S_k^* is −2.49 (see also Figure 7.3) and occurs at around the year 1984. It corresponds to a positive shift of the mean value, whereas a smaller negative shift occurs at around the year 1999. The $\dfrac{R}{\sqrt{N}}$ value is 1.18, which is less than the critical value of 1.50 for $N = 30$ at the 5% significance level. On the basis of this test, the data series can be considered as homogeneous.

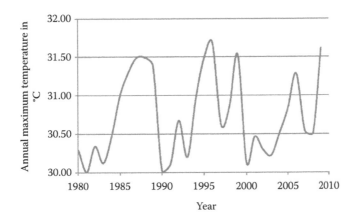

Figure 7.2 Annual maximum temperature in Dhaka City, Bangladesh.

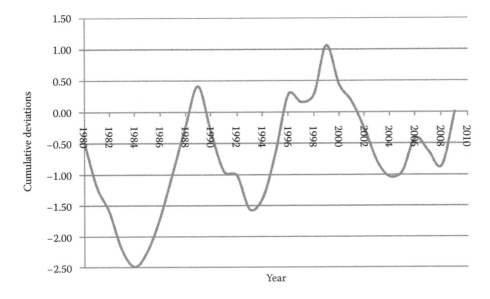

Figure 7.3 Cumulative deviations test (Equation 7.14b) for Dhaka City annual maximum temperature data.

Table 7.4 Critical values of U and A (1% and 5% significance levels)

N	10	20	30	40	50	100	∞
1%	0.575	0.662	0.691	0.698	0.718	0.712	0.743
	(3.14)	**(3.50)**	**(3.70)**	**(3.66)**	**(3.78)**	**(3.82)**	**(3.86)**
5%	0.414	0.447	0.444	0.448	0.452	0.457	0.461
	(2.31)	**(2.44)**	**(2.42)**	**(2.44)**	**(2.48)**	**(2.48)**	**(2.49)**

Source: Buishand, T.A., *Journal of Hydrology* 1982, 58, 11.

Note: A is in bold face inside parenthesis.

7.5.3 Bayesian statistics

In the Bayesian procedure, the shift in mean level is tested using the statistics U and A, which are defined as

$$U = \frac{1}{N(N+1)} \sum_{k=1}^{N-1} \left(S_k^{**}\right)^2 \quad \left(S_k^{**} \text{ is defined in Equation 7.13}\right) \tag{7.15}$$

and

$$A = \sum_{k=1}^{N-1} \frac{\left(S_K^{**}\right)^2}{k(N-k)} \tag{7.16}$$

Large values of U and A are indications of departure from homogeneity. Their critical values, based on 19,999 synthetic Gaussian random numbers obtained by Buishand (1982), are shown in Table 7.4.

7.5.4 Ratio test

In this test (Alexandersson, 1986), which is also known as the standard normal homogeneity test, the ratio series is obtained by dividing the series under consideration by a reference series. It is then standardized to have zero mean and unit variance. There are many forms of reference series that can be used. Ideally, it should be homogeneous and highly correlated to the series under consideration. The null hypothesis to be tested on the standardized (normalized) series of ratios, z_t, is that it has a normal distribution with zero mean and unit variance. The alternative hypothesis is that for some value of k ($0 \leq k \leq N$), the partial series of z_t for $t \leq k$ has a normal distribution with a mean of μ_1 and unit variance, and the remaining partial series for $t \geq k$ has a normal distribution with a mean μ_2 and unit variance ($\mu_1 \neq \mu_2$). Hypothesis testing is done using the statistic T_0 defined as

$$T_0 = \max_{1 \leq k \leq N}(T_k) \tag{7.17}$$

where

$$T_k = k\bar{z}_1^2 + (N-k)\bar{z}_2^2 \tag{7.18}$$

Table 7.5 Critical values of T_0 for the ratio test (1% significance level based on simulations carried out by Jaruškova [1994], and 5% significance level based on simulations carried out by Alexandersson and Moberg [1997])

N	20	30	40	50	70	100
1%	9.56	10.45	11.01	11.38	11.89	12.32
5%	6.95	7.65	8.10	8.45	8.80	9.15

Source: Jaruškova, D., *Monthly Weather Review* 1994, 124, 1535; Alexandersson, H. and Moberg, A., *International Journal of Climatology* 1997, 17, 25.

and

$$\bar{z}_1 = \frac{1}{k} \left(\sum_{t=1}^{k} z_t \right) \tag{7.19}$$

$$\bar{z}_2 = \frac{1}{(N-k)} \left(\sum_{t=k+1}^{N} z_t \right) \tag{7.20}$$

High values of T_0 indicate the non-homogeneous character of the data series. Critical values of T_0 have been given by Alexanderson (1986), Jaruškova (1994), and Alexandersson and Moberg (1997). Table 7.5 shows these values for different sample sizes at two significant levels. This test also permits the evaluation of the position of a possible break in the series, which corresponds to the maximum value of T_k.

For the data set of Example 7.2, the test statistic T_0 has a value of 5 at around the year 1984, which is less than the corresponding critical value of 7.65 for N = 30 at the 5% significance level. Also by this test, the data set can be considered as homogeneous although a break seems to occur in or around the year 1984 (Figure 7.4).

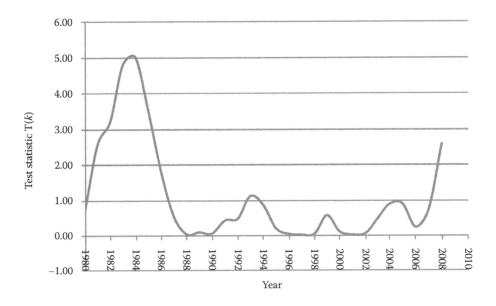

Figure 7.4 Ratio test (Equation 7.17) for Dhaka City annual maximum temperature data.

7.5.5 Pettit test

In the Pettit test (Pettit, 1979), which is non-parametric and based on the Wilcoxon test, the statistics S_k are calculated from ranks $r_1, r_2,...,r_N$ of the series $x_1, x_2,...,x_N$:

$$S_k = 2 \sum_{i=1}^{k} r_i - k(N+1) \quad k = 1,2,...,N \tag{7.21}$$

Test statistic S_{k*} is defined as

$$S_{k*} = \max_{1 \leq k \leq N} |S_k| \tag{7.22}$$

Critical values for S_{k*} are given in Table 7.6.

In this test, if a break occurs in year k^*, then the statistic attains its maximum or minimum near the year $k = k^*$.

> For the data set of Example 7.2, the test statistic S_{k*} (Equation 7.22) has a minimum value of −84 at around the year 1984, which is less than the corresponding critical value of −107 for N = 30 at the 5% significance level. It also shows a slightly smaller maximum value of 25 at around the year 1999 (Figure 7.5). Also by this test, the data set can be considered as homogeneous.

Homogeneity can also be tested graphically by the double mass curve when applied to hydrological data. It is done by plotting the cumulative sums of the variable at the station

Table 7.6 Critical values of S_{k*} (1% and 5% significance levels) based on simulations carried out by Pettit (1979) and reported by Wijngaard et al. (2003)

N	20	30	40	50	70	100
1%	71	133	208	293	488	841
5%	57	107	167	235	393	677

Source: Pettit, A., Journal of the Royal Statistical Society, Series C (Applied Statistics) 1979, 28, 126; Wijngaard, J.B., Klein Tank, A.M.G. and Können, G.P., International Journal of Climatology 2003, 23, 679.

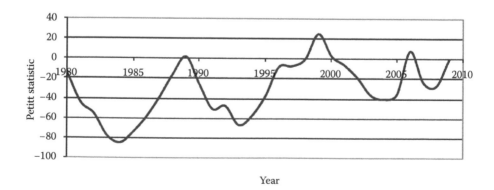

Figure 7.5 Pettit statistic (Equation 7.22) for Dhaka City annual maximum temperature data.

under consideration against the cumulative sums of a number of nearby stations. A high degree of correlation of the data set of the suspect station with those of the nearby stations is implied in this method. The plot tends to be linear if the data are homogeneous, and depart from linearity if they are non-homogeneous. This method also enables the detection of the time at which the data begin to show departure from homogeneity.

In the context of climatology and hydrology, several researchers have carried out homogeneity tests on various climatic and hydrological variables. Examples include the studies on European daily temperature and precipitation data (Wijngaard et al., 2003), rainfall data (Alexandersson, 1986; Buishand et al., 1982; Jayawardena and Lau, 1990b), Bulgarian temperature data (Syrakova and Stefanova, 2009), Turkish meteorological data (Sahin and Cigizoglu, 2010), annual air temperature data in Croatia (Pandžíc and Likso, 2009), monthly Spanish temperature data (Staudt et al., 2007), and Swedish temperature data (Alexandersson and Moberg, 1997).

7.6 COMPONENTS OF A TIME SERIES

A time series normally has a deterministic component, the outcome of which can be predicted with certainty (at least in theory), and a stochastic component, the outcome of which is attributed to chance and cannot be explained physically and therefore cannot be predicted with certainty. The deterministic component may be in the form of a long-term trend, or of a periodic nature or a combination. Therefore, in general, a time series may be considered as a linear combination of a trend, a periodic component, a dependent stochastic component, and an independent residual component.

It is also possible for the three effects to be expressed in a multiplicative form; however, such representations can be easily transformed into linear form by taking the logarithms.

7.7 TREND ANALYSIS

If the mean value of a time series is not time invariant, then the time series may have a trend. Trends may be either increasing or decreasing. They can be removed if detected. It is best to test for trends after carrying out homogeneity tests. Whether a given time series is credible for carrying out trend analysis is determined on the basis of the statistical results of the homogeneity tests discussed in Section 7.5. Wijngaard et al. (2003) classifies data sets as 'useful' if one or zero tests reject the null hypothesis at the 1% significance level, 'doubtful' if two tests reject the null hypothesis at the 1% significance level, and 'suspect' if three or four tests reject the null hypothesis at the 1% significance level. 'Useful' implies no clear signal of non-homogeneity and the data set can be used for trend analysis. 'Doubtful' implies that the trend analysis should be carried out and interpreted with caution, and 'suspect' implies that the data set lacks credibility for trend analysis.

7.7.1 Tests for randomness and trend

When there is no clear visual indication of the presence of a trend, it is necessary to carry out statistical tests to determine whether an apparent trend is significant or not. There are several tests that can be carried out.

7.7.1.1 Turning point test for randomness

In this test, the presence of high and low values is examined by determining the number of turning points in the series. A turning point in a time series x_t ($t = 1, N$) occurs at i if

$$x_i > x_{i-1} \text{ and } x_{i+1} \quad \text{or} \quad x_i < x_{i-1} \text{ and } x_{i+1}$$

For three unequal observations, there are six possible orders of magnitudes:

$x_{i-1} > x_i > x_{i+1};$	$x_{i-1} > x_{i+1} > x_i;$	$x_i > x_{i-1} > x_{i+1};$
$x_i > x_{i+1} > x_{i-1};$	$x_{i+1} > x_{i-1} > x_i;$	$x_{i+1} > x_i > x_{i-1}$

They have equal probabilities of occurrence in the case of a random series. Turning points occur in all except the first and the last. Therefore, the probability of having turning points is 4/6 (or 2/3). In a series, the first and the last terms are not turning points. Therefore, the expected number of turning points, p, in a random series is given by

$$E(p) = \frac{2}{3}(N - 2) \tag{7.23}$$

To establish whether the difference between the observed and the expected number of turning points is significant, it is necessary to calculate the variance of the expected number of turning points. This has been shown to be (Kendall, 1973)

$$\text{Var}(p) = \frac{(16N - 29)}{90} \tag{7.24}$$

A two-tailed test of significance is carried out on p, which is expressed as a standard normal deviate z where

$$z = \frac{p - E(p)}{\sqrt{\text{Var}(p)}} \tag{7.25}$$

which should lie within ±1.96 at the 5% significance level if the series is random. Too many or too few turning points indicate non-randomness.

The turning point test is illustrated using the data set given in Example 7.1. The results are as follows:

Number of turning points in the series, p = 37
Expected number of turning points if the series is random = 37.33
Variance of the expected number of turning points = 9.989
Number of turning points expressed as a standard normal variate, z = −0.106

The hypotheses are

H_0: p = 37.33, and the series is random
H_1: $p \neq$ 37.33, and the series is not random

A two-tailed test is carried out. The hypothesis H_0 (that the series is random) is accepted since z lies between ±1.96 at the 5% significance level.

7.7.1.2 Kendall's rank correlation[5] test (τ test)

Rank correlation (Kendall, 1948; Abdi, 2007) can be used to establish whether an apparent trend in a series is significant or not. The number of times p in all pairs of observations x_i, x_j; $j > i$ that $x_j > x_i$ is determined (i.e., for $i = 1, N - 1$ how many times $x_j > x_i$ for $j = i + 1, i + 2, ..., N$). The ordered subsets in such a series are

$$x_1 : x_2, x_3, x_4, ..., x_N$$

$$x_2 : x_3, x_4, x_5, ..., x_N$$

$$x_3 : x_4, x_5, ..., x_N$$

...

...

$$x_{N-1} : x_N$$

The maximum number of such pairs occurs in a continuously rising curve when all $x_j > x_i$ for $j = i + 1, N; i = 1, N - 1$. Then,

$$p = (N - 1) + (N - 2) + (N - 3) + ... + 1 = \frac{1}{2} N(N - 1) \tag{7.26}$$

In a falling trend, all $x_j < x_i$, and therefore $p = 0$. Therefore, for a trend-free series,

$$E(p) = \frac{1}{2} \left(0 + \frac{1}{2} N(N - 1) \right) = \frac{1}{4} N(N - 1) \tag{7.27}$$

The test is carried out using the statistic τ (known as Kendall's τ and which varies between ±1) defined as

$$\tau = \frac{4p}{N(N - 1)} - 1 \tag{7.28}$$

For a random series, $E(\tau) = 0$, and its variance is given as

$$\text{Var}(\tau) = \frac{2(2N + 5)}{9N(N - 1)} \tag{7.29}$$

As N increases, $\dfrac{\tau - E(\tau)}{\sqrt{\text{Var}(\tau)}}$ converges to a standard normal distribution. It may also be possible to carry out a test using $E(\tau)$ that takes values of −1 and 1, leading to the inference that there is a rising or falling trend.

---

[5] Kendall rank correlation coefficient, commonly referred to as Kendall's tau (τ), is also used to measure the association between two data sets.

Kendall's rank correlation test is illustrated using the data set given in Example 7.1. The results are as follows:

Number of times $x_{i+1} > x_i = 857$
Statistic $\tau = 0.0369$
$\text{Var}(\tau) = 0.008133$
$z = 0.4092$

which again is within the interval ± 1.96 at the 5% significance level. Hence, the hypothesis H_0 (that the series is random) is accepted.

7.7.1.3 Regression test for linear trend

A linear trend may be approximated as

$$y_c = a + bx_i + \zeta_i \text{ (error)} \tag{7.30}$$

The hypothesis to be tested is that $b \neq 0$. A linear regression of the form $y_c = a + bx_i$ is first fitted to the data. The regression coefficients a and b are obtained by minimizing the squared error. For pairs of dependent and independent variables y_i and x_i, it can be shown that

$$a = \bar{y} - b\bar{x} \tag{7.31a}$$

$$b = \frac{\sum_{i=1}^{N}(x_i - \bar{x})(y_i - \bar{y})}{\sum_{i=1}^{N}(x_i - \bar{x})^2} \tag{7.31b}$$

where the overbar indicates the mean.

For computational purposes, it is convenient to write b as

$$b = \frac{\sum_{i=1}^{N} x_t y_t - \dfrac{\sum_{i=1}^{N} x_i \sum_{i=1}^{N} y_i}{N}}{\sum_{i=1}^{N} x_i^2 - \dfrac{1}{N}\sum_{i=1}^{N} x_i} \tag{7.31c}$$

Estimate of the variance of b is given by

$$s_b^2 = \frac{s_{yx}^2}{\sum_{i=1}^{N}(x_i - \bar{x})^2} \tag{7.32}$$

where, s_{yx} is the standard error of regression (or estimate), which is given by

$$s_{yx}^2 = \frac{1}{(N-2)} \sum_{i=1}^{N} (y_c - (a+bx_i))^2 = \frac{\sum_{i=1}^{N} (\text{error}_i)^2}{N-2} \tag{7.33}$$

Student's t is given by the ratio $\dfrac{(b - E(b))}{s_b}$. (The statistic Student's t is used here because σ_b is unknown and s_b is an estimate of σ_b.)

A regression test for linear trend is illustrated using the data set given in Example 7.1. The results are as follows:

The Student's t statistic from the calculations is 0.06063. The hypotheses to be tested are

H_0: $b = 0$, and there is no linear regression
H_1: $b \neq 0$, there is a linear regression

At the 5% level of significance, accept H_0 if t lies in the interval $-t_{0.975}$ to $t_{0.975}$, which for 57 (= 58 – 1) degrees of freedom is the interval –2.003 to 2.003. Hence, H_0 is accepted at the 5% significance level.

7.7.1.4 Mann–Kendall test

The Mann–Kendall test (Mann, 1945; Kendall, 1975) is a non-parametric test to detect trends in time series data. For a time series $x_1, x_2,...,x_N$, the Mann–Kendall statistic S is defined as follows:

$$S = \sum_{i=1}^{N-1} \sum_{j=i+1}^{N} \text{Sgn}(x_j - x_i) \tag{7.34}$$

where

$$\text{Sgn}(x_j - x_i) = \begin{cases} 1 & \text{if } x_j > x_i \\ 0 & \text{if } x_j = x_i \\ -1 & \text{if } x_j < x_i \end{cases} \tag{7.35}$$

If a data set displays a consistently increasing trend, the statistic S will be positive, whereas if it displays a decreasing trend, S will be negative. A larger magnitude S indicates that the trend is more consistent in its direction. The Sgn function enables detection of trends featuring either large or small increase steps from year to year equally.

Under the null hypothesis that there is no trend displayed by the time series, the distribution of S is expected to have a zero-mean and variance

$$\text{Var}(S) = \frac{1}{18} \left[N(N-1)(2N+5) - \sum_{p=1}^{g} t_p(t_p - 1)(2t_p + 5) \right] \tag{7.36a}$$

where N is the number of data under consideration, g is the number of tied[6] groups, and t_p is number of data in the p-th tied group. If there are no tied groups in the series, Equation 7.36a simplifies to

$$\text{Var}(S) = \frac{N(N-1)(2N+5)}{18} \tag{7.36b}$$

The test statistic z is calculated as

$$z = \begin{cases} \dfrac{S-1}{\sqrt{\text{Var}(S)}} & \text{if } S > 0 \\[2ex] 0 & \text{if } S = 0 \\[2ex] \dfrac{S+1}{\sqrt{\text{Var}(S)}} & \text{if } S < 0 \end{cases} \tag{7.37}$$

and tested against the critical value of the normal distribution at a chosen level of significance. With no tied groups, the test is reported to be valid for $N > 10$, and normally valid for $N > 40$ (Gilbert, 1987).

> Mann–Kendall test is illustrated using the data set given in Example 7.2. This data set has three tied groups; tied group 1 has two data points, tied group 2 has two data points, and tied group 3 has two data points. The statistic S (Equation 7.34) works out to be 70 with a variance of 3139, giving a standard normal value (Equation 7.37) of 1.23, which is less than 1.96 at the 5% significance level. Hence, the series in Example 7.2 can be considered as having an insignificant positive trend.

Trend analyses of climatological data have been carried out using one or more of the above methods by several researchers from several geographical regions. Examples of data used include daily maximum and minimum temperatures in Catalonia, Spain (Martínez et al., 2010); summer temperature in Tuscany, Italy (Bartolini et al., 2008); precipitation in the Kalu Ganga basin in Sri Lanka (Ampitiyawatta and Guo, 2009); Turkish precipitation (Partal and Kahya, 2006); low flows in the United States (Douglas et al., 2000); monthly water quality data in the United States (Hirsch et al., 1981, 1991); temperature and precipitation in the Yangtze River basin in China (Su et al., 2006); annual and seasonal rainfall data in the Mediterranean area (Longobardi and Villani, 2009); and rainfall in Amazonia (Paiva and Clarke, 1995).

7.7.2 Trend removal

A trend can be linear or non-linear. Removal in the former case is quite simple by fitting a regression line to the data set. The slope and the intercept of the straight line can be estimated by the least squares method. In the latter case of a non-linear trend, it can be removed by higher-order polynomial regression. The parameters in this case can also be estimated by

---

[6] If the time series has values 10, 15, 10, 20, 15, 18, 20, 30, 20, and 5, then the number of tied groups is 3 and the number of data in tied group 1 (corresponding to 10) is 2, group 2 (corresponding to 15) is 2, and group 3 (corresponding to 20) is 3.

the method of least squares. Trends may also be removed by differencing; that is, by transforming a non-stationary series to a stationary one.

However, higher-order polynomial interpolation does not necessarily lead to better approximations. It has been shown by Runge (as cited by Kreyszig, 1999, p. 861) that the maximum error even approaches infinity as the order of the polynomial tends to infinity. An alternative to global polynomial regression is piecewise regression fitting. This is called spline regression that has the ability to avoid such numerical oscillations. Instead of a single higher-order polynomial, several lower-order polynomials can be used in a piecewise manner.

The piecewise interpolation function $g(x)$ of $f(x)$ should be such that at each end point of the interval

$$g(x_0) = f(x_0); \ g(x_1) = f(x_1); \ldots; g(x_N) = f(x_N) \tag{7.38}$$

In each interval, the function $g(x)$ should be several times differentiable (e.g., for a cubic spline, it should be twice differentiable; for a fourth-order spline, it should be three times differentiable). A linear spline is the simplest but it has sharp corners and the derivatives of adjoining splines cannot be matched easily. A necessary condition for the spline to be continuous at the end points of each segment is that their derivatives computed from either side of the intersection should match.

7.7.2.1 Splines

Spline in the ordinary sense refers to a flexible strip used in drafting to draw a smooth curve through a set of points. Also in the mathematical sense, it has the same meaning. Spline methods can be used for interpolation and regression. The basic objective in spline interpolation is to connect a set of data points by a smooth curve via a combination of several piecewise low-order curves. The interpolation (or regression) functions can be polynomial, sinusoidal, exponential, or their combinations. The cubic spline, which consists of N polynomial functions each of which has order not greater than three, is by far the most widely used. It is of the form

$$y = a_i + b_i x + c_i x^2 + d_i x^3 \tag{7.39}$$

where x,y are the independent and dependent variables, respectively; a_i, b_i, c_i, d_i are parameters for the i-th interval. Thus, in a cubic spline, there are four parameters to be estimated for each interval that require four equations; that is, $4(N-1)$ equations in total. These consist of $2N-2$ equations obtained by conditions of satisfying Equation 7.38 at all data points (twice at interior points and once at end points), $2(N-2)$ equations by equating the first and second derivatives (i.e., continuity at each interior point) of the interpolation function at each interior point, and two more equations by setting the second derivatives of the function at the two end points to zero, adding up to $4(N-1)$ equations.

Cubic spline regression is similar to cubic spline interpolation in all aspects, except that a fewer number of 'knots' ('knot' refers to the intersection of two adjacent splines; k knots link $k + 1$ splines) is used on the basis of the curvature changes of the data trajectory. The cubic regression spline equation has the form

$$y = a_i + b_i x + c_i x^2 + d_i x^3 + \sum_{j=1}^{k} D_j e_j (x_i - x_j)^3 \tag{7.40}$$

where x_i is the location of the i-th knot, k is the number of knots, D_j is a dummy variable (zero or one), and e_j is an additional parameter. For example,

$$y = a_i + b_i x + c_i x^2 + d_i x^3 \text{ (for the first interval; all } D_j\text{'s} = 0) \qquad (7.41a)$$

$$y = a_i + b_i x + c_i x^2 + d_i x^3 + e_1(x - x_1)^3 \text{ (for the second interval; } D_1 = 1) \qquad (7.41b)$$

$$y = a_i + b_i x + c_i x^2 + d_i x^3 + e_1(x - x_1)^3 + e_2(x - x_2)^3 \text{ (for the third interval; } D_1 = D_2 = 1) \qquad (7.41c)$$

and so on (Luo and White, 2005).

In these expressions, the continuity of $g(x)$ and its derivatives is automatically satisfied. If the number of knots and their locations are decided, the parameters a_i, b_i, c_i, d_i, e_i can be obtained by the linear least squares method. The locations can be equally spaced between the end points, in which case it may not be the optimum. If the locations of the knots are also to be optimized from a random allocation, then the parameters can only be determined by non-linear least squares method. More details of the method, including an application, is given by Luo and White (2005).

The smoothing spline function has been shown to be equivalent to minimizing the functional (Reinsch, 1967)

$$S(\lambda) = \sum_{i=1}^{N} \{(y_i - g(x_i)\}^2 + \lambda \int_a^b \{g''(x)\}^2 \, dx \qquad (7.42)$$

where λ is a positive prespecified parameter, and a and b are the beginning and end values of x. The first term in Equation 7.42 represents the residual sum of squares and it penalizes the lack of fit of the function $g(x)$. The second term, which is weighted by λ, represents a roughness penalty that accounts for the curvature of the function $g(x)$. The λ in Equation 7.42 is a smoothing parameter that may vary from zero to infinity. If λ is close to zero, $g(x)$ will not be smooth but will fit the data more closely. If λ is large, $g(x)$ will be smooth but will not fit the data closely. The parameter λ therefore controls the trade-off between the closeness of fit to the data as measured by the residual sum of squares $\sum_{i=1}^{N} \{y_i - g(x_i)\}^2$ and the smoothness as measured by the integral part $\int_a^b \{g''(x)\}^2 \, dx$.

For a given λ, the function $g(x)$, which minimizes $S(\lambda)$, is a cubic spline (Reinsch, 1967). It has the following properties:

$g(x)$ has a continuous derivative everywhere.
$g(x)$ is linear for $x < x_1$ and $x > x_N$.
$g(x)$ is cubic between each successive values of x_i.

The minimization problem can be solved using variational calculus.

In the case of spline regression in a non-parametric sense, the function can be defined as

$$g(x_i) = \sum_{j=1}^{N} w_{ij} y_j \qquad (7.43)$$

where the weights w_{ij} are functions of the distance in the x-space, and are given in the form of kernel functions that can take many forms. For example, Nadaraya (1964) and Watson (1964), commonly referred to as Nadaraya–Watson kernel, used the form

$$w_{ij} = \frac{K(u)}{\sum\limits_{j=1}^{N} K(u)}; \quad u = \frac{x_i - x_j}{h} \tag{7.44}$$

The kernel function $K(u)$ can be a probability distribution function such as the Gaussian as well as several other functions. The following sinusoidal function with $h = \lambda^{0.25}$ and $W_{ij} \cong \frac{1}{h} K\left(\frac{x_i - x_j}{h}\right)$ has also been used:

$$K(u) = 0.5e^{-\frac{|u|}{\sqrt{2}}} \sin\left(0.25\pi + \frac{|u|}{\sqrt{2}}\right) \tag{7.45}$$

7.8 PERIODICITY

Periodicity is a common feature in most natural phenomena where the effects of the revolution of the earth around the sun, rotation of the earth about its own axis, and the rotation of the moon around the earth produce well-defined cyclic patterns. If the period of the periodicity is known, then it can be removed by harmonic analysis. If it is not well defined, the periodicities could be identified by autocorrelation and spectral analysis.

7.8.1 Harmonic analysis – cumulative periodogram

If, for the annual cycle, $\tau = 1, 2, \ldots, p$ are the periods of the harmonics (e.g., in the case of monthly data, $p = 12$; weekly data, $p = 52$; daily data, $p = 365$), the harmonically fitted means, m_τ, for each period τ, can be represented by a series of sine and cosine terms (at this stage, the trend component is assumed to have been removed) as follows:

$$m_\tau = \mu + \sum_{i=1}^{h} \left\{ A_i \cos\left(\frac{2\pi i \tau}{p}\right) + B_i \sin\left(\frac{2\pi i \tau}{p}\right) \right\} \tag{7.46}$$

where μ is the population mean, A_i and B_i are the Fourier coefficients, and h the total number of harmonics. Sample estimates of m_τ from the series are given by

$$\bar{x}_\tau = \frac{p}{N} \sum_{i=1}^{N/p} x_{\tau + p(i-1)} \tag{7.47}$$

The estimates of the population mean, and the Fourier coefficients A_i and B_i are obtained by minimizing the sum of squares of the differences, $\sum_{\tau=1}^{p} (\bar{x}_\tau - m_\tau)^2$; that is,

$$E = \sum_{\tau=1}^{p} (\bar{x}_\tau - m_\tau)^2 = \sum_{\tau=1}^{p} \left[\bar{x}_\tau - \mu - \sum_{i=1}^{h} \left\{ A_i \cos\left(\frac{2\pi i\tau}{p}\right) + B_i \sin\left(\frac{2\pi i\tau}{p}\right) \right\} \right]^2 \tag{7.48}$$

which is done by setting $\dfrac{\partial E}{\partial \mu} = 0$, $\dfrac{\partial E}{\partial A_i} = 0$, and $\dfrac{\partial E}{\partial B_i} = 0$.

The first condition gives the estimate of the population mean μ as

$$\hat{\mu} = \frac{1}{p} \sum_{\tau=1}^{p} \bar{x}_\tau \tag{7.49}$$

and setting $\dfrac{\partial E}{\partial A_i} = 0$ and $\dfrac{\partial E}{\partial B_i} = 0$, and using the orthogonal properties[7] of the sine and cosine functions, it can be shown (Kottegoda, 1980, pp. 37–38) that

$$A_i = \frac{2}{p} \sum_{\tau=1}^{p} \bar{x}_\tau \cos\left(\frac{2\pi i\tau}{p}\right) \quad \text{for } i = 1, 2, ..., h \tag{7.50a}$$

$$B_i = \frac{2}{p} \sum_{\tau=1}^{p} \bar{x}_\tau \sin\left(\frac{2\pi i\tau}{p}\right) \quad \text{for } i = 1, 2, ..., h \tag{7.50b}$$

When p is even, the last two coefficients are given by

$$A_h = \frac{1}{p} \sum_{\tau=1}^{p} \bar{x}_\tau \cos\left(\frac{2\pi h\tau}{p}\right) \tag{7.51a}$$

$$B_h = 0 \tag{7.51b}$$

In the above equations, i refers to the number of harmonics, and h the total number of harmonics,[8] which is equal to $\dfrac{p}{2}$ (or $\dfrac{p-1}{2}$ if p is odd).

For example,

for monthly data $p = 12$, therefore $h = 6$
for weekly data $p = 52$, therefore $h = 26$
for daily data $p = 365$, therefore $h = 182$

---

[7] See Appendix 7.1.

[8] The harmonics have wavelengths (or periods) of $\dfrac{p}{1}, \dfrac{p}{2}, \dfrac{p}{3}, ..., \dfrac{p}{h}$, where $\dfrac{p}{h} \geq 2\Delta t$, and Δt is the time interval between data. The shortest wavelength is $2\Delta t$, because, at least three points are needed to define a curve. If $\Delta t = 1$, $\dfrac{p}{h} = 2$, and, hence $h = \dfrac{p}{2}$ (or $\left(\dfrac{p}{2} - 0.5\right)$ if p is odd).

When m_τ is determined by considering all the harmonics, then it will be the same as the actual periodic means $\bar{x}_\tau$ for all the values of τ. However, in practice only a few harmonics ($h^* < h$) are sufficient. These are called the significant harmonics – those that contribute significantly to the variability of x.

Experience in using the Fourier series representation indicates that for small time interval (daily, weekly, etc.) series, only the first few harmonics are sufficient (about 4–5 out of a total of 182 for daily series). For monthly series, only about one or two harmonics would be sufficient. A more accurate estimate of the number of significant harmonics could be obtained by plotting the cumulative periodogram (Salas et al., 1980, p. 80), which is defined as

$$P_j = \sum_{i=1}^{j} \frac{\text{Var}(h_i)}{\text{Var}(x)} \tag{7.52}$$

where $\text{Var}(h_i)$ is the mean square deviation of m_τ around $\hat{\mu}$ for each harmonic $i = 1, 2, ..., h$, and $\text{Var}(x)$ is the mean square deviation of $\bar{x}_\tau$ around $\hat{\mu}$. They are given as[9]

$$\text{Var}(h_i) = \frac{1}{2}\left(A_i^2 + B_i^2\right); i = 1,2,...,h \tag{7.53}$$

$$\text{Var}(x) = \frac{1}{p}\sum_{\tau=1}^{p}(\bar{x}_\tau - \hat{\mu})^2 \tag{7.54}$$

It can also be shown that

$$\sum_{i=1}^{h} \text{Var}(h_i) = \text{Var}(x) \tag{7.55}$$

that is, $P_j = 1$ when $j = h$, in which case the explained variance is equal to the total variance. It should be noted that in the computation of the cumulative periodogram, the $\text{Var}(h_i)$ must be arranged in decreasing order of magnitude.

The criterion for testing the significance is based on the concept that the variation of P_j versus j is composed of a fast increasing periodic part and a slow increasing sampling part. The point of intersection of these two parts corresponds to the critical harmonic h^*. For an independent series, the sampling part is a straight line, whereas for a linearly dependent series, it is a curve. The periodogram is also identified as the sample spectrum, or the line spectrum, or the discrete spectrum. In the plot, j represents the frequency.

Example 7.3: Periodicity analysis

Monthly rainfall data measured at the Observatory of Hong Kong (Hong Kong Observatory, Meteorological Results Part I) for the period 1884–1939 ($N = 672$) are used to illustrate various tests and analysis of periodic data. The data series, shown in Figure 7.6, has a mean value of 178.4 mm, a standard deviation of 193.4 mm, a skewness coefficient of 1.496, and a kurtosis of 5.216. Typical periodogram results for this data set are shown in Table 7.7 and Figure 7.7.

---

[9] The variance of m_τ around $\hat{\mu}$ is composed of variances of each harmonic, $\text{Var}(h_i)$. It is considered as the part of the variance of m_τ, which is contributed by the harmonic i. Therefore, P_j would represent the ratio of the explained variance by the first j harmonics to the total variance.

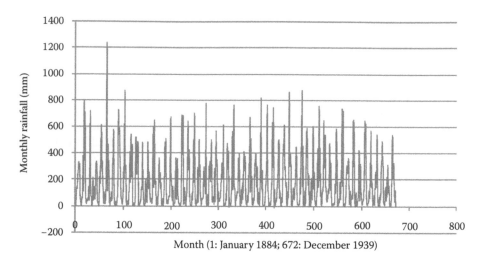

Figure 7.6 Monthly rainfall data in Hong Kong (1884–1939).

As shown in Table 7.7 and Figure 7.7, the cumulative periodogram of the monthly means shows that the first harmonic is very significant, explaining 94.85% of the variance followed by second, third, fourth, fifth, and sixth, explaining 4.292%, 0.4076%, 0.4062%, 0.04498%, and 0.001476% of the variance, respectively. Hence, the first, second, third, and fourth would be sufficient (they together explain 99.9% of the variance). However, all six harmonics have been retained for Fourier series fitting.

As shown in Table 7.8 and Figure 7.7, the cumulative periodogram of the monthly standard deviations shows that the first harmonic explains 89.73% of the variance followed by third, fifth, fourth, second, and sixth explaining 5.277%, 2.484%, 1.348%, 1.024%, and 0.06896% of the variance, respectively. On the basis of the periodogram, five harmonics (they together explain 99.9% of the variance) would be sufficient to describe the monthly standard deviations, but all six harmonics have been retained.

The comparisons of the sample monthly means and standard deviations with the Fourier fitted means and standard deviations for different numbers of harmonics up to the maximum are illustrated in Tables 7.9 and 7.10, and Figures 7.8 and 7.9. It is seen that with the maximum numbers of harmonics, the sample monthly means and the standard deviations and the harmonically fitted monthly means and the standard deviations are exactly the same.

The significance of each harmonic can also be tested by the *F*-test. Analysis of variance is then required.

Table 7.7 Cumulative periodogram for monthly means of Hong Kong rainfall data: 1884–1939 (56 years)

j	A_i	B_i	$\frac{1}{2}\left(A_i^2 + B_i^2\right)$	$\sum_{i=1}^{j} \frac{1}{2}\left(A_i^2 + B_i^2\right)$	$\dfrac{\frac{1}{2}\left(A_i^2 + B_i^2\right)}{\textit{Variance}}$	$\dfrac{\sum_{i=1}^{h} \frac{1}{2}\left(A_i^2 + B_i^2\right)}{\textit{Variance}}$
1	−174.5	−80.33	18,450	18,450	0.9485	0.9485
2	21.46	34.78	835.2	19,290	0.04292	0.9914
3	−2.407	12.36	79.30	19,370	0.004076	0.9955
4	6.421	−10.81	79.04	19,450	0.004062	0.9995
5	−1.378	3.950	8.752	19,460	0.0004498	1.0000
6	−0.7577	0.000	0.2871	19,460	0.00001476	1.0000

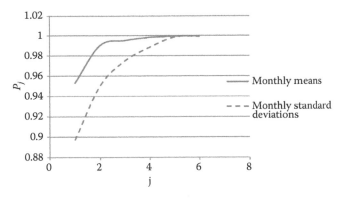

Figure 7.7 Cumulative periodogram for monthly means and standard deviations of Hong Kong rainfall data (See Equation 7.52).

Table 7.8 Cumulative periodogram for monthly standard deviations of Hong Kong rainfall data: 1884–1939 (56 years)

J	A_i	B_i	$\frac{1}{2}\left(A_i^2 + B_i^2\right)$	$\sum_{i=1}^{j} \frac{1}{2}\left(A_i^2 + B_i^2\right)$	$\dfrac{\frac{1}{2}\left(A_i^2 + B_i^2\right)}{Variance}$	$\dfrac{\sum_{i=1}^{h} \frac{1}{2}\left(A_i^2 + B_i^2\right)}{Variance}$
1	−82.37	−38.98	4152	4152	0.8973	0.8973
3	−7.407	20.82	244.2	4396	0.05277	0.9501
5	8.254	12.72	114.9	4511	0.02101	0.9749
4	−5.107	−9.935	62.39	4574	0.01348	0.9884
2	3.979	−8.885	47.39	4621	0.01024	0.9986
6	−2.526	0.000	3.191	4624	0.0006896	0.9993

Table 7.9 Sample and Fourier series fitted means

Month	Sample means	Fourier series fitted means for different numbers of harmonics					
		I Harmonic	*2* Harmonics	*3* Harmonics	*4* Harmonics	*5* Harmonics	*6* Harmonics
January	31.72	−12.85	28	40.36	27.79	30.96	31.72
February	44.7	21.62	41.01	43.42	49.57	45.46	44.7
March	75.42	98.12	76.66	64.29	70.72	74.67	75.42
April	136.8	196.1	155.3	152.9	140.3	137.6	136.8
May	290.1	289.4	270	282.4	288.5	289.3	290.1
June	383.9	353	374.4	376.8	383.3	384.6	383.9
July	383.3	369.7	410.6	398.2	385.7	382.5	383.3
August	361.8	335.3	354.7	352.2	358.4	362.5	361.8
September	252.9	258.8	237.3	249.7	256.1	252.1	252.9
October	111.7	160.8	119.9	122.3	109.7	112.5	111.7
November	41.85	67.48	48.09	35.73	41.88	41.1	41.85
December	27.27	39.32	25.4	22.99	29.41	28.03	27.27

Table 7.10 Sample and Fourier series fitted standard deviations

Month	Sample standard deviations	Fourier series fitted standard deviations for different numbers of harmonics					
		1 Harmonic	*2 Harmonics*	*3 Harmonics*	*4 Harmonics*	*5 Harmonics*	*6 Harmonics*
January	36.55	25.75	46.57	45.78	39.73	34.02	36.55
February	41.10	41.63	49.04	42.15	53.31	43.63	41.10
March	62.93	77.59	56.77	69.49	64.38	60.40	62.93
April	98.58	124.0	116.6	101.5	95.40	101.1	98.58
May	226.1	168.4	189.2	202.7	213.9	223.6	226.1
June	194.4	198.9	206.3	198.1	193.0	197.0	194.4
July	178.1	207.4	186.6	187.4	181.3	175.6	178.1
August	189.9	191.5	184.1	191.0	202.1	192.5	189.9
September	157.1	155.6	176.4	163.7	158.5	154.6	157.1
October	128.8	109.1	116.6	131.7	125.6	131.3	128.8
November	53.77	64.73	26.80	30.40	41.56	51.24	53.77
December	31.40	34.20	26.80	35.05	29.94	33.92	31.40

7.8.2 Autocorrelation analysis

Many time series have some form of dependence between events separated by small time steps. A measure of the degree of linear dependence is the autocovariance function c_k, which represents the covariance between x_t and x_{t+k}, where k is the time lag. For a stationary continuous series, it is defined as

$$c_k = \lim_{T \to \infty} \frac{1}{2T} \int_{-T}^{T} x_t x_{t+k} \, dt \tag{7.56}$$

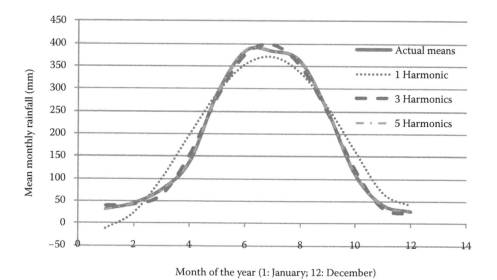

Figure 7.8 Actual and harmonically fitted monthly means of rainfall in Hong Kong.

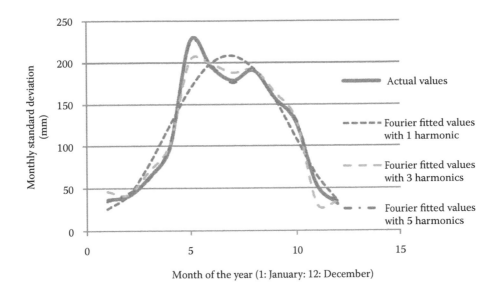

Figure 7.9 Actual and harmonically fitted monthly standard deviations of rainfall in Hong Kong.

For a weakly stationary series, it is defined as

$$c_k = E(x_t x_{t+k})$$ (7.57)

Both these definitions give the same result for a stationary ergodic series.
    The autocorrelation function ρ_k is defined as

$$\rho_k = \frac{c_k}{E(x_t^2)}$$ (7.58)

where $-1 \leq \rho_k \leq 1$; $\rho_{-k} = \rho_k$ for all k; $\rho_0 = 1$. Estimates of ρ_k from observed samples are called
the serial autocorrelation coefficients, r_k, and are obtained from the following equation:

$$r_k = \frac{\dfrac{1}{N-k} \sum_{t=1}^{N-k} (x_t - \bar{x})(x_{t+k} - \bar{x})}{\dfrac{1}{N} \sum_{t=1}^{N} (x_t - \bar{x})^2}$$ (7.59)

where N is the total number of observations, and $\bar{x}$ is the mean of x_t.
    A plot of r_k versus k, called the autocorrelogram, is a useful tool for investigating the existence of periodicities in time series. In a stationary series, the autocorrelogram dies down to zero as k increases. In the computation of r_k, it is assumed that the trend component has already been removed.
    For periodic time series, the autocorrelogram will also be periodic. For non-periodic-dependent series, the strength of correlation depends on the dependence of the series. The serial correlation coefficient usually increases with closeness of data. For example, daily data will show a stronger correlation compared with monthly or annual data.

The variance of the serial autocorrelation coefficient is given approximately by (Bartlett, 1946)

$$\text{Var}(r_k) \approx \left(\frac{1}{N}\right)\left(1 + 2\sum_{i=1}^{q} r_i^2\right) \quad \text{for } k > q \tag{7.60a}$$

This is based on the assumption that $r_k = 0$ for $k > q$ and $r_k \neq 0$ for $k \leq q$. For $k > q$, the r_k values tend to a normal distribution with zero mean. For a dependent series, it is complicated even for large samples from a normal population.

For independent time series $r_k = 0$, except for $k = 0$. For $k = 0$, $r_k = 1$. For $k \neq 0$, the r_k values are normally distributed with zero mean. The variance, from Equation 7.60a, is then given by

$$\text{Var}(r_k) \approx \frac{1}{N} \tag{7.60b}$$

The probability limits for the autocorrelogram of an independent series are given by (Anderson, 1941)

$$r_k(95\%) = \frac{\left\{-1 \pm \left(1.96\sqrt{(N - k - 1)}\right)\right\}}{N - k} \tag{7.61a}$$

$$r_k(99\%) = \frac{\left\{-1 \pm \left(2.326\sqrt{(N - k - 1)}\right)\right\}}{N - k} \tag{7.61b}$$

Figures 7.10 and 7.11 show the autocorrelograms for annual and monthly rainfall data, respectively, measured at the Observatory of Hong Kong. For the annual data, the autocorrelation oscillates within the 95% confidence limits indicating the absence of any clear periodicity. On the other hand, the monthly data clearly shows the annual periodicity.

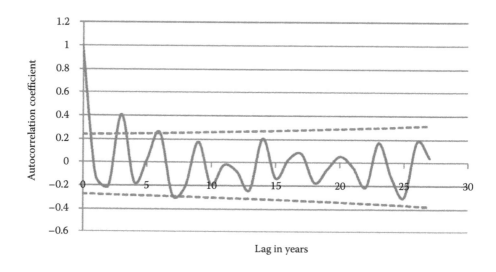

Figure 7.10 Autocorrelation of annual rainfall data in Hong Kong.

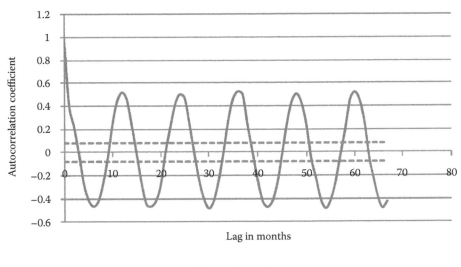

Figure 7.11 Autocorrelation of monthly rainfall data in Hong Kong.

7.8.3 Spectral analysis

An alternative to the time domain analysis of the dependence of a time series by the autocorrelation method is the spectral analysis, which is carried out in the frequency domain. Like the decomposition of a periodic time series into its harmonics of different periods, the objective of spectral analysis is the decomposition on a frequency basis and then the estimation of the frequencies and their relative amplitudes.

By analogy with the spectral analysis in optics, a time series can be considered as a combination of basic frequencies of occurrence of random variables. The white light when passed through a prism is decomposed into fundamental colours violet, indigo, blue, green, yellow, orange, and red, which have specific wavelengths (or frequencies), with the violet having the highest frequency and the red having the lowest frequency. Similarly, the spectrum of the time series decomposes it into components on a frequency basis.

The general definition of spectral density function, which is useful for mathematical calculations, has been shown to be the Fourier transform of the correlation function and the relationship is known as the Wiener–Khintchine equations, named in honour of the two mathematicians N. Wiener of the United States and A.I. Khintchine of the Union of Soviet Socialist Republics in the early 1930s. The equations form a Fourier transform pair and are given as

$$S(f) = \int_{-\infty}^{\infty} \rho(\tau)e^{-j2\pi f\tau} \, d\tau \tag{7.62a}$$

$$\rho(\tau) = \int_{-\infty}^{\infty} S(f)e^{j2\pi f\tau} \, df \tag{7.62b}$$

$$S_{xy}(f) = \int_{-\infty}^{\infty} \rho_{xy}(\tau) e^{-j2\pi f \tau} \, d\tau \qquad (7.62c)$$

$$\rho_{xy}(\tau) = \int_{-\infty}^{\infty} S_{xy}(f) e^{j2\pi f \tau} \, df \qquad (7.62d)$$

In Equation 7.62a, $S(f)$ is the autospectral density function, $S_{xy}(f)$ is the cross-spectral density function, $\rho(\tau)$ is the autocorrelation function, $\rho_{xy}(\tau)$ is the cross-correlation function, f the frequency, and $j^2 = -1$. Equation 7.62a, which is valid in the frequency range $(-\infty, \infty)$, refers to the two-sided power spectral density.

Stationary correlation functions have the following symmetric properties:

$$\rho(-\tau) = \rho(\tau); \quad \rho_{xy}(-\tau) = \rho_{yx}(\tau) \qquad (7.63)$$

Therefore,

$$S(-f) = S(f) \qquad (7.64a)$$

$$S_{xy}(-f) = S_{yx}(f) \qquad (7.64b)$$

indicating that the autospectral density function $S(f)$ is a real-valued even function of f, whereas the cross-spectral density function $S_{xy}(f)$ is a complex valued function of f. Using the cosine transformation,[10] the Fourier transform pair of the two-sided spectral density function $S(f)$ and the autocorrelation function is given as

$$S(f) = \int_{-\infty}^{\infty} \rho(\tau) \cos(2\pi f \tau) \, d\tau = 2 \int_{0}^{\infty} \rho(\tau) \cos(2\pi f \tau) \, d\tau \qquad (7.65a)$$

$$\rho(\tau) = \int_{-\infty}^{\infty} S(f) \cos(2\pi f \tau) \, df = 2 \int_{0}^{\infty} S(f) \cos(2\pi f \tau) \, df \qquad (7.65b)$$

This definition is for physically realizable processes in the frequency range $(0, \infty)$. The one-sided spectral density functions, $G(f)$, defined in the frequency range $(0, \infty)$, are

$$G(f) = 2S(f) \qquad (7.66a)$$

$$G_{xy}(f) = 2S_{xy}(f) \qquad (7.66b)$$

Therefore, in terms of the one-sided spectral density functions, the relationship between the stationary correlation function and the spectral density function can be written as

---

[10] The cosine transformation of the complex term is possible because $\rho(\tau)$ is an even function of τ $[\rho(-\tau) = \rho(\tau)]$. Therefore, $G(f)$ is a real function of f and is also even.

$$G(f) = 4 \int_0^\infty \rho(\tau) \cos(2\pi f \tau) \, d\tau \qquad (7.66c)$$

$$\rho(\tau) = \int_0^\infty G(f) \cos(2\pi f \tau) \, df \qquad (7.66d)$$

$$G_{xy}(f) = 2 \int_{-\infty}^\infty \rho_{xy}(\tau) e^{-j2\pi f \tau} \, d\tau = P_{xy}(f) - jQ_{xy}(f) \qquad (7.66e)$$

$$\rho_{xy}(\tau) = \int_0^\infty \left\{ P_{xy}(f) \cos(2\pi f \tau) + Q_{xy}(f) \sin(2\pi f \tau) \right\} df \qquad (7.66f)$$

where $P_{xy}(f)$ is the real part (called the co-spectrum) and $Q_{xy}(f)$ is the imaginary part of the complex function but real-valued odd function of f (called the quadrature spectrum). The normalized two-sided power spectral density function, $S^*(f)$, is given by

$$S^*(f) = \frac{S(f)}{\sigma_x^2} \qquad (7.67)$$

The power spectral density function may also be calculated via the fast Fourier transform (FFT) method in which a direct Fourier transform of the original data is carried out. This method is more efficient computationally for series involving large amounts of data. However, when the volume of data is not excessive and when the autocorrelation function is anyway determined, the FFT method does not offer a great advantage.

The approximate discrete representation of the power spectral density function given in Equation 7.66c may be evaluated from the following equation[11]

$$G(f) = 2\Delta t \left\{ r_0 + 2 \sum_{k=1}^{M-1} r_k \cos(2\pi f k) + r_M \cos 2\pi f M \right\} \qquad (7.68)$$

where Δt is the time interval between two events, r_k is the sample autocorrelation coefficient, and M is the maximum lag considered in the correlogram. For an independent series, the expected summation in the above equation is zero.

Similar to harmonic analysis in the time domain where the shortest period a harmonic could have is $2\Delta t$, where Δt is the sampling interval (because at least three points are needed to define a curve), the highest frequency in cycles/unit time for which information is sought

---

[11] By the trapezoidal rule of numerical integration,

$$\int r(\tau) \cos(2\pi f \tau) \, d\tau = \sum_{k=0}^{M} r(k) \cos(2\pi f k) \Delta t$$

$$= \Delta t \left\{ 0.5 r_0 + \sum_{k=1}^{M-1} r_k \cos(2\pi f k) + 0.5 r_M \cos(2\pi f M) \right\}.$$

is $\dfrac{1}{2\Delta t}$ in cycles/unit time. This frequency is called the Nyquist frequency (sometimes called the cut-off frequency, f_c). In the computation of the serial autocorrelation coefficient, the maximum lag corresponds to this limiting frequency. The maximum lag is usually chosen to be around $\dfrac{N}{5}$ to $\dfrac{N}{10}$ (subjectively). For example, if $\Delta t = 1$ month, $f_c = 0.5$ cycles/month. Therefore, the frequency range is $(0,0.5)$. This range may be divided into frequency intervals as follows:

$$\left(0, \frac{1}{M}, \frac{2}{M}, \frac{3}{M}, ..., \frac{M}{M} \right) f_c$$

where M is usually taken as the maximum lag (not necessarily). In terms of the frequencies, $G(f)$ can be written as

$$G(f) = 2\Delta t \left\{ r_0 + 2 \sum_{k=1}^{M-1} r_k \cos \frac{\pi k f}{f_c} + r_M \cos \frac{\pi M f}{f_c} \right\} \tag{7.69}$$

In the normalized form (making the area under the power spectral density function equal to unity), $G(f)$ is given by

$$G(f) = \frac{2\Delta t}{2M} \left\{ r_0 + 2 \sum_{k=1}^{M-1} r_k \cos \frac{\pi k f}{f_c} + r_M \cos \frac{\pi M f}{f_c} \right\} \tag{7.70a}$$

$$= \frac{1}{M} \left\{ r_0 + 2 \sum_{k=1}^{M-1} r_k \cos \frac{\pi k f}{f_c} + r_M \cos \frac{\pi M f}{f_c} \right\} \quad \text{if } \Delta t = 1 \tag{7.70b}$$

In this equation, k is an index for the frequency, and varies from 0, 1, 2, ..., M. Frequency, f, is given by

$$f = \frac{k}{2\Delta t M} \tag{7.71}$$

Therefore, $G(f)$ can be expressed as $G(k)$.

If frequency is measured in radians/unit time, M in the above equation should be replaced by π.

Low values of M in the above equation will result in loss of information in the correlogram, leading to excessive smoothing of power spectral density function (fewer points in the power spectral density function). High values of M will lead to a spectrum that is not clear. The sensitivity of the power spectral density function to variations of M may be carried out using different values of M.

The power spectral density function so obtained is called the raw estimate of the power spectral density. It is the convolution of the true power spectral density function with a window function of the box-car type. It has the net effect of causing leakage by spreading the main lobes of the power spectral density function and by adding an infinite number of smaller side lobes. This leads to the possibility of obtaining negative values for the power spectral density function.

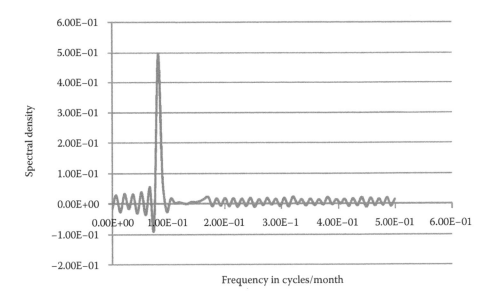

Figure 7.12 Raw spectral density function for monthly rainfall data.

An example of the raw spectral density function for the rainfall data in Example 7.1 is shown in Figure 7.12. Leakage can be reduced by the following procedures:

7.8.3.1 Hanning method (after J. von Hann)

The Hanning method was named after J. von Hann (Blackman and Tukey, 1959). This window function (see Equations 7.74 through 7.76), which is of the form

$$\gamma_k = \frac{1}{2}\left(1 + \cos\frac{\pi k}{M}\right)$$

is equivalent to a weighted moving average of three terms as shown in Equation 7.72 in which GS is the smoothed function.

$$
\begin{aligned}
GS(1) &= 0.5G(1) + 0.5G(2) \quad \text{for } k = 1 \\
GS(k) &= 0.25G(k-1) + 0.5G(k) + 0.25G(k+1) \quad \text{for } k = 2 \text{ to } M-1 \\
GS(M) &= 0.5G(M-1) + 0.5G(M) \quad \text{for } k = M
\end{aligned}
\tag{7.72}
$$

7.8.3.2 Hamming method (after R.W. Hamming, 1983)

$$
\begin{aligned}
GS(1) &= 0.54G(1) + 0.46G(2) \quad \text{for } k = 1 \\
GS(k) &= 0.23G(k-1) + 0.54G(k) + 0.23G(k+1) \quad \text{for } k = 2 \text{ to } M-1 \\
GS(M) &= 0.54G(M-1) + 0.46G(M) \quad \text{for } k = M
\end{aligned}
\tag{7.73}
$$

7.8.3.3 Lag window method (after Tukey, 1965)

As the lag increase, r_k becomes less reliable because there are fewer observations. Therefore, it is multiplied by unequal weights called lag windows γ_k of the form

$$r_k = \frac{1}{2}\left(1 + \cos\frac{\pi k}{M}\right) \tag{7.74}$$

One such form after Tukey (1965) is given by

(a) $D(k) = \dfrac{1}{2}(1 + \cos(2\pi f k))$ (7.75)

$\quad D(k) = 1 - 6\left(\dfrac{k}{M}\right)^2 + 6\left(\dfrac{k}{M}\right)^3 \quad \text{for } k \le \dfrac{M}{2}$

(b) $D(k) = 2 - \left(\dfrac{k}{M}\right)^3 \quad \text{for } \dfrac{M}{2} < k < M$ (7.76)

$\quad D(k) = 0 \quad \text{for } k > M$

An illustration of the smoothed spectral density function by the Hanning method for the rainfall data in Example 7.3 is shown in Figure 7.13. The annual periodicity is shown as a peak in the spectral density function at a frequency of 0.0827 cycles/month $\left(\approx \dfrac{1}{12}\right)$.

The spectrum usually does not give any more information than the autocorrelogram. It, however, indicates where the high and low frequencies are concentrated.

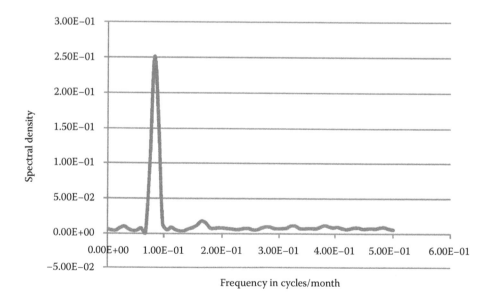

Figure 7.13 Smoothed spectral density function for monthly rainfall data.

Peaks in the power spectral density function (of series where trends are already removed) indicate periodicities. Sharp peaks indicate regular periodicities. A perfect periodicity such as a sinusoidal curve will produce a vertical line in the spectrum. The spectrum in addition to revealing the above characteristics of the series also helps guide the choice of the mathematical model that generates the series. The autocorrelation and spectral density are complementary to each other and should be interpreted together.

Confidence limits can be set for spectral density functions based on a known population.

7.8.4 Cross correlation

Cross correlation is used to determine the dependence between two series x_t and y_t, not only at the same time level but also for one series leading or lagging the other. For x_t and y_t with zero mean, the cross covariance $c_{xy}(\tau)$ is defined as

$$c_{xy}(\tau) = E[x_t y_{t+\tau}] \tag{7.77}$$

$$c_{yx}(\tau) = c_{xy}(-\tau) = E[x_{t+\tau} y_t] \tag{7.78}$$

(Note that $c_{xy}(\tau) \neq c_{xy}(-\tau)$, because the cross covariance is not an even function.)

Cross correlation $\rho_{xy}(\tau)$ is defined as

$$\rho_{xy}(\tau) = E\left\{ \frac{\left[x_t - E(x_t)\right]\left[y_{t+\tau} - E(y_t)\right]}{\sqrt{\mathrm{Var}(x_t)\mathrm{Var}(y_t)}} \right\} \tag{7.79}$$

Sample cross-correlation coefficient $r_{xy}(k)$ is defined as

$$r_{xy}(k) = \frac{\displaystyle\sum_{t=1}^{N-k}(x_t - \bar{x})(y_{t+k} - \bar{y})}{\left\{ \left[\displaystyle\sum_{t=1}^{N}(x_t - \bar{x})^2\right]\left[\displaystyle\sum_{t=1}^{N}(y_t - \bar{y})^2\right]\right\}^{1/2}} \tag{7.80}$$

7.8.5 Cross-spectral density function

The cross-spectral density function and the cross correlation were defined in Equations 7.66e and 7.66f.

An estimate of the sample cross-spectral density functions $s_{xy}(\omega)$ can be obtained as

$$s_{xy}(\omega) = \frac{1}{\pi}\left\{ \sum_{\tau=-M}^{M} \gamma(\tau) r_{xy}(\tau)\cos(\omega\tau)\right\} \tag{7.81}$$

where $\gamma(\tau)$ is a lag window and $\omega = 2\pi f$.

It should be noted that the cross-spectral density function has a real part, which is in phase, and an imaginary part, which is out of phase. The real (in phase) part, known as the normalized sample co-spectrum, is given by

$$P_{xy}(\omega) = \frac{1}{\pi}\left\{ \gamma(0)r_{xy}(0) + \sum_{\tau=1}^{M} \gamma(\tau)\left[r_{xy}(\tau) + r_{xy}(-\tau)\right]\cos(\omega\tau)\right\} \tag{7.82}$$

and the imaginary (quadrature spectrum or out of phase part) is given by

$$Q_{xy}(\omega) = \frac{1}{\pi}\left\{\sum_{\tau=1}^{M}\gamma(\tau)\left[r_{xy}(\tau) - r_{xy}(-\tau)\right]\sin(\omega\tau)\right\} \tag{7.83}$$

where $s_{xy}(\omega) = p_{xy}(\omega) - jQ_{xy}(\omega)$.

Normalized cross-amplitude spectrum, which measures the association between the amplitudes in the two series at the same frequency, is given by

$$\beta_{xy}(\omega) = \left[P_{xy}^2(\omega) + Q_{xy}^2(\omega)\right]^{1/2} \tag{7.84}$$

Phase spectrum is defined as

$$\varphi_{xy}(\omega) = \tan^{-1}\left\{-\frac{Q_{xy}}{P_{xy}}\right\} \tag{7.85}$$

Coherence spectrum, h_{xy}, is defined as

$$h_{xy}(\omega) = \frac{\beta_{xy}^2}{s_x(\omega)s_y(\omega)} \tag{7.86}$$

The coherence that lies between 0 and 1 measures the interdependence between frequency components in the two series at frequency ω. Phase and coherence are not considered in the autocorrelation or spectral analysis. Coherence is analogous to ρ^2.

A plot of functions such as coherence and phase against frequency enables an assessment of the non-linear nature of the time series, which would help in the model formulation. Strong coherence occurs at low frequencies (e.g., annual cycle), whereas at high frequencies, the coherence is small. This means that the periodicities with higher frequencies are spurious.

7.9 STOCHASTIC COMPONENT

After the trend and the periodic component have been removed, the remaining stochastic component, which is assumed to be covariance stationary, may consist of a dependent (correlated) part and an independent (uncorrelated) random part. Four different types of stochastic models can be used to describe the dependent part, namely,

- Autoregressive (AR)
- Moving average (MA)
- Autoregressive moving average (ARMA)
- Autoregressive integrated moving average (ARIMA)

In all the above four types of models, the present value of the stochastic variable is linearly related to the past values in some form. They are therefore identified as linear stochastic models. Stationarity is also implied in the AR-, MA-, and ARMA-type models.

7.9.1 Autoregressive (AR) models

In AR models, the current value of the variable is linearly related to the weighted sum of a number of past values and an independent random value.

The general p-th order AR model has the form

$$z_t = \varphi_{p,1}z_{t-1} + \varphi_{p,2}z_{t-2} + \ldots + \varphi_{p,p}z_{t-p} + \eta_t = \sum_{i=1}^{p} \varphi_{p,i}z_{t-i} + \eta_t \tag{7.87}$$

where the $\varphi_{p,i}$'s are called the autoregressive coefficients, η_t is an independent (uncorrelated) random number (sometimes referred to as 'white noise'), and z_t is the stochastic component, which is obtained by subtracting any trends and periodicities from the original time series. For convenience, z_t is reduced to zero mean and unit variance (normalized). An $AR(p)$ model has $p + 2$ parameters: the mean of the stochastic series, the p autoregressive coefficients, and the variance of the random component.

7.9.1.1 Properties of autoregressive models

The following properties form the basis of model development:

$$\left.\begin{aligned}
E(z_t) &= E(\eta_t) = 0 \\
\mathrm{Var}(z_t) &= E\left(z_t^2\right) = \sigma_z^2 \\
\mathrm{Var}(\eta_t) &= E\left(\eta_t^2\right) = \sigma_\eta^2 \\
\rho_k &= \frac{E(z_t z_{t-k})}{\sigma_z^2} \\
E(\eta_t \eta_{t-k}) &= E(\eta_t z_{t-k}) = 0 \quad \text{for } k = 1,2,3\ldots \\
\sigma_\eta^2 &= \sigma_z^2 \left(1 - \sum_{i=1}^{p} \phi_{p,i}\rho_i\right) \\
\sigma_z^2 &= \frac{\sigma_\eta^2}{1 - \rho_1\phi_{1,1} - \rho_2\phi_{2,2} - \ldots \rho_p\phi_{p,p}}
\end{aligned}\right\} \tag{7.88}$$

In the above equations for σ_η^2, the summation (RHS) is called the coefficient of determination, R^2. It is the square of the multiple correlation coefficient. An unbiased estimate of σ_η^2 may be obtained by multiplying by $\frac{N}{N-p}$. An AR process is completely known if $\varphi_{p,i}$'s and σ_η^2 are known.

7.9.1.2 Estimation of parameters

Multiplying the general autoregressive equation by $z_{t-1}, z_{t-1},...,z_{t-p}$, in turn, and taking expectations,[12] the following p equations can be obtained. They are called the Yule–Walker equations (after Yule, 1927; Walker, 1931).

$$
\begin{bmatrix}
\rho_1 \\
\rho_2 \\
\rho_3 \\
\cdot \\
\cdot \\
\cdot \\
\cdot \\
\rho_p
\end{bmatrix}
=
\begin{bmatrix}
1 & \rho_1 & \rho_2 & \cdot & \cdot & \cdot & \cdot & \rho_{p-1} \\
\rho_1 & 1 & \rho_1 & \cdot & \cdot & \cdot & \cdot & \rho_{p-2} \\
\rho_2 & \rho_1 & 1 & \cdot & \cdot & \cdot & \cdot & \rho_{p-3} \\
\cdot & \cdot & \cdot & \cdot & \cdot & \cdot & \cdot & \cdot \\
\cdot & \cdot & \cdot & \cdot & \cdot & \cdot & \cdot & \cdot \\
\cdot & \cdot & \cdot & \cdot & \cdot & \cdot & \cdot & \cdot \\
\cdot & \cdot & \cdot & \cdot & \cdot & \cdot & \cdot & \cdot \\
\rho_{p-1} & \rho_{p-2} & \rho_{p-3} & \cdot & \cdot & \cdot & \cdot & 1
\end{bmatrix}
\begin{bmatrix}
\varphi_{p,1} \\
\varphi_{p,2} \\
\varphi_{p,3} \\
\cdot \\
\cdot \\
\cdot \\
\cdot \\
\varphi_{p,p}
\end{bmatrix}
\tag{7.89}
$$

The diagonal term in the above matrix corresponds to ρ_0. In a different form, these equations can be written as

$$
\rho_k = \sum_{i=1}^{p} \varphi_{p,i}\rho_{k-i} \quad \text{for } k > 0
\tag{7.90}
$$

A necessary condition for stationarity of the AR model is that the autocorrelation matrix $[\rho]$ is positive definite.[13] This means that the determinant and all its minors are positive; that is,

$$
\rho_0 > 0; \quad
\begin{bmatrix}
\rho_0 & \rho_1 \\
\rho_1 & \rho_0
\end{bmatrix} > 0; \quad
\begin{bmatrix}
\rho_0 & \rho_1 & \rho_2 \\
\rho_1 & \rho_0 & \rho_1 \\
\rho_2 & \rho_1 & \rho_0
\end{bmatrix} > 0, \text{ etc.}
$$

The positive definiteness of the autocorrelation matrix also implies that for AR(1) models, $-1 < \rho_1 < 1$, and for AR(2) models, $-1 < \rho_1 < 1$, $-1 < \rho_2 < 1$, and $-1 < \dfrac{\rho_2 - \rho_1^2}{1 - \rho_1^2} < 1$.

The stationarity condition is also satisfied if the roots of the characteristic equation given below lie within the unit circle given by $u^2 = 1$ (u is a dummy variable).

$$
u^p - \varphi_{p,1}u^{p-1} - \varphi_{p,2}u^{p-2} - ...\varphi_{p,p} = 0
\tag{7.91}
$$

---

[12] Multiplying Equation 7.87 by $z_{t-k}z_{t-k}$ gives $z_t z_{t-k} = \varphi_{p,1} z_{t-1}z_{t-k} + \varphi_{p,2}z_{t-k}z_{t-k} + \cdots + \eta_{p,p}z_{t-p}z_{t-k} + \eta_t z_{t-k}$. Taking expectations term by term (using the definition that $\rho_k = \dfrac{E[z_t z_{t-k}]}{\sigma_z^2}$ and the assumption that $\sigma_z^2 = 1$), we have $\rho_k = \varphi_{p,1}\rho_{k-1} + \varphi_{p,2}\rho_{k-2} + ... + \varphi_{p,p}\rho_{k-p} = \sum_{i=1}^{p} \varphi_{p,i}\rho_{k-i}$ for $k > 0$, and also using $\rho(-k) = \rho(k)$.

[13] Positive semidefinite means ≥ 0.

The AR coefficients can be determined from the estimates of ρ, or if $\varphi_{p,i}$'s are known, then ρ's may be determined.

7.9.1.3 First-order model (lag-one Markov model)

The first-order autoregressive model (AR(1)) is given by

$$z_t = \varphi_{1,1} z_{t-1} + \eta_t \tag{7.92}$$

From the Yule–Walker equations, it can be seen that

$$
\left.
\begin{array}{l}
\rho_1 = \varphi_{1,1} \\
\rho_2 = \varphi_{1,1}\, \rho_1 \\
\rho_3 = \varphi_{1,1}\, \rho_2 \\
\quad \cdot \qquad \cdot \\
\quad \cdot \qquad \cdot \\
\quad \cdot \qquad \cdot \\
\quad \cdot \qquad \cdot \\
\rho_p = \varphi_{1,1}\, \rho_{p-1}
\end{array}
\right\}
\tag{7.93}
$$

The $\rho - \varphi$ relationship can be expressed either as

$$\rho_0 = 1; \quad \rho_k = \varphi_{1,1}^k; \; k \geq 0 \tag{7.94}$$

or as

$$\rho_k = \rho_1^k \tag{7.95}$$

This equation decays exponentially to zero for positive values of $\varphi_{1,1}$. For negative $\varphi_{1,1}$, the autocorrelation function is damped and oscillates around zero.

The variance of the independent component η_t is given by

$$\sigma_\eta^2 = \left(1 - \rho_1^2\right) \; \left(R^2 = \rho_1^2\right) \text{ because } \varphi_{1,2},\ \varphi_{1,3},\ \text{etc., are zero.} \tag{7.96}$$

$$\sigma_z^2 = \frac{\sigma_\eta^2}{1 - \rho_1 \varphi_{1,1}} = \frac{\sigma_\eta^2}{1 - \varphi_{1,1}^2} \tag{7.97}$$

The stationarity condition for[14] AR(1) models can be shown to be $-1 < \rho_1 < 1$.

---

[14] The characteristic equation for an AR(1) model is $u - \varphi_{1,1} = 0$, which gives $u - \varphi_{1,1} = \rho_1$ and hence $|u| < 1$.

7.9.1.3.1 Special properties of an AR(1) model

(a) An AR(1) model can be written as an infinite series as follows:

$$
\begin{aligned}
z_t &= \varphi_{1,1} z_{t-1} + \eta_t \\
&= \varphi_{1,1} (\varphi_{1,1} z_{t-2} + \eta_{t-1}) + \eta_t \\
&= \varphi_{1,1} \left(\varphi_{1,1} (\varphi_{1,1} z_{t-3} + \eta_{t-2}) + \varphi_{1,1} \eta_{t-1} \right) + \eta_t \\
&= \eta_t + \varphi_{1,1} \eta_{t-1} + \varphi_{1,1}^2 \eta_{t-2} + \varphi_{1,1}^3 \eta_{t-3} + \cdots \\
&= \sum_{j=0}^{\infty} \varphi_{1,1}^j \eta_{t-j}
\end{aligned}
\tag{7.98}
$$

This is an MA(∞) process (see also MA(1) model).

(b) An AR(1) model can be used to transform a dependent series into an independent series

$$
z_t = \varphi_{1,1} z_{t-1} + \eta_t
$$

which gives

$$
\eta_t = z_t - \varphi_{1,1} z_{t-1}
\tag{7.99}
$$

which is an independent series.

For example, if z_t versus z_{t-1} from a AR(1) series is plotted, a trend can be seen. On the other hand, no trend or correlation will be seen if z_t versus z_{t-1} is plotted for the process $z_t = \eta_t$ (i.e., $\varphi_{1,1} = 0$), which is white noise.[15]

(c) For z_t to be finite, $|\varphi_{1,1}| < 1$. Otherwise,

$$
z_t = \varphi_{1,1} z_{t-1} + \eta_t
$$

will increase infinitely as t increases, thereby making z_t non-stationary.

(d) Random walk as a limiting case of AR(1)

If $\varphi_{1,1} \to 1$, the AR(1) model becomes

$$
z_t = z_{t-1} + \eta_t
$$

or

$$
\eta_t = z_t - z_{t-1} = \nabla z_t
$$

z_t can also be written as

---

[15] White noise has $\rho_k = \begin{cases} 1 & \text{for } k = 0 \\ 0 & \text{for } k \neq 0 \end{cases}$, $\varphi_{k,k} = \begin{cases} 1 & \text{for } k = 0 \\ 0 & \text{for } k \neq 0 \end{cases}$, and the one-sided spectral density function has a constant value for all frequencies.

$$z_t = z_{t-1} + \eta_t$$
$$= \eta_t + z_{t-1}$$
$$= \eta_t + z_{t-2} + \eta_{t-1}$$
$$= \eta_t + \eta_{t-1} + z_{t-3} + \eta_{t-2}$$
$$\cdots \tag{7.100}$$
$$\cdots$$
$$= \eta_t + \eta_{t-1} + \eta_{t-2} + \cdots$$
$$= \sum_{j=0}^{\infty} \eta_{t-j}$$

which is a random walk process.

In a random walk process, the best forecast at the present time level is the outcome of the process at the previous time level.

7.9.1.4 Second-order model (lag-two model)

The second-order autoregressive model (AR(2)) has the form

$$z_t = \varphi_{2,1} z_{t-1} + \varphi_{2,2} z_{t-2} + \eta_t \tag{7.101}$$

From the Yule–Walker equations,

$$\left\{ \begin{array}{l} \rho_1 = \varphi_{2,1} + \varphi_{2,2}\rho_1 \\ \rho_2 = \varphi_{2,1}\rho_1 + \varphi_{2,2} \\ \rho_3 = \varphi_{2,1}\rho_2 + \varphi_{2,2}\rho_1 \\ \quad \cdot \qquad \quad \cdot \\ \quad \cdot \qquad \quad \cdot \\ \quad \cdot \qquad \quad \cdot \\ \rho_p = \varphi_{2,1}\rho_{p-1} + \varphi_{2,2}\rho_{p-2} \end{array} \right. \tag{7.102}$$

By solving the first two equations simultaneously,

$$\varphi_{2,1} = \frac{\rho_1(1-\rho_2)}{\left(1-\rho_1^2\right)} \tag{7.103a}$$

$$\varphi_{2,2} = \frac{\left(\rho_2-\rho_1^2\right)}{\left(1-\rho_1^2\right)} \tag{7.103b}$$

or alternatively,

$$\rho_0 = 1; \quad \rho_1 = \frac{\varphi_{2,1}}{1-\varphi_{2,2}}; \quad \rho_2 = \varphi_{2,2} + \frac{\varphi_{2,1}^2}{1-\varphi_{2,2}} \tag{7.104}$$

For $k \geq p$, the ρ_k's can be obtained recursively by the Yule–Walker equations:

$$\rho_k = \varphi_{2,1}\rho_{k-1} + \varphi_{2,2}\rho_{k-2} \tag{7.105}$$

The variance of the independent component η_t is given by

$$\sigma_\eta^2 = \frac{1+\varphi_{2,2}}{1-\varphi_{2,2}}\left\{(1-\varphi_{2,2})^2 - \varphi_{2,1}^2\right\} \tag{7.106}$$

$$\sigma_z^2 = \frac{\sigma_\eta^2}{1-\rho_1\varphi_{2,1}-\rho_2\varphi_{2,2}} \tag{7.107}$$

The stationarity condition[16] for AR(2) models can be shown to be

$$-1 < \rho_1 < 1; \, -1 < \rho_2 < 1; \, \text{and} \, \rho_1^2 < \frac{1}{2}(1+\rho_2)$$

or

$$\varphi_{2,1} + \varphi_{2,2} < 1; \quad \varphi_{2,2} - \varphi_{2,1} < 1; \quad -1 < \varphi_{2,2} < 1$$

Depending on the values of $\varphi_{2,1}$ and $\varphi_{2,2}$, the autocorrelation function ρ_k of an AR(2) model may have different forms.

For example, if $\varphi_{2,1}^2 + 4\varphi_{2,2} \geq 0$, ρ_k decays exponentially to zero when $\varphi_{2,1} > 0$, but oscillates around zero if $\varphi_{2,1} < 0$.

If $\varphi_{2,1}^2 + 4\varphi_{2,2} < 0$, the autocorrelation function is pseudo-periodic or damped.

7.9.1.5 Partial autocorrelation function (PAF)

The set of parameters $\varphi_{1,1}, \varphi_{2,2}, \varphi_{3,3}, \dots, \varphi_{k,k}$ of the AR models of order 1, 2, 3, ..., p constitute the partial autocorrelation function. They may be estimated by the method of least squares or by the Yule–Walker equations. The resulting relationship is of the form

$$\rho_j = \varphi_{k,1}\rho_{j-1} + \varphi_{k,2}\rho_{j-2} + \dots \varphi_{k,k}\rho_{j-k}; \, j = 1,2,\dots,k \tag{7.108}$$

The set of parameters $\varphi_{1,1}, \varphi_{2,2}, \varphi_{3,3}, \dots, \varphi_{k,k}$ can then be determined successively for $k = 1,2,\dots$ from Equation 7.108 as

$$\varphi_{1,1} = \rho_1 \tag{7.109a}$$

$$\varphi_{2,2} = \frac{\begin{vmatrix} 1 & \rho_1 \\ \rho_1 & \rho_2 \end{vmatrix}}{\begin{vmatrix} 1 & \rho_1 \\ \rho_1 & 1 \end{vmatrix}} \tag{7.109b}$$

---

[16] The characteristic equation for an AR(2) model is $u2 - \varphi_{2,1} u - \varphi_{2,2} = 0$, the roots of which depend on the values of $\varphi_{2,1}$ and $\varphi_{2,2}$, and should lie outside the unit circle for stationarity.

$$\varphi_{3,3} = \frac{\begin{vmatrix} 1 & \rho_1 & \rho_1 \\ \rho_1 & 1 & \rho_2 \\ \rho_2 & \rho_1 & \rho_3 \end{vmatrix}}{\begin{vmatrix} 1 & \rho_1 & \rho_2 \\ \rho_1 & 1 & \rho_1 \\ \rho_2 & \rho_1 & 1 \end{vmatrix}}$$

(7.109c)

$$\varphi_{k,k} = \frac{\begin{vmatrix} 1 & \rho_1 & \rho_2 & \cdots & \rho_{k-2} & \rho_1 \\ \rho_1 & 1 & \rho_1 & \cdots & \rho_{k-3} & \rho_2 \\ \cdots & \cdots & & \cdots & \cdots & \\ \cdots & \cdots & & \cdots & \cdots & \\ \rho_{k-1} & \rho_{k-2} & \rho_{k-3} & \cdots & \rho_1 & \rho_k \end{vmatrix}}{\begin{vmatrix} 1 & \rho_1 & \rho_2 & \cdots & \rho_{k-1} \\ \rho_1 & 1 & \cdots\rho_1 & \cdots & \rho_{k-2} \\ \cdots & \cdots & \cdots & \cdots & \cdots \\ \cdots & \cdots & \cdots & \cdots & \cdots \\ \rho_{k-1} & \rho_{k-2} & \rho_{k-3}\cdots & \cdots & 1 \end{vmatrix}}$$

(7.109d)

It is a useful tool for determining the order of a model because of the fact that $\varphi_{k,k}$ are theoretically zero for $k > p$ ($\psi_{k,k}$, etc., are the last coefficients in an AR model). It gives the correlation between x_i and x_{k+i}, adjusted for the intervening observations $x_{t+1},...,x_k$.

For AR(2) model, the PAF from Equation 7.108 would be

$$\varphi_{1,1} = \rho_1; \quad \varphi_{2,2} = \frac{\rho_2 - \rho_1^2}{1 - \rho_1^2}; \quad \varphi_{k,k} = 0 \text{ for } k > 2$$

(7.110)

In general, $\varphi_{k,k} \neq 0$ for $k \leq p$ and $\varphi_{k,k} = 0$ for $k > p$.
The variance of the estimated PAF for an AR(p) model is given by

$$\text{Var}(\varphi_{k,k}) \approx \frac{1}{N} \text{ for } k > p$$

(7.111)

where N is the sample length.

The confidence limits can be obtained by assuming that the distribution of the PAF with zero expected value is approximately normal. However, the distribution may be significantly different from normal if $N < 30$ (Anderson, 1941; Kottegoda, 1980, p. 120). Then, the small sampling distribution theory may be used.

7.9.2 Moving average (MA) models

In a moving average model of order q, the current value of the variable is considered as the weighted sum of $q + 1$ independent residuals. It takes the form

$$z_t = \eta_t - \theta_{q,1}\eta_{t-1} - \theta_{q,2}\eta_{t-2} - \ldots \theta_{q,q}\eta_{t-q} = \eta_t - \sum_{i=1}^{q}\theta_{q,i}\eta_{t-i} \tag{7.112}$$

in which $\theta_{q,i}$ are the moving average coefficients. The weight assumed for the current value of the residual is unity. As before, the stochastic component is assumed to be normalized. An MA(q) model has $q + 2$ parameters: the mean of the stochastic series, the q moving average coefficients, and the variance of the random component.

7.9.2.1 Properties of MA models

$$E(z_t) = E(\eta_t) = 0 \tag{7.113a}$$

$$E\left(z_t^2\right) = \sigma_z^2 = 1 \tag{7.113b}$$

$$E(\eta_t\eta_{t-k}) = 0 \text{ for } k \neq 0 \tag{7.113c}$$

By squaring Equation 7.112, taking expectations, and substituting the properties given by Equation 7.113, the variance of the residuals can be shown to be equal to

$$\sigma_\eta^2 = \frac{1}{\left(1 + \theta_{q,1}^2 + \theta_{q,2}^2 + \ldots + \theta_{q,q}^2\right)} \tag{7.114}$$

A finite MA process is always stationary.

7.9.2.2 Parameters of MA models

Multiplying the left hand side (LHS) of Equation 7.112 by z_{t-k} and the right hand side (RHS) by $\eta_{t-k} - \theta_{q,1}\eta_{t-1-k} - \theta_{q,2}\eta_{t-2-k} - \ldots\theta_{q,q}\eta_{t-q-k}$ (this is the same as z_{t-k}), taking expectations, and substituting the properties given by Equation 7.113, gives

$$\rho_k = \sigma_\eta^2\left(-\theta_{q,k} + \theta_{q,1}\theta_{q,k+1} + \theta_{q,2}\theta_{q,k+2} + \cdots + \theta_{q,q-k}\theta_{q,q}\right) \text{ for } k = 1, 2, \ldots, q \tag{7.115}$$

Substituting for σ_η^2 from Equation 7.114, gives

$$\rho_k = \frac{\left[-\theta_{q,k} + \sum_{i=1}^{q-k}\theta_{q,i}\theta_{q,i+k}\right]}{\left[1 + \sum_{i=1}^{q}\theta_{q,i}^2\right]} \text{ for } k = 1, 2, \ldots, q \tag{7.116}$$

and

$$\rho_k = 0 \text{ for } k > q.$$

Equation 7.116 is non-linear and needs to be solved numerically, except for MA(1).

7.9.2.3 MA(1) model

An MA(1) model is given by

$$z_t = \eta_t - \theta_{1,1}\eta_{t-1} \tag{7.117}$$

From Equation 7.116,

$$\rho_1 = -\frac{\theta_{1,1}}{\left(1+\theta_{1,1}^2\right)}; \; \rho_k = 0 \text{ for } k \geq 2 \tag{7.118}$$

Hence, $\theta_{1,1}$ can be estimated.

It can be seen that one root of this equation is the reciprocal of the other. The root that satisfies the invertibility condition $|\theta_{1,1}| > 1$ is the initial estimate, which is not necessarily the best. A better estimate, which will make the generated data look more like the historical data, can be obtained by minimizing the sum of squares of the residuals ε_i where

$$\varepsilon_1 = z_1^{obs}$$
$$\varepsilon_2 = z_2^{obs} + \theta_{1,1}\varepsilon_1 \tag{7.119}$$
$$\varepsilon_3 = z_3^{obs} + \theta_{1,1}\varepsilon_2, \text{ etc.}$$

(z_t^{obs} for $t = 1, 2, \ldots N$, is an observed series). They can also be determined by the more difficult maximum likelihood estimate method.

7.9.2.3.1 Invertibility

By writing the equation for MA(1) recursively, it can be seen that

$$\begin{aligned}
\eta_t &= z_t + \theta_{1,1}\eta_{t-1} \\
&= z_t + \theta_{1,1}(z_{t-1} + \theta_{1,1}\eta_{t-2}) \\
&= z_t + \theta_{1,1}z_{t-1} + \theta_{1,1}^2(z_{t-2} + \theta_{1,1}\eta_{t-3}) \\
&\quad\ldots \\
&\quad\ldots \\
&= z_t + \theta_{1,1}z_{t-1} + \theta_{1,1}^2 z_{t-2} + \ldots
\end{aligned} \tag{7.120}$$

which converges for $|\theta_{1,1}| < 1$, or

$$z_t = -\theta_{1,1}z_{t-1} - \theta_{1,1}^2 z_{t-2} + \ldots + \eta_t \tag{7.121}$$

This can be considered as an AR(∞) process provided that $|\theta_{1,1}| < 1$. Thus, the MA(1) process is an inverted AR(∞) process.

Similarly, an AR(1) process can be represented as an MA(∞) process, as shown in Equation 7.98.

Because of the invertibility condition between MA and AR models, the autocorrelation and the PAF of an MA process behave like the PAF and the autocorrelation of an AR process and vice versa. The autocorrelation function of an MA(1) has a cut-off at lag 1, whereas that of an AR(1) has an exponential decay. The PAF of an MA(1) has an exponential decay, whereas that of an AR(1) has a cut-off after lag 1. A finite-order stationary AR(p) process corresponds to an MA(∞) process and a finite-order invertible MA(q) process corresponds to an AR(∞) process. This property is known as duality. For an MA(1) model, the PAF is given by

$$\phi_{k,k} = -\frac{\theta_{1,1}^k \left(1 - \theta_{1,1}^2\right)}{\left(1 - \theta_{1,1}^{2(k+1)}\right)} \qquad (7.122)$$

and alternates in sign if $\rho_1 > 0$ (i.e., $\phi_{1,1} < 0$), and negative if $\rho_1 < 0$ (i.e., $\phi_{1,1} > 0$).

7.9.2.4 MA(2) model

An MA(2) model is given by

$$z_t = \eta_t - \theta_{2,1}\eta_{t-1} - \theta_{2,2}\eta_{t-2} \qquad (7.123)$$

For stationarity, the parameters should satisfy

$$\theta_{2,2} + \theta_{2,1} < 1$$
$$\theta_{2,2} - \theta_{2,1} < 1$$
$$-1 < \theta_{2,2} < 1$$

which is similar to the stationarity condition of the AR(2) model. The autocorrelation function from Equation 7.116 is

$$\rho_1 = \frac{-\theta_{2,1}(1 - \theta_{2,2})}{1 + \theta_{2,1}^2 + \theta_{2,2}^2} \qquad (7.124a)$$

$$\rho_2 = \frac{-\theta_{2,2}}{1 + \theta_{2,1}^2 + \theta_{2,2}^2} \qquad (7.124b)$$

$$\rho_k = 0 \text{ for } k > 2 \qquad (7.124c)$$

The PAF for MA(2) decays exponentially or is damped with oscillations depending on the sign and magnitude of the moving average parameters. It behaves like the autocorrelation function of an AR(2) process. Duality exists between the autocorrelation function and the PAF of AR(2) and MA(2) processes.

The PAF from Equation 7.109 is

$$\varphi_{1,1} = \rho; \quad \varphi_{2,2} = \frac{\rho_2 - \rho_1^2}{1 - \rho_1^2}; \quad \varphi_{3,3} = \frac{\rho_1^3 - \rho_1\rho_2(2 - \rho_2)}{1 - \rho_2^2 - 2\rho_1^2(1 - \rho_2)}; \qquad (7.125)$$

7.9.3 Autoregressive moving average (ARMA) models

ARMA models constitute a combination of AR and MA models. The ARMA(p,q) model is a combination of an AR(p) model and an MA(q) model. It is of the form

$$z_t = \sum_{i=1}^{p} \varphi_{p,i} z_{t-1} + \eta_t - \sum_{i=1}^{q} \theta_{q,i} \eta_{t-i} \tag{7.126}$$

7.9.3.1 Properties of ARMA(p,q) models

The properties of ARMA(p,q) are given in terms of the cross covariance between z and η:

$$c_k^{(z\eta)} = \mathrm{Cov}[z_{t-k}\eta_t] \quad \begin{array}{l} \neq 0 \quad \text{for } k \leq 0 \\ = 0 \quad \text{for } k > 0 \end{array} \tag{7.127}$$

(because z depends only on the present and past values of η).

An ARMA(p,q) model has $p + q + 2$ parameters: the mean of the stochastic series, p autoregressive coefficients, q moving average coefficients, and the variance of the random component.

Multiplying Equation 7.126 by z_{t-k} and taking expectations of products of the form $z_t z_{t-k}$, (assuming $\theta_{q,0} = -1$) as before,

$$c_k = \sum_{i=1}^{p} \varphi_{p,i} c_{k-i} - \sum_{i=0}^{q} \theta_{q,i} c_{k-i}^{(z\eta)} \text{ for } k < q+1$$

$$= \sum_{i=1}^{p} \varphi_{p,i} c_{k-i} \text{ for } k \geq q+1 \tag{7.128}$$

$$\mathrm{Var}(z_t) = \sigma_\eta^2 + \sum_{i=1}^{p} \varphi_{p,i} c_i - \sum_{i=1}^{q} \theta_i c_{-i}^{(z\eta)} \tag{7.129}$$

$$\rho(k) = \sum_{i=1}^{p} \varphi_{p,i} \rho_{k-i} \text{ for } k \geq q+1 \tag{7.130}$$

For stationarity of the AR part, the roots of the characteristic equation $\phi(B) = 0$ should lie outside the unit circle, and for invertibility of the MA part, the roots of the characteristic equation $\theta(B) = 0$ should lie outside the unit circle.

7.9.3.2 ARMA(1,1) model

An ARMA(1,1) model is given by

$$z_t = \varphi_{1,1} z_{t-1} + \eta_t - \theta_{1,1} \eta_{t-1} \tag{7.131}$$

which, by repeated substitution, may be written as

$$z_t = \eta_t + b_{1,1} \eta_{t-1} + b_{1,2} \eta_{t-2} + \dots \tag{7.132}$$

where

$$b_{1,i} = (\varphi_{1,1} - \theta_{1,1})\varphi_{i,1}^{i-1}$$

This represents an MA process of infinite order. Similarly, it may also be written as an AR process of infinite order as follows:

$$z_t = d_{1,1} + z_{t-1} + d_{1,2}z_{t-2} + \ldots\ldots\ldots\ldots + \eta_t \tag{7.133}$$

where

$$d_{1,i} = (\varphi_{1,i} - \theta_{1,1})\theta_{1,1}^{i-1}$$

The convergence criteria for the two cases are $|\phi_{1,1}| < 1$ and $|\theta_{1,1}| < 1$, respectively. The autocorrelation function is given as

$$\rho_k = 1 \text{ for } k = 0; \; \rho_k = \frac{(\varphi_{1,1} - \theta_{1,1})(1 - \varphi_{1,1}\theta_{1,1})}{1 + \theta_{1,1}^2 - 2\varphi_{1,1}\theta_{1,1}} \text{ for } k = 1; \; \rho_k = \varphi_{1,1}\rho_{k-1} \text{ for } k \geq 2 \tag{7.134}$$

The PAF for ARMA(1,1) is complicated and its shape depends on the magnitudes of the AR and MA coefficients, and therefore cannot be generalized. It behaves like the PAF of an MA(1) process and decays off exponentially like the autocorrelation function. It is dominated by smooth damped exponential decay when $\theta_{1,1}$ is positive, and dominated by oscillating exponential decay when $\theta_{1,1}$ is negative. The stationarity and invertibility conditions are

$$|\rho_2| < |\rho_1|$$

$$\rho_2 > \rho_1(2\rho_1 + 1) \text{ for } \rho_2 < 0$$

$$\rho_2 > \rho_1(2\rho_1 - 1) \text{ for } \rho_1 > 0$$

7.9.4 Backshift operator

ARMA-type models are usually represented using the backshift operator B. The notation is as follows:

$$\begin{aligned}
Bz_t &= z_{t-1} \\
B\eta_t &= \eta_{t-1} \\
B^2 z_t &= z_{t-2} \\
B^2 \eta_t &= \eta_{t-2} \\
&\ldots \\
&\ldots \\
B^p z_t &= z_{t-p} \\
B^q \eta_t &= \eta_{t-q}
\end{aligned} \tag{7.135}$$

Using this notation, Equation (7.126) may be written as

$$\varphi_p(B)z_t = \theta_q(B)\eta_t \tag{7.136}$$

where

$$\varphi_p(B) = 1 - \varphi_{p,1}\,B - \varphi_{p,2}\,B^2 - \ldots - \varphi_{p,p}\,B^p \tag{7.137}$$

and

$$\theta_q(B) = 1 - \theta_{q,1}\,B - \theta_{q,2}\,B^2 - \ldots - \theta_{q,q}\,B^q \tag{7.138}$$

7.9.5 Difference operator

The difference operator ∇ has the following meaning:

$$\begin{aligned}
\nabla z_t &= (1-B)z_t = z_t - z_{t-1} \\
\nabla^2 z_t &= (1-B)^2 z_t = (1-2B+B^2)z_t = z_t - 2z_{t-1} + z_{t-2} \\
&\ldots \\
&\ldots \\
\nabla^d z &= (1-B)^d z_t
\end{aligned} \tag{7.139}$$

7.9.6 Autoregressive integrated moving average (ARIMA) models

ARIMA models are used for non-stationary time series where the periodicities are represented by several parameters. Annual series that are stationary can usually be represented by ARMA models. Representation of non-stationary series such as monthly, weekly, and daily by ARMA-type models requires a large number of parameters. For example, ARMA(1,1) for monthly series requires 27 parameters (12 monthly means, 12 monthly variances, $\varphi_{1,1}$, $\theta_{1,1}$, and σ_η^2). Non-periodic ARIMA(p,d,q) is an alternate way of transforming a time series into a stationary series requiring fewer parameters. The transformation is done by differencing. For example, the d-th-order differenced ARMA model is given by

$$\varphi_p(B)\nabla^d z_t = \theta_q(B)\eta_t \tag{7.140a}$$

which can also be written as

$$\varphi_p(B)(1-B)^d z_t = \theta_q(B)\eta_t \tag{7.140b}$$

where $\varphi_p(B)$, $\theta_q(B)$ are polynomials of degree p, q with all roots of the polynomial equations

$$\varphi_p(B) = 0 \tag{7.141a}$$

$$\theta_q(B) = 0 \tag{7.141b}$$

outside the unit circle. This is necessary for stationarity and invertibility. For example, the random walk model $(1 - B)z_t = \eta_t$ is an ARIMA(0,1,0) model. An ARIMA(p,d,q) model has $p + d + q + 2$ parameters: the mean of the stochastic series, p autoregressive coefficients, orders of differencing d, q moving average coefficients, and the variance of the random component.

Identification of ARIMA models include

- Checking for stationarity
- Taking differences if not stationary, once, twice, etc.
- Checking to see if the autocorrelogram decays to zero
- Finding a cut-off value of lag q, for which $\rho_k = 0$ for $k > q$
- Suggesting an MA(q) model
- Determining the PAF, and finding a cut-off value p for which $\varphi_{p,p} = 0$ for $k > p$
- Suggesting an AR(p) model

7.10 RESIDUAL SERIES

Residual series refer to the series represented by η_t in equations that define the stochastic component. It can be obtained recursively by back substitution once the parameters of the model are determined. For example, if an AR(p) model is considered, η_t is given by Equation 7.87. This, however, requires the past values of z_t, which are not known. For lower-order AR models, neglecting the first p terms of the residual series would lose very little information. Therefore, η_t is determined from $t = p + 1$ (for AR(1) models, $\eta_2, \eta_3, \eta_4, \ldots$; for AR(2) models, $\eta_3, \eta_4, \eta_5, \ldots$; for MA models, assume $\eta_t = 0$, for $t < 0$, then $z_1 = \eta_1 \Rightarrow \eta_1 = z_1$; $z_2 = \eta_2 - \theta_{q,1}\eta_1 \Rightarrow \eta_2 = z_2 + \theta_{q,1}\eta_1$, etc.).

7.10.1 Test of independence

The dependence of the residual series can be tested either by the residual correlogram or by the Porte Manteau lack of fit test (Box and Pierce, 1970). The statistic Q for the Porte Manteau lack of fit test is defined as

$$Q = (N - d)\sum_{k=1}^{M} r_k^2(\eta) \tag{7.142}$$

where d is the number of differences in a general ARIMA(p,d,q) model, and M is the maximum lag, which is taken to be about $\dfrac{N}{5}$. In the case of AR models, $d = 0$. If η_t is independent, then Q, estimated from Equation 7.142, is approximately χ^2 distributed with $(M - p - q)$ degrees of freedom (Ljung and Box, 1979). Therefore, if $Q < \chi^2 (M - p - q)$ at a given level of significance, then the residual series η_t can be considered as independent implying that the assumed model is adequate.

7.10.2 Test of normality

If the residual series is normally distributed, the subsequent analysis and synthesis becomes relatively simple. Normality of the residual series can be tested by a number of methods. For instance, a plot of the empirical distribution on normal probability paper should give a straight line if the residual series is normally distributed. Alternatively, the empirical cumulative frequency distribution of the series could be compared with the theoretical cumulative frequency distribution for a normal series. A third method uses a skewness test based on the criterion that the skewness coefficient for a normal distribution is zero. Snedecor and Cochran (1967) suggested a method using the skewness coefficient based on the condition that if the series is normal, the skewness coefficient is asymptotically normal with zero mean and variance $\frac{6}{N}$. For large values of N (>150), the hypothesis of normality can be accepted at the 5% level of significance if the standard normal variate, z, defined as follows, lies within ± 1.96:

$$z = \frac{\gamma - E(\gamma)}{\sigma_\gamma} \tag{7.143}$$

In Equation 7.143, γ is the skewness coefficient of the residual series, $E(\gamma)$ is the expected value of the skewness coefficient if the series is normal ($= 0$), and σ_γ is $\sqrt{\frac{6}{N}}$.

The above test is sufficiently accurate for $N > 150$. For smaller values of N, the limiting values of the skewness coefficients for different probability levels are given in Table 7.11 (Snedecor and Cochran, 1967, p. 552; or Salas et al., 1980, p. 93). If the computed skewness coefficients fall within these limits, the hypothesis of normality can be accepted.

If the residual series is not normal, the original series could be normalized by a transformation (e.g., a logarithmic transformation). Alternatively, a non-normal distribution may be fitted to the residual series.

7.10.3 Other distributions

For other distributions such as Gamma and log-normal, a test of goodness of fit can be carried out using the χ^2. The test statistic $\hat{\chi}^2$ is defined as

$$\hat{\chi}^2 = \sum_{i=1}^{k} \frac{(O_i - E_i)^2}{E_i} \tag{7.144}$$

Table 7.11 Critical values for the skewness for normality test

N	Significance level, α	
	0.02	0.01
25	1.061	0.711
30	0.986	0.662
35	0.923	0.621
40	0.870	0.587
45	0.825	0.558
50	0.787	0.534
60	0.723	0.492
70	0.673	0.459
80	0.631	0.432
90	0.596	0.409
100	0.567	0.389
125	0.508	0.350
150	0.464	0.321
175	0.430	0.298

where k is the number of classes in the sample, O_i is the observed frequency of the class i, and E_i is the expected[17] frequency under the null hypothesis. The χ^2 goodness of fit test is carried out by comparing the statistic $\hat{\chi}^2$ with the critical values of the χ^2-distribution (with $k - 1 - a$ degrees of freedom) under the null hypothesis at a chosen significance level. For two-parameter distributions, $a = 2$, and therefore, the degree of freedom is $k - 3$. Before comparison, the variates of the theoretical distributions under the null hypothesis and those of the residual series must be standardized.

7.10.4 Test for parsimony

Finding a model with the minimum number of parameters, which will adequately reproduce the statistics that will have the smallest variances (or those statistics, which need to be preserved), is called the principle of parsimony. This has its origin in the principle referred to as the Occam's (or, Ockham) razor, which states that 'entities must not be multiplied beyond necessity', and which is attributed to the 14th century English logician, theologian, and Franciscan friar William of Ockham.

7.10.4.1 Akaike information criterion (AIC) and Bayesian information criterion (BIC)

The parsimony of the assumed model is established by using the Akaike information criterion (AIC), which for ARMA(p,q) models is defined as (Akaike, 1970, 1973, 1974)

$$\text{AIC}(p,q) = N \ln\left(\sigma_\eta^2\right) + 2(p+q) \tag{7.145}$$

where σ_η^2 is the maximum likelihood estimate of the residual variance. The model that gives the minimum AIC value is the one to be chosen.

A more recent criterion is the Bayesian information criterion (BIC) (Akaike, 1978, 1979), which takes the form

$$\text{BIC}(p,q) = N \ln\left(\sigma_\eta^2\right) - (N-p-q)\ln\left\{1 - \frac{(p+q)}{N}\right\} + (p+q)\ln(N) + (p+q)\ln\left[\frac{\left(\frac{\sigma_z^2}{\sigma_\eta^2} - 1\right)}{(p+q)}\right] \tag{7.146}$$

7.10.4.2 Schwartz Bayesian criterion (SBC)

The Schwartz Bayesian criterion (SBC) (Schwartz, 1978) is defined as

$$\text{SBC}(p,q) = N \ln\left(\sigma_\eta^2\right) + (p+q)\ln(N) \tag{7.147}$$

There are other criteria that can be used for model selection (e.g., Parzen's criterion for autoregressive transfer functions, or CAT [Parzen, 1977], among several others). In all these criteria, $p + q$ should be replaced by $p + q + d$ if an ARIMA(p,d,q) model is considered.

---

[17] The expected frequency depends on the assumed distribution.

Fitting ARMA models can be accomplished using several standard software such as IMSL, SPSS, SAS, and MINITAB. Each of such software may have different methods of parameter estimation and therefore the results may differ appreciably from one another.

7.11 FORECASTING

Forecasting is the process of predicting the outcome of a future event before its actual occurrence. Weather forecasting is a typical example in everyday life. Flood forecasting is another example in which the magnitude and time of occurrence of impending floods are used as the bases of early warning systems for disaster mitigation. The methods and tools used for forecasting can also be used for hindcasting, which is a process of reproducing events that have already occurred. The time of occurrence of the event measured from the time the forecast is made is known as the lead time. It is a fact that the reliability of any forecast decreases with increasing lead time; however, its usefulness from the point of view of taking necessary mitigative measures to avert a disaster, such as, for example, in the case of a flood, increases with increasing lead time. Very often, a compromise is needed in satisfying these two conflicting requirements. Forecasting can be made for any length of lead time from a fixed time origin or, one step at a time from a moving time origin. In the former case, the information that goes into the forecasting system is static. The system only uses the information available up to the time of the fixed origin. In the latter case, new information as and when they become available can be made use of when the forecasts are updated at the new origin. Real-time forecasting systems operate in the latter mode.

Forecasts are made using mathematical models, which can be empirical, conceptual, stochastic, deterministic, or their combinations. In all these types, some assumptions are necessary, and the forecasts should be interpreted giving due consideration to such assumptions as well as to the uncertainties associated with model formulation, implementation, and accuracy of data. In the context of time series analysis, stochastic models are widely used after removing the deterministic components. Of the different approaches, the minimum mean square error type difference equation is by far the simplest.

7.11.1 Minimum mean square error type difference equation

The minimum mean square error forecast (Box and Jenkins, 1976) type difference equation takes the form

$$
\hat{z}_{t+l} = \varphi_1[z_{t+l-1}] + \dots + \varphi_p[z_{t+l-p}] - \theta_1[\eta_{t+l-1}] - \dots
$$
$$
- \theta_q[\eta_{t+l-q}] + [\eta_{t+l}]
\tag{7.148}
$$

where $\hat{z}_{t+l}$ is the forecast at origin t for lead time l, z_{t+l-i} is the observed sequence, η_{t+l} is the residual sequence, and φ_i, θ_j are the autoregressive and moving average coefficients. Square brackets, [ ], indicate the conditional expectations (conditional on the past values), which can be calculated as follows:

$$
[z_{t-j}] = E_t[z_{t-j}] = z_{t-j} \quad j = 0,1,2,\dots
\tag{7.149a}
$$

$$
[z_{t+j}] = E_t[z_{t+j}] = \hat{z}_t(j) \quad j = 1,2,\dots
\tag{7.149b}
$$

$$[\eta_{t-j}] = \underset{t}{E}[\eta_{t-j}] = \eta_{t-j} = z_{t-j} - \hat{z}_{t-j-1}(1) \quad j = 0,1,2,\ldots \tag{7.149c}$$

$$[\eta_{t+j}] = \underset{t}{E}[\eta_{t+j}] = 0 \quad j = 1,2,\ldots \tag{7.149d}$$

where $E [ ]$ denotes the expectation operator.

Note:

z_{t-j} $j = 0,1,2,\ldots$ that have already happened are left unchanged.
z_{t+j} $j = 1,2,\ldots$ that have not yet happened are replaced by their forecasts $\hat{z}_t(j)$ at origin t.
η_{t-j} $j = 0,1,2,\ldots$ that have happened are available from $z_{t-j} - \hat{z}_{t-j-1}(1)$.
η_{t+j} $j = 0,1,2,\ldots$ that have not yet happened are replaced by zeros.

Forecast values are updated when new values are available using the following relationship (as a linear function of present and past shocks):

$$\hat{z}_{t+1}(l) = \psi_l \eta_{t+1} + \psi_{l+1} \eta_t + \psi_{l+2} \eta_{t-1} + \ldots \tag{7.150}$$

$$\hat{z}_t(l+1) = \psi_{l+1} \eta_t + \psi_{l+2} \eta_{t-1} + \ldots \tag{7.151}$$

Subtracting, Equation 7.151 from 7.150,

$$\hat{z}_{t+1}(l) = \hat{z}_t(l+1) + \psi_l \eta_{t+1} \tag{7.152}$$

where

$\hat{z}_{t+1}(l)$ = the forecast for lead time l made at time $t + 1$

$\hat{z}_t(l+1)$ = the forecast for lead time $l + 1$ made at time t

$\eta_{t+1} = z_{t+1} - \hat{z}_t(1)$ = the one-step-ahead forecast error, and the ψ_l weights used for updating the forecasts are given by the recursive equation

$$\psi_l = \varphi_1 \psi_{l-1} + \ldots + \varphi_p \psi_{l-p} - \theta_l \tag{7.153}$$

in which $\psi_0 = 1$, $\psi_l = 0$ for $l < 0$ and $\theta_l = 0$ for $l > q$.

That is, that the forecasts for lead time l at origin $t + 1$ is an update of forecast at lead time $l + 1$ at origin t.

$$\psi_1 = \varphi_1 - \theta_1 \tag{7.154a}$$

$$\psi_2 = \varphi_1 \psi_1 + \varphi_2 \psi_0 - \theta_2 \tag{7.154b}$$

$$\psi_3 = \varphi_1 \psi_2 + \varphi_2 \psi_1 + \varphi_3 \psi_0 - \theta_3, \text{ etc.} \tag{7.154c}$$

[The forecast at time t with lead time l, that is, $\hat{z}_t(l)$, is obtained by writing down the equation for z_{t+l}.]

7.11.2 Confidence limits

The 95% confidence limits for lead time forecasts are given by

$$
z_{t+l}(\pm) = \hat{z}_t(l) \pm z_{\alpha/2} \left\{ 1 + \sum_{j=1}^{l-1} \psi_j^2 \right\}^{1/2} \sigma_\eta
\tag{7.155}
$$

This means that given the information up to time t, the actual observation will lie within this confidence band with $(1 - \alpha)$ probability. The quantity $z_{\alpha/2}$ in Equation 7.155 is the z value of a standard normal distribution and should not be confused with the stochastic variable z_t.

7.11.3 Forecast errors

Forecast errors can be estimated by

$$
\text{MSE} = \frac{1}{N} \sum_{l=1}^{N} \left[z_{t+l} - \hat{z}_t(l) \right]^2
\tag{7.156a}
$$

or by

$$
\text{MAPE} = \frac{1}{N} \sum_{l=1}^{N} \left| \frac{z_{t+l} - \hat{z}_t(l)}{z_{t+l}} \right| \times 100
\tag{7.156b}
$$

or by

$$
\text{MAE} = \frac{1}{N} \sum_{l=1}^{N} \left| z_{t+l} - \hat{z}_t(l) \right|
\tag{7.156c}
$$

where
    MSE = mean square error
    MAPE = mean absolute percentage error
    MAE = mean absolute error

7.11.4 Numerical examples of forecasting

Example 7.4: AR(2) model

Consider the AR(2) model defined by $(1 - 1.5B + 0.5B^2)z_{t+1} + \eta_{t=1}$, which can be written as

$$z_{t+l} = 1.5z_{t+l-1} - 0.5z_{t+l-2} + \eta_{t+1}$$

Forecasts at origin t are given by

$\hat{z}_t(1) = 1.5z_t - 0.5z_{t-1} + 0$
$\hat{z}_t(2) = 1.5z_{t+1} - 0.5z_t = 1.5\hat{z}_t(1) - 0.5z_t$
$\hat{z}_t(3) = 1.5z_{t+2} - 0.5z_{t+1} = 1.5\hat{z}_t(2) - 0.5\hat{z}_t(1)$
$\hat{z}_t(4) = 1.5z_{t+3} - 0.5z_{t+2} = 1.5\hat{z}_t(3) - 0.5\hat{z}_t(2)$
...

...

$\hat{z}_t(l) = 1.5z_{t+l-1} - 0.5z_{t+l-2} = 1.5\hat{z}_t(l-1) - 0.5\hat{z}_t(l-2)$

Assume the z_t series to be 23.7, 23.4, 23.1, 22.9, and 22.8 for $t = 1, 2, 3, 4,$ and 5. Taking $t = 2$ as the origin (present time level),

$\hat{z}_2(1) = 1.5z_2 - 0.5z_1 = 1.5 \times 23.4 - 0.5 \times 23.7 = 23.25$
$\hat{z}_2(2) = 1.5\hat{z}_2(1) - 0.5z_2 = 1.5 \times 23.25 - 0.5 \times 23.4 = 23.175$
$\hat{z}_2(3) = 1.5\hat{z}_2(2) - 0.5\hat{z}_2(1) = 1.5 \times 23.175 - 0.5 \times 23.25 = 23.178$
$\hat{z}_2(4) = 1.5\hat{z}_2(3) - 0.5\hat{z}_2(2) = 1.5 \times 23.178 - 0.5 \times 23.175 = 23.119$

After z_3 is observed,

$e_2(1) = \eta_3 = z_3 - \hat{z}_2(1) = 23.1 - 23.25 = -0.15$

Now the origin can be shifted to $t = 3$; then, the updated forecasts are (Equation 7.152)

$\hat{z}_{2+1}(1) = \hat{z}_2(2) + \psi_1\eta_3$
$\hat{z}_{2+1}(2) = \hat{z}_2(3) + \psi_2\eta_3$
$\hat{z}_{2+1}(3) = \hat{z}_2(4) + \psi_3\eta_3$
...
...

The ψ values for updating are given by

$\psi_0 = 1$
$\psi_1 = 1.5 - 0 = 1.5$
$\psi_2 = 1.5 \times 1.5 - 0.5 \times 1 - 0 = 1.75$
$\psi_3 = 1.5 \times 1.75 - 0.5 \times 1.5 - 0 - 0 = 1.875$
...
...

Updated forecasts at $t = 3$ are given by

$\hat{z}_{2+1}(1) = 23.175 + 1.50(-0.15) = 22.95$
$\hat{z}_{2+1}(2) = 23.178 + 1.75(-0.15) = 22.92$
$\hat{z}_{2+1}(3) = 23.119 + 1.875(-0.15) = 22.84$
...
...

Therefore, a new series of lead time forecasts are obtained.

If, on the other hand, one-step-ahead forecasts are required, the procedure is as follows: Again from Equation 7.152

$$\hat{z}_{t+1}(1) = \hat{z}_t(2) + \psi_1\eta_{t+1} \text{ and } \eta_{t+1} = z_{t+1} - \hat{z}_t(1)$$
$$\hat{z}_{t+2}(1) = \hat{z}_{t+1}(2) + \psi_1\eta_{t+2} \text{ and } \eta_{t+2} = z_{t+2} - \hat{z}_t(2)$$
...
...

where z_t are the observed values and $\hat{z}_t(1)$ are the forecast values. In this example,

$$\eta_{2+1} = z_{2+1} - \hat{z}_2(1) = 23.1 - 23.25 = -0.15$$
$$\hat{z}_{2+1}(1) = \hat{z}_2(2) + \psi_1\eta_{2+1} = 23.175 + 1.5(-0.15) = 22.95$$
$$\eta_{2+2} = z_{2+2} - \hat{z}_2(2) = 22.9 - 23.175 = -0.275$$
$$\hat{z}_{2+2}(1) = \hat{z}_3(2) + \psi_1\eta_{2+2} = 22.92 + 1.5(-0.275) = 22.51$$

Therefore, one-step-ahead forecasts can be obtained as and when new data become available. The confidence limits for lead time forecasts at time $t = 2$ are

For $\hat{z}_2(2)$: $\hat{z}_2(2) \pm 1.96\{1 + 1.5^2\}^{1/2}\sigma_\eta = 23.175 \pm 1.803\sigma_\eta$

For $\hat{z}_2(3)$: $\hat{z}_2(3) \pm 1.96\{1 + 1.5^2 + 1.75^2\}^{1/2}\sigma_\eta = 23.178 \pm 2.512\sigma_\eta$

...
...

Example 7.5: ARMA(2,2) model

Consider the ARMA(2,2) model defined by $\nabla^2 z_{t+1} = (1 - 0.5B + 0.5B^2)\eta_{t+1}$, which can be written as

$$z_{t+l} - 2z_{t+l-1} + z_{t+l-2} = (1 - 0.5B + 0.5B^2)\eta_{t+1}$$

$$z_{t+l} = 2z_{t+l-1} - z_{t+l-2} + \eta_{t+1} - 0.5\eta_{t+1-1} + 0.5\eta_{t+1-2}$$

Forecasts are given by

$$\hat{z}_t(1) = 2z_t - z_{t-1} - 0.5\eta_t + 0.5\eta_{t+1} + \eta_{t+1}$$
$$\hat{z}_t(2) = 2z_{t+1} - z_t - 0.5\eta_{t+1} + 0.5\eta_t + \eta_{t+2} = 2\hat{z}_t(1) - z_t + 0.5\eta_t$$
$$\hat{z}_t(3) = 2z_{t+2} - z_{t+1} - 0.5\eta_{t+2} + 0.5\eta_{t+1} + \eta_{t+3} = 2\hat{z}_t(2) - \hat{z}_t(1)$$

For shorter lead times, the η's may be set to their unconditional expected value of zero; that is, up to $\hat{z}_t(q)$, which involve η_t and η_{t-1}. Similarly, for all lead times greater than 2, there will be no η component; that is,

$$\hat{z}_t(4) = 2\hat{z}_t(3) - \hat{z}_t(2)$$
$$\hat{z}_t(5) = 2\hat{z}_t(4) - \hat{z}_t(3)$$
...
...

For the MA(q) part, the forecasts will depend on η's up to $\hat{z}_t(q)$; for longer lead times, the forecasts will not depend on the η's. In Section 7.11.4.1, there is no MA part. Therefore, the forecasts are independent of the η's. In Section 7.11.4.2, the MA part is of order 2.

Therefore, the forecasts up to lead $l = 2$ depend on the η's. Thereafter, the forecasts are independent of the η's.

Using the same data as in Section 7.11.4.1

$$\hat{z}_t(1) = 2 \times 23.4 - 23.7 = 23.1$$
$$\hat{z}_t(2) = 2 \times 23.1 - 23.4 = 22.8$$
$$\hat{z}_t(3) = 2 \times 22.8 - 23.1 = 22.5$$
$$\hat{z}_t(4) = 2 \times 22.5 - 22.9 = 22.1$$

7.12 SYNTHETIC DATA GENERATION

The 'end product' of time series analysis is the independent random component η_t. It is ideal for this to be white noise. If so, independent standard normal variates can be generated using a number of procedures. The dependent structure and the deterministic components are then added to obtain the synthetically generated series.

Random sequences can be generated using the Monte Carlo technique, or tossing a coin. Box and Muller (1958) proposed a method of generating standard normal random numbers ε_1 and ε_2:

$$\varepsilon_1 = \left(\ln\left(\frac{1}{u_1} \right) \right)^{1/2} \cos(2\pi u_2) \tag{7.157a}$$

$$\varepsilon_2 = \left(\ln\left(\frac{1}{u_1} \right) \right)^{1/2} \sin(2\pi u_2) \tag{7.157b}$$

where u_1 and u_2 are random numbers of the uniform distribution $(0, 1)$. Normality and independence tests must be carried out for the generated numbers because they are often pseudo-normal or pseudo-independent.

The stochastic component of an AR process is obtained as follows:

$$\eta_t = \hat{\sigma}_\eta \varepsilon_t \tag{7.158a}$$

$$\hat{z}_t = \varphi_{p,i} \hat{z}_{t-i} + \ldots + \varphi_{p,p} \hat{z}_{t-p} + \eta_t \tag{7.158b}$$

where $\hat{\sigma}_\eta$, $\varphi_{p,i}$'s, and $\hat{z}_t$ are the standard deviation of the residual series, the AR parameters, and the generated stochastic series, respectively. The synthetic series is obtained by adding the deterministic components to $\hat{z}_t$.

However, the estimation of $\hat{z}_t$ requires $\hat{z}_{t-p}$, which is unknown. This problem is solved by assuming 'zero initial condition' and having a sufficient 'warm-up length'. For example, if N_g values of generated data are required, the procedure is as follows:

• Generate the standard normal random variates ε_1 and calculate $\hat{z}_1$ assuming that $\hat{z}_0$, $\hat{z}_{-1}$, ... are zeros.

- In the same manner, generate a second standard normal variate ε_2 and calculate $\hat{z}_2$ based on $\hat{z}_1$ already generated and assuming $\hat{z}_0$, $\hat{z}_{-1}$ .... are zeros.
- Repeat the procedure until the series $\hat{z}_1, \hat{z}_2, \hat{z}_3, ... \hat{z}_{N'}$ is generated, where $N' = N_g + N_w$, and N_w is the warm-up length. (The length N_w is necessary to remove the effect of the starting condition.)
- Delete the first N_w values and re-initialize the last N_g values.

7.13 ARMAX MODELLING

ARMAX models take the form

$$Q_t = \sum_1^p A_i Q_{t-i} + \sum_0^q B_j I_{t-j} \tag{7.159}$$

where I_t and Q_t are the input and output variables at time t; A_i and B_j are the weighting function coefficients for Q_{t-i} and I_{t-j}; p and q are the orders of the models for Q and I. The coefficients A_i and B_j can be estimated by the least squares method. The normal equations are obtained by minimizing the squared errors as follows:

$$E = \left(Q_t - \sum_1^p A_i Q_{t-i} - \sum_0^q B_j I_{t-j} \right)^2 \tag{7.160}$$

Differentiating with respect to the coefficients,

$$\frac{\partial E}{\partial A_1} = -2 \left(Q_t - \sum_1^p A_i Q_{t-i} - \sum_0^q B_j I_{t-j} \right)^2 Q_{t-1} \tag{7.161a}$$

$$\frac{\partial E}{\partial A_2} = -2 \left(Q_t - \sum_1^p A_i Q_{t-i} - \sum_0^q B_j I_{t-j} \right)^2 Q_{t-2} \tag{7.161b}$$

...

...

$$\frac{\partial E}{\partial B_0} = -2 \left(Q_t - \sum_1^p A_i Q_{t-i} - \sum_0^q B_j I_{t-j} \right)^2 I_t \tag{..}$$

...

Setting Equation 7.161 to zero,

$$
\begin{bmatrix}
Q_{t-1}Q_{t-1} & Q_{t-1}Q_{t-2}\cdots Q_{t-1}I_t & Q_{t-1}I_{t-1}\cdots \\
Q_{t-2}Q_{t-1} & Q_{t-2}Q_{t-2}\cdots Q_{t-2}I_t & Q_{t-2}I_{t-1}\cdots \\
Q_{t-3}Q_{t-1} & Q_{t-3}Q_{t-2}\cdots Q_{t-3}I_t & Q_{t-3}I_{t-1}\cdots \\
\cdots & & \\
I_tQ_{t-1} & I_tQ_{t-2}\cdots I_tI_t & I_tI_{t-1}\cdots \\
I_{t-1}Q_{t-1} & I_{t-1}Q_{t-2}\cdots I_{t-1}I_t & I_{t-1}I_{t-1}\cdots \\
I_{t-2}Q_{t-1} & I_{t-2}Q_{t-2}\cdots I_{t-2}I_t & I_{t-2}I_{t-1}\cdots \\
\cdots & &
\end{bmatrix}
\begin{bmatrix}
A_1 \\ A_2 \\ A_3 \\ \cdot \\ B_0 \\ B_1 \\ B_3 \\ \cdot
\end{bmatrix}
=
\begin{bmatrix}
Q_{t-1}Q_t \\ Q_{t-2}Q_t \\ Q_{t-3}Q_t \\ \cdots \\ I_tQ_t \\ I_{t-1}Q_t \\ I_{t-2}Q_t \\ \cdot
\end{bmatrix}
$$

which is of the form

$$[A][\phi] = [B] \tag{7.162}$$

which can be solved for the coefficients A_i and B_j.

7.14 KALMAN FILTERING

Kalman filter (Kalman, 1960; Kalman and Bucy, 1961) recursively estimates the state of a linear dynamic system in a way that minimizes the mean of the squared error. Only the estimate of the state at the previous time step and the current measurement are needed to compute the estimate of the present state. In comparison, batch estimates require the entire history of estimates and measurements that lead to solution of sets of simultaneous equations.

Linear Kalman filter algorithm starts with two equations:

$$x_{t+1} = Ax_t + Bi_t + w_t \quad \text{(state equation)} \tag{7.163}$$

$$y_t = Cx_t + v_t \quad \text{(measurement equation)} \tag{7.164}$$

where x_t is the state that may include more than one variable. In general, it is a vector of the form

$$
x_t =
\begin{bmatrix}
x_t \\
x_{t-1} \\
x_{t-2} \\
\cdots \\
\cdots
\end{bmatrix}
\tag{7.165}
$$

In a one-dimensional system, x_t is a scalar quantity. The term i_t in Equation 7.163 is an input variable (controllable), which may or may not exist; w_t is an independently distributed random variable that is assumed to be normally distributed with zero mean and variance R_w. It accounts for the error in describing the process (or state) by Equation 7.163; y_t is the measurement that is related to the state x_t with an error v_t, which is also assumed to be normally distributed with zero mean and variance R_v; that is,

$$w_t \sim N(0, R_w)$$
$$v_t \sim N(0, R_v)$$

It is further assumed that the two error terms are independent of each other and independent of the state variable x_t and the measurement y_t; that is,

$$E(w_t v_t) = 0$$
$$E(w_t x_t) = E(w_t y_t) = E(v_t x_t) = E(v_t y_t) = 0 \tag{7.166a}$$

$$E\left(w_t w_t^T\right) = R_w$$
$$E\left(v_t v_t^T\right) = R_v \tag{7.166b}$$

A, B, and C are matrices that are generally assumed to be time invariant, that is, constants. However, they can also be time dependent. In a linear Kalman filter, they are assumed to be constants.

The Kalman filter is a predictor–corrector type of algorithm that minimizes the error covariance; that is, min $(x_t - \hat{x}_t)^2$. The Kalman filter does it optimally compared with other error minimization techniques. The equations can be written in two stages:

At prediction:

$$x_{t+1|t} = A x_{t|t} + B i_t$$
$$P_{t+1|t} = A P_{t|t} A^T + R_w \tag{7.167}$$

Since the Kalman filter is a recursive algorithm, it is necessary to have initial conditions; in this case, $x_{t|t}$, $P_{t|t}$, and R_w.

$x_{t|t}$ = State value using the information up to and including the time level t
$P_{t|t}$ = State (or process) error covariance using the information up to and including the time level t
R_w = Variance of the random error component w_t

To start the algorithm, these initial conditions must be provided. Estimates of these can be made in an offline mode using past information.

At correction:

Once a new measurement is made, there is new information brought that allows a correction to be made to the previous prediction. It is done as follows:

$$P_{t+1|t} = A P_{t|t} A^T + R_w \tag{7.168}$$

$$K_{t+1} = P_{t+1|t} C^T \{ C P_{t+1|t} C^T + R_v \}^{-1} \tag{7.169}$$

$$\hat{x}_{t+1|t+1} = \hat{x}_{t+1|t} + K_{t+1} \{ y_{t+1} - C\hat{x}_{t+1|t} \} \tag{7.170}$$

and the updated error covariance

$$P_{t+1|t+1} = \{ I - K_{t+1} C \} P_{t+1|t} \tag{7.171}$$

The term K is known as the Kalman gain, and the quantity $\{y_{t+1} - C\hat{x}_{t+1|t}\}$ is known as the innovation (or, sometimes the residual). I is the identity matrix.

If R_v (variance of measurement error) is very large, then from Equation 7.169,

$$K_{t+1} \to 0 \tag{7.172}$$

Therefore, the state estimate update (Equation 7.170) becomes

$$\hat{x}_{t+1|t+1} = x_{t+1|t} \tag{7.173}$$

that is, the state estimate is equal to the predicted value from the mathematical model.

The updated error covariance (Equation 7.171) then becomes

$$P_{t+1|t+1} = P_{t+1|t} = AP_{t|t}A^T + R_w \tag{7.174}$$

This depends only on the initial error covariance $P_{0|0}$ and the variance of the model error. The effect of $P_{0|0}$ is insignificant.

On the other hand, if R_v is small, then Equation 7.169 becomes

$$K_{t+1} = P_{t+1|t}C^T \{CP_{t+1|t}C^T + 0\}^{-1} = C^{-1} \tag{7.175}$$

Therefore, the state estimate update (Equation 7.170) becomes

$$\hat{x}_{t+1|t+1} = \hat{x}_{t+1|t} + K_{t+1}\{y_{t+1} - C\hat{x}_{t+1|t}\} = C^{-1}y_{t+1} \tag{7.176}$$

that is, the updated state estimate at $t + 1$ is equal to the observed state at $t + 1$.

If C is the identity matrix, that is, $y_t = x_t + v_t$, then

$$x_{t+1|t+1} = y_{t+1} \tag{7.177}$$

Therefore,

$$x_{t+2|t+1} = Ay_{t+1} + Bi_{t+1} \tag{7.178}$$

(When R_v is large, $K_{t+1} \to 0$, and when R_v is small, $K_{t+1} \to 1$.)

The model error and the measurement error covariance matrices take the form

$$R_w = \begin{bmatrix} \sigma_w^2 & 0 & 0\dots \\ 0 & \sigma_w^2 & 0\dots \\ 0 & 0 & \sigma_w^2 \\ \dots \\ 0 & 0 & 0 & \sigma_w^2 \end{bmatrix} \tag{7.179}$$

$$R_v = \begin{bmatrix} \sigma_v^2 & 0 & 0\dots \\ 0 & \sigma_v^2 & 0\dots \\ 0 & 0 & \sigma_v^2 \\ \dots \\ 0 & 0 & 0 & \sigma_v^2 \end{bmatrix} \tag{7.180}$$

and the model error variance is given by

$$\sigma_w^2 = \frac{1}{N} \sum_1^N (x_t - \hat{x}_t)^2 \tag{7.181}$$

With a knowledge of the initial estimates of $x_{0|0}$, $P_{0|0}$, R_v, and R_w,

$$x_{1|0} \to P_{1|0} \to K_1 \to x_{1|1} \to P_{1|1}; x_{2|1} \to P_{2|1} \to K_2 \to x_{2|2} \to P_{2|2};$$
$$x_{3|2} \to P_{3|2} \to K_3 \to x_{3|3} \to P_{3|3};$$

...

...

can be recursively estimated in that sequence.

The Kalman filtering algorithm has many variables that need to be determined. The model parameters A, B, and C can be estimated using any parameter estimation technique. The error estimates R_w and R_v, however, are unknowns that cannot be estimated from the available input and output data. The state of the process x_t is an unknown; only the observation y_t is known. Therefore, it is not possible to separate the two error terms w_t and v_t. They are therefore assigned arbitrary values. In particular, $R_v = 0$ make $K = 1$, which gives the best correction of the previous forecast that is the observed value itself.

The performance of the predictor is judged by comparing $\hat{x}_{t+1|t}$ with y_{t+1} and, not with x_{t+1}. The latter is not available.

If, on the other hand, information about the statistics of the measurement error is available *a priori*, then R_v should be assigned a value. Otherwise, it is not unreasonable to set the measurement error equal to zero.

Underestimation of the model error R_w may reduce the Kalman gain, and hence the filtering of previous forecast $\hat{x}_{t+1|t}$ using the new observation y_{t+1}. Correction strongly depends on Kalman gain and hence upon R_w. The propagation of the forecast error to the prediction of $\hat{x}_{t+1|t+1}$ can be prevented by choosing the right value of R_w. It is therefore best to have a scheme whereby R_w is re-estimated at each time level (adaptive Kalman filtering), which can be made on the basis of the comparison of y_{t+1} with $C\hat{x}_{t+1|t}$, y_{t+2} with $Cx_{t+2|t+1}$, etc.

The quantity $\left[y_{t+1} - C\hat{x}_{t+1|t} \right]$ is called the innovation – the new information brought about by the current observation. The new observation can be used for

- Filtering the state vector x_t
- Correcting the error covariance R_w
- Updating the model parameters A, B, and C

If the measurement noise R_v is large, K_t will be small and therefore the measurement will have little or no impact on future state estimate $\hat{x}_{t+1}$. If, on the other hand, R_v is small, then K_t will be large and the measurement will have an impact on the future state estimate $\hat{x}_{t+1}$. The state vector update, Kalman gain, and the error covariance equations represent an asymptotically stable system and therefore the estimates of state x_t become independent of the initial estimates of $x_{t|t}$ and $P_{t|t}$ as t increases.

7.15 PARAMETER ESTIMATION

All the models described above have parameters that need to be estimated by some technique. The poor fitting of the model results with the actual observations that may result in some situations may not necessarily be due to inadequacies in the model, but more likely on the inaccuracies in the parameter estimation. Several methods for model parameter estimation are available but their estimates could sometimes differ appreciably. Initial estimates of parameters are usually obtained by the method of moments by solving the Yule–Walker equations. For MA and ARMA models, this procedure is complicated and the moment estimators are also sensitive to round-off errors. They should not be used as final estimates of parameters. MA and ARMA model parameters are estimated iteratively.

A better but more complicated alternative is the maximum likelihood method in which the log likelihood function of the parameters is maximized. With an assumed initial values of z_t and η_t, the method leads to the conditional maximum likelihood estimates. It is also possible to define an unconditional log likelihood function and an exact likelihood function. More details of these procedures can be found in several textbooks (e.g., Box and Jenkins, 1976, Chapter 7; Wei, 1990, Chapter 7).

Example 7.6: Kalman filtering

Use the following process and measurement equations to forecast the series starting from the origin $t = 3$:

> Process equation: $x_{t+1} = 1.8x_t - 0.8x_{t-1} + w_{t+1}$
> Measurement equation: $y_t = x_t + v_t$; ($y_t = 23.7, 23.4, 23.1, 22.9, 22.8$ for $t = 1, 2,..., 5$)

INITIAL ESTIMATES

Since the state variable x_t and x_{t-1} are not known, the best estimate of these past values will be the measurement. Therefore,

$$\hat{x}_{2|2} = 23.4$$
$$\hat{x}_{1|1} = 23.7$$

The initial error covariance $P_{2|2}$ is also unknown. It measures the error between the actual value of the state variable and its estimate; that is,

A priori: $e_{t|t-1} = x_t - \hat{x}_{t|t-1}$; $P_{t|t-1} = E\left(e_{t|t-1}e_{t|t-1}^T\right)$

A posteriori: $e_{t|t} = x_t - \hat{x}_{t|t}$; $P_{t|t} = E\left(e_{t|t}e_{t|t}^T\right)$

The initial value of $P_{t|t}$ we assume is not crucial as long as it is non-zero, because its value quickly converges as the forecasting progresses in time. Therefore, let $P_{2|2} = 100$.

We also need values of R_w and R_v to start the Kalman filtering algorithm. If they can be estimated from prior information, then they could be assigned those values. Since there is not much prior information available, let them be arbitrarily chosen as $R_w = 10$ and $R_v = 5$.

KALMAN FILTER ESTIMATES

The Kalman filter equations (for this example) are

$$\hat{x}_{t+1|t} = A\hat{x}_{t|t}$$
$$P_{t+1|t} = AP_{t|t}A^T + R_w$$

$$K_{t+1} = P_{t+1|t}C^T\{CP_{t+1|t}C^T + R_v\}^{-1}$$
$$\hat{x}_{t+1|t+1} = \hat{x}_{t+1|t} + K_{t+1}\{y_{t+1} - C\hat{x}_{t+1|t}\}$$
$$P_{t+1|t+1} = \{I - K_{t+1}C\}P_{t+1|t}$$

At time $t = 3$,
A priori estimate:

$$\hat{x}_{3|2} = (1.8 - 0.8)\begin{bmatrix}\hat{x}_{2|2}\\\hat{x}_{1|1}\end{bmatrix} = 1.8 \times 23.4 - 0.8 \times 23.7 = 23.16$$

$$P_{3|2} = (1.8 - 0.8)\begin{bmatrix}1.8\\-0.8\end{bmatrix}P_{2|2} + R_w = (1.8^2 + 0.8^2)100 + 10 = 398$$

$$K_3 = 398(398 + R_v)^{-1} = 398/(398 + 5) = 0.987$$

A posteriori estimate:

$$\hat{x}_{3|3} = \hat{x}_{3|2} + K_3(y_3 - \hat{x}_{3|2}) = 23.16 + 0.987(23.1 - 23.16) = 23.10$$

$$P_{3|3} = (I - K_3) P_{3|2} = 398 (1 - 0.987) = 5.174$$

At time $t = 4$,
A priori estimate:

$$\hat{x}_{4|3} = (1.8 - 0.8)\begin{bmatrix}\hat{x}_{3|3}\\\hat{x}_{2|2}\end{bmatrix} = 1.8 \times 23.10 - 0.8 \times 23.4 = 22.86$$

$$P_{4|3} = (1.8 - 0.8)\begin{bmatrix}1.8\\-0.8\end{bmatrix}P_{3|3} + R_w = (1.8^2 + 0.8^2)5.174 + 10 = 30.07$$

$$K_4 = 30.07(30.07 + R_v)^{-1} = 30.07/(30.07 + 5) = 0.857$$

A posteriori estimate:

$$\hat{x}_{4|4} = \hat{x}_{4|3} + K_4(y_4 - \hat{x}_{4|3}) = 22.86 + 0.857(22.9 - 22.86) = 22.89$$

$$P_{4|4} = (I - K_4) P_{4|3} = 30.07 (1 - 0.857) = 4.30$$

At time $t = 5$,

A priori estimate:

$$\hat{x}_{5|4} = (1.8 - 0.8)\begin{bmatrix} \hat{x}_{4|4} \\ \hat{x}_{3|3} \end{bmatrix} = 1.8 \times 22.89 - 0.8 \times 23.1 = 22.72$$

$$P_{5|4} = (1.8 - 0.8)\begin{bmatrix} 1.8 \\ -0.8 \end{bmatrix} P_{4|4} + R_w = (1.8^2 + 0.8^2)4.30 + 10 = 26.68$$

$$K_5 = 26.68(26.68 + R_v)^{-1} = 26.68/(26.68 + 5) = 0.842$$

A posteriori estimate:

$$\hat{x}_{5|5} = \hat{x}_{5|4} + K_5(y_5 - \hat{x}_{5|4}) = 22.72 + 0.842(22.8 - 22.72) = 22.79$$

$$P_{5|5} = (I - K_5)\, P_{5|4} = 26.68\,(1 - 0.842) = 4.21$$

7.16 APPLICATIONS

Time series analysis and modelling have diverse applications in many disciplines. It is used in statistics, actuarial sciences, banking and insurance, signal processing, control engineering, economics, life sciences, and in many areas of science and engineering. In the context of this chapter, the interest is restricted to applications in the hydro-environment. A comprehensive list of relevant literature is beyond the scope of this chapter, but the studies carried out by Thomann (1967), Lohani and Wang (1987), Jayawardena and Lai (1989), and Huck and Farquhar (1974) in water quality modelling; Srikanthan and McMahon (1982) in annual and monthly rainfall simulation; McMicheal and Hunter (1972) in modelling temperature and flows in rivers; Jayawardena and Lau (1990a) on stochastic modelling of evaporation data; Lawrance and Kottegoda (1977) in river flow time series modelling; Gupta and Chanhan (1986) in irrigation requirement modelling; and Delleur and Kavvas (1978) in synthetic generation of monthly rainfall data, among many others, may lead the readers to more applications in the field.

7.17 CONCLUDING REMARKS

In this chapter, an attempt has been made to describe the various techniques of time series analysis and forecasting in the context of hydrological and environmental time series. It should be noted that time series analysis and modelling are iterative processes that involve diagnostic checking and model selection. The content of this chapter is restricted to techniques and analysis under the assumptions of linearity and stationarity. Non-stationary series can be analysed by transforming them into stationary ones. Also, seasonal models have not been dealt with explicitly but they can be taken care of by the various techniques of periodicity analysis. Examples to illustrate the techniques have been given where possible. The reader is, however, encouraged to refer to relevant literature given for further understanding and consolidation.

APPENDIX 7.1

FOURIER SERIES REPRESENTATION OF A PERIODIC FUNCTION

Any periodic function $f(x)$ can be represented as a combination of sine and cosine function as follows:

$$f(x) = a_0 + \sum_{n=1}^{\infty} \left\{ a_n \cos(nx) + b_n \sin(nx) \right\} \tag{A1-1}$$

The coefficients a_n and b_n are determined by integration Equation A1-1 as

$$\int_{-\pi}^{\pi} f(x)\,dx = \int_{-\pi}^{\pi} a_0\,dx + \sum_{n=1}^{\infty} \left\{ a_n \int_{-\pi}^{\pi} \cos(nx)\,dx + b_n \int_{-\pi}^{\pi} \sin(nx)\,dx \right\} \tag{A1-2}$$

which leads to (because all sine and cosine integrals are zero)

$$a_0 = \frac{1}{2\pi} \int_{-\pi}^{\pi} f(x)\,dx \tag{A1-3a}$$

If in Equation A1-1, $\frac{a_0}{2}$ is used instead of a_0, then the corresponding equation will be

$$a_0 = \frac{1}{\pi} \int_{-\pi}^{\pi} f(x)\,dx \tag{A1-3b}$$

The coefficients a_n's can be obtained by multiplying Equation A1-1 by $\cos(mx)$ and integrating from $-\pi$ to π as

$$a_n = \frac{1}{\pi} \int_{-\pi}^{\pi} f(x) \cos(mx)\,dx \tag{A1-4}$$

Similarly, the coefficients b_n's can be obtained by multiplying Equation A1-1 by $\sin(mx)$ and integrating from $-\pi$ to π as

$$b_n = \frac{1}{\pi} \int_{-\pi}^{\pi} f(x) \sin(mx)\,dx \tag{A1-5}$$

Equations A1-3, A1-4, and A1-5 are referred to as Euler equations for determining the Fourier coefficients (Kreszig, 1999, pp. 530–532).

The above formulation is valid for a periodic function with period 2π. For any other function with period $p = 2L$, Equation A1-1 can be written as (Kreszig, 1999, p. 537)

$$f(x) = a_0 + \sum_{n=1}^{\infty} \left\{ a_n \cos\left(\frac{n\pi x}{L}\right) + b_n \sin\left(\frac{n\pi x}{L}\right) \right\} \tag{A1-6}$$

and the corresponding Fourier coefficients are

$$a_0 = \frac{1}{2L} \int_{-L}^{L} f(x)\,dx$$
(A1-7)

$$a_n = \frac{1}{L} \int_{-L}^{L} f(x) \cos\left(\frac{n\pi x}{L}\right) dx \quad \text{for } n = 1, 2, 3, \dots$$
(A1-8)

$$b_n = \frac{1}{L} \int_{-L}^{L} f(x) \sin\left(\frac{n\pi x}{L}\right) dx \quad \text{for } n = 1, 2, 3, \dots$$
(A1-9)

In the integration of Equation A1-1, a key property used in deriving Euler formulas is the orthogonality conditions of trigonometric functions (Kreszig, 1999, p. 534). Orthogonality implies that the integral of the products of any two different combinations of these functions is zero over the interval $-\pi$ to π. The period 2π can be easily replaced by any other period $2L$. The orthogonality conditions are (Kreszig, 1999, p. 537) given as follows:
For integers m and n,

$$\int_{-\pi}^{\pi} f(x) \cos(mx) \cos(nx)\,dx = \begin{cases} 0 & \text{if } m \neq n \\ \pi & \text{if } m = n \neq 0 \end{cases}$$
(A1-10)

$$\int_{-\pi}^{\pi} f(x) \sin(mx) \sin(nx)\,dx = \begin{cases} 0 & \text{if } m \neq n \\ \pi & \text{if } m = n \neq 0 \end{cases}$$
(A1-11)

and for any integers m and n,

$$\int_{-\pi}^{\pi} f(x) \cos(mx) \sin(nx)\,dx = 0$$
(A1-12)

It is also important to note that the Fourier series of an even function of period $2L$ is a cosine function, whereas that of an odd function is a sine function; that is,

$$f(x) = a_0 + \sum_{n=1}^{\infty} \left\{ a_n \cos\left(\frac{n\pi x}{L}\right) \right\}$$
(A1-13)

$$f(x) = \sum_{n=1}^{\infty} \left\{ b_n \sin\left(\frac{n\pi x}{L}\right) \right\}$$
(A1-14)

REFERENCES

Abdi, H. (2007): Kendall rank correlation, In: Salkind, N.J. (Ed.), *Encyclopedia of Measurement and Statistics*. Sage, Thousand Oaks, CA.

Akaike, H. (1970): Statistical predictor identification. *Annals of the Institute of Statistical Mathematics*, 22, 203–217.

Akaike, H. (1973): Information theory and an extension of the maximum likelihood principle. In: Petrov, B.N. and Csaki, F. (Eds.), *Proceedings, 2nd International Symposium on Information Theory*, Akademiai Kaido, Budapest, pp. 267–281.

Akaike, H. (1974): A new look at the statistical model identification. *IEEE Transactions on Automatic Control*, 19, 716–723.

Akaike, H. (1978): A Bayesian analysis of the minimum AIC procedure. *Annals of the Institute of Statistical Mathematics*, 30A, 9–14.

Akaike, H. (1979): A Bayesian analysis of the minimum AIC procedure of auto-regressive model fitting. *Biometrica*, 66, 237–242.

Alexandersson, H. (1986): A homogeneity test applied to precipitation data. *Journal of Climatology*, 6, 661–675.

Alexandersson, H. and Moberg, A. (1997): Homogenization of Swedish temperature data, part 1: Homogeneity test for linear trends. *International Journal of Climatology*, 17, 25–34.

Ampitiyawatta, A.D. and Guo, S. (2009): Precipitation trends in the Kalu Ganga basin in Sri Lanka. *The Journal of Agricultural Sciences*, 4(1), 10–18.

Anderson, R.L. (1941): Distribution of serial correlation coefficients. *Annals of Mathematical Statistics*, 8(1), 1–13.

Bartlett, M.S. (1946): On the theoretical specification of sampling properties of auto-correlated time series. *Journal of the Royal Statistical Society*, B8, 27.

Bartolini, G., Marco, M., Alfonso, C., Daniele, G., Tommaso, T., Martina, P., Giampiero M. and Simone, O. (2008): Recent trends in Tuscany (Italy) summer temperature and indices of extremes. *International Journal of Climatology*, 28, 1751–1760.

Blackman, R.B. and Tukey, J. (1959): Particular pairs of windows. In: *The Measurement of Power Spectra, From the Point of View of Communications Engineering*. Dover, New York, pp. 98–99.

Box, G.E.P. and Jenkins, G.M. (1976): *Time Series Analysis: Forecasting and Control*, Revised Edition. Holden-Day, Oakland, CA, p. 575.

Box, G.E.P. and Muller, M.E. (1958): A note on generation of random normal deviates. *Annals of Mathematical Statistics*, 29, 610–611.

Box, G.E.P. and Pierce, D.A. (1970): Distribution of residual autocorrelations in auto-regressive integrated moving average time series models. *Journal of the American Statistical Association*, 65, 1509–1526.

Buishand, T.A. (1982): Some methods for testing the homogeneity of rainfall records. *Journal of Hydrology*, 58, 11–27.

Delleur, F.W. and Kavvas, M.L. (1978): Stochastic models for monthly rainfall forecasting and synthetic generation. *Journal of Applied Meteorology*, 17(10), 1528–1536.

Douglas, E.M., Vogel, R.M. and Kroll, C.N. (2000): Trends in floods and low flows in the United States: Impact of spatial correlation. *Journal of Hydrology*, 240(1–2), 90–105.

Gilbert, R.O. (1987): *Statistical Methods for Environmental Pollution Monitoring*. John Wiley & Sons Inc., New York.

Gupta, R.K. and Chanhan, H.S. (1986): Stochastic model of irrigation requirements. *Journal of Irrigation and Drainage Engineering, ASCE*, 112(1), 65–76.

Hamming, R.W. (1983): *Digital Filters*, 2nd Edition. Prentice Hall, Englewood Cliffs, NJ.

Hirsch, R.M., Alexander, R.B. and Smith, R.A. (1991): Election of methods for the detection and estimation of trends in water quality. *Water Resources Research*, 27(5), 803–813.

Hirsch, R.M., Slack, J.R. and Smith, R.A. (1981): Techniques of trend analysis for monthly water-quality data. US Geological Survey Open-File Report No. 81-488.

Hong Kong Observatory (1884–1939): *Meteorological Results Part I (1884–1939)*.

Huck, P.M. and Farquhar, G.J. (1974): Water quality models using Box-Jenkins method. *Journal of Environmental Engineering, ASCE*, 100(3), 733–753.

Jaruškova, D. (1994): Change-point detection in meteorological measurement. *Monthly Weather Review*, 124, 1535–1543.

Jayawardena, A.W. and Lai, F. (1989): Time series analysis of water quality data in Pearl River, China. *Journal Environmental Engineering, ASCE*, 115(3), 590–607.

Jayawardena, A.W. and Lau, W.H. (1990a): Stochastic analysis and generation of monthly and 14-day evaporation data. *Journal of the Japan Society of Hydrology and Water Resources*, 3(3), 56–67.

Jayawardena, A.W. and Lau, W.H. (1990b): Homogeneity tests for rainfall data. *Journal of the Hong Kong Institution of Engineers*, 22–25.

Kalman, R.E. (1960): A new approach to linear filtering and prediction problems. *ASME Journal of Basic Engineering, Series*, D82, 35–45.

Kalman, R.E. and Bucy, R.S. (1961): New results in linear filtering and prediction theory. *ASME Journal of Basic Engineering, Series*, D83, 95–108.

Kendall, M. (1975): *Rank Correlation Methods*. Charles Griffin, London.

Kendall, M.G. (1948): *Rank Correlation Methods*. Charles Griffin & Company Limited, London.

Kendall, M.G. (1973): *Time-series*. Charles Griffin, London.

Kottegoda, N.T. (1980): *Stochastic Water Resources Technology*. John Wiley, New York.

Kreyszig, E. (1999): *Advanced Engineering Mathematics*, 8th Edition. John Wiley & Sons, New York.

Lawrance, A.J. and Kottegoda, N.T. (1977): Stochastic modelling of river flow time series. *Journal of Royal Statistical Society, Series A*, 140, 1–47.

Ljung, G. M. and Box, G.E.P. (1978): On measure of lack of fit in time series models. *Biometrica*, 65(2), 297–303.

Ljung, G.M. and Box, G.E.P. (1979): The likelihood function of stationary autoregressive moving average models. *Biometrica*, 66, 265–270.

Lohani, B.N. and Wang, M.M. (1987): Water quality data analysis in Chung Kang River. *Journal of Environmental Engineering, ASCE*, 113(1), 186–195.

Longobardi, A. and Villani, P. (2009): Trend analysis of annual and seasonal rainfall time series in the mediterranean area. *International Journal of Climatology*, 30(10), 1538–1546.

Luo, Q. and White, R.E. (2005): Cubic spline regression for the open-circuit potential curves of a lithium-ion battery. *Journal of the Electrochemical Society*, 152(2), A343–A350.

Mann, H.B. (1945): Non parametric test against trend. *Econometrica*, 13(3), 245–259.

Martínez, M.D., Serra, C., Burgueño, A. and Lana, X. (2010): Time trends of daily maximum and minimum temperatures in Catatonia (NE Spain) for the period 1975–2004. *International Journal of Climatology*, 30, 267–290.

McMicheal, F.C. and Hunter, J.S. (1972): Stochastic modelling of temperature and flow in rivers. *Water Resources Research*, 8(1), 87–98.

Nadaraya, E.A. (1964): On estimating regression. *Theory of Probability and its Applications*, 9(1), 141–142.

Neumann, J. von (1941): Distribution of the ratio of the mean square successive difference to the variance. *The Annals of Mathematical Statistics*, 12(4), 367–395.

Owen, D.B. (1962): *Handbook of Statistical Tables*. Addison Wesley, Reading, MA.

Paiva, E.M.C.D. de and Clarke, R.T. (1995): Time trends in rainfall records in Amazonia. *Bulletin of the American Meteorological Society*, 76(11), 2203–2209.

Pandžíc, K. and Likso, T. (2009): Homogeneity of average annual air temperature time series for Croatia. *International Journal of Climatology*, 30(8), 1215–1225.

Partal, T. and Kahya, E. (2006): Trend analysis in Turkish precipitation data. *Hydrological Processes*, 20(9), 2011–2026.

Parzen, E. (1977): Multiple time series: Determining the order of approximating autoregressive schemes, In: Krishnaiah, P. (Ed.), *Multivariate Analysis IV*. North Holland, Amsterdam, 283–296.

Pettit, A. (1979): A non-parametric approach to the change-point detection. *Journal of the Royal Statistical Society, Series C (Applied Statistics)*, 28(2), 126–135.

Reinsch, C. (1967): Smoothing by spline functions. *Numerische Mathematic*, 10, 177–183.

Sahin, S. and Cigizoglu, H.K. (2010): Homogeneity analysis of Turkish meteorological data set. *Hydrological Processes*, 24, 981–992.

Saifuddin, A.K.M. (2010): Homogeneity and trend analysis of temperature for urban and rural areas. Thesis submitted in partial fulfilment of the requirement for the Master's Degree in Disaster Management, National Graduate Institute for Policy Studies, Tokyo, Japan.

Salas, J.D., Delleur, J.W., Yevjevich, V. and Lane, W.L. (1980): *Applied Modelling of Hydrologic Time Series*. Water Resources Publications, Littleton, CO.

Schwartz, G. (1978): Estimating the dimension of a model. *Annals of Statistics*, 6, 461–464.

Snedecor, G.W. and Cochran, W.G. (1967): *Statistical Methods*. The Iowa State University Press, Ames, IA.

Srikanthan, R. and McMahon, T.A. (1982): Simulation of Annual and Monthly Rainfalls—A preliminary study at five Australian stations. *Journal of Applied Meteorology*, 21, 1472–1479.

Staudt, M., Esteban-Parra, M.J. and Castro-Díez, Y. (2007): Homogenization of long-term monthly Spanish temperature data. *International Journal of Climatology*, 27, 1809–1823.

Su, B.D., Jiang, T. and Jin, W.B. (2006): Recent trends in observed temperature and precipitation extremes in the Yangtze River basin, China. *Theoretical and Applied Climatology*, 83, 139–151.

Syrakova, M. and Stefanova, M. (2009): Homogenization of Bulgarian temperature series. *International Journal of Climatology*, 29, 1835–1849.

Thomann, R.V. (1967): Time series analysis of water quality data. *Journal of Sanitary Engineering Division, ASCE*, 93(1), 1–23.

Tukey, J.W. (1965): Data analysis and frontiers of geophysics. *Science*, 148, 1283–1289.

Walker, G. (1931): On periodicity in series of related terms. *Proceedings of the Royal Society of London, Series A*, 131, 518–532.

Watson, G.S. (1964): Smooth regression analysis. *Sankhya, Series A*, 26, 359–372.

Wei, W.W.S. (1990): *Time Series Analysis, Univariate and Multivariate Methods*. Reprinted with corrections, 1994, Addison-Wesley, USA.

Wijngaard, J.B., Klein Tank, A.M.G. and Können, G.P. (2003): Homogeneity of 20th century European daily temperature and precipitation series. *International Journal of Climatology*, 23, 679–692.

Yule, G.U. (1927): On a method of investigating periodicities in disturbed series, with special reference to Wolfer's sunspot numbers. *Philosophical Transactions of the Royal Society of London, Series A*, 226, 267–298.

Chapter 8

Artificial neural networks

8.1 INTRODUCTION

In recent years, artificial neural networks, or ANNs, have found applications in many areas of science, engineering, medicine, and finance. They emulate the brain, which can be considered as a biological neural network. The processors operate on the data received via the connections. The transformation of an input to a corresponding output by a single neuron is relatively simple. The complexity arises as a result of the interactions of many neurons.

The first artificial neuron was the Threshold Logic Unit (McCulloch and Pitts, 1943), used as a transfer function for binary classification. Since then, advances have been made over the years in the technique itself as well as in extending the application areas into diverse fields. A major milestone in the theoretical development has been the introduction of the perceptron, which is a collection of simple neurons into a network (Rosenblatt, 1962). With the development of the multilayer perceptron, which is a cascade of single-layer perceptrons (SLPs) arranged in layers, ANNs have found applications in modelling many types of complex systems.

There are several types of problems that can be solved using ANNs. Among them are classification problems that include pattern (e.g., face and object identification) and sequence (e.g., speech and handwriting) recognition; medical diagnosis; regression (or function approximation) problems that include time series modelling and prediction; and data processing problems that include filtering, clustering, and compression. In classification problems, input variables are classified into different classes depending on the outcome of the network. The simplest is when there are only two outcomes that lead to binary classification. In regression or function approximation problems, the objective is to estimate the value of the output variable when the values of the input variables are presented. This is also the focus of this chapter, particularly applicable to hydrological and environmental systems.

A system in which a large number of neurons are interconnected with exposure to the external environment is called an artificial neural network. The power of ANNs lies in their large number of interconnections. Very often the neurons are arranged in layers with each layer having a number of neurons. This chapter gives a general overview of ANNs, highlighting the different types, activation functions, learning paradigms, and application areas particularly relevant to hydrology and environmental systems. For more detailed theoretical aspects, the reader is referred to relevant textbooks (e.g., Haykin, 1999; Schalkoff, 1997). Examples of applications in other areas can be found in the relevant literature.

8.2 ORIGIN OF ARTIFICIAL NEURAL NETWORKS

8.2.1 Biological neuron

The brain is a biological neural network with over 10 billion neurons and over 60 trillion synapses. Each neuron (sometimes called neurone or nerve cell) is a cell that receives, processes, and transmits information through biochemical processes. A biological neuron has a cell body, which is the central part, of typical sizes ranging from about 10 to 80 μm; dendtrites, which are root-like extensions from the cell body with typical size a few micrometers; and an axon, which is a single tubular fibre projecting from the cell body with typical diameters a few micrometers and lengths ranging from about 0.1 mm to 1 m. The dendrites act as receptors for signals from nearby neurons, whereas the axons transmit the generated neuron activity to other neurons or muscle fibres. When a neuron fires, an electric charge is received by a dendrite. Such charges are summed spatially and temporally and passed onto the cell body. Spatial summation takes place when several weak signals are converted into a single strong signal, whereas temporal summation takes place when a series of weak signals from one source is converted into a single strong signal. If the strength of the summed signal is greater than the threshold value of the axon, then the neuron fires and an output is transmitted via the axon. Otherwise, the neuron does not fire and there is no output transmitted. The strength of the output is constant regardless of the strength of the input signal as long as it is above the threshold value. The axon has several terminal buttons, which are, in turn, connected to other neurons across a small gap called the synapse whose characteristics determine the strength and polarity of the new input signal. Over a period, the strengths of the synapses are adjusted through experience or exposure to various signals, such as vision, hearing, emotions, etc.

Neurons in the brain are slower than those in an artificial neuron. The time taken for a typical operation in a silicon chip is of the order of nanoseconds (10^{-9} s), whereas in the brain it is of the order of milliseconds (10^{-3} s). However, the brain is much more energy efficient when compared to computers. Typical energy consumption in the brain is of the order of 10^{-16} J/operations, whereas the best computers consume about 10^{-6} J/operations (Faggin, 1991).

8.2.2 Artificial neuron

An artificial neuron can be thought of as a mathematical model of the biological neuron. It has five components: input(s), connection weights, threshold (or bias), activation functions, and output(s) (Figure 8.1). It receives one or more inputs (analogous to dendrites) from the

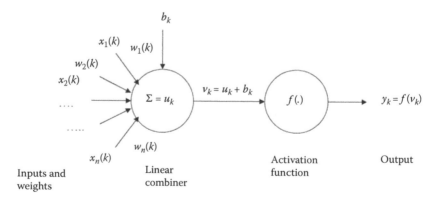

Figure 8.1 Artificial non-linear neuron.

external environment and sums them up to produce an output analogous to the axon of a biological neuron. This output then acts as the input to the next layer via a synapse. In the same way as a biological neural network has interconnected neurons, artificial neurons also have connections to other neurons. It is also possible for the output of the first layer to exit as the final output. Figure 8.1 corresponds to a non-linear neuron. If the activation function is linear, or if there is no activation function (only inputs, weights, linear combiner, and outputs), then it becomes a linear neuron.

8.2.2.1 Bias/threshold

The threshold is a real number that is subtracted from the weighted sum of the inputs. When it is positive, it is referred to as the bias, which is a real random number between 0 and 1. In artificial neurons, the bias term is added to the weighted sum. It gives a base level of activity to the output neuron. If the bias b is negative, the weighted sum must be a positive value ($>-b$) for the output neuron to produce an output. For simplicity, it can be regarded as another input/weight pair, where $w_0 = b$ and $x_0 = 1$.

The bias can also be thought of as equivalent to the intercept in a linear regression model. In the same way as a linear regression model will pass through the origin without an intercept, the hyperplane in the multidimensional space will also pass through the origin if there is no bias. Bias is an external parameter of the neuron that has the effect of applying an 'affine transformation'[1] to the output u_k of the linear combiner. It can be introduced in one of two ways:

$$y_k = f(v_k) = f(u_k + b_k) = f\left(\sum_{j=1}^{N} w_{kj}x_j + b_k\right) \tag{8.1}$$

or

$$y_k = f\left(\sum_{j=0}^{N} w_{kj}x_j\right) \tag{8.2}$$

with $x_0 = 1$ and $w_{k0} = b_k$; b_k is the bias for neuron k; v_k is referred to as the induced local field or activation potential, and $f(.)$ is the activation function. Adding a bias value is equivalent to subtracting a threshold value. The bias has the effect of increasing (when the bias is positive) or decreasing (when the bias is negative) the net input of the activation function. The output y_k may become an input to another neuron or may be exposed to the external environment.

There are many types of activation functions (see Section 8.5) but the simplest (for binary classification) is the hardlimiter defined as

$$f(v) = \begin{cases} 1 & \text{if } v \geq 0 \\ 0 & \text{if } v < 0 \end{cases} \tag{8.3}$$

---

[1] Affine transformation preserves collinearity; that is, all points initially lying on a straight line will still lie on a line after the transformation. In addition, it preserves the ratios of distances (e.g., the midpoint of a line segment remains the midpoint after transformation). It does not preserve angles and lengths.

Thus, the basic model performs a weighted sum of its inputs and compares it to the internal threshold level. If the threshold is exceeded, the neuron will turn on. Otherwise, it will not. This kind of system where the information flow is in the forward direction is known as a feed-forward type of neural network. A single neuron can only perform binary classification. This can be illustrated with a simple artificial neuron (Figure 8.2), which has two inputs x_1 and x_2, weights w_1 and w_2, and a threshold T.

The computational procedure for a simple neuron for binary classification is illustrated in Table 8.1. For this example, the input variables x_1 and x_2 can take any value. The weights are obtained after exposing the network to a set of training data, which will be pairs of inputs and corresponding target outputs. Once trained, the network can be used to obtain the outputs corresponding to unseen inputs. Columns 1 and 2 give the unseen inputs, columns 3 through 5 the calibrated weights and threshold, and the last column the neuron output. The value of y is either 0 or 1 regardless of the range of values for the inputs x_1 and x_2. The activation function in this case is the hardlimiter, which produces an output if the threshold value is equalled or exceeded and no output otherwise.

An artificial neuron also has other names; for example, binary neuron, linear threshold function, McCulloch–Pitts neuron, etc.

Two or more single neurons in the same layer can classify linearly separable inputs into multiple classes. Single neuron is the basis of adaptive filter that is widely used in signal processing. It uses the least mean square (LMS) algorithm, which is also known as the delta rule (Widrow and Hoff, 1960). In a linear dynamical system that has inputs defined by $x_1(n)$, $x_2(n)$, $x_3(n)$,..., $x_N(n)$, which may be time or space dependent, and is characterized by a set of synaptic weights $w_k(n)$, the estimation procedure consists of a filtering process in which the outputs $y(n)$ are processed in response to the inputs $x(n)$. The errors $e(n)$ are obtained by comparing with the target outputs, and the synaptic weights are adjusted according to the errors by an adaptive process. These two processes constitute a feedback loop. The errors are given by

$$e(n) = d(n) - y(n) \tag{8.4}$$

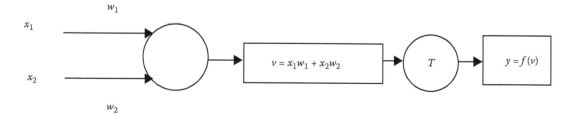

Figure 8.2 Schematic diagram of an artificial neuron.

Table 8.1 Computations in a simple artificial neuron

x_1	x_2	w_1	w_2	T	y
10	5	0.6	0.4	1	1
5	8	0.4	0.5	1	1
3	10	0.8	0.7	0	1
0.5	−8	0.5	0.5	0.5	0

where $d(i)$'s are the target or desired outputs. The objective (cost) function is normally defined as

$$E = \frac{1}{2} \sum_{n=1}^{N} e^2(n) \qquad (8.5)$$

where the factor 1/2 is introduced for simplification of subsequent analysis. The optimization of the weights can be done using several techniques some of which are briefly described in the next section.

8.3 UNCONSTRAINED OPTIMIZATION TECHNIQUES

Adaptive filtering is based on the concept of *local iterative descent*, which aims at reducing the error at each iteration. Starting with an initial guess $w(0)$, the method generates a sequence of weights $w(1)$, $w(2)$,..., such that the cost function $E(w)$ is reduced at each iteration; that is,

$$E(w(n + 1)) < E(w(n))$$

It is expected that the algorithm will converge, but there is no guarantee. There are three unconstrained optimization techniques that follow the concept of *local iterative descent*. They are applicable to linear adaptive filters as well as to neural networks in general.

8.3.1 Method of steepest descent

In this method, adjustments are made in the direction of steepest descent; that is, in the direction opposite to the gradient vector $\Delta E(w)$, which is defined as

$$\mathbf{g} = \nabla E(w) = \left[\frac{\partial E}{\partial w_1}, \frac{\partial E}{\partial w_2}, \frac{\partial E}{\partial w_3} \cdots \frac{\partial E}{\partial w_N} \right]^T \qquad (8.6)$$

The weight adjustment then is

$$\Delta w(n) = w(n + 1) - w(n) = -\eta g(n) \qquad (8.7)$$

where η is the learning rate parameter and $g(n)$ is the gradient vector. To satisfy the condition given in Equation 8.6, a first-order Taylor series expansion (which is justified for small η) around $w(n)$ to approximate $E(w(n + 1))$ is used:

$$E(w(n + 1)) \cong E(w(n)) + g^T(n)\Delta w(n) \qquad (8.8)$$

Substituting from Equation 8.7 gives

$$E(w(n+1)) \cong E(w(n)) - \eta g^T(n)g(n)$$
$$= E(w(n)) - \eta \|g(n)\|^2 \qquad (8.9)$$

When η is small, the algorithm is overdamped, and the trajectory is smooth; when η is large, it is underdamped, and therefore zigzag; when η is greater than a critical value, it is unstable. Typically, η ranges from about 0.2 to about 0.8. This method converges to the optimal solution slowly.

8.3.2 Newton's method (quadratic approximation)

This method uses a second-order Taylor series expansion and minimizes a quadratic approximation of the cost function $E(w)$ around the current point $w(n)$ as shown below.

$$\Delta E(w(n)) = E(w(n+1)) - E(w(n)) = g^T(n)\Delta w(n) + \frac{1}{2}\Delta w^T(n)H(n)\Delta w(n) \tag{8.10}$$

where $H(n)$ is the Hessian matrix of $E(w)$ defined as

$$H = \nabla^2 E(w) = \begin{bmatrix} \dfrac{\partial^2 E}{\partial w_1^2}, \dfrac{\partial^2 E}{\partial w_1 \partial w_2}, \dfrac{\partial^2 E}{\partial w_1 \partial w_3}, \cdots \dfrac{\partial^2 E}{\partial w_1 \partial w_M} \\[2ex] \dfrac{\partial^2 E}{\partial w_2 \partial w_1}, \dfrac{\partial^2 E}{\partial w_2^2}, \dfrac{\partial^2 E}{\partial w_2 \partial w_3}, \cdots, \dfrac{\partial^2 E}{\partial w_2 \partial w_M} \\[2ex] \cdots \\[1ex] \cdots \\[1ex] \dfrac{\partial^2 E}{\partial w_M \partial w_1}, \dfrac{\partial^2 E}{\partial w_M \partial w_2}, \dfrac{\partial^2 E}{\partial w_M \partial w_3}, \cdots, \dfrac{\partial^2 E}{\partial w_M^2} \end{bmatrix} \tag{8.11}$$

where M is the number of weights. For this method, the $E(w)$ needs to be twice continuously differentiable with respect to w. Weight adjustment is done according to

$$w(n + 1) = w(n) + \Delta w(n) = w(n) - H^{-1}(n)g(n) \tag{8.12}$$

It converges fast but asymptotically. For the method to work, H must be positive definite[2] for all n, although it is not guaranteed all the time. Hessian matrix has significant influence on the dynamics of the back-propagation (BP) and LMS algorithms.

8.3.3 Gauss–Newton method

This method is applicable to a cost function expressed in the form of Equation 8.5. Weight update is given as

$$w(n + 1) = w(n) - (J^T(n)J(n))^{-1}(n)e(n) \tag{8.13}$$

where the Jacobian matrix $J(n)$ of $e(n)$ is defined as (evaluated at $w = w(n)$)

---

[2] An $M \times M$ matrix R is said to be non-negative if $a^T R a \geq 0$ for any vector a, and positive definite if $a^T R a > 0$. For a positive definite matrix, R^{-1} exists.

$$J(n) = \begin{bmatrix} \dfrac{\partial e(1)}{\partial w_1}, \dfrac{\partial e(1)}{\partial w_2}, \dfrac{\partial e(1)}{\partial w_3}, \cdots \dfrac{\partial e(1)}{\partial w_M} \\[2ex] \dfrac{\partial e(2)}{\partial w_1}, \dfrac{\partial e(2)}{\partial w_2}, \dfrac{\partial e(2)}{\partial w_3}, \cdots \dfrac{\partial e(2)}{\partial w_M} \\[2ex] \cdots \\[1ex] \cdots \\[1ex] \dfrac{\partial e(n)}{\partial w_1}, \dfrac{\partial e(n)}{\partial w_2}, \dfrac{\partial e(n)}{\partial w_3}, \cdots \dfrac{\partial e(n)}{\partial w_M} \end{bmatrix} \tag{8.14}$$

For this method to work, the matrix $(J^T(n)J(n))$ must be non-singular. Each row of the Jacobian matrix corresponds to one example of the training data set, and each column corresponds to one of the parameters to be determined. Methods described in Sections 8.3.1 through 8.3.3 are suited to linear adaptive filtering.

8.3.4 LMS algorithm

The LMS algorithm is the most widely used adaptive filtering algorithm. The objective of weight updating is to reduce the difference between the desired output and the network output. A measure of this difference is the mean square error (MSE) and the adjustment is done in a direction opposite to the direction of the error gradient. The LMS algorithm operates by performing an approximate steepest descent on the MSE surface in the weight space. In the weight space, the steepest descent follows a quadratic trajectory; however, in the LMS algorithm, the weight vector follows a random trajectory and hence it is sometimes called 'stochastic gradient algorithm'. It is also sometimes identified as the stochastic implementation of the steepest descent algorithm. The LMS algorithm produces only an estimate of the weight vector.

The LMS algorithm is based on the instantaneous value of the objective function instead of its expected value, which is defined as

$$e(w) = \frac{1}{2} e^2(n) \tag{8.15}$$

at time n.

When differentiated,

$$\frac{\partial e(w)}{\partial w} = e(n) \frac{\partial e(n)}{\partial w}$$

and the instantaneous error is given as

$$e(n) = d(n) - x^T(n)w(n) \tag{8.16}$$

from which, $\dfrac{\partial e(n)}{\partial w} = -x(n).$

Therefore,

$$\frac{\partial e(w)}{\partial w} = -x(n)e(n) \tag{8.17}$$

Using Equation 8.17 as an estimate of the gradient vector, the weight updating is given as

$$\hat{w}(n+1) = \hat{w}(n) + \eta x(n)e(n) \qquad (8.18)$$

where $\hat{w}$ and x are vectors. In the LMS algorithm, the initial weights are normally set to zero or some small random values. The procedure consists of initializing the weights, choosing the next training pair (x, d), computing the outputs, computing $y_j = w^T x$, computing the errors (Equation 8.4), and updating the weights according to Equation 8.18. The LMS algorithm is a special case of the BP algorithm. The steps involved in the LMS algorithm are

- Initialize the weights to small random numbers.
- Choose the next training pair (x, d).
- Compute outputs.
- Compute $y_j = w^T x$.
- Compute errors $e_j = d_j - y_j$.
- Update weights according to $w_j(n + 1) = w_j(n) + \eta e_j x$.

One complete cycle of training with all input training data is called an 'epoch', and the training is complete when it goes through without error. If a new input data set unseen in the training set is presented to the network, the network will tend to generalize with an output value similar to the target values for input values in the neighbourhood of the unseen data.

8.4 PERCEPTRON

Perceptron (the term 'perceptron' was coined by Frank Rosenblatt in 1962 and is used to describe the connection of simple neurons into networks) is a neural network with one layer of input nodes connected to an output layer. It is the simplest feed-forward neural network and is built around a non-linear neuron. It has a linear summation, an externally applied bias, and a hardlimiter activation function, thereby making it non-linear. Linear neuron has only the linear summation. Perceptron is a binary classifier that uses the rule given in Equation 8.3. The hardlimiter activation function produces an output of +1 if the hardlimiter input is positive and zero if it is negative. For a single linear neuron, the output is the same as the induced field; that is, $y_k = v_k$.

SLP, which has only one layer of activity is necessarily feed-forward. The input layer that does not have any activity is not counted. The goal of the SLP is to correctly classify the inputs into two classes (binary classification), which is possible only when the two classes are linearly separable. It consists of one or more artificial neurons arranged in parallel. These networks are simplified versions of the real nervous system where some properties are exaggerated while others are ignored. The connection weights and the bias in an SLP can be determined by presenting a set of input and corresponding output data. Once trained, the perceptron can be used to produce the correct output when an unseen input is presented. Since they can generalize from training data, SLPs are specially suited to simple classification problems. The output is related to the inputs by Equation 8.1 or 8.2.

The SLP is a binary classifier that uses the hardlimiter activation function and is applicable only to linearly separable classes. The classes can be considered as two decision regions separated by the hyperplane

$$\sum_{i=1}^{N} w_i x_i + b = 0 \qquad (8.19)$$

The bias merely shifts the decision boundary from the origin (affine transformation). Adding more neurons allows the classification to extend to more than two classes. SLP can be used to represent many Boolean functions. The training of the SLP is carried out according to the perceptron learning algorithm, which uses an error-correcting rule as summarized below:

- Start with an initial random set of weights, usually small. Setting $w(0) = 0$ would satisfy this condition.
- At time step n, activate the perceptron by applying the input vector $x(n)$ and desired output vector $d(i)$.
- Compute the actual response $y(n) = \mathrm{Sgn}[w^T(n)x(n)]$, where Sgn, the Signum function, is given as $\mathrm{Sgn}(v) = \begin{cases} +1 & \text{if } v > 0 \\ -1 & \text{if } v < 0 \end{cases}$.
- If the perceptron output matches with the desired output, no weight adjustment is done. Otherwise,
- Update weights according to $w(n + 1) = w(n) + \eta\,[d(n) - y(n)]\,x(n)$, where

$$d(n) = \begin{cases} +1 & \text{if } x(n) \text{ belongs to class 1} \\ -1 & \text{if } x(n) \text{ belongs to class 2} \end{cases}, \text{ and } \eta \text{ is a learning rate parameter.}$$

- $n = n + 1$ and repeat until the stopping criterion is met.

The perceptron convergence theorem guarantees convergence to a solution within a finite number of steps for linearly separable data sets. When the data are not linearly separable, the learning rule may fail. For a neuron that consists of a linear summation followed by a non-linear unit, the SLP can perform pattern classifications that are linearly separable regardless of the type of non-linearity (i.e., whether it is hardlimiter or sigmoid, etc.).

SLP has limitations: it can have only two outcomes as the output and that it is applicable only to linearly separable sets of data. Minsky and Papert (1969) proved mathematically that the SLP cannot solve classification problems that are linearly inseparable. Perceptron is built around non-linear neurons. LMS algorithm is built around linear neurons.

8.4.1 Linear separability

When two classes of inputs can be separated by a straight line, they are said to be linearly separable, as in Figure 8.3a. When they can be separated by a non-linear function, or by more than one linear function, they are said to be non-linearly separable, as in Figure 8.3b. Consider for example the points (0,0), (0,1), (1,0), (1,1) in the x_1–x_2 plane (Figure 8.3c). Any line with $x_1 < 1$ and $x_2 < 1$, such as the one drawn, can separate the points (0,1), (1,1), (1,0) from the point (0,0). It is a linear classifier and the set of points is linearly separable. Not all sets of points are linearly separable. There are many non-linearly separable input patterns in the real world. For example, consider the same four points but the group containing the points (0,0) and (1,1) cannot be separated by a line from the group containing the points (1,0) and (0,1). They can, however, be separated by two lines similar to the ones shown in Figure 8.3d. In an n-dimensional space, two groups are said to be linearly separable if they can be separated by a $(n - 1)$ dimensional hyperplane.

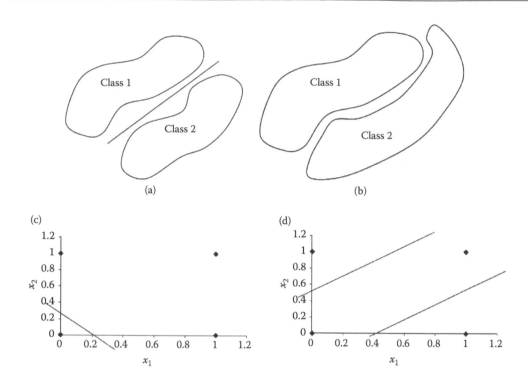

Figure 8.3 (a) Linearly separable classes, (b) Non-linearly separable classes, (c) Linearly separable points, and (d) Non-linearly separable points.

8.4.2 'AND', 'OR', and 'XOR' operations

Perceptrons can be used to represent some basic Boolean operators such as 'AND' and 'OR'. A Boolean operator can be thought of as a box with many inputs and one output. They are used for classifying input variables that produce two outcomes, one complement of the other. The operator 'AND' implies that the output will be on (meaning 1, or high) if and only if all inputs are on. The operator 'OR' implies that the output will be on (meaning 1, or high) if any input is on. The exclusive 'OR', or 'XOR', will be on if the inputs are different. With the 'AND' operation, the output neuron will fire only if both inputs are firing (meaning high or 1). For example, consider an SLP with two neurons at the input layer and a single output with equal connection weights of 1 and a threshold of 1.5 (Figure 8.4a). The output from the perceptron will match the target outputs given in Table 8.2 (also called the 'truth table') for all input conditions (0,0), (0,1), (1,0), and (1,1).

In the 'OR' operation, the output neuron will fire only if it receives input from either or both neurons. The 'OR' perceptron will also match with the target outputs given in Table 8.2 for the same input conditions and equal connection weights, but with a threshold value of 0.9 (Figure 8.4b). The main difference between the 'AND' and 'OR' operations is the magnitude of the threshold value. In the exclusive 'OR', or 'XOR', problem, the input conditions (0,0), (0,1), (1,0) can be matched with the target outputs, but not the input condition (1,1). In other words, the SLP cannot solve the 'XOR' problem. It can be solved by the addition of an extra layer of neurons called the hidden layer. This is illustrated with the network shown in Figure 8.4c in which the connection weights from input to hidden layers take equal values of 1, and those from the hidden layer to the output layer take values of −1 and 1. The

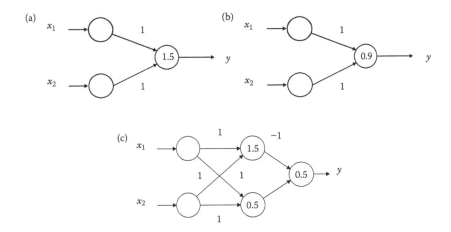

Figure 8.4 (a) 'AND' operation, (b) 'OR' operation, and (c) 'XOR' operation.

Table 8.2 Truth table for 'AND', 'OR', and 'XOR' operations

x_1	x_2	x_1 AND x_2	x_1 OR x_2	x_1 XOR x_2
0	0	0	0	0
0	1	0	1	1
1	0	0	1	1
1	1	1	1	0

threshold values, respectively, are 1.5 and 0.5 at the hidden layer and 0.5 at the output layer. All target values can be matched with the output values of the network for all the input conditions given in Table 8.2.

The solutions shown above are not the only possible ones for the 'AND', 'OR', or 'XOR' problems. In a simple network such as the one shown above, the detailed estimation of the parameters can be easily demonstrated. However, in a network with a large number of inputs, it is not so simple. Since the inputs, weights, and the thresholds can take any real value, positive (excitatory) or negative (inhibitory), there can be an infinite number of possible combinations. How these values can be determined is by a process of training by providing a set of input and expected output values.

SLPs cannot classify input patterns that are not linearly separable. The simplest example is the exclusive OR, or XOR, Boolean function.

8.4.3 Multilayer perceptron (MLP)

The standard MLP (Figure 8.5) is a cascade of SLPs consisting of a layer of input nodes, a layer of output nodes, and one or more intermediate layers called 'hidden layers'. It is the most popular neural network architecture in practical applications. Each layer in an MLP uses a linear combination function. The inputs are fully connected to the first hidden layer, each hidden layer is fully connected to the next, and the last hidden layer is fully connected to the outputs. In exceptional cases, however, there can be direct connections from input layer to output layer as well as skipping of layers. Each node in an MLP network has a

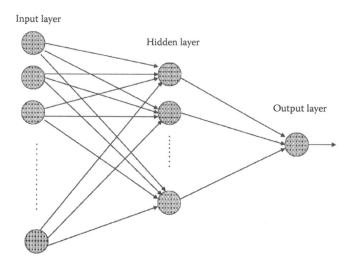

Figure 8.5 Schematic diagram of an MLP network.

response $f(w^T x)$ where x is the vector of output activations from the preceding layer, w is a vector of weights, and f is a bounded non-decreasing, non-linear activation function such as the sigmoid. The activation functions in the hidden layer(s) introduce the non-linearity into the network. The activation function should be smooth, differentiable, and preferably bounded if the BP algorithm is to work, as opposed to the hardlimiter used in Rosenblatt's perceptron.

It has been proven (Cybenko, 1989) that an MLP network with a single hidden layer can approximate arbitrarily well any functional continuous mapping from one finite-dimensional space to another, provided the number of hidden units is sufficiently large. MLP networks are said to possess *universal approximation capability*. This result is important because it offers a theoretical justification for using single hidden layer MLP for function approximation.

The main problem in MLP function approximation is how to determine the number of nodes (neurons) in the hidden layer. Too few nodes may not model the process adequately and too many will require a long computational time as well as resulting in overfitting. The objective of ANNs is to model the signal; however, overfitting will fit into the noise as well, producing a very good fit that lacks generalization properties when presented with unseen data. The optimal number of nodes in the hidden layer is determined by cross validation using different numbers of nodes, which is a trial-and-error approach.

8.4.4 Optimal structure of an MLP

One of the difficult problems in the MLP formulation is the choice of the optimal network structure as well as the number of hidden layers and the number of neurons in each hidden layer. The number of hidden layers can be relatively easily fixed on the basis of the *universal approximation theorem* (Haykin, 1999, p. 208–209), which states that a single hidden layer is sufficient for an MLP to compute a uniform ε (> 0) approximation to a given training data set represented by inputs x_i; $i = 1, 2,...,N$, and corresponding target output of d_i; $i = 1, 2,...,N$). The theorem, however, does say that a single hidden layer is optimum in the sense of the ease of implementation and generalization. Notwithstanding the above theorem, it is a fact that with more hidden layers, a network is able to model more complex problems. The aim should, however, be generalization rather than overfitting.

Two approaches could be followed to fix the number of neurons in the hidden layer: a building process and a pruning process. In the former, a network with the minimum number of neurons in a single hidden layer is first attempted, and if it does not work, the complexity of the network is increased by gradually adding more neurons into the hidden layer, and if necessary, by adding more hidden layers. Haykin (1999, p. 193) has demonstrated in a computer experiment based on success probability that two hidden nodes are sufficient for pattern classification. In the pruning process, a complex network with a large number of neurons in one or more hidden layers is first attempted, and the network is gradually pruned by removing neurons in the hidden layers gradually until satisfactory outcomes are attained. It is, of course, more logical to follow the building process, since in the pruning process there is no objective criterion to choose the starting network. The building process is also in conformity with the principle of parsimony. A problem that may occur in the case of a single hidden layer is that the neurons tend to interact with each other globally, resulting in improvements made at one point being offset by deterioration at another point.

Another issue that needs to be resolved is the size of the training data set. It has been recommended (Haykin, 1999, p. 208) that the training data size should be of the order of $\frac{W}{\varepsilon}$, where W is the total number of free parameters in the network and ε is the fraction of classification errors permitted in the test data. For example, a 10% error means $\varepsilon = 0.1$.

8.5 TYPES OF ACTIVATION FUNCTIONS

The activation function in the hidden layers(s) introduces the non-linearity into the network, thereby making it a powerful input–output model. An activation function should have the property that it is continuously differentiable and preferably bounded. There are many possible types of activation functions.

8.5.1 Linear activation function (unbounded)

It is also sometimes referred to as the identity function (Figure 8.6a). Mathematically, it is of the form

$$f(x) = x \tag{8.20}$$

8.5.2 Saturating activation function (bounded)

$$f(x) = \begin{cases} 0 & \text{for } x < 0 \\ x & \text{for } 0 \le x \le 1 \\ 1 & \text{for } x > 1 \end{cases} \tag{8.21}$$

It is similar to the linear activation function except that it is bounded between 0 and 1 (Figure 8.6b).

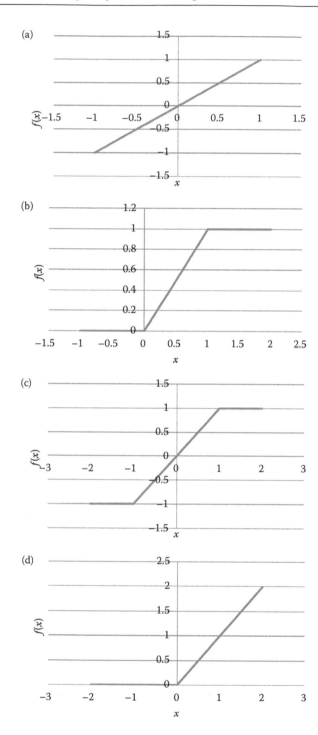

Figure 8.6 (a) Linear activation function (unbounded), (b) Saturating activation function (bounded), (c) Symmetric saturating function (bounded), (d) Positive linear function, (e) Hardlimiter (Heaviside; McCulloch–Pitts) function, (f) Symmetric hardlimiter function, (g) Triangular function, (h) Sigmoid function, (i) Tanh function, (j) Multiquadratic radial basis function, (k) Inverse multiquadratic radial basis function, (l) Gaussian radial basis function, (m) Polyharmonic spline function, (n) Polyharmonic function (for positive values of x), and (o) Softmax function for the data shown in Table 8.3.

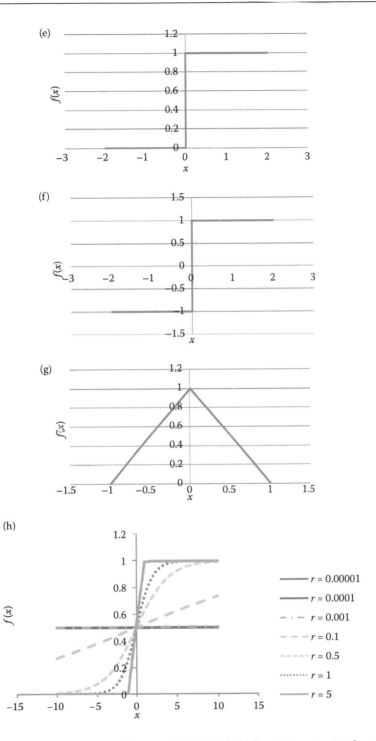

Figure 8.6 (Continued) (a) Linear activation function (unbounded), (b) Saturating activation function (bounded), (c) Symmetric saturating function (bounded), (d) Positive linear function, (e) Hardlimiter (Heaviside; McCulloch–Pitts) function, (f) Symmetric hardlimiter function, (g) Triangular function, (h) Sigmoid function, (i) Tanh function, (j) Multiquadratic radial basis function, (k) Inverse multiquadratic radial basis function, (l) Gaussian radial basis function, (m) Polyharmonic spline function, (n) Polyharmonic function (for positive values of x), and (o) Softmax function for the data shown in Table 8.3. (*continued*)

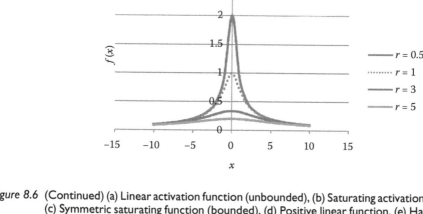

Figure 8.6 (Continued) (a) Linear activation function (unbounded), (b) Saturating activation function (bounded), (c) Symmetric saturating function (bounded), (d) Positive linear function, (e) Hardlimiter (Heaviside; McCulloch–Pitts) function, (f) Symmetric hardlimiter function, (g) Triangular function, (h) Sigmoid function, (i) Tanh function, (j) Multiquadratic radial basis function, (k) Inverse multiquadratic radial basis function, (l) Gaussian radial basis function, (m) Polyharmonic spline function, (n) Polyharmonic function (for positive values of x), and (o) Softmax function for the data shown in Table 8.3.

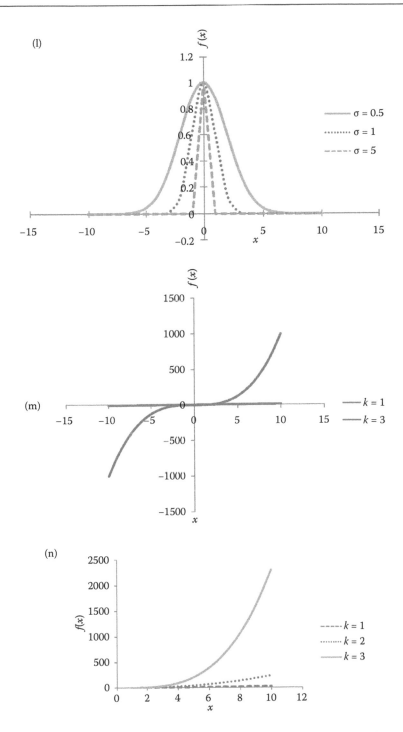

Figure 8.6 (Continued) (a) Linear activation function (unbounded), (b) Saturating activation function (bounded), (c) Symmetric saturating function (bounded), (d) Positive linear function, (e) Hardlimiter (Heaviside; McCulloch–Pitts) function, (f) Symmetric hardlimiter function, (g) Triangular function, (h) Sigmoid function, (i) Tanh function, (j) Multiquadratic radial basis function, (k) Inverse multiquadratic radial basis function, (l) Gaussian radial basis function, (m) Polyharmonic spline function, (n) Polyharmonic function (for positive values of x), and (o) Softmax function for the data shown in Table 8.3. (*continued*)

(o)

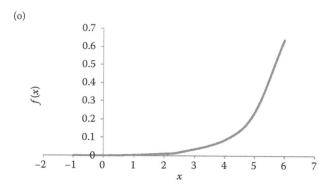

Figure 8.6 (Continued) (a) Linear activation function (unbounded), (b) Saturating activation function (bounded), (c) Symmetric saturating function (bounded), (d) Positive linear function, (e) Hardlimiter (Heaviside; McCulloch–Pitts) function, (f) Symmetric hardlimiter function, (g) Triangular function, (h) Sigmoid function, (i) Tanh function, (j) Multiquadratic radial basis function, (k) Inverse multiquadratic radial basis function, (l) Gaussian radial basis function, (m) Polyharmonic spline function, (n) Polyharmonic function (for positive values of x), and (o) Softmax function for the data shown in Table 8.3.

Table 8.3 Inputs and outputs for a softmax function

x	f(x)
−1	0.000581
4	0.086194
2.5	0.019233
−0.5	0.000958
0.4	0.002355
6	0.636894
2	0.011665
1.5	0.007075
−0.75	0.000746
5	0.2343

8.5.3 Symmetric saturating activation function (bounded)

$$f(x) = \begin{cases} -1 & \text{for } x < -1 \\ x & \text{for } -1 \leq x \leq 1 \\ 1 & \text{for } x > 1 \end{cases} \tag{8.22}$$

This is similar to the saturating activation function and is bounded between ±1 (Figure 8.6c).

8.5.4 Positive linear activation function

$$f(x) = \begin{cases} 0 & \text{for } x < 0 \\ x & \text{for } x > 0 \end{cases} \tag{8.23}$$

8.5.5 Hardlimiter (Heaviside function; McCulloch–Pitts model) activation function

The hardlimiter transfer function (Heaviside function) gives a perceptron the ability to classify input vectors by dividing the input space into two regions (Figure 8.6e). Specifically, outputs will be 0 if the net input is <0, or 1 if the net input is ≥0. It is also called the McCulloch–Pitts model or the threshold function. Mathematically, it is of the form

$$f(x) = \begin{cases} 0 & \text{if } x < 0 \\ 1 & \text{if } x \geq 0 \end{cases} \tag{8.24}$$

8.5.6 Symmetric hardlimiter activation function

$$f(x) = \begin{cases} -1 & \text{if } x \leq 0 \\ 1 & \text{if } x > 0 \end{cases} \tag{8.25}$$

8.5.7 Signum function

$$f(x) = \begin{cases} 1 & \text{for } x > 0 \\ 0 & \text{for } x = 0 \\ -1 & \text{for } x < 0 \end{cases} \tag{8.26}$$

This is also called the sign function, $f(x) = \text{Sign}(x)$, and it is similar to the symmetric hardlimiter function.

8.5.8 Triangular activation function

$$f(x) = \begin{cases} 0 & \text{for } x < -1 \\ 1 + x & \text{for } -1 \leq x \leq 0 \\ 1 - x & \text{for } 0 \leq x \leq 1 \\ 0 & \text{for } x > 1 \end{cases} \tag{8.27}$$

8.5.9 Sigmoid logistic activation function

The sigmoid is perhaps the most popular family of activation functions (Figure 8.6h). It derives the name from the S-shape it has. It is bounded, continuous, and continuously differentiable. The sigmoid function attains its maximum (1) and minimum (0) values asymptotically. For practical purposes, neurons that use the sigmoid function should be considered fully activated at around 0.9 and turned off at around 0.1. An example of the sigmoid family is the logistic function, which is of the form

$$f(x) = \frac{1}{1 + e^{-rx}} \tag{8.28}$$

In this equation, r is the steepness parameter, and as $r \to \infty$, sigmoid $\to$ hardlimiter function. As $r \to 0$, $f(x) \to 0.5$. For the implementation of the BP algorithm, which is used in MLPs, it is necessary to ensure that the activation function is continuously differentiable. The first derivative of the sigmoid function is

$$f'(x) = f(x)(1 - f(x)) \tag{8.29}$$

8.5.10 Sigmoid hyperbolic tangent function

The hyperbolic tangent (tanh) is also continuous, bounded between ±1, continuously differentiable, and belongs to the sigmoid family. The tanh function also attains its maximum (1) and minimum (–1) values asymptotically. For practical purposes, neurons that use of the tanh function should be considered fully activated at around 0.9 and turned off at around –0.9. The hyperbolic tangent, which yields both positive and negative values, is known to train faster than those that have only positive values. Mathematically, it is of the form

$$f(x) = \tanh(rx) = \frac{1 - e^{-rx}}{1 + e^{rx}} \tag{8.30}$$

The tanh function also tends to the symmetric hardlimiter function as $r \to \infty$, and approaches zero as $r \to 0$. The first derivative of the tanh function is

$$f'(x) = 1 - [\tanh(rx)]^2 \tag{8.31}$$

The tanh function, which is 'antisymmetric',[3] is preferred for training purposes (Figure 8.6i). A preferred tanh function proposed by LeCun (1989, 1993) as referenced in Haykin (1999, p. 179) is of the form

$$f(x) = a\tanh(rx) \tag{8.32}$$

with $a = 1.7159$ and $r = 2/3$, and has the following properties: $f(1) = 1$; $f(-1) = -1$; $f(0) = ar = 1.1439$, and $f''(x)$ attains its maximum value at $x = 1$. This function has an almost linear part and a saturation part. When $|x| > 1$, the function lies in the saturation part, and when $|x| < 1$, it lies in the almost linear part.

8.5.11 Radial basis functions

There are several types of radial basis functions, such as the multiquadratic, inverse multiquadratic, polyharmonic spline, thin plate spline, and the Gaussian functions. They are respectively given as follows:

8.5.11.1 Multiquadratic

$$f(x) = (x^2 + r^2)^{1/2}; \ r > 0 \tag{8.33a}$$

---

[3] $f(-x) = -f(x)$.

8.5.11.2 Inverse multiquadratic

$$f(x) = \frac{1}{(x^2 + r^2)^{1/2}}; r > 0 \qquad (8.33b)$$

8.5.11.3 Gaussian

$$f(x) = e^{-\frac{x^2}{2\sigma^2}}; \sigma > 0 \qquad (8.33c)$$

The multiquadratic function is unbounded, whereas the inverse multiquadratic and Gaussian functions are bounded with $f(x) \to 0$ as $x \to \pm\infty$.

8.5.11.4 Polyharmonic spline function

$$f(x) = \begin{cases} x^k; & k = 1, 3, 5, \ldots \\ x^k \ln(x); & k = 2, 4, 6, \ldots \end{cases} \qquad (8.33d)$$

It can be seen that the polyharmonic spline function (Figure 8.6m) for even values of k is defined only for positive values of x.

8.5.11.5 Thin plate spline function

Thin plate spline is a special case of the polyharmonic spline (Figure 8.6n) and is given as

$$f(x) = x^2 \ln(x) \qquad (8.33e)$$

8.5.12 Softmax activation function

$$f(x) = \text{softmax}(x) = \frac{e^{x_i}}{\sum\limits_{j=1}^{n} e^{x_j}} \qquad (8.34)$$

where x_i's are the inputs, $f(x)$ are the outputs, and n is the number of nodes. For example, in a network with three inputs and two outputs,

$$f(x_1) = \frac{e^{x_1}}{e^{x_1} + e^{x_2}}; \quad f(x_2) = \frac{e^{x_2}}{e^{x_1} + e^{x_2}} \text{ and } f(x_1) + f(x_2) = 1$$

With four inputs $x_1 = 0$, $x_2 = 1$, $x_3 = -0.5$, and $x_4 = 0.5$, it is easy to see that the outputs are $f(x_1) = 0.17$, $f(x_2) = 0.46$, $f(x_3) = 0.10$, and $f(x_4) = 0.28$.

The softmax activation function squashes unevenly distributed data into an appropriate range. The degree to which a value is squashed depends on its distance from the mean and the standard deviation of the data set. The further the data point from the mean, the greater the squashing. It enforces the constraints that the outputs lie between 0 and 1, and add up to unity. If the data are evenly linearly distributed, the softmax function has the same effect as a linear squashing function. The softmax function is particularly useful in situations where the outputs need to be interpreted as posterior probabilities for a target variable. Sometimes the softmax is also known as a 'multiple logistic' function (Figure 8.60).

Activation functions are also needed at the output layer. When the target output values are bounded, a bounded activation such as the sigmoid is necessary. When the target output values are unbounded, an unbounded activation function is preferred. Very often, an identity function, which is equivalent to an activation function, is used in the latter case. Exponential-type activation functions are used when the target output values are positive but unbounded. There are also studies carried out by some investigators where a linear function has been used in the output layer (e.g., Safavi et al., 2003; Karul et al., 2000). It has been reported that the linear transfer function gives faster training for regression type of ANN models (Tian, 2007). There is also no proof that the non-linear transfer function in the output layer can always produce better solutions than its linear counterpart.

8.6 TYPES OF ARTIFICIAL NEURAL NETWORKS

A system in which a large number of neurons are interconnected with exposure to the external environment is called an artificial neural network. The power of ANNs lies in their large number of interconnections. Very often the neurons are arranged in layers, with each layer having a number of neurons. ANNs constitute a type of non-parametric regression models. This means that the parameters in ANNs have no meaning unlike in parametric type of regression models where the parameters convey some intrinsic meaning.

ANNs can be classified according to criteria such as learning strategies, direction of information flow, architecture, learning algorithms, type of basis function used, etc. On the basis of learning strategies, they can be fixed or adaptive. Adaptive networks can be further subdivided into error feedback and *self-organizing maps* (*Kohonen networks*). On the basis of the direction of information flow, ANNs can be divided into *feed-forward* networks in which information flows only in the forward direction, and feedback (or *recurrent*) networks in which the information flows in both directions. *Product unit neural network* (PUNN) is a relatively new type of network where non-linear basis functions are used.

ANNs can also be classified on the basis of their architecture. In the feed-forward type, information flows unidirectionally from input to output. The output therefore is a function of inputs only. In recurrent (or feedback) type, information can flow forwards and backwards. In this type, the output from any neuron can be an input to the same neuron or to a neuron in the preceding layer. The architecture of a neural network is determined by specifying the number of layers, the number of neurons in each layer, the layout of the connections, the connection (synaptic) weights, and the biases. The network size has a major effect on the performance. A smaller size network is preferred for better generalization.

They can also be classified according to the three basic learning algorithms – supervised, reinforced, and unsupervised. In the supervised learning mode, the synaptic weights are adjusted according to a measure of the error between the network output and the target or expected output for a given input. This is the most common approach of training neural networks. The delta rule and generalized delta rule (back-propagation) belong to this category. In reinforced learning, the network weights are adjusted so as to develop an input/output behaviour that maximizes the probability of receiving a reward and minimizes that of receiving a penalty. In the unsupervised learning mode, target or expected outputs are not required. The synaptic weights are adjusted by clustering the input patterns into similar groups.

8.6.1 Feed-forward neural networks

In a feed-forward-type neural network, the induced field at the hidden layer can be expressed as

$$v_{pj}^h = \theta_j^h + \sum_{i=1}^{N} w_{ji}^h x_{pi} \tag{8.35}$$

where w_{ji}^h is the connection weight from the i-th input node to the j-th hidden node, superscript h refers to quantities in the hidden layer, θ_j^h is the bias term (bias weight) for the hidden layer, p is an index for the number of input data patterns, and N is the dimension of the input vector. There can be single layer as well as multilayer feed-forward neural networks.

The output from the hidden layer i_{pj} is obtained by applying the activation function to the input. Then,

$$i_{pj} = f_j^h \left(v_{pj}^h \right) \tag{8.36}$$

In the same way, the input to the output layer is the summation of the outputs from the hidden layer, which can be written as

$$v_{pk}^o = \theta_k^o + \sum_{j=1}^{L} w_{kj}^o i_{pj} \tag{8.37}$$

where superscript 'o' refers to quantities in the output layer, θ_k^o is the bias term (bias weight) for the output layer, and L is the dimension of the hidden layer (number of nodes). The output from the output layer is then of the form

$$o_{pk} = g_k^o \left(v_{pk}^o \right) \tag{8.38}$$

The activation function g at the output layer may be the same as the activation function f in the hidden layer. However, in certain situations, it may be desirable to use a linear activation function at the output layer.

The above sequence completes the feed-forward propagation stage. The BP starts at the output layer and incrementally adjusts the connection weights between the output and hidden layers, and hidden and input layers, in that order. The weight updating is done according to the BP algorithm (described in Section 8.8) as follows:

$$w_{ij}(k+1) = w_{ij}(k) - \eta \frac{\partial E_p}{\partial w_{ij}} + \alpha \{ w_{ij}(k) - w_{ij}(k-1) \} \tag{8.39}$$

where η and α are coefficients known as learning rate and momentum term, respectively; k is epoch counter; and E_p is an objective function (error). The learning rate is an indicator of the rate of convergence. If it is too small, the rate of convergence will be slow owing to the large number of steps needed to reach the minimum error. If it is too large, the convergence initially will be fast, but will produce undue oscillations, and may not reach the minimum error. The momentum term is a parameter that adds inertia to the weight change depending on the direction of its previous change. It has the tendency to increase the speed of convergence. It also improves the learning capability when the training data contain noise and its value lies between zero and unity. A value close to zero gives more weight to the current error than the previous errors, and vice versa. The objective function (error) can be defined as

$$E_p = \frac{1}{2} \sum_{k=1}^{M} (y_{pk} - o_{pk})^2 \qquad (8.40)$$

where y_{pk} is the target output and M is the dimension of the output vector.

8.6.2 Recurrent neural networks

In feed-forward neural networks, there is no feedback from one layer to a preceding layer. In recurrent neural networks, information can flow forwards and backwards. They are different from feed-forward networks in that they operate not only on an input space but also on an internal state space, which is a trace of the history of the network. Recurrent networks consist of a 'context' layer, which retains information between observations. At each time step, new inputs as well as the previous contents of the hidden layer are fed into the hidden layer in the next time step. These 'context units' may receive information from the hidden layer, as in the case of Elman-type recurrent networks (Elman, 1990), or from the output layer as in the case of Jordan-type networks (Jordan, 1986). These two types are otherwise similar in architecture. Another type known as Hopfield network (Hopfield, 1982) has symmetric connections and is a binary classifier that has a threshold function at the output layer.

In recurrent neural networks, there are connections from forward to back as well as feedback to itself (same neuron); that is, output of a neuron fed back into its own inputs (self-feedback), or each neuron in the layer feeding its outputs to the inputs of all other neurons (no self-feedback). The latter is introduced with a time lag (type of Hopfield network). The sequence is important, making them amenable to time series prediction. In MLPs, sequence is not important. The patterns may be presented in any order.

For a neural network to be dynamic, it should have a memory, which may be short term or long term. Long-term memory is stored in the synaptic weights. Short-term memory is introduced into the network by using time delays. The role of the memory is to transform a static neural network into a dynamic one.

Time-lagged recurrent networks (TLRNs) have the advantages that they have smaller network size, low sensitivity to noise, faster learning rate, and better generalization. They are a kind of non-linear moving average models that can be extended to non-linear ARMA models. TLRNs are dynamic systems, whereas feed-forward networks are static systems. The two types may be expressed by the following equations:

For a feed-forward network:

$$v_j^h(n) = \sum_{i=1}^{N} w_{ji}^h x_i(n) + b_j^h \quad \text{(induced field from input layer to hidden layer)} \qquad (8.41a)$$

$$y_j^h(n) = f\left(v_j^h(n)\right) \quad \text{(output from the hidden layer)} \qquad (8.41b)$$

$$v_k^o(n) = \sum_{j=1}^{M} w_{kj}^o y_j^h(n) + b_k^o \quad \text{(induced field from hidden layer to output layer)} \qquad (8.41c)$$

$$y_k^o(n) = g\left(v_k^o(n)\right) \quad \text{(output from the output layer)} \tag{8.41d}$$

For a recurrent network,

$$v_j^h(n) = \sum_{i=1}^{N} w_{ji}^h x_i(n) + \sum_{h=1}^{M} u_{ji}^h y_h^h(n-1) + b_j^h \tag{8.42a}$$

$$y_j^h(n) = f\left(v_j^h(n)\right) \tag{8.42b}$$

$$v_k^o(n) = \sum_{j=1}^{M} w_{kj}^o y_j^h(n) + b_k^o \tag{8.42c}$$

$$y_k^o(n) = g\left(v_k^o(n)\right) \tag{8.42d}$$

In Equations 8.41 and 8.42, the index i refers to the input layer, j and h refer to the hidden layer, and k refers to the output layer; w_{ji}^h is the synaptic weight from the input layer to the hidden layer; w_{ji}^o is the synaptic weight from the hidden layer to the output layer; b_j^h and b_k^o are the biases in the hidden and output layers, respectively; u_{ji}^h are the recurrent weights; N is the number of inputs; M is the number of hidden nodes; n is an iteration counter; and f and g are non-linear differentiable functions (which can be the same). In the case of the recurrent network, the input vector is propagated through a weight layer w_{ji}^h, and combined with previous state activations through an additional recurrent weight layer u_{ji}^h.

Simple recurrent neural networks (Elman, 1990) have short-term memory. The hidden layer receives inputs from the input layer as well as feedback from previous hidden state activated by a certain weight. The feedback is modified by a set of weights so as to enable automatic adaptation through learning. Simple recurrent networks (SRN) (Elman, 1990) contain 'context units' consisting of unit delays. They store the outputs of hidden units for one time step and feed them back to the input layer.

Generic recurrent neural networks originate from MLPs. They take the form

$$y(n + 1) = f\{y(n), y(n - 1),...y(n - q + 1), x(n), x(n - 1),...x(n - q + 1)\} \tag{8.43}$$

where q is the number of memory units. The larger the number of memory units used, the greater the memory requirements because past errors as well as activations need to be stored. Very often, a large memory is not needed because the error gets smaller and smaller through each layer.

8.6.2.1 Back-propagation through time (BPTT)

Learning of recurrent neural networks can be accomplished by back-propagation through time (BPTT) (Rumelhart et al., 1986) in the offline mode, real-time recurrent learning in the online mode, and by extended Kalman filtering. BPTT, which is popular and similar to standard BP, is a powerful tool applicable to pattern recognition, dynamic system modelling, and control systems, in addition to neural networks. Normally, errors are back-propagated at the same time level. Through BPTT, errors can be back-propagated even further. In the same way as the error is back-propagated in space in a feed-forward network, the error in

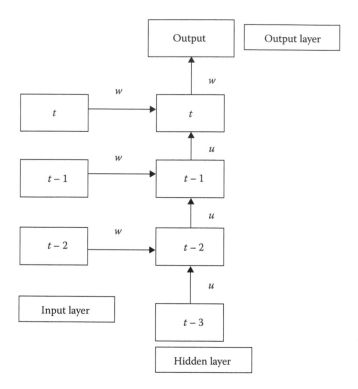

Figure 8.7 Unfolding in BPTT with a memory of three units (*w* is the weight from the input layer to hidden layer and from hidden layer to output layer; *u* is the weight from the previous hidden layer to the present hidden layer).

a recurrent network in addition is back-propagated through time, which is an 'unfolding' process (Figure 8.7). Training of recurrent neural networks is rather slow.

The basic principle of BPTT is that of 'unfolding'. Errors are back-propagated according to the gradient descent method, and the weight change is proportional to the negative gradient of the error function with respect to the weight. A summary of the relevant equations are given below (Bodèn, 2001):

The total error is defined as

$$E = \frac{1}{2} \sum_{p}^{N} \sum_{k}^{M} (d_{pk} - y_{pk})^2 \tag{8.44}$$

where *d* is the desired output and *y* is the network output; *N*, the total number of input samples; and *M*, the total number of output nodes. Subscript *p* is an index for the number of input data patterns. The weight change is given by

$$\Delta w = -\eta \frac{\partial E}{\partial w} \tag{8.45}$$

More specifically, the error component δ_{pk} for an output node is

$$\delta_{pk} = -\frac{\partial E}{\partial y_{pk}} \frac{\partial y_{pk}}{\partial v_{pk}} = (d_{pk} - y_{pk}) g'(y_{pk}) \tag{8.46}$$

and for a hidden node, δ_{pj} is

$$\delta_{pj} = -\left(\sum_{k=1}^{M} \frac{\partial E}{\partial y_{pk}} \frac{\partial y_{pk}}{\partial v_{pk}} \frac{\partial v_{pk}}{\partial y_{pk}}\right)\left(\frac{\partial y_{pj}}{\partial v_{pj}}\right) = \sum_{k=1}^{M} \delta_{pk} w_{kj}^{b} f'(y_{pj}) \tag{8.47}$$

Since $\dfrac{\partial v}{\partial w}$ is equal to the input activation, the weight change for an output node is

$$\Delta w_{kj}^{o} = \eta \sum_{p=1}^{N} \delta_{pk} y_{pj} \tag{8.48}$$

and for a hidden node (input weights),

$$\Delta w_{ji}^{b} = \eta \sum_{p=1}^{N} \delta_{pj} x_{pi} \tag{8.49}$$

For recurrent networks, the error component can be modified by adding a time index as

$$\delta_{pj}(t-1) = \sum_{b=1}^{M} \delta_{pb}(t) u_{bj} f'(y_{pj}(t-1)) \tag{8.50}$$

and the corresponding weight change as

$$\Delta u_{jb} = \eta \sum_{p=1}^{N} \delta_{pj}(t) y_{pb}(t-1) \tag{8.51}$$

where b is the index for the activation receiving node and j for the sending node (one time step back).

As the number of memory steps increase, the computer memory requirements as well as computing time increase. However, a large value of memory steps is not necessary because the error from one layer to another becomes smaller and smaller until it is reduced to zero.

Applications of recurrent neural networks include pattern recognition, time series prediction, system identification, and signal processing. In the water sector, TLRN have been used recently for monthly reservoir inflow modelling (Kote and Jothiprakash, 2009), suspended sediment load forecasting (Wang and Traore, 2009), reconstruct rainfall–runoff processes reconstruction (Pan and Wang, 2005), stream-flow forecasting (Chang et al., 2002), and rainfall forecasting (Patil and Ghatol, 2009).

8.6.3 Self-organizing maps (Kohonen networks)

Self-organizing maps (SOMs), also known as Kohonen networks (Kohonen, 1982, 2001), use a neighbourhood function to preserve the features of the input space. The objective is to map higher-dimensional input patterns into a one- or two-dimensional discrete map. Patterns close to one another in the input space should also be close to one another in the map. They are characterized by clustering of input patterns in which the locations of the neurons are indicative of their statistical features. SOM is inherently non-linear and may be thought of as a generalization of principal component analysis (PCA). The method is useful for visualizing a low-dimensional view of a high-dimensional input space.

SOMS are inspired by the neurological organization of the brain that follows the principle of topographic map formation. There are basically two types of SOM models, the Willshaw–von der Malsburg (1976) and the Kohonen (1982). The former is particularly suited for situations where the input dimension is the same as the output dimension, whereas the latter is more general and suitable for dimension reduction, such as in data compression.

In SOMs, each input pattern consists of a localized active region or position against a quiet background. The location of this region or position varies from one input pattern to another. Therefore, the network should be exposed to as many different input patterns as possible to ensure that self-organization has a chance of reaching maturity.

The SOM computational process starts by weight initialization in which the weights are randomly set to some small values. For an input vector $[x_1, x_2, ..., x_N]$, with associated weights $[w_{j1}, w_{j2}, ..., w_{jN}]$, $j = 1, 2, ..., M$, where M is the total number of neurons in the network, the inner products $w_j^T x$ for $j = 1, 2, ..., M$ are calculated and compared to select the largest one. This determines the location where the centre of the topological neighbourhood of the excited neurons has to be placed. Maximizing the inner product is equivalent to minimizing the Euclidean distance between x and w_j. Next, define an index $i(x)$ as

$$i(x) = \arg\ \min_j \left\| x - w_j \right\| \text{ for } j = 1, 2, ..., M \tag{8.52}$$

The neuron that satisfies this condition is the 'winning' neuron for the input vector x. The output of the network may be either the index of the 'winning' neuron, or, the weight vector closest to the input vector as measured by the Euclidean distance.

The winning neuron locates the centre of the topological neighbourhood. It is also logical to think that excited neurons in the neighbourhood of the winning neuron also influence the network output. To accommodate this feature, a neighbourhood function h_{ji} of the distance between the winning neuron and other neurons in the neighbourhood, d_{ji}, is defined subject to the requirements that the neighbourhood is symmetric around the winning neuron (i.e., when $d_{ji} = 0$), and that the function decreases monotonically with increasing distance d_{ji}. An obvious choice that satisfy these conditions is the Gaussian function, which is also used in radial basis function neural networks, and which is defined as

$$h_{ji(x)} = \exp\left(-\frac{d_{ji}^2}{2\sigma^2}\right) \tag{8.53}$$

where the parameter σ can be considered as the width of the neighbourhood (same interpretation as the receptive field in radial basis functions). Another requirement of SOMs is that the topological neighbourhood should shrink with time, implying that the width σ should be a function that decreases with time. This feature can be achieved by having the width dependent on time discretely as

$$\sigma(n) = \sigma(0)\exp\left(-\frac{n}{\tau_1}\right) \quad n = 0, 1, 2, ... \tag{8.54}$$

where τ_1 is a time constant and n is an iteration counter. Then, Equation 8.53 can be rewritten as

$$h_{ji(x)}(n) = \exp\left(-\frac{d_{ji}^2}{2\sigma^2(n)}\right) \quad n = 0,1,2,\ldots \tag{8.55}$$

Alternatively, it is possible to have a constant width and a gradually increasing number of neurons in the neighbourhood. The new neurons are added half way between the old ones.

The training of the SOM requires the weights to be adjusted as and when new input vectors are presented. Unlike in feed-forward-type networks, Kohonen networks have no target output values to compare. Weight adjustment is therefore done in a different way. The adjustment takes place according to

$$\Delta w_j = \eta y_j x - g(y_j) w_j \tag{8.56}$$

where $g(y_j)$ is some positive scalar function of the output y_j with

$$g(y_j) = 0 \quad \text{for } (y_j) = 0$$

Setting $g(y_j) = \eta y_j$ and $y_j = h_{ji(x)}$, the weight change expression (Equation 8.56) can be simplified to

$$\Delta w_j = \eta h_{ji(x)} (x - w_j) \tag{8.57}$$

The updated weight is then given as

$$w_j(n + 1) = w_j(n) + \eta(n) h_{ji(x)} (n)(x - wj(n)) \tag{8.58}$$

Equation 8.58 is applied to all the neurons in the neighbourhood of the winning neuron. To complete the training process, it is necessary to determine the learning rate parameter, η, which in this case is exponentially decreasing, but never attains a zero value. Heuristically, it can be assigned according to

$$\eta(n) = \eta_0 \exp\left(-\frac{n}{\tau_2}\right) \quad n = 0,1,2,\ldots \tag{8.59}$$

where τ_2 is another constant of the SOM algorithm. The suggested values for these parameters are (Haykin, 1999, p. 474) $\eta_0 = 0.1$, $\tau_1 = \frac{1000}{\log \sigma_0}$, and $\tau_2 = 1000$. The initial width σ_0 is set equal to the radius of the lattice (input neurons are arranged in the form of a lattice).

8.6.4 Product unit–based neural networks (PUNN)

In most of the widely used neural networks, the sigmoid-type activation (or basis) function is used to describe the relationship between the inputs and outputs. An often-criticized drawback in this kind of representation is their lack of transparency. Among the several alternative basis functions suggested, multiplicative (instead of additive)-type neural networks may perhaps be the most promising. PUNN, introduced by Durbin and Rumelhart (1989), is a type of multiplicative network that has the advantage of increased information capacity and the ability to form higher-order combinations of the inputs. Durbin and Rumelhart (1989) empirically showed, by using the capacity of learning random Boolean patterns, that the information capacity of a product unit is approximately $3N$, where N is the number of input variables, compared to $2N$ of an additive unit. Despite such advantages, one major drawback of this type of network is the difficulty in training with the BP algorithm. Parametric learning such as genetic algorithms (Janson and Frenzel, 1993), and simulated annealing (Leerink et al., 1995) have been proposed to overcome this difficulty. The algorithm proposed by Martinez-Estudillo et al. (2006) enables the determination of both the architecture and the parameters of a PUNN.

A model of evolutionary programming for automatically obtaining the architecture and the weights of a PUNN is that proposed by Martinez-Estudillo et al. (2006). Their method is relatively easy to implement and does not require the kind of expertise needed for the design and training of a standard neural network. This model has been shown to work in several benchmark problems of synthetic data and a real data set of microbial growth, and is considered to be robust. It has also been shown that the product units they have used can approximate any function to a given accuracy.

Evolutionary algorithm to train PUNNs has several advantages over other techniques (Martinez-Estudillo et al., 2006). It can obtain both the architecture and the weights of the neural network, which normally is very difficult and requires a long process of trial and error. It can also be viewed as a global optimization algorithm, although the convergence to a global optimum is guaranteed only in a weak probabilistic sense. Another positive feature of evolutionary algorithms is that they perform well on noisy functions where there may be multiple local optima. Evolutionary algorithms tend not to get trapped in local minima and can often find globally optimal solutions. This feature is important as networks based on product units are prone to be trapped in local minima. The algorithm does not require a derivable error function and therefore can be applied to problems where such error functions do not exist.

The PUNN considered by Martinez-Estudillo et al. (2006) has three-layers: an input layer, a single hidden layer of product units, and an output layer of additive units (Figure 8.7). It is similar to the architecture of an MLP with a single hidden layer except for the output function. The output function of the PUNN is of the form

$$y = f(x_1, x_2, \ldots, x_K) = \beta_0 + \sum_{j=1}^{M} \beta_j \left(\prod_{i=1}^{k} x_i^{w_{ji}} \right) \tag{8.60}$$

where x_i: $i = 1, 2, \ldots, K$ are the input variables, β_j: $= 0, 1, \ldots, M$ are scalar coefficients, and w_{ji}: $i = 1, 2, \ldots, K$; $j = 1, 2, \ldots, M$ are the exponents. It should be noted that when a negative number is raised to a non-integer power, it produces a complex number. This can happen when negative inputs are used in Equation 8.60. Since neural networks with complex outputs are rarely used in applications, Durbin and Rumelhart (1989) suggested using only the real part for further processing. In such cases, no finite VC dimension bounds can be derived (Schmitt, 2001). The focus is therefore restricted to input domains with non-negative sets only.

In evolutionary algorithms, a feasible solution for the problem of interest is coded in an entity called an individual. A group of such individuals is called a population. The population is progressed through generations by applying certain rules called operators. Operators are defined such that the population approaches towards better solutions over generations. In the context of PUNN, an individual consists of the parameters, that is, w_{ji}: $i = 1, 2, \ldots, K$; $j = 1, 2, \ldots, M$, and β_j: $j = 0, 1, \ldots, M$, and its architecture, that is, the connections from inputs to hidden layer and the number of hidden nodes. A typical PUNN architecture consists of

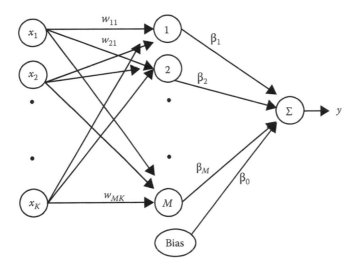

Figure 8.8 Typical architecture of product unit–based neural network.

connections from inputs to hidden layer and hidden to output layer, as denoted by arrows in Figure 8.8. In an individual in the evolutionary algorithm, a connection is denoted by 1 if the connection exists, or zero if the connection does not exist. Unlike in MLPs where all the nodes/inputs are fully connected, PUNN may have nodes that are not connected to other nodes/inputs.

The evolutionary algorithm of Martinez-Estudillo et al. (2006) starts with an initial population that is updated with some operators in a series of subsequent generations. The population consists of individuals representing the parameters and the architecture of the product unit–based networks; that is, one individual represents one neural network based on product units. The algorithm uses the replication operation and two types of mutation operations: parametric mutation and structural mutation. The parametric mutation modifies the coefficients and the exponents of the network using the simulated annealing algorithm. In structural mutation, the architecture of the network is modified through addition, deletion and fusion of nodes, and connections. Therefore, the structural mutation modifies the function represented by the network and allows exploration of different regions of the search space. The crossover operator is not used in the algorithm due to its potential disadvantages in evolving neural networks (Angeline et al., 1994). The population is evolved through generations until predefined stopping criteria are met. The main steps of the algorithm of Martinez-Estudillo et al. (2006) includes the generation of an initial population of size of about $10N_R$ where N_R is the number of individuals in the evolving population and repeating the following steps until the stopping criteria are fulfilled:

a. Calculation of the fitness of every individual in the population
b. Ranking of the individuals according to their fitness
c. Copying of the best individual into the new population (The copy of the elite individual will not go through any mutation in steps (e) and (f))
d. Replacement of the 10% worst individuals with the replications of the 10% of best individuals
e. Application of parametric mutation to the 10% best individuals
f. Application of structural mutation to the remaining 90% of individuals

8.6.4.1 Generation of the initial population

The number of neurons in the hidden layer for each network in the initial population is chosen within $\left[0, \dfrac{M}{2}\right]$, where M is the maximum number of hidden nodes allowed in subsequent populations as proposed by Martinez-Estudillo et al. (2006), or within the interval $[1, M]$ from a uniform distribution as proposed and used by Karunasingha et al. (2011). The number of connections between a hidden node and the inputs is chosen from a discrete uniform distribution in the interval $[1, K]$, where K is the number of independent variables. All the hidden neurons may or may not have connections to the output node.

Once the architecture is determined, values are assigned to the exponents and the coefficients, whose connections are defined from continuous uniform distributions in the predefined intervals $[-E, E]$ and $[-C, C]$ respectively. Here, $-E$, E, $-C$, and C are the lower and upper bounds for the values for exponents and coefficients in the first generation. By this procedure, the exponents are determined, the individuals are evaluated, and the best N_R solutions among the $10\,N_R$ individuals are chosen as the initial population.

8.6.4.2 Fitness function

The fitness function should be a strictly decreasing function of the form

$$A(k) = \frac{1}{(1 + \mathrm{MSE}(k))} \tag{8.61}$$

where $\mathrm{MSE}(k)$ is the MSE of the individual k. Martinez-Estudillo et al. (2006) used this function for an independent and identically distributed training data set. Karunasingha et al. (2011) used the normalized root mean square error (NRMSE) in place of $\mathrm{MSE}(k)$ in Equation 8.61. NRMSE is defined as

$$\mathrm{NRMSE} = \sqrt{\frac{\displaystyle\sum_{i=1}^{N}(x_i - \hat{x}_i)^2}{\displaystyle\sum_{i=1}^{N}(x_i - \overline{x})^2}} \tag{8.62}$$

where x_i is the observed value, $\hat{x}_i$ is the predicted value, N is the number of points predicted, and $\overline{x}$ is the average value of the observed time series. A value of zero for NRMSE denotes a perfect prediction and a value greater than 1 indicates that the predictions are no better than the long-term average. The MSE depends on the range of output (observed) values. The advantage of using NRMSE as the fitness function is that it does not depend on the range of observed values.

8.6.4.3 Parametric mutation

In parametric mutation, the exponents (w_{ji}) and the coefficients (β_j) of the product unit networks are updated. This stage can also be seen as an exploitation stage of the evolutionary algorithm compared with the structural mutation stage where the main thrust is on exploration. Martinez-Estudillo et al. (2006) used a simulated annealing algorithm (Geyer and Thompson, 1996; Kirkpatrik et al., 1983; Otten and Ginneken, 1989) in parametric mutation. Simulated annealing is a strategy that advances through jumps from a current state to another state according to some user-defined mechanism. The closeness of the actual function to any solution of the problem is represented by a fictitious 'temperature'. The severity of mutation of an individual k depends on this 'temperature', $T(k)$, which is given by

$$T(k) = 1 - A(k), \quad 0 \leq T(k) \leq 1 \tag{8.63}$$

For parametric mutation, the nodes and connections are selected sequentially in the given order, with probability $T(k)$ of the network in a certain generation. Thus, once an individual approaches towards a solution, the chances of its exponents/coefficients being selected for parametric mutation are less. From one generation to another, each parameter w_{ji}, β_j of a product unit–based network selected for parametric mutation is changed by an amount of Gaussian noise in which the variance depends on the 'temperature' of the network. This allows an initial coarse-grained search, and as the model approaches towards a solution, a finer-grained search. Exponents (w_{ji}) update of a single hidden unit is performed in one batch. The exponents, w_{ji}, of a network are updated as follows:

$$w_{ji}(t + 1) = w_{ji}(t) + \xi_1(t), \quad i = 1, 2,..., K, \quad j = 1, 2,..., M \tag{8.64a}$$

where $\xi_1(t) \in N(0, \alpha_1(t)T(k))$ represents a normally distributed random variable with zero mean and variance $\alpha_1(t)T(k)$, and $\alpha_1(t)$ is an adaptive parameter that determines the severity of a mutation. Similarly the coefficients β_j of a network are updated as follows:

$$\beta_j(t + 1) = \beta_j(t) + \xi_2(t), \quad j = 1, 2,...,M \tag{8.64b}$$

where $\xi_2(t) \subset N(0, \alpha_2(t)T(k))$ represents a normally distributed random variable with zero mean and variance $\alpha_2(t)T(k)$. Modification of each β_j is made one at a time. Once all the other parameters are known, β_0, the only remaining unknown, is taken to be the average of the training error as

$$\beta_0 = \frac{1}{N} \sum_{i=1}^{n} \left(y_i - \hat{y}_i \right) \tag{8.65}$$

where y_i is the actual value, $\hat{y}_i$ is the simulated value from Equation 8.60 without taking β_0, and N is the number of training samples.

After the mutation is performed, the fitness of each individual k is calculated and a simulated annealing criterion is applied. If ΔA is the difference in fitness function before and after a random step (i.e., either updating w_{ji} as in Equation 8.64a or updating β_j as in Equation 8.64b), then following the simulated annealing strategy, the new step is accepted with probability 1 if $\Delta A > 0$ and with a probability $\exp(\Delta A / T(k))$ if $\Delta A \leq 0$.

The adaptive parameter of w_{ji} is much smaller than that of β_j (i.e., $\alpha_1(t) \ll \alpha_2(t)$). These parameters attempt to avoid the evolutionary process getting trapped in local minima, and are changed in every generation using the following rule (Karunasingha et al., 2011):

$$\alpha_k(t) = \begin{cases} (1+\lambda)\alpha_k(t), & \text{if } A(g_s) > A(g_{s-1}), \; \forall s \in \{t, t-1, ..., t-\rho\} \\ (1+\lambda)\alpha_k(t), & \text{if } A(g_s) = A(g_{s-1}), \; \forall s \in \{t, t-1, ..., t-\rho\} \\ \alpha_k(t), & \text{otherwise} \end{cases} \quad k = 1, 2. \tag{8.66}$$

In Equation 8.66, $A(k_s)$ is the fitness of the best individual k_s in the generation s. The values of λ and ρ have to be defined by the user. Martinez-Estudillo et al. (2006) have used values of $\lambda = 0.1$ and $\rho = 10$. A generation is considered successful if the best individual is better than the best individual of the previous generation.

The parameters $\alpha_1(t)$, $\alpha_2(t)$, and the 'temperature' change throughout the evolutionary process. They determine the severity of mutation and drive the individuals towards better solutions by changing the solution severely when circumstances are good for exploration (i.e., when there are many consecutive successes) or by changing them by small amounts when larger steps produce poor results (i.e., when fitness of the best individual does not change over several generations).

8.6.4.4 Structural mutation

Structural mutation modifies the structure of the networks. This encourages exploration of the search space and maintains diversity of the population. Possible structural mutations include node addition, node deletion, connection addition, connection deletion, and node fusion. These five mutations can be applied sequentially to each network. For each mutation, the nodes and connections are selected randomly, and the number of nodes or connections selected for mutation (except for node fusion) is chosen within the range of a minimum (Δ_{min}) and a maximum (Δ_{max}) number of mutations defined by the user. Thus, the number of elements chosen for mutation may be calculated as

$$\Delta_{min} + [uT(g)(\Delta_{max} - \Delta_{min})] \tag{8.67}$$

where u is a random value in the interval $[0, 1]$. Values of Δ_{min} and Δ_{max} as well as other parameters used by Martinez-Estudillo et al. (2006), which are claimed to be robust, are given in Table 8.4.

8.6.4.4.1 Node addition, deletion, fusion, connection addition, and deletion

Nodes may or may not have connections from all input variables. Whether a connection exists between a newly added node and a particular input variable is randomly decided. The w_{ji} values associated with these connections are also generated randomly. The β_j values associated with the connections to the output node are similarly assigned. Node addition, node deletion, connection addition, and connection deletion are carried out randomly. Node fusion is done by selecting two nodes, a and b, randomly and replacing them by a node c,

Table 8.4 Parameters used in the applications

Population parameters		Parametric mutation parameters		Structural mutation parameters interval [Δ_{min}, Δ_{max}]	
Parameter	Value	Parameter	Value	Parameter	Value
Maximum number of hidden nodes in a population	8	$\alpha_1(0)$	1	Node addition	[1, 2]
Exponent interval [$-E, E$]	[−5, 5]	$\alpha_2(0)$	5	Node deletion	[1, 5]
Coefficient interval [$-C, C$]	[−5, 5]	λ	0.1	Connection addition	[1, 2k]
		ρ	10	Connection deletion	[1, 2k]

which is a combination of the original two nodes. When there are connections from a certain input variable to these two nodes, such connections are replaced by one connection to node c with weight given by

$$w_{ci} = \frac{w_{ai} + w_{bi}}{2} \tag{8.68}$$

Other connections from inputs to node a or b are included in c with a probability of 0.5 and their weights are unchanged. The coefficient of the connection from node c to the output node is modified as

$$\beta_c = \beta_a + \beta_b \tag{8.69}$$

The accuracy and robustness of the PUNN model has been demonstrated using synthetic data generated by some benchmark functions. It has also been successfully applied to predict daily river flow discharges measured at the Pakse gauging station in the Mekong River (Karunasingha et al., 2011).

8.6.5 Wavelet neural networks

Much of the hydrological and environmental data recorded in the field are as functions of time and therefore constitute time series. Such time series data can be analysed in the time domain as well as in the frequency domain (see also Chapters 2 and 7). Depending on the purpose of analysis, time domain analysis can be more useful than frequency domain analysis, and vice versa. Although the basic extractable information is more or less the same, the two approaches have their unique features. In some situations, important information is hidden in the frequency content of the signal, which can be extracted by the Fourier transform of the time domain signal. Fourier transform enables the signal to be broken down into spectral components. The Fourier-transformed signal gives how much each frequency exists in the signal but not the times of occurrences of these frequencies. The Fourier transform approach can be used in situations where the times of occurrences of various frequency components are not of interest but the only interest is in determining what frequency components exist.

The short time (or term) Fourier transform (STFT), or windowed Fourier transform, which is a variant of the normal Fourier transform, and which is capable of giving the times of occurrences of a band of frequencies rather than the exact frequency, can be considered as an improved version, but it still has problems. STFT calculates the Fourier transform of the signal, which is divided into small segments by a window function. The Fourier transform is performed on each windowed part of the signal and the window is shifted over the signal. The result of using the fixed window length is the time–frequency content of a signal with a constant frequency and time resolution that is often not the most desired resolution. A good frequency resolution is required for the signal with low-frequency components, while a good time resolution is required for the signal with high-frequency components. It is not possible to obtain a good resolution in both time and frequency at the same time by using the fixed window as is done in STFT. Wavelet transform can capture both the short-duration, high-frequency and the long-duration, low-frequency information simultaneously.

Wavelets, meaning small waves, which can be thought of as a spinoff from STFT, overcome many such problems. Unlike in Fourier analysis, the wavelet analysis does not require assumptions about stationarity and periodicity of data. The basic approach in Fourier as well as wavelet decomposition is to convolute the signal function by a basis function. In the case of Fourier approach, the basis function is a combination of sines and cosines. In the case of wavelet approach, there exist a number of different basis functions, such as the Haar wavelet

(Haar, 1910), the Daubechies wavelet (Daubechies, 1988, 1992), the Mexican Hat wavelet (normalized second derivative of a Gaussian function), and the Morlet wavelet (Goupillaud et al., 1984). Of these, the Haar wavelet has the major advantages of being conceptually simple, computationally fast, and exactly reversible. It has a rectangular shape and has been used in many fields of study. One of the shortcomings of the Haar wavelet is its discontinuity, which makes it impossible to directly apply to solve differential equations. The most popular wavelet in signal processing is the Daubechies wavelet (Daubechies, 1988), which is continuous and symmetric. Other wavelets that have been widely used include the Morlet wavelet (Goupillaud et al., 1984), which is symmetric and has the advantage of minimizing the Heisenberg uncertainty principle, and the Mexican hat wavelet, which is the second derivative of the Gaussian function and is also symmetric. Among the many popular uses of the Haar wavelet is its application in the JPEG format of digital image compression.

Wavelet analysis is a kind of data preprocessing technique. One of its attractive features is its flexibility to be integrated with other techniques of analysis. For example, when it is integrated with ANNs, the resulting network is referred to as wavelet neural networks (WNN); when it is integrated with neuro-fuzzy systems, it is referred to as wavelet-neuro-fuzzy networks.

Wavelet transform can be used when signals are characterized by localized high-frequency components for short duration or when they are characterized by low-frequency components for long duration. For low scaling parameter ($\ll 1$), the wavelet function is concentrated with high-frequency components (compressed), whereas when it is high ($\gg 1$), the wavelet is stretched and concentrated with low-frequency components (dilated). Thus, small scales give a high resolution, whereas large scales give the signal structure.

The wavelet transforms work by repeatedly filtering the signal with a short wave called wavelet at different scales and positions. Continuous wavelet transform is defined as

$$\text{CWT}(s,\tau) = \frac{1}{\sqrt{s}} \int_{-\infty}^{\infty} x(t)\psi\left(\frac{t-\tau}{s}\right) dt \tag{8.70}$$

where $\psi(t)$ is the basic wavelet (sometimes also referred to as the mother wavelet) with effective length t, which is much shorter than the target time series, $x(t)$; s, the scaling (dilation, shape) parameter $(-\infty,\infty)$; and $\tau(0, \infty)$ the location (shifting, translation) parameter that represents the sliding of the wavelet over $x(t)$. As the value of s changes, the wavelet is dilated or compressed to cover a large interval of frequencies or scales. As t changes, the centre of the wavelet moves.

To avoid the difficulty in the calculation of wavelet coefficients at every possible scale with the continuous wavelet transformation, the discrete wavelet transformation (DWT) in which the scale and location are expressed as powers of two (dyadic scales and locations) can be used. It is of the form

$$\text{DWT}(a,b) = \frac{1}{\sqrt{2^a}} \sum_{k=1}^{N} x(k)\psi\left(\frac{k}{2^a} - b\right) \tag{8.71}$$

where $DWT(a, b)$ is the wavelet coefficient for the discrete wavelet function with scale $s = 2^a$ and location $\tau = 2^a b$. For the Haar wavelet,

$$\psi(t) = \begin{cases} 1 & \text{for } t \in (0,0.5) \\ -1 & \text{for } t \in (0.5,1) \\ 0 & \text{for } t \notin (0,1) \end{cases} \tag{8.72}$$

In the implementation of discrete wavelet transform, the original signal $x(t)$ passes through a high-pass filter and a low-pass filter and emerges as two signals (Figure 8.9), the approximate component (A) and the detailed components (D) (Mallat, 1989, 1997). The approximate component (global view) corresponds to high-scale low-frequency, whereas the detailed component corresponds to low-scale high-frequency resolution. Of the two resolutions, the approximate component that identifies the signal is more important. The decomposition process can be continued into many lower resolutions (Figure 8.10). The output of the low-pass filter consists of the average of every two samples and the output of the high-pass filter contains the difference of every two samples. The output of the low-pass filter contains much of the information.

Each step of a wavelet transformation produces a set of averages and a set of differences, thereby halving the size of the input data. With an input data size of 2^N (almost all wavelet algorithms work with data expressed as a power of two), recursive repetition of this process leaves with one (2^0) sum and $2^N - 1$ differences at different scales. The differences are referred to as 'wavelet coefficients'. The approximate and detailed coefficients for level 1 decomposition are respectively given as

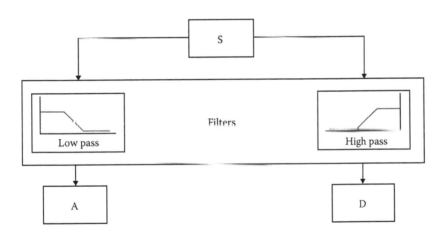

Figure 8.9 Decomposition of the signal by high-pass and low-pass filters.

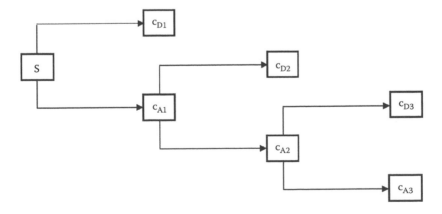

Figure 8.10 Tree-like structure of wavelet decomposition.

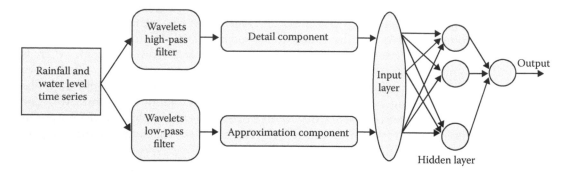

Figure 8.11 Architecture of wavelet neural network. (From Amnatsan, S., Water level prediction in Nan River, Thailand using wavelet neural networks. Master's degree Thesis, National Graduate Institute for Policy Studies, Tokyo, Japan and International Centre for Water Hazard and Risk Management, Tsukuba, Japan, unpublished, 2010.)

$$C_A(n) = \frac{1}{\sqrt{2}}[x(2n-1) + x(2n)], \quad n =, 1, 2, ..., \frac{N}{2} \tag{8.73}$$

$$C_D(n) = \frac{1}{\sqrt{2}}[x(2n-1) - x(2n)], \quad n =, 1, 2, ..., \frac{N}{2} \tag{8.74}$$

The decomposed series has the same length as the original series, which can be reconstructed by adding the two components. When N is odd, it can be made even by adding a zero.

In WNNs, the decomposed data are used as inputs of the neural network. Applications of wavelet analysis in the hydrological field include the studies carried out to observe the periodicity and trend of long-term time series of climatic and hydrological data (Andreo et al., 2006), multiscale analysis in hydrology to characterize daily stream flows (Smith et al., 1998; Saco and Kumar, 2000), daily reservoir inflows (Coulibaly et al., 2000), to generate annual and monthly streamflows (Bayazit and Aksoy, 2001), and daily flows for intermittent streams (Aksoy and Bayazit, 2000). The wavelet approach has also been used to generate annual and monthly rainfall data (Unal et al., 2004; Aksoy and Unal, 2007) and streamflow data (Bayazit et al., 2001), and in karstic hydrology (Labat et al., 1999, 2005), rainfall data simulation by hidden Markov model and discrete wavelet transformation (Jayawardena et al., 2009). Better results of water level time series prediction have also been obtained by using WNNs than with ordinary MLP-type neural networks (Amnatsan, 2010). In this work, the architecture of the WNN used is shown in Figure 8.11.

8.7 LEARNING MODES AND LEARNING

8.7.1 Learning modes

Haykin (1999), citing from Mendel and McLaren (1970), defines learning in the context of neural networks as

> Learning is a process by which the free parameters of a neural network are updated through a process of stimulation by the environment in which the network is embedded. The type of learning is determined by the manner in which the parameter changes take place.

The external environment is the training data sets. Learning can take place under a supervised mode or unsupervised mode. Supervised learning is like learning with a teacher. The connectivity (synaptic) weights are adjusted according to a measure of the difference between the neural network output and the target or expected output for a given input. Examples include the delta rule and the BP algorithm, which is a generalized delta rule. The BP method is the most widely used supervised learning algorithm and it belongs to the error correction learning category. In the unsupervised mode, there is no equivalent of a teacher, meaning that there is no target output available. The connectivity weights are adjusted automatically by clustering the input patterns into similar groups. Unsupervised learning often refers to clustering algorithms, for example, the k-means algorithm, which can be used for training radial basis function–type neural networks.

8.7.2 Types of learning

8.7.2.1 Error correction learning (optimum filtering)

In this method of learning, it is first necessary to define a cost (error) function. Usually, it is of the form

$$E = \frac{1}{2}\sum_k e_k^2(n) = \frac{1}{2}\sum_k \left(d_k(n) - y_k(n)\right)^2 \tag{8.75}$$

where d_k and y_k are the target and network outputs at the k-th neuron, and n is an iteration counter. The weights are adjusted incrementally until they reach a steady-state condition according to

$$\Delta w_{kj}(n) = \eta e_k(n)x_j(n) \tag{8.76}$$

$$w_{kj}(n + 1) = w_{kj}(n) + \Delta w_{kj}(n) \tag{8.77}$$

Here, w_{kj} is the weight of neuron k excited by $x_j(n)$ of input vector $x(n)$ at time (or iteration) n. This kind of learning rule is known as the delta rule or the Widrow–Hoff rule (Widrow and Hoff, 1960). It is a form of closed-loop feedback system. The learning rate parameter η plays an important role in the performance of the system.

8.7.2.2 Memory-based learning

In this type of learning, all past experiences (inputs and corresponding target outputs) are stored in a large memory. When an unseen input x_t is presented to the network, the algorithm retrieves and analyses all stored data in a 'local neighbourhood' of the input x_t. In this context, it is first necessary to define the meaning of 'local neighbourhood', which in most cases uses the nearest neighbourhood rule.

A vector $x_N' = (x_1, x_2, x_3, \ldots, x_N)$ is said to be a nearest neighbour of x_t if

$$\min\left(d(x_i, x_t)\right) = d\left(x_N', x_t\right) \text{ for all } i$$

where $d(x_i, x_t)$ is the Euclidean distance between the vectors x_i and x_t

8.7.2.3 Hebbian learning (Hebb, 1949) (unsupervised)

Hebbian learning is one of the oldest forms of learning rules. It was postulated by Hebb (1949) in the context of neurophysiology to describe the modification of synaptic connections between neurons. Specifically, in his book (1949, p. 62), it is stated as '*When an axon of cell A is near enough to excite a cell B and repeatedly or persistently takes part in firing it, some growth process or metabolic change takes place in one or both cells such that A's efficiency, as one of the cells firing B, is increased*'. When translated into ANNs, it implies that the weights will increase if the two connecting neurons are activated simultaneously and decrease if they are activated separately. Weight adjustment are done according to

$$\Delta w_{kj}(n) = f(y_k(n), x_j(n)) \tag{8.78}$$

where the function f can take many forms, x_j and y_k are pre- and postsynaptic signals (dimensionless) of the k-th neuron at time n (iteration counter). The simplest form of Hebb's hypothesis takes the form

$$\Delta w_{kj}(n) = \eta y_k(n) \, x_j(n) \tag{8.79}$$

Hebbian learning has limitations, and several of its variations are currently available. One such variation, according to the covariance hypothesis (Sejnowski, 1977a,b), takes the form

$$\Delta w_{kj}(n) = \eta(y_k - \bar{y})(x_j - \bar{x}) \tag{8.80}$$

where η is a learning parameter, and, $\bar{x}$, $\bar{y}$ are the time-averaged values of x, y. In this case, $\Delta w_{kj} > 0$ if $y_k > \bar{y}$, $x_j > \bar{x}$ and $\Delta w_{kj} < 0$ if $y_k > \bar{y}$, $x_j < \bar{x}$, or $y_k < \bar{y}$, $x_j > \bar{x}$.

8.7.2.4 Competitive learning (unsupervised)

In competitive learning (Rumelhart and Zipser, 1985; Kia and Coghill, 1994), the output neurons compete with each other to become active. Compared with Hebbian learning where several output neurons may be active simultaneously, in competitive learning, only a single output neuron is active at any instant of time. Mathematically,

$$y_k = \begin{cases} 1 & \text{if } v_k > v_j \text{ for all } j, \, j \neq k \\ 0 & \text{otherwise} \end{cases} \tag{8.81}$$

where v_k, which is the induced local field, is the combined effect of all forward and feedback inputs to neuron k. If each neuron is allotted a fixed amount of synaptic weight, which is distributed among all input nodes, then

$$\sum_j w_{kj} = 1 \quad \text{for all } k \tag{8.82}$$

that is, neurons learn by shifting the weights from inactive (those that do not respond to a particular input) ones to active (those that respond) ones. Inactive neurons do not learn. The standard competitive learning rule is

$$\Delta w_{kj} = \begin{cases} \eta(x_j - w_{kj}) & \text{if neuron } k \text{ wins} \\ 0 & \text{if neuron } k \text{ loses} \end{cases} \tag{8.83}$$

For a neuron k to be winning, its induced local field v_k for a given input pattern x must be the largest among all neurons in the network. If a neuron wins, each input node of that neuron gives up some proportion of its weight, which is distributed equally among the active input nodes.

The main form of competitive learning is the vector quantization, sometimes referred to as 'Kohonen networks' (Kohonen, 1982, 1984, 2001), which is similar to 'k'-means clustering. Kohonen's SOMs combine competitive learning with dimensionality reduction.

8.7.2.5 Boltzmann learning

According to the Boltzmann learning rule (Hinton and Sejnowski, 1986), which is stochastic, the weight change is given by

$$\Delta w_{kj} = \eta\left(\rho_{kj}^+ - \rho_{kj}^-\right), \quad j \neq k \tag{8.84}$$

where ρ_{kj}^+ is the correlation between the states of neurons k and j in the *clamped* condition (neurons exposed to the external environment are clamped into specific states), ρ_{kj}^- is the correlation between the states of neurons k and j in the free-running condition (neurons exposed to the external environment as well as those hidden are free to take any value). The correlations are averaged over all possible states $\left(-1 < \rho_{kj}^+, \rho_{kj}^- < 1\right)$.

8.7.2.6 Reinforced learning (unsupervised)

The network receives a global reward/penalty signal. Weights are changed so as to develop an input/output behaviour that maximizes the probability of receiving a reward and minimizes that of receiving a penalty.

For a neuron to be rewarded, its summed input for a given input pattern should be the largest among all neurons in the network. The output of this neuron is set to unity, while the outputs of all other neurons are zero; that is,

$$y_k = \begin{cases} 1 & \text{if } v_k > v_j \text{ for all } j, j \neq k \\ 0 & \text{otherwise} \end{cases} \qquad \text{(same as Equation 8.81)}$$

The weight updating is done according to

$$\Delta w_{kj} = \begin{cases} \eta(x_j - w_{kj}) & \text{if neuron } k \text{ wins the competition} \\ 0 & \text{if neuron } k \text{ loses the competition} \end{cases} \qquad \text{(same as Equation 8.83)}$$

with $\sum_j w_{kj} = 1$ for all k

8.7.2.7 Hybrid learning

Hybrid learning combines supervised and unsupervised learning, and is particularly applicable to radial basis function–type networks. The width, location of centres, and the number of hidden nodes of the radial basis function are determined in an unsupervised mode, whereas the connection weights from the hidden layer to the output layer are determined in a supervised mode.

8.7.3 Learning rate (η) and momentum term (α)

The learning rate is an indicator of the rate of convergence. If it is too small, the rate of convergence will be slow due to the large number of steps needed to reach the minimum error. If it is too large, the convergence initially will be fast, but will produce undue oscillations, and may not reach the minimum error. The momentum term is a parameter that adds inertia to the weight change depending on the direction of its previous change. It has the tendency to increase the speed of convergence. It also improves the learning capability when the training data contains noise and its value lies between zero and unity. A value close to zero gives more weight to the current error than the previous errors, and vice versa.

The optimum values of learning rate and momentum term are defined (Haykin, 1999, pp. 193–194) as those values that on average yield convergence to a local minimum in the error surface with the least number of epochs, or, for the worst case or on average yield convergence to the global minimum with the least number of epochs, or, on average yield convergence to the network configuration that has the best generalization over the entire input space with the least number of epochs. For classification problems, Haykin (1999, p. 197) using a computer simulation reports optimal values of $\eta = 0.1$ and $\alpha = 0.5$ in the context of convergence to a local minimum. For $\eta \to 0$, α tends to a faster speed of convergence. For $\eta = 1$, $\alpha \to 0$ is required to ensure learning stability.

8.8 BP ALGORITHM

The BP algorithm, which is based on the application of the chain rule of calculus, has several steps. The first step is the forward pass in which the input patterns are presented to the network using an initially assumed set of random weights, usually small (in the range from −0.5 to 0.5, or sometimes in the range 0 to 1). The input signals are transmitted in the forward direction from layer to layer and from neuron to neuron until they reach the output layer in a sequential manner. The BP algorithm can also be applied to networks with no hidden layers. Hidden layers allow more complex systems to be analysed but at a higher computational cost. For a typical input pattern from the training data set, the following equations can be written for a typical neuron j:

For the weighted linear summation,

$$v_j(n) = \sum_{i=0}^{N} w_{ji}(n)x_i(n) \tag{8.85}$$

Output after passing through the activation function f,

$$y_j(n) = f_j(v_j(n)) \tag{8.86}$$

In these two equations, w_{ji} refers to the connection weight from neuron j to neuron i, subscript n refers to the n-th pattern of the training data set, and N refers to the number of input variables reaching neuron j. If the neuron j is in the hidden layer of a three-layer MLP, the inputs will be from the external environment, whereas if j is in the output layer, the inputs will be the outputs from the hidden layer nodes weighted by the connection weights from hidden layer nodes to the output layer nodes. In the weighted summation of inputs (Equation 8.85), $i = 0$ corresponds to the bias b_j with weight w_{j0} and input $x_0(n) = 1$. After computing the output y_j at an output layer neuron, it is compared with the corresponding target output $d_j(n)$, which comes from the training data set, and the instantaneous error $e_j(n)$ is computed as follows:

$$e_j(n) = d_j(n) - y_j(n)$$ (same as Equation 8.4)

When the instantaneous errors at all the output layer nodes are calculated, the total squared error, $E(n)$, is calculated as

$$E(n) = \frac{1}{2} \sum_{j=1}^{L} e_j^2(n)$$ (8.87)

where L is the number of nodes in the output layer, and the factor 1/2 is introduced to make subsequent analysis simpler. The mean squared error can also then be defined as

$$F_{ave} = \frac{1}{N} \sum_{n=1}^{N} E(n)$$ (8.88)

where N is the total number of patterns in the training data set. The error terms $E(n)$ and E_{ave} are functions of the free parameters of the network, namely the connection weights and the biases. The weights, however, are updated on a sequential basis, layer by layer starting from the output layer, neuron by neuron, by presenting each pattern of the training data set one after the other. The completion of presenting all the patterns of the training data set constitutes one *epoch*. To implement the BP algorithm, it is necessary to calculate the error gradient $\dfrac{\partial E(n)}{\partial w_{ji}(n)}$ at each neuron. If increasing the weight leads to more error, then the weighs are adjusted downwards and vice versa.

The instantaneous error gradients with respect to the weights, at a typical neuron j by the chain rule of calculus, can be written as

$$\frac{\partial E(n)}{\partial w_{ji}(n)} = \frac{\partial E(n)}{\partial e_j(n)} \frac{\partial e_j(n)}{\partial y_j(n)} \frac{\partial y_j(n)}{\partial v_j(n)} \frac{\partial v_j(n)}{\partial w_{ji}(n)}$$ (8.89)

From Equations 8.4, 8.86, 8.87, and the fact that for a neuron $j, v_j(n) = \sum_{i=0}^{M} w_{ji}(n)y_i(n)$, it can be seen that

$$\frac{\partial e_j(n)}{\partial y_j(n)} = -1$$ (8.90a)

$$\frac{\partial E(n)}{\partial e_j(n)} = e_j(n)$$ (8.90b)

$$\frac{\partial y_j(n)}{\partial v_j(n)} = f'\{v_j(n)\}$$ (8.90c)

$$\frac{\partial v_j(n)}{\partial w_{ji}(n)} = y_i(n)$$ (8.90d)

Substituting Equation 8.90 in Equation 8.89 yields

$$\frac{\partial E(n)}{\partial w_{ji}(n)} = -e_j(n)f'\{v_j(n)\}y_i(n) \tag{8.91}$$

By the delta rule, the weight correction is

$$\Delta w_{ji}(n) = -\eta \frac{\partial E(n)}{\partial w_{ji}(n)} = \eta e_j(n)f'\{v_j(n)\}y_i(n) = \eta \delta_j(n)y_i(n) \tag{8.92}$$

where $\delta_j(n)$ is the local gradient given by

$$\delta_j(n) = -\frac{\partial E(n)}{\partial v_j(n)} = -\frac{\partial E(n)}{\partial e_j(n)}\frac{\partial e_j(n)}{\partial y_j(n)}\frac{\partial y_j(n)}{\partial v_j(n)} = e_j(n)f'\{v_j(n)\} \tag{8.93}$$

To implement the above algorithm, it is necessary to know the error at each node. In the case of nodes at the output layer, these can be found because the network output is compared with the target output (Equation 8.4). In the case of nodes in the hidden layer, there are no target values to compare with and therefore the errors are not known explicitly. Since the error at the hidden layer node contributes to the error at the output node, a share of the output error is assigned to the hidden node in accordance with the *credit-assignment problem*[4] (Minsky, 1961). Continuous differentiability of the activation function is also necessary to implement the BP algorithm. For a node j in the hidden layer, the local gradient is defined as follows:

$$\delta_j(n) = -\frac{\partial E(n)}{\partial v_j(n)} = -\frac{\partial E(n)}{\partial y_j(n)}\frac{\partial y_j(n)}{\partial v_j(n)} = -\frac{\partial E(n)}{\partial y_j(n)}f'(v_j(n)) \tag{8.94}$$

The term $\dfrac{\partial E(n)}{\partial y_j(n)}$ is calculated as follows:

For an output layer node k, Equation 8.87 can be rewritten as (replacing subscript j by k)

$$E(n) = \frac{1}{2}\sum_{k=1}^{L} e_k^2(n)$$

which when differentiated gives

$$\frac{\partial E(n)}{\partial y_j(n)} = \sum_{k=1}^{L} e_k(n)\frac{\partial e_k(n)}{\partial y_j(n)} = \sum_{k=1}^{L} e_k(n)\frac{\partial e_k(n)}{\partial v_k(n)}\frac{\partial v_k(n)}{\partial y_j(n)} \tag{8.95}$$

For an output layer node,

$$e_k(n) = d_k(n) - y_k(n) = d_k(n) - f_k\{v_k(n)\} \tag{8.96}$$

---

[4] Weights are adjusted to effect the change needed to obtain the desired system performance by giving 'credits' (rewards) or 'debits' (penalties) as the case may be.

which when differentiated gives

$$\frac{\partial e_k(n)}{\partial v_k(n)} = -f_k'\{v_k(n)\} \tag{8.97}$$

For the neuron k

$$v_k(n) = \sum_{j=0}^{M} w_{kj}(n)y_j(n) \tag{8.98}$$

where M is the total number of inputs applied to neuron k (equal to the number of hidden layer neuron), and, which when differentiated gives

$$\frac{\partial v_k(n)}{\partial y_j(n)} = w_{kj}(n) \tag{8.99}$$

It should also be noted that w_{k0} is the bias $b_k(n)$ with $y_0(n) = 1$. Substituting Equations 8.97 and 8.99 in Equation 8.95,

$$\frac{\partial E(n)}{\partial y_j(n)} = -\sum_{k=1}^{L} e_k(n)f'\{v_k(n)\}w_{kj}(n) = -\sum_{k=1}^{L} \delta_k(n)w_{kj}(n) \tag{8.100}$$

where $\delta_k(n)$ has the same definition as before (Equation 8.94), excepting that the subscript j is replaced by k; that is,

$$\delta_k(n) = -\frac{\partial E(n)}{\partial v_k(n)} = e_k(n)f'\{v_k(n)\} \tag{8.101}$$

Because $\delta_j(n) = -\dfrac{\partial E(n)}{\partial y_j(n)}f'\{v_j(n)\}$, the local gradient for a hidden node j is given as

$$\delta_j(n) = \sum \delta_k(n)w_{kj}(n)f'\{v_k(n)\} \tag{8.102}$$

As before, by the delta rule,

$$\Delta w_{ji}(n) = \eta\delta_j(n)\,y_i(n) \qquad \text{(same as Equation 8.92)}$$

Having completed one cycle of back-propagating the error to all the neurons in the network, the procedure is repeated by presenting the remaining patterns in the training data set one after the other. The total error can now be computed using Equations 8.87 and 8.88. If this error is zero (very unlikely), the training process stops. If not, the procedure is repeated using a different combination of the input patterns until a prescribed acceptable error is reached. Randomizing the presentation of input data tend to make the search for the optimum weights stochastic. This kind of implementation of BP is known as sequential mode, or online mode, or stochastic mode. In the alternative, which is referred to as the batch mode, weight updating is done after presenting all training samples to the network, or in other words, epoch by epoch basis. The error estimation and weight updating in the latter case is according to the following equations:

$$E_{ave} = \frac{1}{2N} \sum_{n=1}^{N} \sum_{j} e_j^2(n) \tag{8.103}$$

$$\Delta w_{ji} = -\eta \frac{\partial E_{ave}}{\partial w_{ji}} = \frac{\eta}{N} \sum_{n=1}^{N} e_j(n) \frac{\partial e_j(n)}{\partial w_{ji}} \tag{8.104}$$

A description of BP in the context of hydrological time series forecasts has been given by Lachtermacher and Fuller (1994).

8.8.1 Generalized delta rule

The learning rate parameter η ($0 \le \eta \le 1$) controls the speed of attaining the optimum solution. A small value allows the weight change to be small and smooth but takes a longer computation time. A large value, on the other hand, accelerates the process, but may give oscillatory solutions and also may become unstable. It also need not be a constant value, although ideally all neurons in the network should learn at the same rate. Usually, neurons in the output layer have larger gradients and therefore their learning rates should be smaller. Neurons with many inputs should have a relatively small learning rate compared with those with fewer inputs. It is difficult to determine the learning rate parameter in a systematic way, but a rule of thumb is that it is proportional to the average magnitude of the vectors feeding into the network (Hush and Horne, 1993). The addition of a momentum term, α (Rumelhart et al., 1986), can increase the learning rate and yet avoid oscillations. It is introduced as follows:

$$\Delta w_{ji}(n) = \alpha \Delta w_{ji}(n - 1) + \eta \delta_j(n) x_i(n) \tag{8.105}$$

This is referred to as the generalized delta rule, which becomes the delta rule when $\alpha = 0$. The momentum term lies between 0 and 1. It also has the tendency to avoid getting trapped in local optima. If the gradient $\frac{\partial E}{\partial w_{ji}}$ has the same sign in two consecutive iterations, the weight adjustment is relatively large. If, on the other hand, they have opposite signs, the adjustment would be small. On the basis of the optimality criterion that η and α on average yield convergence to a local optimum in the error surface with the least number of epochs, Haykin (1999) recommends optimal values of $\eta = 0.1$ and $\alpha = 0.5$. Optimality can also be defined using other criteria. Although the BP is by far the most popular learning algorithm in neural networks, it may get trapped in local minima at times. The sequential learning mode has an advantage over the batch mode from the speed and storage points of view.

8.9 ANN IMPLEMENTATION DETAILS

8.9.1 Data preprocessing: Principal Component Analysis (PCA)

When more and more time-lagged variables are added into the input layer, the dimensionality of the problem becomes larger and larger without a proportionate contribution to the output. In other words, some input variables tend to become redundant. PCA aids to achieve parsimony of a data set consisting of a large number of interrelated variables by extracting the smallest number components that account for most of the variation in the original multivariate data and to summarize the data with little loss of information (Jolliffe, 2002). For the BP algorithm to work efficiently, it is also important that the input variables in the training set should be uncorrelated. This can be achieved by the PCA. They should then be scaled so that their covariances are approximately equal.

PCA is aimed at reducing the dimensionality of a problem that involves many inter-related variables while retaining the important variations present in the data. Dimension reduction can also be achieved by 'factor analysis', which makes an assumption about the causal relationship among variables that PCA does not. PCA transforms original variables into a new set of variables, or principal components (PCs), which are uncorrelated and ordered so that the first few account for most of the variations in the original data. Mathematically, it can be defined as an orthogonal linear transformation that transforms the data into a new coordinate system such that the largest variance by any projection of the data will lie on the first coordinate, the second largest on the second coordinate, and so on. Each PC is a linear transformation of the entire original data set that maximizes the variance in the lower-dimensional space. In other words, it is a domain transformation that enables identification of significant elements, which cannot be done in the original domain. In least squares sense, PCA is the optimal linear transform for compressing a given high-dimensional data set to a lower-dimensional one. PCA is non-parametric; the answer it gives is unique and independent of any hypothesis about the probability distribution of data. However, no prior knowledge can be incorporated into PCA and there is always some loss of information. The objective is to retain as much information as possible; that is, minimize the error. PCA is reversible in the sense that the original data can be recovered exactly. PCA is sometimes referred to as the Karhunen–Loêve transformation in communication theory. An example of PCA application to monitor spatial and temporal variations in water quality is given by Bengraine and Marhaba (2003).

PCA can be carried out using covariance matrices or correlation matrices. Covariance-based analysis is unit dependent, whereas correlation-based analysis uses standardized variables and is easier to implement. PCs depend on the ratios of correlations and not on the absolute values of correlation. When there are widely differing variances, care should be taken to interpret the covariance-based PCA. The lower-dimensional space is determined by the eigenvectors of the covariance matrix of the normalized original data set that correspond to the largest eigenvalues. These eigenvectors are called the PCs. A vector V is an eigenvector of A if and only if

$$AV = \lambda V \tag{8.106}$$

where V has a dimension N; A is a square matrix of dimension $N \times N$; and λ, a scalar, is the eigenvalue. Equation 8.106 is called the eigenvalue equation. For each eigenvalue,

$$(A - \lambda_i I)V = 0 \tag{8.107}$$

where I is the identity matrix.

The matrix A can be factorized as

$$A = Q\Lambda Q^{-1} \tag{8.108}$$

where Q is a square matrix ($N \times N$) whose i-th column is the eigenvector q_i of A and Λ is the diagonal matrix with $\Lambda_{ii} = \lambda_i$. q_i's are usually normalized.

Two approaches can be used for PCA: covariance method and singular value decomposition. With the covariance method, the goal is to transform the given data set X of dimension $M \times N$ to a lower-dimension $L \times N$ data set Y ($L \le M$). It is equivalent to the Karhunen–Loêve transform (KLT) of the data matrix X.

This method starts by organizing the data set, which involves arranging the data X_1, X_2, ..., X_N (N is the number of data vectors) as column vectors, each of which has M rows and placing them into a single matrix X of dimension $M \times N$. Then, the empirical means along each dimension $m = 1$, M, where M is the number of elements in each vector (dimension), are calculated and subtracted from each data value in each row. The covariance matrix C_X ($M \times M$) is then calculated as

$$C_X = \frac{1}{N} X X^T \tag{8.109}$$

The diagonal terms of this matrix are in fact the variances. The covariance matrix is positive definite. The computation of the eigenvectors and eigenvalues of the covariance matrix is iterative and non-trivial, and is best obtained by standard mathematical software (e.g., MATLAB®). The outputs of many such software have eigenvectors of unit length. The highest eigenvalue corresponds to the first PC, the second highest the second PC, and so on. Usually the PCs corresponding to the smaller eigenvalues are ignored, thereby reducing the dimensionality of the problem.

The eigenvectors matrix, which diagonalizes the covariance matrix, is computed as

$$P^{-1}CP = D \tag{8.110}$$

where D is the diagonal matrix of eigenvalues of C. The elements of the diagonal matrix give the eigenvalues

$$D_{pq} = \lambda_m \text{ for } p = q = m; \text{ and } D_{pq} = 0 \text{ for } p \ne q \tag{8.111}$$

The eigenvalues and eigenvectors are then ordered in decreasing order of magnitude of eigenvalue and paired; that is, the m-th eigenvalue corresponds to the m-th eigenvector. The subset of eigenvectors is selected using the criterion given in Equation 8.112.

$$\frac{\sum_{i=1}^{L} \lambda_i}{\sum_{i=1}^{M} \lambda_i} > \text{threshold (usually about } 0.90 - 0.95); \quad L \le M \tag{8.112}$$

Normally, PCs with eigenvalues greater than 1 are retained. It can be shown that the error due to dimension reduction is $\frac{1}{2} \sum_{i=L+1}^{M} \lambda_i$. When $L = M$, all the PCs are retained, and there is no dimension reduction and no loss of information.

Some assumptions implied in the PCA are linearity (i.e., that the data set is a linear combination of a certain basis), Gaussian distribution of eigenvectors, and that the observed data has a high signal-to-noise ratio (SNR). The SNR is defined as

$$\text{SNR} = \frac{\sigma_{\text{signal}}^2}{\sigma_{\text{noise}}^2} \tag{8.113}$$

8.9.1.1 *Eigenvalue decomposition*

Eigenvalue decomposition and PCA are equivalent. Let X (x_i are the columns of X) be the original data set, an $M \times N$ matrix, where M is the number of measurement types and N is the number of samples. Each column is a single sample. Y (y_i are the columns of Y) is a new $M \times N$ matrix related to X by a linear transformation P (p_i are the rows of P).

The goal is to find an orthogonal matrix P in

$$Y = PX \tag{8.114}$$

such that

$$C_Y = \frac{1}{N} YY^T \tag{8.115}$$

is a diagonal matrix (uncorrelated). In Equation 8.115 C_Y is the correlation matrix of Y. In expanded form,

$$PX = \begin{bmatrix} p_1, p_2, \ldots, p_m \end{bmatrix} \begin{bmatrix} x_1 \\ x_2 \\ . \\ . \\ x_n \end{bmatrix} \tag{8.116}$$

$$Y = \begin{bmatrix} p_1.x_1, p_1.x_2, \ldots, p_1.x_n \\ p_2.x_1, p_2.x_2, \ldots, p_2.x_n \\ \ldots \\ \ldots \\ p_m.x_1, p_m.x_2, \ldots p_m.x_n \end{bmatrix} \quad y_i = \begin{bmatrix} p_1.x_i \\ p_2.x_i \\ \ldots \\ \ldots \\ p_m.x_i \end{bmatrix} \tag{8.117}$$

In Equation 8.117, each column y_i is given as a dot product of p_i and x_i.

The rows p_i's of P are the PCs of X. C_Y can be written in terms of the unknown variables as

$$\begin{aligned} C_Y &= \frac{1}{N} YY^T \\ &= \frac{1}{N} (PX)(PX)^T \\ &= \frac{1}{N} PXX^T P^T \\ &= P\left(\frac{1}{N} XX^T\right) P^T \\ &= PC_X P^T \end{aligned} \tag{8.118}$$

Note that C_X is the covariance matrix of X, which is symmetric and square ($M \times M$) and whose diagonal terms give the variances of particular measurements. The off-diagonal terms give the covariance between measurement types. Any symmetric matrix can be diagonalized by an orthogonal matrix of its eigenvectors. By selecting P where each row p_i is an eigenvector of C_X, the following relationship can be written:

$$C_X = PDP^T \tag{8.119}$$

where D is a diagonal matrix and P is a matrix of eigenvectors of C_X arranged as columns. Matrix P is chosen to be one where each row p_i is an eigenvector of C_X. The goals are to minimize the redundancy as measured by the covariance and maximize the signal measured by the variance; that is, C_Y is a diagonalized matrix as shown in Equation 8.120 implying Y is decorrelated. It also means all off-diagonal terms of C_Y are zero.

$$
\begin{aligned}
C_Y &= PC_XP^T \\
&= P(PDP^T)P^T \\
&= (PP^T)D(PP^T) \\
&= (PP^{-1})D(PP^{-1}) \\
&= D
\end{aligned}
\tag{8.120}
$$

In Equation 8.120, the relationship that $P^{-1} = P^T$ (the inverse of an orthogonal matrix is its transpose) is also used. It can also be seen that the PCs of X are the eigenvectors of C_X and that the i-th diagonal value of C_Y is the variance of X along p_i.

8.9.1.2 Deriving the new data set

After the PCA, the final data set can be obtained as $(\lambda_1, \lambda_2, ..., \lambda_M)X^T$. If all the PCs are taken into account, there is no loss of information. However, if only the significant PCs are considered, there will be some loss of information due to truncating the number of eigenvectors. The original data can be retrieved by inverse transformation of the above relationship. When all the eigenvectors are taken into account, the inverse of the vector of eigenvalues is equal to its transpose. This is true because the elements of the matrix are all unit eigenvectors of the data set.

MLP architectures are good at ignoring irrelevant inputs. PCA is one way of selecting linear subspaces of reduced dimensionality. Since the first hidden layer is formed by a linear combination of the inputs, the network attention is confined to the subspace of the weight vectors. Therefore, addition of irrelevant inputs to the training data does not increase the number of hidden units required, but increases the amount of training data required.

8.9.2 Data normalization

One of the definitions used in the context of data normalization is 'rescaling' a vector, which is meant to add or subtract a constant and then multiply or divide by a constant, as in the case of changing the units of measurement of the data, such as to convert a temperature from Celsius to Fahrenheit. A second definition is 'normalizing' a vector, which most often means dividing by a norm of the vector, such as to make the Euclidean length of the vector equal to unity. In the neural networks literature, 'normalizing' also often refers to rescaling by the minimum and range of the vector, to make all the elements lie between 0 and 1. A third definition is 'standardizing' a vector, which most often means subtracting a measure of location and dividing by a measure of scale. For example, if the vector contains random values with a Gaussian distribution, subtracting the mean and dividing by the standard deviation coverts the data to 'standard normal' random variables with zero mean and unit standard deviation.

Normalization is done by reducing the data range to fixed limits. The commonly used fixed limits are (–1, 1) or (0, 1). In ANN models, the input data, which may take any value ranging from $-\infty$ to ∞, are first normalized to the range (–1, 1) by a linear transformation. This is desirable although not a requirement because the activation function anyway squashes the summed input data to the range (–1, 1) or (0, 1). With the sigmoid transfer function at the output layer, the model outputs will also be within the range (0, 1). The target outputs should therefore be transformed to the same range in the same way for comparison purposes. With an unbounded linear transfer function ($f(x) = x$) in the output layer, the model output values as well as the target values are linearly transformed into the range (–1, 1) for comparison purposes. Since the activation functions reach their maximum and minimum values asymptotically, the theoretical maximum and minimum values are replaced by the more practical values of (–0.9, 0.9) or (0.1, 0.9). The commonly adopted linear transformation is of the form

$$x_i' = \text{low} + \frac{(x_i - x_{\min})(\text{up} - \text{low})}{(x_{\max} - x_{\min})} \tag{8.121}$$

where x_i' denote the normalized data, x_i the original data, $x_{\max}$ and $x_{\min}$ the maximum and minimum values of the original data, and 'low' and 'up' the smallest and largest output values allowed in the network. The low and up values are usually preset to (0.1, 0.9) or (–0.9, 0.9). Another way is to use a zero mean and a unit variance transformation, which is referred to as 'standardization' of data.

Normalizing the input data, although not required, has its benefits. In this respect, it is better to have the data centred around zero, rather than confining to the interval (0, 1). Standardizing the input data tend to make the training process more efficient computationally and reduces the chances of getting trapped in local optima. For example, when considering inputs with values that differ by several orders of magnitude, the contributions to the output, if measured by the Euclidean distance as in the case of MLPs, the smaller inputs will become insignificant relative to the larger inputs. It also removes the scale dependence of initial connection weights. However, it should be done with caution as it discards information. If the discarded information is irrelevant, then standardization has advantages, whereas in the contrary case, it has the opposite effect. Standardizing input variables can have different effects when used with different training algorithms. MLPs with BP algorithm are sensitive to scaling and slow down the convergence when the data are ill conditioned. Scaling under such circumstances is an important consideration.

At the output layer, the target variables also need rescaling to reflect their relative importance. If they are of equal importance, they should be standardized to the same range or to the same standard deviation. In the former case, it is important to use the lower and upper bounds for the range rather than the minimum and maximum values in the training data set. In the latter case, it is important to use zero mean and unit standard deviation. With unbounded target data, an unbounded activation function should be used in the output layer.

When dealing with noisy data, it is sometimes necessary to transform the target non-linearly, or approximately linearly. In such situations, smooth functions are preferred as they have better generalization properties.

8.9.3 Choice of input variables

A rule of thumb is that the number of samples in the training set should at least be greater than the number of synaptic weights (Rogers and Dowla, 1994; Tarassenko, 1998). This gives the upper limit of the number of nodes for the network.

8.9.4 Heuristics for implementation of BP

When implementing the BP algorithm, the sequential (or online) mode of processing is preferable as it is computationally faster. Batch mode requires the computation of the Hessian matrix, which is time consuming. To maximize the information content, it is desirable to use examples that result in large training errors. Large local gradient leads to faster learning. Use of examples radically different from those previously used, and randomizing the presentation of patterns are desirable.

The choice of the activation function also has influence on efficient BP. Of the two widely used activation functions, the *logistic* function is non-symmetric and the tanh function is antisymmetric about zero; that is, $f(-v) = -f(v)$. An antisymmetric function is better for efficient training. Because of the asymptotic nature of the sigmoid function, the target values modified by a small value are preferred. For example, $d_j = a - \varepsilon$ for a limiting value of a, and $d_j = -a + \varepsilon$ for a limiting value of $-a$ (ε is a small quantity).

In normalization of input data, it is better to get the patterns from both positive and negative values. This can be achieved by mean removal. Also, decorrelation of inputs can be achieved by transforming the data from time domain to frequency domain, as well as by PCA. Another preprocessing that can be done is to carry out covariance equalization. Learning will be better with the above normalizations.

Initialization of the weights should be such that they are neither too small nor too large. Small initial values tend to produce induced fields close to the near-linear part of the sigmoid, implying that the network operates on a flat area of the error surface, which is not desirable. On the other hand, large initial values will produce induced fields close to the saturation part of the sigmoid where the gradients are approaching zero, thereby slowing down the learning process.

8.9.5 Stopping criteria

Since the BP algorithm is an iterative process, some criterion to stop the weight updating should be given. This may be after a prescribed number of epochs. One drawback in this stopping criterion is that the algorithm may stop without achieving the optimum solution; on the other hand, it will prevent the computations continuing in an endless cycle. Another criterion is cross-validation (Stone, 1974). In the latter case, the training data set is divided into a training set and a testing set. The training set is further subdivided into an estimation (or calibration) subset and a validation subset. Approximately 80% of the training set is assigned to the estimation subset and the remaining 20% to the validation subset. Generalization performance is measured on the testing set. To implement the cross-validation, training is stopped after a certain number of epochs, and the thus-far trained network is used in the forward direction with the validation data. This procedure is repeated with further training and validation until the mean squared error begins to increase with increasing number of epochs (Figure 8.12). In general, the mean squared error drops rapidly during the early stages of learning (for small number of epochs), and continues to decrease rather slowly, and somewhat asymptotically. The point at which the validation error changes direction corresponds to the early stopping criterion (Amari et al., 1996).

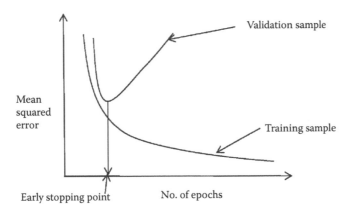

Figure 8.12 Illustration of cross-validation.

When the network gives the correct output for unseen input data, it is said to have attained generalization characteristics.

8.9.6 Performance criteria

The performance of ANN models can be measured by several error indicators. The most commonly used ones are the mean absolute error (MAE), root mean square error (RMSE), relative root mean square error with respect to the expected mean value (RRMSE), the Nash–Sutcliffe coefficient of efficiency (NSE), modified coefficient of efficiency (MNSE), coefficient of correlation r, and the coefficient of determination (CD), which is different from R^2 in statistics. The MAE and RMSE have the units of the variable, whereas RRMSE and NSE are dimensionless. Their definitions and optimum values are given below:

$$\text{MAE} = \frac{1}{N}\sum_{i=1}^{N}\left|(O_i - P_i)\right|; \text{ optimum} = 0, \text{ units same as for data} \tag{8.122}$$

$$\text{RMSE} = \sqrt{\frac{1}{N}\sum_{i=1}^{N}(O_i - P_i)^2}\text{ ; optimum} = 0, \text{ units same as for data} \tag{8.123}$$

$$\text{RRMSE} = \frac{\sqrt{\dfrac{1}{N}\sum_{i=1}^{N}(O_i - P_i)^2}}{\bar{O}}\text{ ; optimum} = 0, \text{ dimensionless} \tag{8.124}$$

$$\text{NSE} = 1 - \frac{\sum_{i=1}^{N}(O_i - P_i)^2}{\sum_{i=1}^{N}(O_i - \bar{O})^2}\text{ ; optimum} = 1, \text{ dimensionless} \tag{8.125}$$

$$\text{MNSE} = 1 - \frac{\sum\limits_{i=1}^{N} |(O_i - P_i)|}{\sum\limits_{i=1}^{N} |(O_i - \bar{O})|} \; ; \text{optimum} = 1, \text{dimensionless} \tag{8.126}$$

$$\text{CD} = \frac{\sum\limits_{i=1}^{N} (O_i - \bar{O})^2}{\sum\limits_{i=1}^{N} (P_i - \bar{O})^2} \; ; \text{optimum} = 1, \text{dimensionless} \tag{8.127}$$

$$1 - \text{NSE} = \frac{\sum\limits_{i=1}^{N} (O_i - P_i)^2}{\sum\limits_{i=1}^{N} (O_i - \bar{O})^2} \; ; \text{optimum} = 0, \text{dimensionless} \tag{8.128}$$

$$r = \frac{\sum\limits_{i=1}^{N} (O_i - \bar{O})(P_i - \bar{P})}{\left[\sum\limits_{i=1}^{N} (O_i - \bar{O})^2\right]^{0.5} \left[\sum\limits_{i=1}^{N} (P_i - \bar{P})^2\right]^{0.5}} \; ; \text{optimum} = 1, \text{dimensionless} \tag{8.129}$$

In these equations, N is the number of records; O_i and P_i are the expected and simulated values of the variable, respectively; and $\bar{O}$ is the mean of the expected value within the number of records. RMSE and MAE are absolute error measures, and they are expressed in the units of the variables. In general, RMSE is not less than MAE. The degree to which RMSE exceeds MAE is an indicator of the extent of outliers. For comparison of different time series, RMSE and MAE are inappropriate because of the differences in the scales of data. In such situations, relative error measures that are dimensionless are preferred. Pearson's product-moment correlation coefficient, or simply the correlation coefficient r, is the most widely used measure despite the fact that it is oversensitive to extreme values and insensitive to additive and proportional differences between predicted and targeted values. It also evaluates the linear relationship between the variables.

The coefficient of efficiency (Nash and Sutcliffe, 1970), also known as the Nash–Sutcliffe coefficient of efficiency, is an improvement over the correlation coefficient in that it is sensitive to differences in the variances of the predicted and targeted values. It is widely used in hydrological model evaluations. The modified coefficient of efficiency (Legates and McCabe, 1999) is not sensitive to outliers and is an improvement to the coefficient of determination as well as the correlation coefficient.

8.10 FEEDBACK SYSTEMS

Feedback systems offer great advantages over feed-forward systems for modelling of dynamical systems. A small feedback system may sometimes be equivalent to a large (or even an infinite) feed-forward system. Much of the work associated with feedback systems have been done in the context of 'linear systems theory'; however, there are difficulties that can only be resolved by resorting to non-linear dynamics. Time delay neural networks are commonly referred to as recurrent neural networks. Because of the feedback mechanism, all such neural networks are recursive. Such neural networks do not have general learning algorithms.

8.11 PROBLEMS AND LIMITATIONS

Although the ANN approach has many applications in several diverse fields with non-linearities, it has its own problems and limitations. The type of ANN to be used and the architecture to be adopted are not known *a priori* and are often determined by trial and error. Such an approach may not always lead to the optimum type of ANN or the optimum architecture. It only gives the best of the trials attempted. The learning process, which normally follows the gradient descent search approach, is prone to getting trapped at a local optimum leading to a suboptimal solution. There could also be a situation where the learning reaches a stage where the weights cannot be adjusted any further, a condition known as paralysis. The number of training samples to be used is another issue, although it is more important to have the data range cover extremes. The level of training, or when to stop training, should be decided so as not to overfit. When the training fits to the data too closely, it is likely that the function approximation fits the noise as well rather than the signal only. In such kind of overfitting situations, the networks lack generalization property. In general, a larger network size can potentially produce a better approximation, but at a heavy computational cost. Adding hidden units sequentially as well as pruning from a larger network can be used as ways of achieving an optimal network.

8.12 APPLICATION AREAS

Artificial neural networks belong to a class of data-driven models. They are particularly suitable for data-rich, theory-weak problems (Figure 8.13). They have several attractions compared with traditional methods of systems modelling such as the ability to learn from examples, and not requiring *a priori* knowledge of the system dynamics. ANNs normally process information in a distributed and parallel manner, thereby enabling them to handle large amounts of data. The computations of the individual processors are largely independent of each other. This is clearly a distinct advantage in that the network will function even when a few processing elements are faulty, which is not the case for serial systems. They are also insensitive to noise, which is inherently present in all types of real data, and can handle non-linear problems, which is a distinct advantage. Online processing is also possible by exposing the network to new data as and when they become available. Once trained, ANNs have the ability to generalize.

Because of the above and other attractions, ANNs have been used in diverse fields such as science, engineering, medical sciences, finance, life sciences, etc.

8.12.1 Hydrological applications

Application of ANN techniques to hydrology started perhaps in the early 1980s and within a short span of time, many researchers have produced a large number of publications (e.g., approximately in chronological order: Daniell, 1991; Liong and Chan, 1993; Karunanithi et al., 1994; Maier and Dandy, 1996, 2000; Jayawardena and Fernando, 1995a,b, 1996;

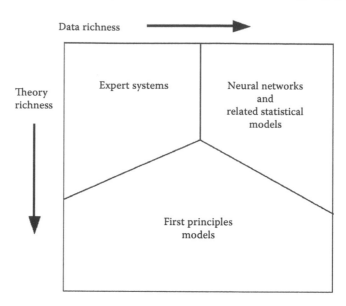

Figure 8.13 Application domains of ANNs. (From Smolensky, P., Mozer M.C. and Rumelhart, D.E. (Eds.), *Mathematical Perspectives on Neural Networks*. Lawrance Erlbaum Associates, Mahwah, NJ, p. 537, 1996.)

Raman and Sunilkumar, 1995; Smith and Eli, 1995; Minns and Hall, 1996; Jayawardena and Fernando, 1996; Jayawardena et al., 1996a,b; Shamseldin, 1997; Khondar et al., 1998; Govindaraju, 2000; Govindaraju and Rao, 2000; Anmala et al., 2000; ASCE Task Committee, 2000a,b; Jayawardena and Fernando, 2000, 2001; Dawson and Wilby, 2001; Sivakumar et al., 2002; Rajukar et al., 2002; Li and Gu, 2003; Jayawardena et al., 2004; Jayawardena and Tian, 2005a,b). Much of the applications have used the MLP type of networks. A comprehensive description of all relevant hydrological applications is beyond the scope of this chapter. Instead, a summary of the studies carried out by the author and his colleagues is presented next.

8.12.1.1 *River discharge prediction*

a. Streamflow prediction at Lo Shue Ling, Hong Kong

The smallest catchment where an MLP-type ANN application has been made is for the prediction of daily streamflow from daily rainfall input data. The application site has been the Lo Shue Ling catchment (catchment area 10.78 km²) in Hong Kong where the present discharge $Q(t)$ is assumed to be dependent on antecedent discharges $Q(t-1)$, $Q(t-2)$, $Q(t-3)$; present rainfall depths measured at Ta Kwu Ling Police Station and Ta Kwu Ling Farm rain gauges $P1(t)$, $P2(t)$; and daily evaporation and $E(t)$ measured at King's Park meteorological station (Jayawardena and Fernando, 1995b). The MLP network trained with the BP algorithm consisted of a six-node input layer, one hidden layer with three nodes, and an output layer of one node. A small learning rate of 0.05 has been used to avoid oscillation of the output. The data used were from 1975 to 1978 for training and from 1979 to 1982 for testing, standardized to the range −0.999 to +0.999. Continuous weight updating has been done and the training has been terminated after 10,000 iterations. The model outputs were compared with those from a conceptual model (Tank model; Sugawara et al., 1984) and with actual observations. The degree of agreement as measured by the RMSE was found to be better (70%) than that by the Tank model. One of the problems in dealing with this data set is the presence of very low or no flows during the dry season. This affects continuous simulation.

b. River discharge prediction in Shaoxie River, China
A similar application with more input data made for the prediction of daily discharges at the Qiaodouchun gauging station (latitude: 30°17′N; longitude: 119°38′E) across Shaoxie River (catchment area = 233 km²) in the Taihu Basin in West Zhejiang Province of China uses daily rainfall measured at Shilin, Yangshanyu, Xiekou, Yanchaotou, Zhaominshi, Nanzhuang, and Qiaodouchun gauging stations and the daily discharges measured at Qiaodouchun gauging station (Jayawardena et al., 1996a). An MLP with a multinode input layer consisting of the antecedent rainfalls $R(t-1)$, $R(t-2)$, and $R(t-3)$ at the seven rain gauges, as well as the antecedent discharges $Q(t-1)$, $Q(t-2)$, and $Q(t-3)$, a single hidden layer with five nodes, and a single output layer have been used. The data, taken from Chinese Hydrologic Almanac covered the period 1980–1988. A hyperbolic tangent activation function and a learning rate of 0.05 have been used. Learning was terminated after 20 error-reducing iterations after which there was no appreciable improvement. The data for the period 1980–1984 have been used for training, while the entire data set from 1980 to 1988 has been used for testing. The results were compared with observations and those from a Tank model simulation. The objective function in this case has been taken as the mean of the MSE and the mean of the logarithms of the squared error. Such a multiobjective criterion is aimed at obtaining agreements in the high-flow part as well as in the low-flow part.

The above two examples were for continuous simulation. Event simulations can also be made by the MLP approach in the same way as illustrated by an example for the Shek Pi Tau gauging station, another small catchment in Hong Kong (Jayawardena and Fernando, 1995b; Fernando, 1997).

c. Evaporation prediction at King's Park Meteorological Station, Hong Kong
Monthly evaporation predictions have also been made using other meteorological data such as wind speed, vapour pressure difference, mean air temperature, and duration of sunshine hours as inputs at the King's Park meteorological station (latitude 22°18′12.82″N; longitude: 114°10′18.75″E) in Hong Kong (Jayawardena and Fernando, 1996). The results, when compared with observations, were found to be better than those obtained from an empirical equation calibrated for the region. In this application, an adaptive MLP has been used.

d. Discharge prediction in Fuji River, Japan
The results of an exercise aimed at studying the selection of input variables, optimization of internal parameters, choice of stopping criteria, and the use of distributed rainfall is reported separately by Fernando (2002) with reference to an application of an MLP–BP network to predict 1-day-ahead discharge Q_{t+1} at the Kitamatsuno gauging station across Fuji River in Japan. Fuji River is one of the steepest rivers in Japan and drains an area of 3432 km² and has a length of 128 km. The data used have been the daily discharges at Kitamatsuno gauging station, and rainfalls measured at 10 weather stations around the basin for the period January 1990 to December 1993. The data set has been divided into training (January 1990–December 1992), validation (January 1993–August 1993), and testing (September 1993–December 1993). This is not in line with the general guideline that the training, validation, and testing samples should have approximately equal sizes. In this case, the extreme events were contained in the training sample, thereby making it relatively long. The input variables have been selected on the basis of autocorrelation and cross-correlation analysis of all the discharge and rainfall data. Time-lagged discharges (Q_t, Q_{t-1}, Q_{t-2}, ..., Q_{t-6}),

and time-lagged rainfalls measured at the 10 weather stations ($R1_t$, $R1_{t-1}$, $R1_{t-2}$,...$R1_{t-6}$, $R2_t$, $R2_{t-1}$, $R2_{t-2}$,...$R2_{t-6}$,..., $R10_t$, $R10_{t-1}$, $R10_{t-2}$,...$R10_{t-6}$), thus amounting to 77 input variables, have been used in the final optimum network that had 12 hidden nodes. The optimal network has been obtained by increasing number of hidden nodes incrementally from 2 to 12. Twelve nodes in the hidden layer is about the maximum that should be used in an MLP network (Tarassenko, 1998). The momentum term and the learning rate that gave best simulations, obtained by trial and error, were 0.7 and 0.05, respectively. Table 8.5 gives the number of iterations at the termination of training. It can be seen that for certain combinations of momentum term and learning rate, the training does not converge. Table 8.6 gives the error indicators for different numbers of neurons in the hidden layer. Two stopping criteria have been used: cross validation and fixed number of iterations set at 2000. The performance of the network has been evaluated with four error indicators: MAE, RMSE, correlation coefficient r, and MNSE (see section 8.9.6 for their definitions). It is to be highlighted here that modelling with time-lagged discharges only has not produced satisfactory results in this example.

e. **Discharge prediction at Nakhon Sawan gauging station in Chao Phraya River, Thailand**

Chao Phraya River, the main waterway in Thailand, is one of the intensely studied rivers in Southeast Asia under the Global Energy Water Cycle Experiment (GEWEX) and the GEWEX Asian Monsoon Experiment (GAME). It begins at the confluence of Nan River and Ping River where the gauging station Nakhon Sawan is located. The

Table 8.5 Number of iterations at the termination of training

Momentum term	Learning rate						
	0.01	0.05	0.10	0.30	0.50	0.70	0.90
0	1926	843	254	NC	NC	NC	NC
0.10	1960	663	1707	45	37	NC	NC
0.30	1989	603	29	NC	NC	NC	NC
0.50	1016	24	599	80	NC	NC	NC
0.70	494	551	109	250	NC	NC	NC
0.90	820	142	11	NC	NC	NC	NC

Source: Fernando, T.M.K.G., Hydrological applications of MLP neural networks with back-propagation. M. Phil Thesis, The University of Hong Kong, 2002.

Note: NC, not converging.

Table 8.6 Error indicators for different numbers of nodes in the hidden layer

No. of nodes in the hidden layer	Error indicator							
	MAE (m³/s)		RMSE (m³/s)		r		MNSE	
	Training	Testing	Training	Testing	Training	Testing	Training	Testing
2	69	90	187	217	0.661	0.415	0.212	0.085
4	57	73	112	206	0.920	0.465	0.354	0.263
6	38	55	78	138	0.950	0.645	0.569	0.440
8	35	60	70	171	0.962	0.522	0.600	0.392
10	42	60	75	151	0.954	0.583	0.526	0.392
12	38	36	88	64	0.942	0.937	0.567	0.639

Source: Fernando, T.M.K.G., Hydrological applications of MLP neural networks with back-propagation. M. Phil Thesis, The University of Hong Kong, 2002.

catchment area at this point is 110,569 km². An attempt to model and forecast 1-, 7-, and 14-day-ahead discharges using an MLP–BP ANN is summarized here, the details of which can be found elsewhere (Jayawardena and Fernando, 2001; Fernando, 2002). The daily data used in the study covers the period April 1978 to March 1994 and are taken from the Global River Data Centre (GRDC) in Koblenz, Germany. Since all the significant events occurred during the first quarter of the data series, only the first 2250 samples were used in the study. They were split up into training (first 1750 samples), validation (next 400 samples), and testing (last 400 samples) subsets. This is in agreement with the general guideline that 10%–20% of training data should be used for validation (Kasabov, 1996). The sigmoid function has been used in both hidden and output layers. All data have been normalized to the range 0.1 to 0.9, and the initial weights were assumed to be normally distributed in the range –1 to 1. Training was done in the pattern mode and the order of presenting the data was also randomized. After carrying out sensitivity analysis using different values of learning rate and momentum term, values of 0.1 and 0.3 respectively have been found to be appropriate for this application (Jayawardena et al., 2000; Jayawardena and Fernando, 2001). Stopping criteria were cross validation and fixed number of iterations set at 1000. Performance was measured by four error indicators: MAE, RMSE, correlation coefficient (r), and modified coefficient of efficiency.

By trial and error, the three networks that gave the optimal results for the cases of 1-day, 7-day, and 14-day lead-time forecasting were 7–4–1 and 3–40–1, 7–6–1, and 14–60–1, respectively (e.g., 7–4–1 imply, seven input variables, four hidden nodes, and one output node). The input variables were time-lagged discharge values ranging from $Q(t)$, $Q(t-1)$,..., $Q(t-13)$ for the prediction of $Q(t+l)$ where l is the lead-time. Although three input variables ($Q(t)$, $Q(t-1)$, and $Q(t-2)$) with 40 hidden nodes gave almost the same performance as the network 7–4–1 for 1-day lead-time, three input variables were not sufficient for 7- and 14-day lead-time predictions. The optimal number of nodes in the hidden layer in each case was obtained by increasing it gradually from 2. In general, the performance of the network improves with increasing number of hidden nodes up to a limit, and thereafter tends to deteriorate. The MSEs decrease rapidly at first and then slowly as the number of iterations increase. Although $Q(t)$ has the strongest correlation with $Q(t+l)$, a network with only $Q(t)$ as the input variable did not produce satisfactory results. In fact, the best result with one input variable and 30 hidden nodes was worse than the worst result with seven input nodes and four hidden nodes. As expected, the reliability of forecasts depend on the lead time – the longer the lead time, the less reliable the forecast but more useful from a practical point of view. Also, increasing the maximum number of iterations from 1000 to 5000 in the case of the network 3–40–1 did not improve the performance.

f. **Discharge prediction at Pakse gauging station in Mekong River, Lao People's Democratic Republic**

The Mekong Basin is geographically bounded between the latitudes 8°–34°N and the longitudes 94°–110°E, and is one of the largest international river basins in the world. The Mekong River is the tenth longest in the world with a mainstream length of about 4620 km and a drainage area of about 795,500 km² (Hori, 2000). It is the most important and most controversial international river in Asia.

Originating from the snow-covered Tang-Ku La mountain of the Tibet highland (elevation 4969 m) of China, the river flows through Tsinghai and Yunnan provinces of China,[5] Myanmar, Lao PDR, Thailand, Cambodia, and Vietnam before draining into

---

[5] The reach of the Mekong River within the borders of China is called Langcang Jiang.

South China Sea. Generally, the basin area belonging to China and Myanmar (≈18%) is called the Upper Mekong, and the remaining, the Lower Mekong. The upper basin is narrow, while the lower part is wide. The climate in the basin varies from temperate to tropical. Averaged with daily data from 1973 to 1980, the mean annual rainfall is 1672 mm and the mean daily discharge over the period of record is 9900 m^3/s at Pakse. The Tonle Sap Lake in Cambodia is the largest body of freshwater in Southeast Asia and a key part of the Mekong hydrological system due to the flow reversal in Tonle Sap River, which links the lake and the mainstream of the Mekong River. The lake retains about 80% of the sediment and nutrients carried into it by the flow reversal. The lower mainstream divides into a complex and increasingly controlled artificial system of branches and canals, flowing southeast to the South China Sea (Mekong River Commission, 2005).

With the goals of promoting and supporting coordinated sustainable and pro-poor development, enhancing effective regional cooperation, strengthening basin-wide environmental monitoring and impact assessment, and strengthening integrated water resources management capacity and knowledge base of the Mekong River riparian countries and other stakeholders, the Mekong River Commission (MRC) was established on April 5, 1995, by an agreement between the governments of Cambodia, Lao PDR, Thailand, and Vietnam. The four countries signed The Agreement on the Cooperation for the Sustainable Development of the Mekong River Basin, and agreed on joint management of their shared water resources and development of the economic potential of the river. In 1996, China and Myanmar became dialogue partners of the MRC and the countries now work together within a cooperation framework. The MRC Secretariat, which is currently (2012) based in Vientiane, Lao PDR, provides technical and administrative services to the MRC Council and Joint Committee.

The middle reach of the river has 10 gauging stations. Figure 8.14 shows the relative locations of these 10 gauging stations with respective names Chiang Saen (most upstream), Luang Prabang, Chiang Khan, Pa Mong Dam site, Vientiane, Nong Khai, Nakhon Phanom, Mukdahan, Khong Chiam, and Pakse (most downstream). The river, while being the lifeline for the people living in the basin, also can be destructive at times. Flooding is a recurrent phenomenon and flood damages are also increasing with time partly owing to increased population growth in and around the banks of the river. Structural means of mitigating flood damages is quite costly and the riparian countries as well as the MRC have adopted the alternative non-structural means of addressing the problem. Early warning is an essential component of non-structural means of mitigating flood damages where prediction of impending floods plays a crucial role. Several studies on flood forecasting in the Mekong River have been carried

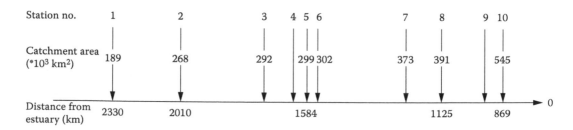

Figure 8.14 Schematic locations of the selected stations. (From Hori, H., *The Mekong: Environment and Development.* United Nations University Press, Tokyo, 249 pp., 2000.)

out in the recent past (e.g. Jayawardena et al., 2004; Jayawardena and Mahanama, 2002; Jirayoot and Al-Soufi, 2000).

For ANN modelling of daily discharges of the Mekong River, two continuous time series (1972–1983 and 1989–1994) were used. These are averaged values compiled from two stage measurements made every day at 7:00 am and 7:00 pm (Khem, 2010). The stage measurements are converted into discharge measurements using a rating curve that makes use of more frequent stage and corresponding flow measurements. The data set was divided into a training set (May 1, 1972–December 31, 1983, and January 31, 1989–December 31, 1989) and a testing set (January 1, 1990–December 31, 1994) with a total sample size of 6423 (72% for the training set and 28% for the testing set). The discharge at Pakse, the most downstream, was chosen as the target. What follows next is a summary of the work reported elsewhere (Jayawardena and Tian, 2005; Jayawardena et al., 2004; Tian, 2007). These data have been taken from the Infrastructure Development Institute (IDI) Water Series No. 10.

The discharge at Pakse is expected to be highly dependent on its own past discharges as well as the discharges at upstream stations. From autocorrelation and cross-correlation analysis, it has been found that the discharge at Pakse, in addition to being related to its own time-lagged discharges, is also highly related to the discharges at Vientiane, Nong Khai, Nakhon Phanom, Mukdahan, and Khong Chiam in decreasing order of correlation coefficient (Table 8.7). The discharge at Pakse is least correlated to the discharge at Pa Mong Dam site. Thus, the inputs for predicting the discharge at Pakse can begin with information at Pakse itself and add on the discharges at upstream stations sequentially. In addition, there is a time delay between the discharges with the highest correlation coefficient between stations. The time delay is consistent with the peak delay between these stations, which reflects the travel time of river flow. The proposed relationship is assumed to be of the form

$$Q_{t+\gamma}^{10} = f\left(Q_t^{10}, \ldots Q_{t-\beta_{10}}^{10}, Q_{t_9}^9, \ldots, Q_{t_9-\beta_9}^9, \ldots, Q_{t_2}^2, \ldots, Q_{t_2-\beta_2}^2, Q_{t_1}^1, \ldots, Q_{t_1-\beta_1}^1\right) + e \tag{8.130}$$

where the superscripts refer to the station number; $Q_{t+\gamma}^{10}$ is the discharge prediction with γ-lead time at Pakse; β is the lag time; Q_t^{10} and $Q_{t-\beta_{10}}^{10}$ are the discharges at Pakse

Table 8.7 Highest cross-correlation coefficient and time delay against Pakse

Station no.	Station name	Relationship	Delay (days)	Highest correlation coefficient
1	Chiang Saen	10 vs. 1	8	0.865
2	Luang Prabang	10 vs. 2	4,5	0.906
3	Chiang Khan	10 vs. 3	3,4	0.918
4	Pa Mong Dam site	10 vs. 4	3	0.757
5	Vientiane	10 vs. 5	2,3	0.925
6	Nong Khai	10 vs. 6	2,3	0.928
7	Nakhon Phanom	10 vs. 7	1	0.977
8	Mukdahan	10 vs. 8	1	0.982
9	Khong Chiam	10 vs. 9	0	0.996
10	Pakse			

Source: Tian, Y., Macro-scale flow modelling of the Mekong River with spatial variance. PhD Thesis, The University of Hong Kong, 2007.

at time t and $t - \beta_{10}$; $Q_{t_9}^9$ and $Q_{t_9 - \beta_9}^9$ are the discharges at Khong Chiam at time t_9 and $t_9 - \beta_9$, etc.; and e is the mapping error to be minimized. The difference of the above equation from other discharge prediction formulae is that it takes into account the travel time between different stations explicitly by varying the corresponding input with flexible subscripts.

In the ANN model, the inputs and outputs were first normalized to the range (–1, 1) by the linear transformation given in Equation 8.121 for the network with a linear activation function at the output layer. Although not a requirement, this is desirable for initially equalizing the importance of variables (Masters, 1993). In this way, all the weights can remain small and in predictable ranges, and the situation in which the error is dominated by outputs with large variabilities can be avoided. With the sigmoid activation function at the output layer, the model outputs will be within the range (0, 1). The target outputs were therefore transformed to the more practical range (0.1, 0.9) in the same way. If the values are scaled to the extreme limits of the activation function, the size of the weight updates can be extremely small and flatspots in training are likely to occur (Maier and Dandy, 2000).

The ANN model has several parameters and input variables that need some trial-and-error work as well as optimization. The momentum term, the learning rate, and the number of neurons in the hidden layer whose values are all problem specific should first be predetermined by trial and error before the connection weights are optimized by the BP algorithm. To fix these parameters, the input variables were chosen to be time-lagged discharges Q_t^{10}, $Q_{t-1}^{10},..., Q_{t-13}^{10}$ at Pakse for the prediction of discharge Q_{t+7}^{10} (at Pakse). Although it is not possible to determine precise values for the momentum term and the learning rate, examination of the performance surfaces of some error indicators with the simultaneous variation of these two parameters indicated that values of 0.3 and 0.5, respectively, for the learning rate and the momentum term seem satisfactory. These values were adopted for subsequent trials. A maximum of 2000 iterations has been used as the stopping criterion.

Table 8.8 illustrates the various performance indicators of the network for 1-day, 7-day, and 14-day lead-time predictions with the non-linear output layer. Table 8.9 shows the corresponding results with the linear output layer. Different input variables have been tried, and their relative performances can be clearly seen. They have some results in common. With seven time-lagged discharges Q_t^{10}, $Q_{t-1}^{10},..., Q_{t-6}^{10}$ at Pakse, the model performance has perhaps reached its optimum for 1-day lead-time prediction. Further addition of input variables does not improve the prediction accuracy significantly. The network is robust for 1-day lead-time prediction. As expected, the reliability of the predictions decreases with increasing lead-time. For 7- and 14-day lead times, there are improvements in the prediction accuracy when upstream discharges are also used as input variables. Additional trials show that there is no need to consider explicitly the time delay between different stations.

This application clearly illustrates (Tables 8.8 and 8.9) the superior performance resulting from using a linear activation function at the output layer instead of the traditional non-linear activation function. For example, with only one input variable, Q_t^{10} at Pakse, even the 1-day lead-time prediction is not satisfactory for the network with the non-linear activation function at the output layer. It is also to be noted that the network with the linear activation function at the output layer converges faster, and that the results are more stable.

Table 8.8 Typical model forecasting performance with non-linear output layer

Network	[]–3–1		[1^I]	[7^{II}]	[14^{III}]	[10^{IV}]	[70^V]	[140^{VI}]
1-day lead time	MAE (m³/s)	6769	1576	1298	1669	1391	1020
	RMSE (m³/s)	8699	2249	2408	3002	2261	1880
	RRMSE	0.86	0.22	0.24	0.30	0.22	0.19
	1 − NSE	0.78	0.05	0.06	0.09	0.05	0.04
7-day lead time	MAE (m³/s)	4544	3153	2875	2133	1762	1576
	RMSE (m³/s)	6164	5144	4535	3694	3190	3015
	RRMSE	0.61	0.51	0.45	0.37	0.32	0.30
	1 − NSE	0.39	0.27	0.21	0.14	0.10	0.09
14-day lead time	MAE (m³/s)	3524	3246	3617	3338	2596	2689
	RMSE (m³/s)	6612	5195	5506	5265	4451	4386
	RRMSE	0.66	0.52	0.55	0.52	0.44	0.44
	1 − NSE	0.45	0.28	0.31	0.29	0.20	0.20

Source: Tian, Y., Macro-scale flow modelling of the Mekong River with spatial variance. PhD Thesis, The University of Hong Kong, 2007.

Note: *I* indicates input Q_t^{10}; *II* indicates inputs $Q_t^{10}, Q_{t-1}^{10},..., Q_{t-6}^{10}$; *III* indicates inputs $Q_t^{10}, Q_{t-1}^{10},..., Q_{t-13}^{10}$; *IV* indicates inputs $Q_t^{10}, Q_t^9,..., Q_t^1$; *V* indicates inputs $Q_t^{10}, Q_{t-1}^{10},..., Q_{t-6}^{10}, Q_t^9, Q_{t-1}^9,..., Q_{t-6}^9,...., Q_t^2, Q_{t-1}^2,..., Q_{t-6}^2, Q_t^1, Q_{t-1}^1,..., Q_{t-6}^1$; and *VI* indicates inputs $Q_t^{10}, Q_{t-1}^{10},..., Q_{t-13}^{10}, Q_t^9, Q_{t-1}^9,..., Q_{t-13}^9,...., Q_t^2, Q_{t-1}^2,..., Q_{t-13}^2$.

Table 8.9 Typical model forecasting performance with linear output layer

Network	[]–3–1	[1^I]	[7^{II}]	[14^{III}]	[10^{IV}]	[70^V]	[140^{VI}]
1-day lead time	MAE (m³/s)	658	477	742	7422	556	649
	RMSE (m³/s)	1302	1028	1454	1478	1174	1349
	RRMSE	0.13	0.10	0.14	0.15	0.12	0.13
	1 − NSE	0.02	0.01	0.02	0.02	0.01	0.02
7-day lead time	MAE (m³/s)	1858	2092	1925	1762	1576	1855
	RMSE (m³/s)	3523	3806	3494	3333	3147	2919
	RRMSE	0.35	0.38	0.35	0.33	0.31	0.29
	1 − NSE	0.13	0.15	0.13	0.11	0.10	0.09
14-day lead time	MAE (m³/s)	2826	2797	2788	2411	2318	2133
	RMSE (m³/s)	4888	4814	4897	4397	4203	3990
	RRMSE	0.49	0.48	0.49	0.44	0.42	0.40
	1 − NSE	0.25	0.24	0.25	0.20	0.18	0.16

Source: Tian, Y., Macro-scale flow modelling of the Mekong River with spatial variance. PhD Thesis, The University of Hong Kong, 2007.

Note: The notation is the same as in Table 8.8.

For 7- and 14-day lead-time predictions, when the time-lagged daily discharges at each station are included into the input layer, there are noticeable improvements in network performance compared with predictions using a single daily discharge at each station. For the network with the linear activation function in the output layer, the improvement is more noticeable for 14-day lead-time predictions. This leads to the conclusion that more time-lagged daily discharge information is needed for longer lead-time predictions.

When time-lagged daily discharges at Pakse itself and some upstream stations are added into the input layer, the dimension of the input vector becomes larger and larger. Some of the inputs may, however, be highly redundant and only add to the dimensionality of the problem. This feature is reflected by autocorrelation and cross-correlation analysis. One way of reducing the dimensionality of the problem is to preprocess the data using PCA, which transforms the input data so that the elements of the input vector become uncorrelated. In this study, the sizes of the input vectors are controlled by retaining only those PCs that contribute more than a specified fraction of the total variation in the data set. By summing up the individual contributions of variations, the cumulative contributions from retained PCs can also be determined.

The results show that satisfactory 1-day lead-time predictions can be obtained without PCA but with a linear activation function at the output layer. Upstream discharges as inputs can increase the accuracy of 7- and 14-day lead-time predictions. Generally, the network with PCA performs a little better than that without PCA. The other merit is that the input vector with PCA is much smaller than that without PCA.

g. Discharge prediction at Pakse gauging station in Mekong River using PUNN, Lao People's Democratic Republic

The new version of ANNs described in section 8.6.4 has been used by Karunasingha et al. (2011) for discharge prediction at Pakse gauging station in the Mekong River with daily discharge data measured at the above referenced 10 stations for the period January 1991 to December 1994. Two different models with different input variable combinations have been considered: (1) predicting 1-day-ahead flow at Pakse with current day's flow and the previous 2 days' flows (i.e., two time-lagged values) of Pakse station, and (2) predicting 1-day-ahead flow at Pakse station with the current records of Pakse and the other nine stations. The two models may be expressed respectively by the following equations:

$$Q_{Pakse}(t+1) = F_1\left(Q_{Pakse}(t),\ Q_{Pakse}(t-1),\ Q_{Pakse}(t-2)\right) \tag{8.131}$$

$$Q_{Pakse}(t+1) = F_2\left(Q_1(t),\ Q_2(t),...,\ Q_{10}(t)\right) \tag{8.132}$$

The two cases are expected to represent two common situations in predicting real world data: (1) predicting a variable of a system with its own time-lagged records (time series data), and (2) predicting a variable with the help of other variables of the system.

The best model obtained by PUNN when only data records of Pakse are used is given as

$$Q_{Pakse} = 0.83472 + 1.2765x_1^{1.4857} + x_2^{-0.5156} - 0.2501x_1^{4.4023}x_2^{4.2687} \tag{8.133}$$

where

$$x_1 = \text{normalized } ST_{\text{Pakse}}(t)$$
$$x_2 = \text{normalized } ST_{\text{Pakse}}(t-1)$$
$$Q_{\text{Pakse}} = \text{scaled flow value at Pakse at time } t+1$$

Prediction errors on Mekong River flow have shown that PUNN has outperformed MLP in prediction accuracy in validation. The results are consistent with all error indicators. Another interesting aspect of PUNN results compared with those of ANN is the transparency in the derived models.

h. **Stage prediction at Sylhet gauging station in Surma River, Bangladesh**
Being the most downstream country of several large international rivers that have origins in the high altitudes of the Himalayas, Bangladesh is perennially faced with many different types of natural water-related disasters. In addition to flooding in the flat areas of Bangladesh, the country is also exposed to storm surges with severe consequences. Structural measures to mitigate such damages do not appear to be economically feasible in a developing country such as Bangladesh, and therefore non-structural measures such as early warning systems seem to be the only viable option. This application in which ANNs have been used is aimed at forecasting daily water levels at the Sylhet gauging station (latitude: 24°42′N; longitude: 91°53′E) across Surma River, which is one of the principal rivers originating from the Assam and Meghalaya hilly areas of India. During heavy rainfall in the upland area, water moves quickly towards the south-western direction through a number of rivers and tributaries and causes flood in *Haor* basin (saucer-shaped shallow depression). The network used is of the MLP–BP type with three layers: an input layer, a hidden layer, and an output layer. The input layer has four nodes representing the water level on the previous day, rainfalls on the same day and 2 preceding days. The output layer gives the current water level. The data used for training covers the period from August 20, 1980, to December 11, 1989, for validation; the period from December 23, 1989, to April 15, 1999; and for application, the period from April 27, 1999, to August 17, 2008. Time series plots of the results using MLP and BP algorithms are shown in Figures 8.15 through 8.17. Scatter diagrams for training, validation, and application data sets have given very

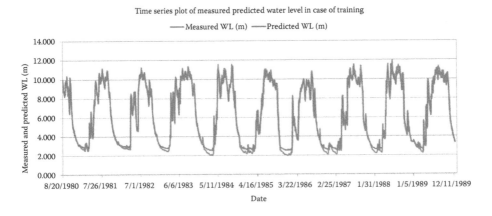

Figure 8.15 Time series plot of scaled measured and predicted water levels (training).

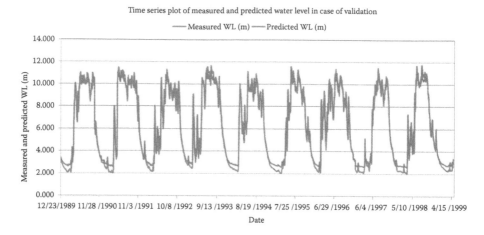

Figure 8.16 Time series plot of scaled measured and predicted water levels (validation).

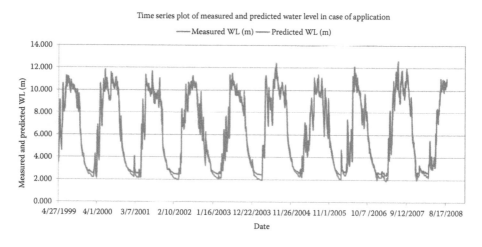

Figure 8.17 Time series plot of scaled measured and predicted water levels (application).

good correlations with respective R^2 values of 0.9893, 0.9895, and 0.9887 (Biswas et al., 2009). In this study, a momentum term of 0.5 and learning rates ranging from 0.1 to 0.5 have been used.

8.12.2 Environmental applications

8.12.2.1 Algal bloom prediction, Hong Kong

Abundance of certain types of phytoplanktons such as *Noctiluca scintillans*, *Skeletonema costatum*, and *Gonyaulax polygramma* cause the waters to turn red to brown in colour, leading to what is commonly referred to as the appearance of 'red tides'. When such an occurrence takes place, the water quality is adversely affected. The red tides block sunlight necessary for photosynthesis and use up the oxygen in the water. Certain types of red tide organisms produce chemicals that can be harmful to fish, shellfish, and humans. For example, *Alexandrium catenella* is toxic to humans, *Gymnodinium* is harmful to fish, and

Dinophysis is harmful to shellfish and therefore also to humans thorough the food chain. It is also known that severe rapid oxygen depletion resulting from their presence causes fish kill. The water discoloration itself can significantly impair recreational uses of coastal waters. If the timing and magnitude of algal blooms could be predicted in the primary stage, then many of these deleterious effects can be prevented or minimized.

The basic requirement for model building is a reasonable understanding of the system dynamics and an appropriate set of data for calibration and testing. In the case of the algal bloom problem, the dynamics are so complex that a quantitative understanding is not available, at least at the present time, although some of the main factors that cause blooms are known qualitatively. It involves a complex and non-linear combination of physical, biological, and chemical processes. With such constraints, a process-based model is difficult, if not impossible, to formulate at the present time. Instead, alternative 'data-driven' approach of modelling could be pursued without the need for *a priori* understanding of the system dynamics.

As the name implies, data-driven approaches try to capture the underlying dynamics from the available data. It is therefore imperative that the data should cover the broad spectrum of possibilities, spatially and temporally. In practice, however, such data are rarely available. In several previous studies, interpolated high-frequency data that have been obtained from low-frequency measured data have been used for model calibration and prediction. Such data inherently has a major drawback when used for prediction purposes because interpolated data implicitly contain information about the future (the value that is to be predicted). The general approach of modelling algal growth involves simulation of chlorophyll-*a* as an indicator for estimating the total biomass of a wide variety of phytoplankton species (Whitehead and Hornberger, 1984). One such approach is the use of ANNs.

Earlier studies of algal blooms, or harmful algal blooms to be more specific, have been carried out in fresh water bodies such as lakes and rivers (e.g., Recknagel et al., 1997, 1998; Yabunaka et al. 1997; Whitehead et al., 1997; Maier et al., 1998; Karul et al., 2000), as well as in coastal waters (Barciela et al., 1999). The study to be presented here summarizes the work carried out in two coastal areas of Hong Kong: Tolo Harbour and Lamma Island. Details of these studies are reported elsewhere (e.g., Lee et al., 2003a,b; Huang, 2001). The algal dynamics are represented by the biweekly chlorophyll-*a* concentration data and the weekly cell count data of the phytoplankton species *Skeletonema* for Tolo Harbour and Lamma Island, respectively. The performance of the network is then presented and discussed along with issues of data interpolation and real-time forecasting. An interpretation of the neural network via examination of the interconnection weights is also attempted.

For Tolo Harbour, the monthly/biweekly water quality data (chlorophyll-*a*, NH_4–N, NO_3–N, NO_2–N, and total N), collected as part of a routine water quality monitoring programme of the Hong Kong Government Environmental Protection Department, is used as the basis to develop the ANN model. The data of 1982–1992 (during a time when algal blooms were most frequent and intense) is used for training the network, while the data of 1993–2000 is used for model testing/validation. Depth-averaged values (obtained from the data for surface, middle, and bottom of the water column) are adopted for analysis. In addition, daily meteorological data such as rainfall, wind speed, solar radiation, and tidal range recorded by the Hong Kong Observatory are also used. For short-term prediction, daily water quality data are obtained through linear interpolation of the biweekly data. For Lamma Island, weekly *Skeletonema* cell count data together with limited water quality data (1996–2000) measured at Mo Tat Wan have been used (Lee et al., 2003a). In all the cases, the objective was to predict the chlorophyll-*a* concentration from other water quality variables and their time-lagged values.

In the ANN models attempted, the initial weights are set randomly. The training process is terminated after a preset stopping criterion such as a fixed number of iterations, a fixed number of error reduction operations, or cross validation. In this study, a learning rate of 0.01, a momentum term of 0.5, and a stopping criterion of 3000 iterations have been used.

One of the drawbacks of earlier ANN model predictions of algal blooms has been the interpolation of biweekly data to obtain daily values. This in effect uses the future information that the model is aimed to forecast, to drive the model. To avoid this drawback, two approaches have been studied by Lee et al. (2003a). One approach was to predict the algal dynamics with a lead time equal to the minimum time interval of original observations (i.e., biweekly forecasts). The second approach predicts 7-day lead-time using interpolated daily data with the testing carried out assuming that no observations are available. For the biweekly prediction, data of 1982–1992 have been used for training, while those of 1993–2000 for testing. Input variables each with two time lags consist of ammonia nitrogen ($NH_4 - N$), nitrate nitrogen ($NO_3 - N$), nitrite nitrogen ($NO_2 - N$), total inorganic nitrogen (TIN), chlorophyll-a, PO_4, turbidity, water temperature, dissolved oxygen (DO), salinity, Secchi disc depth, and solar radiation. In other words, the chlorophyll-a at time t depends on the concentration of the input variables 2 and 4 weeks ago. For the 7-day prediction, instead of using the time-lagged interpolated data, the prediction is made using the time-lagged forecasts.

For Lamma Island, input data consist of *Skeletonema*, water temperature, total inorganic nitrogen, phosphorus, salinity, Secchi disc depth, wind speed, and tidal range with time lags ranging from 7 to 13 days. To avoid large variation of numbers used for network analysis, the logarithmic value of the algal cell concentration is used. The performances of the ANN models, as measured by the RMSE and the correlation coefficient with different input variables are shown in Tables 8.10 and 8.11, respectively, for Tolo Harbour and Lamma Island.

The outcome of the ANN modelling at the two different locations is different and offers some insights. On the one hand, the analysis of algal dynamics in Tolo Harbour is based

Table 8.10 Different tested ANNs for prediction of algal blooms in Tolo Harbour

Scenario	Inputs	RMSE		Correlation coefficient	
		Training	Testing	Training	Testing
I	TIN, chlorophyll-a, PO_4, temperature, DO, Secchi disc depth, solar radiation, rainfall, wind speed, tidal range	1.819	4.481	0.963	0.795
2	TIN, chlorophyll-a, PO_4, temperature, DO, Secchi disc depth, solar radiation, rainfall, wind speed	1.346	4.853	0.980	0.794
3	TIN, chlorophyll-a, PO_4, temperature, DO, Secchi disc depth, solar radiation, rainfall	1.830	4.445	0.962	0.814
4	TIN, chlorophyll-a, PO_4, temperature, DO, Secchi disc depth, solar radiation	2.267	3.520	0.941	0.861
5	TIN, chlorophyll-a, PO_4, temperature, DO, Secchi disc depth	1.458	5.653	0.980	0.683
6	TIN, chlorophyll-a, PO_4, temperature, DO	2.352	3.369	0.937	0.871
7	TIN, chlorophyll-a, PO_4, temperature	1.855	3.248	0.962	0.889
8	TIN, chlorophyll-a, PO_4	2.413	2.875	0.933	0.909
9	TIN, chlorophyll-a	1.865	3.045	0.961	0.901
10	Chlorophyll-a	2.226	3.393	0.943	0.905

Source: Lee, J.H.W., Huang, Y., Dickman, M. and Jayawardena, A.W., *Ecological Modelling*, 2003, 159, 179.

Table 8.11 Different tested ANNs for prediction of algal blooms in Lamma Island

Scenario	Inputs	RMSE		Correlation coefficient	
		Training	Testing	Training	Testing
1	Log(cells/l), water temperature, salinity, Secchi disc depth, TIN, PO$_4$, wind speed, tidal range	0.384	1.705	0.941	0.180
2	Log(cells/l), water temperature, salinity, Secchi disc depth, TIN, PO$_4$, wind speed	0.223	1.796	0.980	0.242
3	Log(cells/l), water temperature, salinity, Secchi disc depth, TIN, PO$_4$	0.445	1.613	0.920	0.281
4	Log(cells/l), water temperature, salinity, Secchi disc depth, TIN	0.150	1.388	0.992	0.613
5	Log(cells/l), TIN, PO$_4$	0.298	1.539	0.966	0.290
6	Log(cells/l), water temperature, salinity	0.272	1.392	0.972	0.523
7	Log(cells/l), water temperature, PO$_4$	0.386	0.927	0.942	0.774
8	Log(cells/l), temperature	0.423	0.853	0.931	0.807
9	Log(cells/l)	0.551	0.784	0.879	0.837

Source: Lee, J.H.W., Huang, Y., Dickman, M. and Jayawardena, A.W., *Ecological Modelling*, 2003, 159, 179.

on an 18-year database that captures all the long-term variations. However, in view of the insufficient sampling frequency of 2 weeks, the predictive ability of the derived ANN is unsatisfactory. On the other hand, the Lamma Island data set has a reasonably fine sampling interval, but the length of data is insufficient to produce a network aimed at reliable forecasts. The use of cell counts instead of chlorophyll-*a* is also expected to lead to more data variability related to the measurement itself. However, the truly predictive forecast at Lamma Island is able to capture general trends, and is most encouraging.

For both these locations, which have different hydrographic and environmental characteristics, it is found that best results are obtained with the simplest network; that is, the network that consists of the time-lagged algal concentration as the only inputs gives the best results when compared with other more complicated networks with many environmental variables, suggesting a non-linear autoregressive type of dynamics.

8.13 CONCLUDING REMARKS

Over the last two decades or so, there have been significant developments in ANN techniques as well as the scope of applications. It is impractical to condense the essence of all such studies into a single book chapter. What has been highlighted here is a glimpse of the basic theoretical background as well as applications in the hydro-environment in which the author had some involvement. The technique is still advancing and the application areas are also expanding.

REFERENCES

Aksoy, H. and Bayazit, M. (2000): A model for daily flows of intermittent streams. *Hydrological Processes*, 14, 1725–1744.

Aksoy, H. and Unal, N.E. (2007): Discussion of 'Comparison of two nonparametric alternatives for stochastic generation of monthly rainfall' by R. Srikanthan et al. *ASCE Journal of Hydrologic Engineering*, 12, 699–702.

Amari, S., Murata, N., Muller, K.R., Finke, M. and Yang, H. (1996): Statistical theory of overtraining – Is cross-validation asymptotically effective? *Advances in Neural Information Processing Systems*, 8, 176–182, MIT Press, Cambridge, MA.

Amnatsan, S. (2010): Water level prediction in Nan River, Thailand using wavelet neural networks. Master's degree Thesis, National Graduate Institute for Policy Studies, Tokyo, Japan and International Centre for Water Hazard and Risk Management, Tsukuba, Japan (unpublished).

Andreo, B., Jemenez, P., Duran, J., Carrasco, F., Vadillo, I. and Mangin, A. (2006): Climatic and hydrological variations during the last 117–166 years in the south of the Iberian Peninsula, from spectral and correlation analyses and continuous wavelet analyses. *Journal of Hydrology*, 324, 24–39.

Angeline, P.J., Saunders, G.M. and Pollack, J.B. (1994): An evolutionary algorithm that constructs recurrent neural networks. *IEEE Transactions on Neural Networks*, 5(1), 54–65.

Anmala, J., Zhang, B. and Govindaraju, R.S. (2000): Comparison of ANNs and empirical approaches for predicting watershed runoff. *Journal of Water Resources Planning and Management*, 126(3), 156–166.

ASCE Task Committee on Application of Artificial Neural Networks in Hydrology (2000a): Artificial neural networks in hydrology I: preliminary concepts. *Journal of Hydrologic Engineering ASCE* 5(2), 115–123.

ASCE Task Committee on Application of Artificial Neural Networks in Hydrology (2000b): Artificial neural networks in hydrology II: Hydrologic applications. *Journal of Hydrologic Engineering ASCE*, 5(2), 124–137.

Barciela, R.M., Garcia, E. and Fernandez, E. (1999): Modelling primary production in a coastal embayment affected by upwelling using dynamic ecosystem models and artificial neural networks. *Ecological Modelling*, 120, 199–211.

Bayazit, M. and Aksoy, H. (2001): Using wavelets for data generation. *Journal of Applied Statistics*, 28, 157–166.

Bayazit, M., Onoz, B. and Aksoy, H. (2001): Nonparametric streamflow simulation by wavelet or Fourier analysis. *Hydrological Sciences Journal*, 46, 623–634.

Bengraine, K. and Marhaba, T.F. (2003): Using principal component analysis to monitor spatial and temporal changes in water quality. *Journal of Hazardous Materials*, 100(1–3), 179–195.

Biswas, R.K., Jayawardena, A.W. and Takeuchi, K. (2009): Prediction of water level in the Surma River of Bangladesh by artificial neural networks. *Proceedings of 2009 Annual Conference, Japan Society of Hydrology and Water Resources*, Kanazawa, Japan, P-17, August 19–21 (in CD ROM).

Bodèn, M. (2001): A guide to recurrent neural networks and back-propagation. http://www.itee.uq.edu.au/~mikael/papers/rn_dallas.pdf.

Chang, F.J., Chang, L.C. and Huang, H.L. (2002): Real-time recurrent learning neural network for stream-flow forecasting. *Hydrological Processes*, 16(13), 2577–2588.

Coulibaly, P., Anctil, F. and Bobee, B. (2000): Daily reservoir inflow forecasting using artificial neural networks with stopped training approach. *Journal of Hydrology*, 230, 244–257.

Cybenko, G. (1989): Approximation by superposition of sigmoidal function. *Mathematics of Control Signals and Systems*, 2, 303–314.

Daniell, T.M. (1991): Neural networks – Applications in hydrology and water resources engineering. *Proc., Int. Hydrology & Water Resour. Symp.*, Perth, October, pp. 797–802.

Daubechies, I. (1988): Orthonormal bases of compactly supported wavelets. *Communications on Pure and Applied Mathematics*, 41, 909–996.

Daubechies, I. (1992): *Ten Lectures on Wavelets*. SIAM, New York.

Dawson, C.W. and Wilby, R.L. (2001): Hydrological modelling using artificial neural networks. *Progress in Physical Geography*, 25(1), 80–108.

Durbin, R. and Rumelhart, D. (1989): Product units: A computationally powerful and biologically plausible extension to back-propagation networks. *Neural Computation*, 1, 133–142.

Elman, J.L. (1990): Finding structure in time. *Cognitive Science*, 14(2), 179–211.

Faggin, F. (1991): VLSI implementation of neural networks. *Tutorial Notes, International Joint Conference on Neural Networks*, Seattle.

Fernando, D.A.K. (1997): On the application of artificial neural networks and genetic algorithms in hydrological modelling. PhD Thesis, The University of Hong Kong, May.

Fernando, T.M.K.G. (2002): Hydrological applications of MLP neural networks with back-propagation. M. Phil Thesis, The University of Hong Kong.

Geyer, C.J. and Thompson, E.A. (1996): Annealing Markov chain Monte Carlo with applications to ancestral inference. *Journal of the American Statistical Association*, 909–920.

Goupillaud, P., Grossman, A. and Morlet, J. (1984): Cycle-octave and related transforms in seismic signal analysis. *Geoexploration*, 23, 85–102.

Govindaraju, R.S. (2000): Artificial neural networks in hydrology II: Hydrologic applications. *Journal of Hydrologic Engineering*, 5(2), 24–137.

Govindaraju, R.S. and Rao, A.R. (2000): *Artificial Neural Networks in Hydrology*. Kluwer, Dordrecht, The Netherlands.

Haar, A. (1910): Zur Theorie der orthogonalen Funktionensysteme. *Mathematische Annalen*, 69, 331–371.

Haykin, S.S. (1999): *Neural Networks: A Comprehensive Foundation*. Prentice Hall, New York.

Hebb, D.O. (1949): *The Organization of Behaviour: A Neuropsychological Theory*. Wiley, New York.

Hinton, G.E. and Sejnowski, T.J. (1986): Learning and re-learning in Boltzmann machines. In: Rumelhart, D.E. and McClelland, J.L. (Eds.), *Parallel Distributed Processing: Explorations in Microstructure of Cognition*. MIT Press, Cambridge, MA.

Hong Kong Environmental Protection Department (1998): *Marine Water Quality in Hong Kong*. Hong Kong Government, Hong Kong, Hong Kong.

Hopfield, J.J. (1982): Neural networks and physical systems with emergent collective computational abilities. *Proceedings of the National Academy of Sciences of the USA*, 79(8), 2554–2558.

Hori, H. (2000): *The Mekong: Environment and Development*. United Nations University Press, Tokyo, 249 pp.

Huang, Y. (2001): Neural network modelling of coastal algal blooms. M. Phil Thesis, The University of Hong Kong.

Hush, D.R. and Horne, B.G. (1993): Progress in supervised neural networks: What's new since Lippmann? *IEEE Signal Processing Magazine*, 10, 8–39.

Infrastructure Development Institute (IDI), Japan (1960–1994): *The Mekong River Hydrological Database* (1960–1994) in CD-ROM.

Janson, D.J. and Frenzel, J.F. (1993): Training product unit neural networks with genetic algorithms. *IEEE Expert*, 8(5), 26–33.

Jayawardena, A.W. and Fernando, D.A.K. (1995a): Hydrological forecasting using artificial neural networks. *Proceedings of the Second International Study Conference on GEWEX in Asia and GAME*, Pattaya, Thailand, pp. 376–379.

Jayawardena, A.W. and Fernando, D.A.K. (1995b): Artificial neural networks in hydrometeorological modelling. In: Topping, B.H.V. (Ed.), *Proc. 4th Int. Conf. on the Application of AI to Civ. and Struct. Engrg.: Developments in Neural Networks and Evolutionary Computing for Civ. and Struct. Engrg*, Civil-Comp Press, UK, pp. 115–120.

Jayawardena, A.W. and Fernando, D.A.K. (1996): Use of artificial neural networks in estimating evaporation. *Proceedings of the International Conference on Water Resources & Environment Research: Towards the 21st Century*, Kyoto, Japan, vol. I, October 29–31, pp. 141–147.

Jayawardena, A.W., Fernando, D.A.K. and Zhou, Y. (1996a): Prediction of daily runoff using an artificial neural networks approach. *Proceedings of International Workshop on Macro-scale Hydrological Modelling*, Hohai University Press, Nanjing, China, May, pp. 145–148.

Jayawardena, A.W., Fernando, D.A.K. and Zhou, M.C. (1996b): Comparison of multilayer perceptron and radial basis function networks as tools for flood forecasting. In: Leavesley, G.L., Lins, H.F., Nobilis, F., Parker, R.S., Schneider, V.R. and van de Ven, F.H.M. (Eds.), Destructive water: Water-caused natural disasters, their abatement and control, *Proc. Conference held at Anaheim, June 1996*, International Association of Hydrological Sciences Publication No. 239, Anaheim, CA, pp. 173–181.

Jayawardena, A.W. and Fernando, T.M.K.G. (2000): Forecasting algal blooms using artificial neural networks. In: Tsaptsinos, D. (Ed.), *Proceedings of the International Conference on Engineering Applications of Neural Networks*, Department of Mathematics, Kingston University, Surrey, UK, July 17–19, pp. 115–122.

Jayawardena, A.W. and Fernando, T.M.K.G. (2001): River flow prediction: An artificial neural network approach. In: Schumann, A.H., Acreman, M.C., Davis, R., Marino, M.A., Rosbjerg, D. and Jun, X. (Eds), Regional Management of Water Resources, *Proceedings of Symposium S2, Sixth Scientific Assembly, IAHS,* Maastricht, The Netherlands, July 18–27, International Association of Hydrological Sciences Publication No. 268, pp. 239–245.

Jayawardena, A.W., Fernando, T.M.K.G., Chan, C.W. and Chan, W.C. (2000): Comparison of ANN, dynamical systems and support vector approaches for river discharge prediction. *Proceedings of the 19th Chinese Control Conference,* Hong Kong, China, December 6–8, 2000, pp. 504–508.

Jayawardena, A.W. and Mahanama, S.P.P. (2002): Meso-scale hydrological modeling: application to Mekong and Chao Phraya basins. *Journal of Hydrologic Engineering,* 7(1), 12–26.

Jayawardena, A.W. and Tian, Y. (2005a): Flow modelling of Mekong River with variance in spatial scale. *Proceedings of the International Symposium on Role of Water Sciences in Transboundary River Basin Management,* Ubon Ratchathani, Thailand, March 10–12, pp. 147–154.

Jayawardena, A.W. and Tian, Y. (2005b): River flow modelling of the Mekong River Basin (Poster). *The 5th International Scientific Conference on the Global Energy and Water Cycle,* Orange County, CA, June 20–24.

Jayawardena, A.W., Tian, Y. and Herath, S. (2004): Prediction of daily discharges of lower Mekong River using artificial neural networks. *Proceedings of the International Conference on Water Sensitive Urban Design: 'Cities and Catchments',* Jakarta, Indonesia, November 22–23, pp. 83–90.

Jayawardena, A.W., Xu, P.C. and Li, W.K. (2009): Rainfall data simulation by hidden Markov model and discrete wavelet transformation. *Stochastic Environment Research and Risk Assessment (SERRA),* 23, 863–877, doi:10.1007/s00477-008-0264-0.

Jirayoot, K. and Al-Soufi, R. (2000): Prediction of daily river discharge by an artificial neural network model. *Mekong River Commission, Workshop on Hydrologic and Environmental Modelling in Mekong Basin,* Phnom Penh, Cambodia, September 11–12, pp. 149–157.

Jolliffe, I.T. (2002): *Principal Component Analysis,* 2nd Edition. Springer, New York.

Jordan, M.I. (1986): Attractor dynamics and parallelism in a connectionist sequential machine. *Proceedings of the Eighth Annual Conference of the Cognitive Science Society,* Erlbaum, Englewood Cliffs, NJ, pp. 531–546 (Reprinted in *IEEE Tutorial Series,* IEEE Publishing Services, New York, 1990).

Karul, C., Soyupak, S., Çilesiz, A.F., Akbay, N. and Germen, E. (2000): Case studies on the use of neural networks in eutrophication modeling. *Ecological Modelling,* 134(2–3), 145–152.

Karunanithi, N., Grenny, W.J., Whitley, D. and Bovee, K. (1994): Neural networks for river flow prediction. *Journal of Computing in Civil Engineering, ASCE,* 8(2), 201–220.

Karunasingha, D.S.K., Jayawardena, A.W. and Li, W.K. (2011): Evolutionary product unit based neural networks for hydrological time series analysis. *Journal of Hydroinformatics,* 134(2011), 825–841.

Kasabov, N.K. (1996): *Foundations of Neural Networks, Fuzzy Systems, and Knowledge Engineering.* MIT Press, Cambridge, MA.

Khem, S. (2010): Personal communication. Mekong River Commission.

Khondker, M.H., Wilson, G. and Klinting, A. (1998): Application of neural networks in real time flash flood forecasting. In: Babovic, V., Larsen, L.C. (Eds.), *Proceedings of the 3rd International Conference on Hydroinformatics,* Copenhagen, Denmark, August 24–25, pp. 777–782.

Kia, S.J. and Coghill, G. (1994): Soft competitive learning in the extended differentiator network. *Proceedings of the 1994 IEEE International Conference on Neural Networks,* 2, 714–718.

Kirkpatrik, S., Jr., Gelatt, C.D. and Vecchi, M.P. (1983): Optimization by simulated annealing. *Science,* 220(4598), 671–680.

Kohonen, T. (1982): Self-organized formation of topologically correct feature maps. *Biological Cybernetics,* 43, 59–69.

Kohonen, T. (1984): *Self-organization and Associative Memory.* Springer Series in Information Science, vol. 8. Springer-Verlag, Berlin.

Kohonen, T. (2001): *Self-Organizing Maps,* 3rd Extended Edition. Springer-Verlag, Berlin.

Kote, A.S. and Jothiprakash, V. (2009): Monthly reservoir inflow modelling using time-lagged recurrent networks. *International Journal Tomography and Statistics,* 12(F09), 64–84.

Labat, D., Ababou, R. and Mangin, A. (1999): Wavelet analysis in karstic hydrology. Part 2: Rainfall–runoff cross-wavelet analysis. *Comptes Rendus de l'Academie des Sciences Series IIA Earth and Planetary Science* 329, 881–887.

Labat, D., Ronchail, J. and Guyot, J.L. (2005): Recent advances in wavelet analyses: Part 2-Amazon, Parana, Orinoco and Congo discharges time scale variability. *Journal of Hydrology*, 314, 289–311.

Lachtermacher, G. and Fuller, J.D. (1994): Back-propagation in hydrological time series forecasting. In: Hipel, K.W., McLeod, A.I., Panu, U.S. and Singh, V.P. (Eds.), *Stochastic and Statistical Methods in Hydrology and Environmental Engineering*, vol. 3, Time Series Analysis in Hydrology and Environmental Engineering. Kluwer Academic Publishers, Dordrecht, Netherlands, pp. 229–242.

LeCun, Y. (1989): Generalization and network design strategies Technical Report, CRG-TR-89-4, Department of Computer Science, University of Toronto, Canada.

LeCun, Y. (1993): *Efficient Learning and Second Order Methods, A tutorial given at Neural Information Processing Systems Conference (NIPS93)*. Denver, CO.

Lee, J.H.W., Huang, Y., Dickman, M. and Jayawardena, A.W. (2003a): Neural network modelling of coastal algal blooms. *Ecological Modelling*, 159(2–3), 179–201.

Lee, J.H.W., Huang, Y. and Jayawardena, A.W. (2003b): Comment on 'Comparative application of artificial neural networks and genetic algorithms for multivariate time-series modelling of algal blooms in freshwater lakes'. *Journal of Hydroinformatics*, 05.1, 2003, 71–74.

Leerink, L.R., Horne, B.G., Giles, C.L. and Jabri, M.A. (1995): Learning with product units. In: *Advances in Neural Information Processing Systems 7*. MIT Press, Cambridge, MA, pp. 537–544.

Legates, D.R. and McCabe, C.G. (1999): Evaluating the use of 'goodness of fit' measures in hydrologic and hydroclimatic model validation. *Water Resources Research*, 35(1), 233–241.

Li, Y. and Gu, R.R. (2003): Modeling flow and sediment transport in a river system using an artificial neural network. *Environmental Management* 31(1), 122–134.

Liong, S.Y. and Chan, W.T. (1993): Runoff volume estimates with neural networks. In: Topping, B.H.V. and Khan, A.I. (Eds.), *Proc., 3rd Int. Conf. in the Application of AI to Civ. and Struct. Engrg: Neural Networks and Combinatorial Optimization in Civ. and Struct. Engrg*, Civil-Comp Press, UK, pp. 67–70.

Maier, H.R. and Dandy, G.C. (1996): Neural network models for forecasting univariate time series. *Neural Network World*, 6(5), 747–772.

Maier, H.R. and Dandy, G.C. (2000): Neural networks for the prediction and forecasting of water resources variables: A review of modelling issues and applications. *Environmental Modelling and Software*, 15(1), 101–124.

Maier, H.R., Dandy, G.C. and Burch, M.D. (1998): Use of artificial neural networks for modelling cyanobacteria *Anabaena* spp. in the River Murray, South Australia. *Ecological Modelling*, 105, 257–272.

Mallat, S. (1997): *A Wavelet Tour of Signal Processing*. Academic Press, San Diego, CA.

Mallat, S.G. (1989): A theory for multi-resolution signal decomposition: The wavelet representation. *IEEE Transactions on Pattern Analysis and Machine Intelligence*, 7, 674–692.

Martinez-Estudillo, A., Martinez-Estudillo, F., Hervas-Martinez, C. and Garcia-Pedrajas, N. (2006): Evolutionary product unit based neural networks for regression. *Neural Networks*, 19, 477–486.

Masters, T. (1993): *Practical Neural Network Recipes in C++*. Academic Press, Boston, pp. 60–61.

McCulloch, W. and Pitts, W. (1943): A logical calculus of the ideas immanent in nervous activity. *Bulletin of Mathematical Biophysics*, 7, 115–133.

Mekong River Commission (2005): *Overview of the Hydrology of the Mekong Basin*. Mekong River Commission, Vientiane, pp. 9–26.

Mendel, J.M. and McLaren, R.W. (1970): Reinforcement-learning control and pattern recognition systems. In: Mendel, J.M. and Fu, K.S. (Eds.), *Adaptive, Learning, and Pattern Recognition Systems: Theory and Applications*, vol. 66. Academic Press, New York, pp. 287–318.

Minns, A.W. and Hall, M.J. (1996): Artificial neural network as rainfall runoff models. *Hydrological Sciences Journal*, 41(3), 399–417.

Minsky, M.L. (1961): Steps towards artificial intelligence. *Proceedings of the Institute of Radio Engineers*, 49, 8–30. Reprinted in Feigenbaum, E.A. and J. Feldman (Eds.). *Computers and Thought*. McGraw Hill, New York, pp. 406–450.

Minsky, M.L. and Papert, S.A. (1969): *Perceptrons*. MIT Press, Cambridge, MA.

Nash, J.E. and Sutcliffe, J.V. (1970): River flow forecasting through conceptual models: A discussion of principals. *Journal of Hydrology*, 10, 282–290.

Otten, R.H.J.M. and Ginneken, L.P.P.P. (1989): *The Annealing Algorithm*. Kluwer, Boston.

Pan, T.-Yi. and Wang, R.-Y. (2005): Using recurrent neural networks to reconstruct rainfall–runoff processes. *Hydrological Processes*, 19(18), 3603–3619.

Patil, Y. and Ghatol, A.A. (2009): Application of time lag recurrent network – Rainfall forecasting. *International Journal of Multi-discipline Research and Advances in Engineering, IJMRAE*, 1(1), 257–270.

Rajurkar, M.P., Kothyari, U.C. and Chaube, U.C. (2002): Artificial neural networks for daily rainfall–runoff modeling. *Hydrological Sciences Journal*, 47(6), 865–878.

Raman, H. and Sunilkumar, N. (1995): Modelling water resources time series using artificial neural networks. *Hydrological Sciences Journal*, 40(2), 145–162.

Recknagel, F., French, M., Harkonen, P. and Yabunaka, K.-I. (1997): Artificial neural network approach for modelling and prediction of algal blooms. *Ecological Modelling*, 96, 11–28.

Recknagel, F., Fukushima, T., Hanazato, T., Takamura, N. and Wilson, H. (1998): Modelling and prediction of phyto- and zooplankton dynamics in Lake Kasumigaura by artificial neural networks. *Lakes & Reservoirs: Research and Management*, 3, 123–133.

Rogers, L.L. and Dowla, F.U. (1994): Optimization of ground water remediation using artificial neural networks with parallel solute transport modelling. *Water Resources Research*, 30(2), 457–481.

Rosenblatt, F. (1958): The perceptron: A probabilistic model for information storage and organization in the brain, Cornell Aeronautical laboratory. *Psychological Review*, 65(6), 386–408.

Rosenblatt, F. (1962): *Principles of Neurodynamics*. Spartan Books, Washington, DC.

Rumelhart, D.E. and Zipser, D. (1985): Feature discovery by competitive learning. *Cognitive Science*, 9, 75–112.

Rumelhart, D.E., Hinton, G.E. and Williams, R.J. (1986): Learning representations by backpropagating errors. *Nature*, 323, 533–536.

Saco, P. and Kumar, P. (2000): Coherent modes in multiscale variability of streamflow over the United States. *Water Resources Research*, 36, 1049–1068.

Safavi, A., Abdollahi, H. and Nezhad, M.R.H. (2003): Artificial neural networks for simultaneous spectrophotometric differential kinetic determination of Co(II) and V(IV). *Talanta*, 59(3), 515–523.

Schmitt, M. (2001): On the complexity of computing and learning with multiplicative neural networks. *Neural Computation*, 14, 241–301.

Sejnowski, T.J. (1977a): Strong covariance with nonlinearly interacting neurons. *Journal of Mathematical Biology*, 4, 303–321.

Sejnowski, T.J. (1977b): Statistical constraints on synaptic plasticity. *Journal of Theoretical Biology*, 69, 385–389.

Shamseldin, A.Y. (1997): Application of a neural network technique to rainfall–runoff modelling. *Journal of Hydrology*, 199, 272–294.

Sivakumar, B., Jayawardena, A.W. and Fernando, T.M.K.G. (2002): River flow forecasting: Use of phase-space reconstruction and artificial neural networks approaches. *Journal of Hydrology*, 265, 225–245.

Smith, J. and Eli, R.N. (1995): Neural network models of rainfall–runoff process. *Journal of Water Resources Planning and Management, ASCE*, 121(6), 499–508.

Smith, L.C., Turcotte, D.L. and Isacks, B.L. (1998): Stream flow characterization and feature detection using a discrete wavelet transform. *Hydrological Processes*, 12, 233–249.

Smolensky, P., Mozer M.C. and Rumelhart, D.E. (Eds.) (1996): *Mathematical Perspectives on Neural Networks*. Lawrance Erlbaum Associates, Mahwah, NJ, p. 537.

Stone, M. (1974): Cross-validatory choice and assessments of statistical predictions. *Journal of the Royal Statistical Society*, B36, 111–133.

Sugawara, M., Watanabe, I., Ozaki, E. and Katsuyama, Y. (1984): Tank model with snow component. Research notes of the National Research Centre for Disaster Prevention, No. 65, November 1984, Science and Technology Agency, Japan.

Tarassenko, L. (1998): *A Guide to Neural Computing Applications*. Wiley, New York.

Tian, Y. (2007): Macro-scale flow modelling of the Mekong River with spatial variance. PhD Thesis, The University of Hong Kong.

Unal, N.E. Aksoy, H. and Akar, T. (2004): Annual and monthly rainfall data generation schemes. *Stochastic Environment Research and Risk Assessment (SERRA)*, 18, 245–257.

Wang, Y.M. and Traore, S. (2009): Time lagged recurrent network for forecasting episodic event suspended load in typhoon area. *International Journal of Physical Sciences*, 4(9), 519–528.

Whitehead, P.G. and Hornberger, G.M. (1984): Modelling algal behaviour in the River Thames. *Water Resources*, 18(8), 945–953.

Whitehead, P.G., Howard, A. and Arulmani, C. (1997): Modelling algal growth and transport in rivers: A comparison of time series analysis, dynamic mass balance and neural network techniques. *Hydrobiologia*, 349, 39–46.

Widrow, B. and Hoff, M.E., Jr. (1960): Adaptive switching circuits. In: *1960 IRE WESCON Convention Record*, Part 4, IRE, New York, pp. 96–104.

Willshaw, D.J. and von der Malsburg, C. (1976): How patterned neural connexions can be set up by self-organisation. *Proceedings of the Royal Society B*, 194, 431–445.

Yabunaka, K., Hosomi, M. and Murakami, A. (1997): Novel application of a back-propagation artificial neural network model formulated to predict algal bloom. *Water Science and Technology*, 36(5), 89–97.

Chapter 9

Radial basis function (RBF) neural networks

9.1 INTRODUCTION

The multilayer perceptron–back-propagation (MLP–BP) neural network can be considered as a type of stochastic approximation. On the other hand, radial basis function (RBF) neural networks are considered as curve-fitting problems in high-dimensional space. Unlike in MLP-type networks, which can have several layers of neurons, RBF-type networks have only three layers: an input layer, a hidden layer, and an output layer. The main distinguishing characteristics of RBF-type neural networks are that the hidden layer is of high dimensionality having a non-linear activation function, and the output layer is always linear. The mathematical justification for the rationale of using a non-linear transformation followed by a linear transformation has been given by Cover (1965) in his theorem on separability of patterns, which states that a complete pattern classification problem cast in a high-dimensional space non-linearly is more likely to be linearly separable than in a low-dimensional space. Therefore, the dimension of the hidden layer is made high, and it can be said that the higher the dimension of the hidden layer, the more accurate the approximation.

RBFs were first introduced by Powell (1985, 1987a,b) in the context of multivariate interpolation problems. In the context of neural networks, they were first used by Broomhead and Lowe (1988). Subsequent contributions have been made by Moody and Darken (1989) and Poggio and Girosi (1990a,b). RBFs are monotonically decreasing (as in the case of Gaussian functions), or monotonically increasing (as in the case of multiquadratic functions) from the centres (see also Section 8.5.11). They give significant responses in the neighbourhood of the centres. The parameters include the centres μ_j and the widths σ_j, which control their radii of influence. Gaussian functions produce significant non-zero responses when the input patterns fall within $\mu_j \pm \sigma_j$, which is referred to as the receptive field.

9.2 INTERPOLATION

RBFs can be used to solve the interpolation problem in multidimensional space. The problem involves finding a function $F(x)$ that satisfies the interpolation condition $F(x_i) = y_i$ for $i = 1, 2, \ldots, N$, given a set of input and corresponding output data points (x_i, d_i). The RBF form of $F(x)$ is defined as (Powell, 1985, 1987a,b)

$$F(x) = \sum_{i=1}^{N} w_i \phi \left\| x - \mu_i \right\| \tag{9.1}$$

where the ϕ's are the RBFs; $\|.\|$ is a norm, usually the Euclidean;[1] w_i's are the weighting parameters; and μ_i's are the centres (same as the data points, x_i's) of the RBFs. Since the approximating function $F(x)$ is linear in weights, they can be determined by solving the resulting set of simultaneous linear equations. For the given set of input and output vectors x_i: $i = 1, 2,..., N$; d_i: $i = 1, 2,..., N$, and a weight vector w_i: $i = 1, 2,..., N$, the RBF relationship in Equation 9.1 can be written in matrix form as

$$\begin{bmatrix} \phi_{11} & \phi_{12} & \phi_{13}...\phi_{1N} \\ \phi_{21} & \phi_{22} & \phi_{23}...\phi_{2N} \\ ... \\ ... \\ \phi_{N1} & \phi_{N2} & \phi_{N3}...\phi_{NN} \end{bmatrix} \begin{bmatrix} w_1 \\ w_2 \\ ... \\ ... \\ w_N \end{bmatrix} = \begin{bmatrix} d_1 \\ d_2 \\ ... \\ ... \\ d_N \end{bmatrix} \qquad (9.2)$$

where $\phi_{ji} = \phi\left(\|x_j - \mu_i\|\right)$. Equation 9.2 is a set of simultaneous linear equations and its solution gives the weights. For a unique solution to be obtained, the interpolation matrix ϕ should be non-singular. This condition is guaranteed for a certain class of RBFs. The Gaussian and inverse multiquadratic, which are localized functions, that is, $\phi(r) \to 0$ as $r \to \infty$, belong to this class. For this class of RBFs with all distinct data, the interpolation matrix ϕ is positive definite,[2] implying that it always has an inverse. It is also symmetric,[3] implying $\phi^T = \phi$. On the other hand, multiquadratic RBF is non-local, that is, $\phi(r) \to \infty$ as $r \to \infty$, in which case ϕ is non-positive definite ($N - 1$ negative eigenvalues and one positive eigenvalue). When this matrix is *ill conditioned* (close to singularity), it can be made *well conditioned* by a process called regularization. It is also reported (Powell, 1985, 1987a,b) that unbounded RBFs give better accuracy than bounded ones for input–output mapping, a somewhat surprising result.

9.3 REGULARIZATION

A *well-posed* problem should satisfy three conditions: existence of an input x and a corresponding output y, uniqueness implying that every pair of values is different, and continuous implying that mapping is continuous. If any of these conditions is not satisfied, then the problem is *ill-posed*. An *ill-posed* problem has a lot of data but not much information. An *ill-posed* problem can be made *well-posed* by a process called regularization, which involves perturbing the *ill-conditioned* matrix ϕ to $\phi + \lambda I$, where λ is a regularization parameter and I is the identity matrix. A brief description of the regularization theory, which was first introduced by Tikhonov (1963) and later developed by Poggio and Girosi (1990a) in the context of RBFs for representing a multivariate continuous function by an approximate function having a fixed number of parameters, is given below.

---

[1] The Euclidean distance between two vectors x_i and x_j: $i, j = 1, 2, ..., m$ is given by $d(x_i, x_j) = \|x_i - x_j\| = \left[\sum_{k=1}^{m}(x_{ik} - x_{jk})^2\right]^{1/2}$, where x_{ik} is the k-th element of x_i and x_{jk} is the k-th element of x_j.

[2] A matrix M ($N \times N$) is non-negative definite if $a^T M a \geq 0$; positive definite if $a^T M a > 0$ for any vector a. A positive definite matrix is non-singular and has all positive eigenvalues. A positive definite matrix always has an inverse.

[3] A symmetric matrix is a square matrix M with $M = M^T$; that is, $M_{ij} = M_{ji}$. Any square diagonal matrix is symmetric. The inner product of a symmetric matrix is positive definite.

The error in a model, which has to be minimized, may be considered as consisting of two components: a standard component and a regularized component. In equation form,

$$E(F(x)) = \frac{1}{2}\sum_{i=1}^{N}(d_i - y_i)^2 + \frac{1}{2}\lambda\|DF(x)\|^2 = \frac{1}{2}\sum_{i=1}^{N}(d_i - F(x_i))^2 + \frac{1}{2}\lambda\|DF(x)\|^2 \tag{9.3}$$

where $E(F(x))$ is a functional of the mapping function $F(x)$, sometimes referred to as the Tikhonov functional; y_i is the response to input x_i; and D is a linear differential operator sometimes referred to as a stabilizer. The operator D contains the *a priori* information about the solution and therefore depends on the particular problem. It stabilizes the solution $F(x)$, making it smooth and continuous. The regularization parameter can also be thought of as a balancing factor for the model bias and the model variance.[4] Minimization of the functional $E(F(x))$ gives the solution to the regularization problem; that is, to find a function $F_\lambda(x)$ that minimizes the functional $E(F(x))$. If $\lambda = 0$, $F(x)$ can be determined from the data samples only. If, on the other hand, $\lambda \to \infty$, the data are unreliable, and *a priori* smoothness condition is sufficient to determine $F(x)$. The function $F_\lambda(x)$ that minimizes the functional $E(F(x))$ is obtained by evaluating the Fréchet differential (best local linear approximation) and setting it to zero. It has been shown to be (Haykin, 1999)

$$F_\lambda(x) = \frac{1}{\lambda}\sum_{i=1}^{N}[d_i - F(x_i)]G(x,\mu_i) = \sum_{i=1}^{N}w_i G(x,\mu_i) \tag{9.4}$$

where $G(x, \mu_i)$ is a Green's function, which has a singularity at $x = \mu_i$ (μ_i's are the centres), and

$$w_i = \frac{1}{\lambda}[d_i - F(x_i)]; \quad i = 1, 2, \ldots, N \tag{9.5}$$

In matrix form,

$$F_\lambda = wG \tag{9.6}$$

$$w = \frac{1}{\lambda}(d - F_\lambda) \tag{9.7}$$

---

[4] Bias and variance. The mean squared error is defined as

$$\text{MSE} = \left\langle (d(x) - F(x))^2 \right\rangle$$

where $F(x)$ is the prediction at x and d is true value corresponding to x. The notation $\langle \cdot \rangle$ refers to the expectation taken over a large number of training samples. This can be split into two parts as follows:

$$\text{MSE} = \left(d(x) - \langle F(x)\rangle\right)^2 + \left\langle \left(F(x) - \langle F(x)\rangle\right)^2 \right\rangle$$

The first term is the bias and the second term is the variance. If $d(x) = \langle F(x)\rangle$, the bias = 0. An unbiased model can still have a variance in which case a small bias is introduced so that the variance can be reduced, keeping the net effect a reduction in the MSE. This is achieved by removing the degrees of freedom either by lowering the order of the polynomial (in case of a polynomial) or reducing the number of parameters in the approximating function. A method of restricting the flexibility of a linear model is to modify the sum of squared error by adding a term that penalizes large weights as follows:

$$E = \sum_{i=1}^{N}(d_i - F(x_i))^2 + \lambda\sum_{j=1}^{M}w_j^2$$

This is called ridge regression, or weight decay, and λ controls the balance between fitting the data and avoiding penalty.

where

$$F_\lambda = [F_\lambda(x_1), F_\lambda(x_2),..., F_\lambda(x_N)]^T \tag{9.8}$$

$$d = [d_1, d_2,..., d_N]^T \tag{9.9}$$

$$w = [w_1, w_2,..., w_N]^T \tag{9.10}$$

$$G = \begin{bmatrix} G(x_1,\mu_1), G(x_1,\mu_2),..., G(x_1,\mu_N) \\ G(x_2,\mu_1), G(x_2,\mu_2),..., G(x_2,\mu_N) \\ ... \\ ... \\ G(x_N,\mu_1), G(x_N,\mu_2),..., G(x_N,\mu_N) \end{bmatrix} \tag{9.11}$$

Eliminating F_λ in Equations 9.6 and 9.7 (Poggio and Girosi, 1990a)

$$w = (G + \lambda I)^{-1}d \tag{9.12}$$

where I is the identity matrix ($N \times N$). Equation 9.12 is a set of simultaneous equations from which w_i's can be determined. The number of Green's functions is equal to the number of samples in the training data set. The matrix G is a Green's matrix, and is symmetric implying $G^T = G$. It has the same role as the matrix ϕ in interpolation and is positive definite for certain classes of Green's functions provided $x_1, x_2,..., x_N$ are distinct. Gaussian and inverse multiquadratic functions belong to this class. The value of λ should be sufficiently large to ensure that $G + \lambda I$ is positive definite. The estimation of the optimum value of λ is not trivial. Generalized cross validation as well as a method based on the study by Craven and Wahba (1979), which defines an estimate of the average squared error as described in Haykin (1999, pp. 284–289), can be used to obtain the optimum value of λ. In practice, a trial-and-error approach is normally used. The Green's functions centred at μ_i depend only on the Euclidean distances between x and μ_i; that is,

$$G(x,\mu_i) = G\left(\left\|x - \mu_i\right\|\right) \tag{9.13}$$

This means that the Green's functions must be RBFs implying

$$F_\lambda(x) = \sum_{i=1}^{N} w_i G\left(\left\|x - \mu_i\right\|\right) \quad \text{(Poggio and Girosi, 1990a)} \tag{9.14}$$

The function space (linear in weights) $F_\lambda(x)$ passes through all data points. When $\lambda = 0$, $F_\lambda(x)$ in Equation 9.14 is the same as $F(x)$ in Equation 9.1.

With the multivariate Gaussian function used as the Green's function:

$$G(x,\mu_i) = \exp\left(-\frac{1}{2\sigma_i^2}\left\|x - \mu_i\right\|^2\right) \tag{9.15}$$

and

$$F_\lambda(x) = \sum_{i=1}^{N} w_i \exp\left(-\frac{1}{2\sigma_i^2}\left\|x - \mu_i\right\|^2\right) \tag{9.16}$$

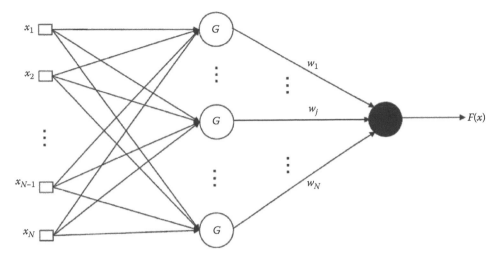

Figure 9.1 Regularized radial basis function network.

Although each centre can have its own width, the condition $\sigma_i = \sigma$ for all i is imposed to simplify the process. Regularized networks are universal approximators that are optimal for multivariate continuous functions (Poggio and Girosi, 1990b) provided there are sufficient numbers of hidden nodes. Figure 9.1 shows the typical architecture of a regularized RBF network. It has an input layer with as many nodes as there are input variables, a hidden layer with one hidden node for each input node and fully connected, and an output layer fully connected to the hidden layer. The output of the hidden layer at the node i is $G(x, \mu_i)$. There is a one-to-one correspondence between the input data x_i and the Green's functions $G(x, \mu_i)$. In Figure 9.1, the output layer has only one node, but it can be expanded to have multiple nodes. The output of the network is a linear weighted sum of the hidden layer outputs.

However, when N is large, the computational cost may be too high and the chances of the matrix becoming ill conditioned are also higher. To avoid this situation, generalized RBFs can be used.

9.4 GENERALIZED RBFs

An approximation (suboptimal) to the regularized solution may be obtained by considering a lower-dimensional space in which a new set of basis functions and a new set of weights are used. With z_i as the new centres, the RBFs may be expressed as

$$\phi_i(x) = G\left(\|x - z_i\|\right) \quad i = 1, 2, \ldots, M(M \leq N; \quad \text{and when } M = N, \; x_i = z_i) \tag{9.17}$$

With $F^*(x)$ as the new approximate interpolation function, Equation 9.1 can be written as

$$F^*(x) = \sum_{i=1}^{M} w_i \phi_i(x) = \sum_{i=1}^{M} w_i G\left(\|x - z_i\|\right) \quad \text{(Poggio and Girosi, 1990a)} \tag{9.18}$$

This particular choice of basis functions guarantees that when $M = N$ and $z_i = x_i$ for $i = 1$, $2,..., N$, the solution to Equation 9.18 converges to that of Equation 9.1. The new functional that has to be minimized as before (similar to Equation 9.3) may be written as

$$E(F^*(x)) = \frac{1}{2} \sum_{i=1}^{N} \left(d_i - \sum_{j=1}^{M} w_j G\left(\left\| x_i - z_j \right\|^2 \right) \right) + \frac{1}{2} \lambda \left\| DF^*(x) \right\|^2 \tag{9.19}$$

The first term on the right-hand side of this equation can be written as the Euclidean distance $\left\| d - Gw \right\|^2$, where

$$d = [d_1, d_2,..., d_N]^T \tag{9.20}$$

$$w = [w_1, w_2,..., w_M]^T \tag{9.21}$$

$$G = \begin{bmatrix} G(x_1, z_1), G(x_1, z_2),..., G(x_1, z_M) \\ G(x_2, z_1), G(x_2, z_2),..., G(x_2, z_M) \\ ... \\ ... \\ G(x_N, z_1), G(x_N, z_2),..., G(x_N, z_M) \end{bmatrix} \tag{9.22}$$

The new matrix G, which corresponds to G in Equation 9.11, is of size $N \times M$ and hence not square. The second term on the right-hand side of Equation 9.19 may be written as (Haykin, 1999)

$$\left\| DF^*(x) \right\|^2 = w^T G_0 w \tag{9.23}$$

where

$$G_0 = \begin{bmatrix} G(z_1, z_1), G(z_1, z_2),..., G(z_1, z_M) \\ G(z_2, z_1), G(z_2, z_2),..., G(z_2, z_M) \\ ... \\ ... \\ G(z_M, z_1), G(z_M, z_2),..., G(z_M, z_M) \end{bmatrix} \tag{9.24}$$

which is an $M \times M$ square matrix. Minimizing the error functional $E(F^*(x))$ given in Equation 9.19 with respect to the weights has been shown (Broomhead and Lowe, 1988) to be equivalent to solving the equation

$$(G^T G + \lambda G_0) \, w = G^T d \tag{9.25}$$

which when $\lambda \to 0$ takes the form (Broomhead and Lowe, 1988)

$$w = G^+ d \tag{9.26}$$

where

$$G^+ = (G^T G)^{-1} G^T \tag{9.27}$$

The quantity $(G^T G)^{-1} G^T$ is the pseudo-inverse of the matrix G. The system defined by Equation 9.25 is overdetermined (more equations than unknowns) and is equivalent to a least squares fitting problem.

In regularized RBFs, the number of hidden nodes is exactly equal to the number of input data samples, whereas in generalized RBFs the number of hidden nodes is less than or equal to the number of input data samples. Also, in regularized RBFs, the weights are unknown but the centres of the Green's functions, which are at the training data points, are known. In generalized RBFs, the weights as well as the centres are unknown. The architecture of a typical generalized RBF, which also has a bias term (equal to its linear weight, and the RBF equal to unity) at the output layer, is shown in Figure 9.2.

The norm in Equation 9.18 is intended to be Euclidean type, but sometimes it is more appropriate to consider a general weighted norm[5] of the form

$$\|x\|_C^2 = (Cx)^T (Cx) = x^T C^T C x \quad \text{(Poggio and Girosi, 1990a)} \tag{9.28}$$

where C is a norm-weighting matrix ($N \times N$) where N is the dimension of the input vector x. Then Equation 9.18 can be rewritten as

$$F^*(x) = \sum_{i=1}^{M} w_i G\left(\|x - z_i\|_C\right) \tag{9.29}$$

A Gaussian RBF $G\left(\|x - z_i\|_C\right)$ centred at z_i and with norm-weighting matrix C can be expressed as

$$
\begin{aligned}
G\left(\|x - z_i\|_C\right) &= \exp\left[-(x - z_i)^T C^T C (x - z_i)\right] \\
&= \exp\left[-\frac{1}{2}(x - z_i)^T \sum{}^{-1} (x - z_i)\right]
\end{aligned}
\tag{9.30}
$$

where the inverse covariance matrix[6] Σ^{-1} is defined as

$$\frac{1}{2}\sum{}^{-1} = C^T C \tag{9.31}$$

When the norm-weighting matrix is an identity matrix, the result is the Euclidean norm.

---

[5] The weighted norm can take one of three forms: $\|x\|_C^2 = \|x\|^2$ in which case $C = I$; $\|x\|_C^2 = $ diagonal matrix $= \sum_{k=1}^{N} c_k x_k^2$ where x_k is the k-th element of input vector x and c_k is the k-th diagonal element of the matrix C; or, $\|x\|_C^2$ is non-diagonal in which case the weighted norm takes a quadratic form; that is, $\|x\|_C^2 = \sum_{k=1}^{N} \sum_{l=1}^{N} a_{kl} x_k x_l$, where a_{kl} is the k-th element of matrix product $C^T C$.

[6] The covariance matrix may be expressed as $\sum = \sigma^2 I$ where σ^2 is the common variance, $\sum = \text{diag}\left(\sigma_1^2, \sigma_2^2, \sigma_3^2, \ldots, \sigma_N^2\right)$ where σ_j^2 is the variance of the j-th element of the vector x_i, $i = 1, 2, \ldots, N$, or $\sum = Q^T \Lambda Q$ where Q is the orthonormal rotation, and Λ is a diagonal matrix.

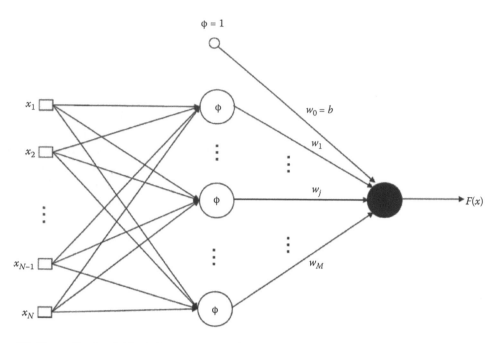

Figure 9.2 Generalized radial basis function network.

9.5 NORMALIZED RADIAL BASIS FUNCTIONS (NRBFs) AND KERNEL REGRESSION

In addition to solving the interpolation problem, RBFs can also be used in kernel regression, which is a non-parametric statistical technique for estimating the conditional expectation of a random variable; that is, to find a non-linear relationship between a pair of random variables x and y. The conditional expectation of a variable y relative to a variable x can be expressed as

$$f(x) = E(y|x) \tag{9.32}$$

Unlike in linear regression where the attempt is to find a function that fits all the data, the objective in kernel regression is to find a function that fits the data locally. No prior assumption about the distribution of the data is necessary. Kernel regression is closely related to nearest-neighbour search.

An estimate of the kernel regression function $f(x)$ can be expressed as (Haykin, p. 318)

$$F(x) = \hat{f}(x) = \frac{\sum_{i=1}^{N} y_i K\left(\frac{x - x_i}{h}\right)}{\sum_{j=1}^{N} K\left(\frac{x - x_j}{h}\right)} \tag{9.33}$$

$F(x)$ in Equation 9.33 is also a universal approximator. The kernel function K, which has a band width h, is real, symmetric, continuous, and bounded with the property that

$$\int K(x)\,\mathrm{d}x = 1 \tag{9.34}$$

The kernel regression function is similar to a probability density function. The kernel estimator may be written as (Nadaraya, 1964; Watson, 1964)

$$F(x) = \sum_{i=1}^{N} w_{N,i}(x) y_i \tag{9.35}$$

where

$$w_{N,i}(x) = \frac{K\left(\dfrac{x - x_i}{h}\right)}{\displaystyle\sum_{j=1}^{N} K\left(\dfrac{x - x_j}{h}\right)}, \quad i = 1, 2, \ldots N \tag{9.36}$$

is defined as the normalized weighting function with

$$\sum_{i=1}^{N} w_{N,i}(x) = 1 \text{ for all } x \tag{9.37}$$

or as

$$F(x) = \sum_{i=1}^{N} w_i \Psi_N(x, x_i) \tag{9.38}$$

where

$$\Psi_N(x - x_i) = \frac{K\left(\dfrac{\|x - x_i\|}{h}\right)}{\displaystyle\sum_{j=1}^{N} K\left(\dfrac{\|x - x_j\|}{h}\right)}, \quad i = 1, 2, \ldots N \tag{9.39}$$

is defined as the normalized RBF with

$$\sum_{i=1}^{N} \Psi_N(x, x_i) = 1 \quad \text{for all } x \tag{9.40}$$

The normalized RBF should satisfy the condition $0 \le \Psi_N(x, x_i) \le 1$ for all x and x_i. The denominator of Equation 9.39, which is the normalization factor, is an estimator of the probability density function of the input vector x and hence satisfies Equation 9.40. For ordinary RBFs, this condition (Equation 9.40) may not be satisfied.

Using Gaussian RBFs, Equations 9.35 and 9.38 can respectively be written as

$$F(x) = \frac{\sum_{i=1}^{N} y_i \exp\left(-\frac{\|x - x_i\|^2}{2\sigma^2}\right)}{\sum_{j=1}^{N} \exp\left(-\frac{\|x - x_j\|^2}{2\sigma^2}\right)} \quad \text{(Specht, 1991)} \tag{9.41}$$

and

$$F(x) = \frac{\sum_{i=1}^{N} w_i \exp\left(-\frac{\|x - x_i\|^2}{2\sigma^2}\right)}{\sum_{j=1}^{N} \exp\left(-\frac{\|x - x_j\|^2}{2\sigma^2}\right)} \tag{9.42}$$

In Equations 9.41 and 9.42, the centres of the normalized RBFs coincide with the data points. For parsimony, a small number of data points chosen heuristically or according to some defined criteria can be used.

In normalized radial basis function (NRBF) networks, the output activity is normalized by the total input activity in the hidden layer. Normalization is done by the hidden nodes before the summation in the output node (Moody and Darken, 1989), thereby making it a non-local operation that requires each node to 'know' about the outputs of the other hidden nodes. As a result, NRBF networks are computationally more resource consuming than ordinary RBF networks. It has novel computational features and has attracted the attention of many researchers from the neural network community (e.g., Moody and Darken, 1989; Servin and Cuevas, 1993; Cha and Kassam, 1995; Rao et al., 1997; Bugmann, 1998a,b).

Owing to normalization, RBF nodes become case indicators rather than basis functions proper. In that sense, NRBF nets are similar to fuzzy inference systems, as discussed in Jang and Sun (1993). It has been pointed out (Bugmann, 1998) that NRBF networks have very good generalization properties that can be exploited to reduce the number of hidden nodes in classification problems. This is due to the property of normalized RBF networks to produce a significant output even for input vectors far from the centre of the receptive field of any of the hidden nodes. A modified learning procedure (Bugmann, 1998) when applied to classification problems results in nodes being added mainly along class boundaries, thus reducing the number of hidden nodes. NRBF classifiers behave as nearest-neighbour classifiers without the curse of dimensionality problem. The widths of basis functions are not critical, which makes them easier to use.

9.6 LEARNING OF RBFs

Since different layers of an RBF network perform different tasks at different rates, it is logical to separate the optimization of the hidden and output layers using different (hybrid) techniques and at different rates. Normally, but not necessarily, the selection of centres is done in an unsupervised mode, whereas the optimization of the weights is done in a supervised mode. Learning is a problem of hyperspace reconstruction, which may become *ill-posed* owing to lack of sufficient information (data) and the presence of noise in the data. *A priori* assumption is then made to make the problem *well-posed*.

9.6.1 Fixed centres selection (random)

The original RBF method of learning required as many centres as data points. In terms of computing cost, this can be prohibitively high when N is large. In the fixed centres selection method, first described by Broomhead and Lowe (1988), the locations of the centres are chosen randomly from the training data set. Once the centres are fixed, a normalized Gaussian RBF whose receptive field width is dependent on the spread of the chosen centres may be employed to represent the non-linearity in the hidden layer. A normalized RBF centred at μ_i is defined as

$$G\left(\left\|x - \mu_i\right\|^2\right) = \exp\left(-\frac{M}{d_{max}^2}\left\|x - \mu_i\right\|^2\right); \quad i = 1, 2, ..., M \tag{9.43}$$

where M is the number of centres and d_{max} is the maximum distance between the chosen centres. The receptive field width (standard deviation of the Gaussian function) is fixed as

$$\sigma = \frac{d_{max}}{\sqrt{2M}} \tag{9.44}$$

which is the same for all RBFs. In such situations, the flatness or peakedness of the RBF is not distinguished. However, too sharp or too flat RBFs are not desirable.

The linear weights are obtained by a pseudo-inverse method (Broomhead and Lowe, 1988) using the relationship given in Equation 9.26; that is, $w = G^+y$, where G^+ is the pseudo inverse defined in Equation 9.27. The elements of the matrix G are

$$g_{ji} = \exp\left(-\frac{M}{d_{max}^2}\left\|x_j - \mu_i\right\|^2\right) \quad j = 1, ... N; \quad i = 1, 2, ..., M \tag{9.45}$$

The pseudo-inverse matrix G^+ can be computed by the singular value decomposition, a method by which any matrix can be decomposed into the product of two orthogonal[7] matrices and a diagonal[8] one as follows:

$$H = UKM \tag{9.46}$$

where H is an $N \times M$ matrix, U is an $N \times N$ matrix, K is an $N \times M$ diagonal matrix, and M is an $M \times M$ matrix.

---

[7] The inverse of an orthogonal matrix M is its own transpose, that is, $M^{-1} = M^T$, implying $M^TM = MM^T = I$ (identity matrix).

[8] A diagonal matrix M (which need not be square) has $m_{ij} = 0$ for $i \neq j$.

9.6.2 Forward selection

One approach of reducing the number of centres in an RBF network is by means of subset selection. It can be done by comparing the performance of models made up of different subsets from the same fixed pool of candidates. The selection of the best subset by this method is quite difficult since there are $2^N - 1$ subsets in a set of size N. The standard practice is to search for a small subset of the space of all subsets heuristically. One such heuristics is the forward selection method, which starts from an empty subset and incrementally adds basis functions, one at a time, until the greatest reduction in the sum of squares of the error is achieved. The other is the backward elimination, which starts with the full subset and removes one basis function at a time until the least increase in the sum of squares of errors is achieved. In the subset selection, the optimization algorithm searches in a discrete space of subsets of a set of hidden units with fixed centres and widths and tries to find the subset with the lowest prediction error. It has the advantage that there is no need to fix the number of hidden nodes in advance and that the required computing resources are relatively low. In forward selection, one basis function is added at each step. The procedure is terminated when the generalized cross-validation criterion is met; that is, when the error that incrementally decreases with the addition of each basis function begins at some point to increase. This is when the model begins to fit to the noise rather than to the signal. Regularized forward selection is a combination of standard ridge regression and forward selection. Ridge regression is a means of controlling the balance between the bias and the variance by varying the effective number of parameters in a linear model with fixed size. It is equivalent to weight decay in neural networks and is used in problems with insufficient information and therefore not having a unique solution.

9.6.3 Orthogonal least squares (OLS) algorithm

The standard approach of training an RBF network involves the selection of centres randomly and the estimation of the weights by single value decomposition. However, the random selection is not ideal and not parsimonious. Random selection may not also be representative of the input domain. If the chosen centres are too close to each other, it is also possible for numerical ill conditioning to occur. The performance of an RBF network depends not much on the choice of the non-linearity (Gaussian, multiquadratic, inverse multiquadratic, or thin plate spline) but more on the choice of centres, which are subsets of the training data. The OLS algorithm (Chen et al., 1989, 1990a,b, 1991) selects the RBF centres sequentially in a rational way until an adequate network has been constructed. The method does not lead to numerical or ill-conditioning problems. In the OLS algorithm, the selection of centres can be considered as a subset selection; that is, selection of a suitable set of centres (regressors) from a large pool of possible candidates. The OLS algorithm, which is superior to the ordinary least squares method, implements the forward selection for subset selection. It is also capable of estimating the parameters. The OLS method and the single value decomposition method can overcome the problem of numerical ill conditioning, which is quite common in the ordinary least squares approach. The algorithm transforms the set of regressors p_i's into orthogonal basis vectors.

Considering RBF as a special case of the linear regressor

$$y(t) = \sum_{i=1}^{M} p_i(t)\theta_i + \varepsilon_i(t) \qquad (9.47)$$

where $y(t)$ are the dependent variables (outputs), $p_i(t)$ are the regressors, θ_i are the parameters, and $\varepsilon_i(t)$ are the errors, which are uncorrelated with $p_i(t)$. The regressors are functions of the independent variable $x(t)$, $t = 1, 2, ..., N$; that is,

$$p_i(t) = p_i(x(t)) \tag{9.48}$$

Bias can be incorporated by adding a constant term and setting the corresponding $p_i(t) = 1$. The regressors may be any one of the possible non-linear functions used in RBFs, but the Gaussian is the most common one. In matrix notation Equation 9.47 may be written as (Chen et al., 1991)

$$Y = P\theta + E \tag{9.49}$$

where

$$Y = [y_1, y_2,..., y_N]^T \tag{9.50}$$

$$P = [p_1, p_2,..., p_M]^T \tag{9.51}$$

$$p_i = \{p_i(1), p_i(2),..., p_i(N)\}^T \quad 1 \leq i \leq M \tag{9.52}$$

$$\theta = [\theta_1, \theta_2,..., \theta_M]^T \tag{9.53}$$

$$E = [\varepsilon_1, \varepsilon_2,..., \varepsilon_N]^T \tag{9.54}$$

The OLS algorithm transforms the set of $p_i(t)$'s into a set of orthogonal basis vectors. The matrix P can be decomposed into

$$P = WA \tag{9.55}$$

where W is an $N \times M$ matrix with the orthogonal columns w_i such that

$$W^T W = H \tag{9.56}$$

and

$$A = \begin{bmatrix} 1, \alpha_{12}, \alpha_{13},..., \alpha_{1M} \\ 0, 1, \alpha_{23}, \alpha_{24},..., \alpha_{2M} \\ 0, 0, 1, \alpha_{34}, \alpha_{35},..., \alpha_{3M} \\ ... \\ ..., \alpha_{M-1,M} \\ 0, 0, 0,..., 0, 0, 0, 0, 1 \end{bmatrix} \tag{9.57}$$

In Equation 9.57, $A(M \times M)$ is an upper triangular matrix and in Equation 9.56, H is a diagonal matrix with element h_i, given by

$$h_i(t) = w_i^T(t)w_i(t) = \sum_{t=1}^{N} w_i(t)w_i(t) \quad 1 \leq i \leq M \tag{9.58}$$

Since the space spanned by the set of orthogonal basis vectors $w_i(t)$ is the same as the space spanned by $p_i(t)$, it is possible to write Equation 9.49 as

$$Y = Wg + E \tag{9.59}$$

and the orthogonal least squares solution of Equation 9.59 is

$$\hat{g} = H^{-1}W^T y \tag{9.60}$$

or

$$g_i = \frac{w_i^T y}{w_i^T w_i} \quad 0 \le i \le M \tag{9.61}$$

In Equations 9.49 and 9.59 $\hat{\theta}$ and $\hat{g}$ satisfy

$$A\hat{\theta} = \hat{g} \tag{9.62}$$

Equation 9.62 can be solved by the Gram–Schmidt method for the least squares estimates of $\hat{\theta}$. The Gram–Schmidt orthogonalization method ensures that each new column added to the design matrix P of the growing subset is orthogonal to all previous columns. A similar decomposition of the matrix P can be done using the Householder[9] transformation method (Golub, 1965). The computational procedure is summarized in the following equations (Chen et al., 1991):

$$w_1 = p_1 \tag{9.63a}$$

$$\alpha_{ik} = \frac{w_i^T p_k}{w_i^T w_i} \quad 1 \le i \le k \tag{9.63b}$$

$$w_k = p_k - \sum_{i=1}^{k-1} \alpha_{ik} w_i \tag{9.63c}$$

for $k = 2,3,...,M$.

In RBFs the number of data points $x(t)$ is often very large and the number of potential regressors M can also be quite large. An adequate model may require only M_s ($\ll M$) significant regressors. These significant regressors are selected using the OLS algorithm operating in a forward regression manner. Since w_i and w_j are orthogonal for $i \ne j$, the sum of squares of $y(t)$ is

$$y^T y = \sum_{i=1}^{M} g_i^2 w_i^T w_i + E^T E \tag{9.64}$$

---

[9] The Householder transformation or Householder reflection is a linear transformation that describes a reflection about a plane or hyperplane containing the origin. Householder transformations are widely used in numerical linear algebra to perform QR decompositions and in the first step of the QR algorithm. QR decomposition of a matrix A is $A = QR$, where R is an upper triangular matrix and Q is an orthogonal matrix; that is, $Q^T Q = I$. The Householder transformation was introduced by Householder (1958).

The variance of $y(t)$ is given by (it is assumed that the mean of $y(t)$ has been removed)

$$N^{-1}y^T y = N^{-1} \sum_{i=1}^{M} g_i^2 w_i^T w_i + N^{-1} E^T E \tag{9.65}$$

$$= \text{explained variance} + \text{unexplained variance}$$

Thus, $\sum_{i=1}^{M} g_i^2 w_i^T w_i$ is the part of output variance that can be explained by the regressors and $\dfrac{g_i^2 w_i^T w_i}{N}$ is the increment to the explained variance introduced by w_i. The error reduction ratio due to w_i can therefore be defined as

$$(\text{err})_i = \frac{g_i^2 w_i^T w_i}{y^T y}; \quad 1 \le i \le M \tag{9.66}$$

The computational procedure for regressor selection is summarized as follows (Chen et al., 1991):

At the first step, compute

$$w_1^i = p_i \tag{9.67a}$$

$$g_1^i = \frac{(w_1^i)^T y}{(w_1^i)^T w_1^i} \tag{9.67b}$$

$$(\text{err})_1^i = \frac{\left(g_1^i\right)^2 \left(w_1^i\right)^T w_1^i}{y^T y} \tag{9.67c}$$

Find $(\text{err})_1^{i_1} = \max\left\{(\text{err}_1^i, \ 1 \le i \le M)\right\}$ and select

$$w_1 = w_1^{i_1} = p_{i_1} \tag{9.67d}$$

At the k-th step ($k \ge 2$ for $1 \le i \le M$, $i \ne i_1, \ldots, i \ne i_{k-1}$), compute

$$\alpha_{jk}^i = \frac{w_j^T p_i}{w_j^T w_j} \quad 1 \le j \le k \tag{9.68a}$$

$$w_k^i = p_i - \sum_{j=1}^{k-1} \alpha_{jk}^i w_j \tag{9.68b}$$

$$g_k^i = \frac{\left(w_k^i\right)^T y}{\left(w_k^i\right)^T w_k^i} \tag{9.68c}$$

$$(\text{err})_k^i = \frac{\left(g_k^i\right)^2 \left(w_k^i\right)^T w_k^i}{y^T y} \tag{9.68d}$$

Find $(\text{err})_1^{i_k} = \max \left\{ \left(\text{err}_k^i, 1 \le i \le M, \; i \ne i_1, \ldots, i \ne i_{k-1} \right) \right\}$ and select

$$w_k = w_k^{i_k} = p_{i_k} - \sum_{j=1}^{k-1} \alpha_{jk} w_j \tag{9.68e}$$

where

$$\alpha_{jk} = \alpha_{jk}^{i_k}, \; 1 \le j \le k \tag{9.68f}$$

The procedure is terminated at the M_s-th time step when $1 - \sum_{j=1}^{M_s} (\text{err})_j < \rho$, where $0 < \rho < 1$ is a chosen tolerance. The OLS procedure will produce a smaller RBF network than a randomly selected one. The parameter ρ balances the accuracy and complexity and is greater than but close to $\dfrac{\sigma_\varepsilon^2}{\sigma_d^2}$, where σ_ε^2 is the variance of the residuals and σ_d^2 is the variance of the measured output. σ_d^2 is known from measured data, whereas σ_ε^2 is estimated during the selection procedure.

One of the drawbacks of the OLS method is that it can only operate in the batch mode. Fun and Hagan (1999) have proposed a recursive orthogonal least squares learning algorithm with automatic weight selection (ROLS-AWS) for a two-layered Gaussian neural network. Their ROLS-AWS algorithm is capable of selecting useful weights suboptimally and recursively. In doing so, they claim that the algorithm will not only reduce the growth of the size of the weights but also minimize the number of weights used. Owing to the recursive nature of this algorithm, it can be applied to any online system, such as in control and signal processing applications. The ROLS-AWS method is based on the batch mode of the OLS method. However, the ROLS-AWS operates recursively without redoing the whole OLS calculation. The potential centres in the ROLS-AWS algorithm are limited by the data points presented to the RBF network, while the batch OLS method has the full range of data points to work with at the beginning. As a result, the ROLS-AWS method selects more centres (still smaller in size) than the batch OLS.

9.6.3.1 Regularized orthogonal least squares (ROLS) algorithm

The OLS algorithm is an efficient procedure for selecting the significant regressors, which leads to a parsimonious RBF network. A parsimonious model generally has better generalization properties. However, owing to the presence of noise in the data, it is possible even for a parsimonious model to overfit the data resulting in lack of generalization. The least squared error criterion used in the OLS algorithm is $e^T e$, which is prone to overfit under certain circumstances. This is prevented by a technique that involves regularized forward selection by considering the zero-order regularized error criterion (Orr, 1996):

$$e^T e + \lambda \theta^T \theta \tag{9.69}$$

where λ (≥ 0) is the regularization parameter and θ is the same as in Equations 9.49 and 9.62. The regularized forward selection algorithm selects one centre from the full model at a time so as to maximize the decrease of the regularized squared error (Equation 9.69). This method is computationally more demanding than the OLS algorithm. By combining the parsimonious principle of OLS with the zero-order regularization method, Chen et al. (1996) have developed a regularized orthogonal least squares (ROLS) algorithm that is capable of constructing small RBF networks that generalize well. It is a forward selection algorithm that is computationally in par with the OLS algorithm. Their procedure is as follows (Chen et al., 1996):

The orthogonal decomposition of the matrix P in Equation 9.55 is described in Equations 9.55 through 9.62. For known A and g, Equation 9.62 can be solved for θ using, for example, the Gram–Schmidt or Householder transformation schemes. A computationally efficient ROLS scheme is obtained by considering the following zero-order regularized error criterion:

$$e^T e + \lambda g^T g \tag{9.70}$$

The criterion in Equation 9.70 is similar to the criterion given in Equation 9.69 due to the relationship in Equation 9.62. The term $\lambda g^T g$ penalizes large g_i, which is equivalent to penalizing large θ_i. After simplification, the regularized error criterion given in Equation 9.70 can be shown to be (Chen et al., 1996)

$$e^T e + \lambda g^T g = y^T y - \sum_{i=1}^{N} \left(w_i^T w_i + \lambda \right) g_i^2 \tag{9.71}$$

which when normalized by dividing by $y^T y$ yields

$$\frac{e^T e + \lambda g^T g}{y^T y} = 1 - \frac{\sum_{i=1}^{N} \left(w_i^T w_i + \lambda \right) g_i^2}{y^T y} \tag{9.72}$$

In the same way as in OLS, the error reduction ratio due to w_i can be defined as

$$(\text{err})_i = \frac{\left(w_i^T w_i + \lambda \right) g_i^2}{y^T y}. \tag{9.73}$$

On the basis of this ratio, significant regressors can be selected in a forward regression procedure exactly as in the case of the OLS algorithm (Chen et al., 1991). The selection is terminated at the M_s-th stage when

$$1 - \sum_{k=1}^{M_s} (\text{err})_k < \rho \tag{9.74}$$

where ρ $(0 < \rho < 1)$ is a chosen tolerance. This procedure gives the subset of M_s regressors for the network.

The solution obtained by the ROLS algorithm is identical to that obtained by the regularized forward selection algorithm of Orr (1996) but requires considerably less computing resources. The solutions obtained by both methods are suboptimal but sufficient for all practical purposes.

The regularization parameter λ is problem dependent but it has been suggested that the performance of the RBF is fairly insensitive to the choice of λ (Bishop, 1991). It can, however, be iteratively determined using the formula (Mackay, 1992)

$$\lambda = \frac{\gamma}{N - \gamma} \frac{e^T e}{g^T g} \tag{9.75}$$

where

$$\gamma = \sum_{i=1}^{M_s} \frac{w_i^T w_i}{w_i^T w_i + \lambda} \tag{9.76}$$

Chen (2006) has pointed out that the subset model selection is carried out in the transformed orthogonal space but the selected subset of the orthogonal bases corresponds to a subset of the original model bases. This is an obvious but often overlooked property of the OLS algorithm. The subset spanned by the selected orthogonal bases is the same space spanned by the corresponding subset of the original model regressors. The one-to-one property of the OLS algorithm ensures that introducing regularization in the orthogonal weight space is equivalent to introducing regularization in the original model weight space.

ROLS algorithm employs a single common regularization parameter for every orthogonal weight in the model. For this reason, it is referred to as the uniformly regularized orthogonal least squares (UROLS) algorithm. Chen (2006) has proposed an individually regularized approach in which each candidate regressor has an associated individual regularization parameter. The method is referred to as the locally regularized orthogonal least squares (LROLS) algorithm. Its computation requirements are simple and the associated optimization process does not suffer from numerical ill conditioning. The LROLS algorithm is reported to have considerable computational advantages, including a well-conditioned solution and faster convergence speed. More recently, Chen et al. (2008) have proposed a complex-valued RBF (CVRBF) network, which has wide applications in non-linear signal processing.

9.6.4 Self-organized selection of centres

This is a kind of hybrid learning technique involving two stages: a self-organized learning stage for selection of centre locations in an unsupervised mode and a supervised learning stage for determining the weights. Both batch and adaptive approaches can be used but adaptive approach is preferred. The first stage involves clustering the training data using an algorithm, such as the k-means algorithm[10] (MacQueen, 1967; Duda and Hart, 1973), which partitions data into different groups by placing the centres of the RBFs in only those regions of the input space where significant data are present. The k-means algorithm assigns each data point to the cluster whose centre is nearest. The centre is the mean of all points in the cluster; that is, its coordinates are the means of the all points in the cluster. It is done so as to minimize the within-cluster sum of squares. It works as follows (Moody and Darken, 1989):

First, the number of clusters k is assumed *a priori*. The initial centres $\mu_k(0)$ are chosen randomly with the only restriction that they are different. It is also desirable to keep the

---

[10]The k-means algorithm is also referred to as the Lloyd's algorithm in the computer science community.

Euclidean norms of the centres small. Next, a sample vector x is drawn from the input space with a certain probability and used as input into the algorithm at iteration n. If $k(x)$ is the best matching centre for input vector x, find $k(x)$ at iteration n by using the minimum distance Euclidean criterion

$$k(x) = \arg \min_{k} \|x(n) - \mu_k(n)\|; \quad k = 1, 2, ..., M ^{11} \tag{9.77}$$

where $\mu_k(n)$ is the centre of the k-th RBF at iteration n. Updating is done according to

$$\mu_k(n+1) = \begin{cases} \mu_k(n) + \eta[(x(n) - \mu_k(n)] & k = k(x) \\ \mu_k(n) & \text{otherwise} \end{cases} \tag{9.78}$$

where η is a learning rate parameter ($0 \le \eta \le 1$). These steps are repeated until there are no noticeable changes in the centres μ_k. The algorithm may also be summarized in the following steps:

- Assume the number of centres k.
- Randomly generate k clusters and determine their centres, or directly generate k random points as cluster centres.
- Assign each point to the nearest cluster centre.
- Recalculate the new cluster centres.
- Repeat the last two steps until some convergence criterion has been met. Normally, the criterion is that the centres do not move.

The k-means algorithm is a special case of competitive learning known as self-organizing maps (SOMs). It has the advantage that it is fast and simple. The algorithm has the limitation that it can achieve only a local minimum depending on the initial choice of the centres. There is also no theoretical solution to find the optimal number of clusters for a given data set. The usual practice would be to try several runs with different k values and compare the results. Chen (1995) proposed an enhanced k-means clustering algorithm that converges to an optimum solution independent of the initial choice of the centre locations. Once the centres have been selected, the RBF widths σ_j are estimated as the average distance between the cluster centres μ_j and the training patterns that belong to that cluster (Hush and Horne, 1993):

$$\sigma_j^2 = \frac{1}{p_j} \sum_{x \in T_j} (x - \mu_j)^T (x - \mu_j) \tag{9.79}$$

where T_j is the set of training patterns grouped with centres μ_j and p_j is the number of patterns in T_j. The next stage of the hybrid learning is to estimate the weights. This is done by the least mean square (LMS) algorithm (see Section 8.3.4), although other learning algorithms for linear mapping can also be used. Outputs from the hidden layer form the input to the LMS algorithm.

---

[11] Argmin $\{f(x)\}$ refers to the mathematical notation for the value of the argument x for which the function $f(x)$ is a minimum.

9.6.5 Supervised selection of centres

In this approach of learning, which is applicable to the most general form of RBFs, all parameters are estimated in the supervised mode. It is done by error correction learning – a gradient descent procedure that is a generalization of the LMS algorithm (see Section 8.3.4). The free parameters w_i, z_i, and $\sum_i^{-1}$ (which is related to the norm-weighting matrix C_i) need to be estimated by minimizing the functional E

$$E = \frac{1}{2}\sum_{j=1}^{N} e_j^2 \tag{9.80}$$

and

$$e_j = d_j - F^*(x_j) = d_j - \sum_{i=1}^{M} w_i G\left(\left\|x_j - z_i\right\|_{C_i}\right) \tag{9.81}$$

Linear weights are obtained by differentiating E with respect to w_i and equating to zero leading to

$$\frac{\partial E(n)}{\partial w_i(n)} = \sum_{j=1}^{N} e_j(n)G\left(\left\|x_i - z_i(n)\right\|_{C_i}\right) \tag{9.82}$$

$$w_i(n+1) = w_i(n) - \eta_1\frac{\partial E(n)}{\partial w_i(n)}; \quad i = 1,\ 2,...,M \tag{9.83}$$

Position centres are obtained by differentiating E with respect to μ_i and equating to zero leading to

$$\frac{\partial E(n)}{\partial z_i(n)} = 2w_i(n)\sum_{j=1}^{N} e_j(n)G'\left(\left\|x_j - z_i(n)\right\|_{C_i}\right)\sum_i^{-1}(x_j - z_i(n)) \tag{9.84}$$

$$z_i(n+1) = z_i(n) - \eta_2\frac{\partial E(n)}{\partial z_i(n)}; \quad i = 1, 2,...,M \tag{9.85}$$

Spreads of centres are obtained by differentiating E with respect to the inverse covariance matrix $\sum_i^{-1}$ and equating to zero leading to

$$\frac{\partial E(n)}{\partial \sum_i^{-1}(n)} = -w_i(n)\sum_{j=1}^{N} e_j(n)G'\left(\left\|x_j - z_i(n)\right\|_{C_i} Q_{ji}(n)\right) \tag{9.86}$$

$$Q_{ji}(n) = [x_j - z_i(n)][x_j - z_i(n)]^T \tag{9.87}$$

$$\sum_i^{-1}(n+1) = \sum_i^{-1}(n) - \eta_3\frac{\partial E(n)}{\partial \sum_i^{-1}(n)} \tag{9.88}$$

In all the above equations, n is an iteration counter and G' is the first derivative of the Green's function G. The inverse covariance matrix $\sum_i^{-1}$ is defined by Equation 9.31 as

$$\frac{1}{2}\sum_i^{-1} = C_i^T C_i \tag{9.89}$$

Updating of the free parameters is done using different learning rate parameters η_1, η_2, and η_3. The problem of getting trapped in a local minimum in the weight space could be avoided by carefully choosing the initial parameter set.

Supervised training optimizes the locations of centres, whereas hybrid learning does not. As a result, supervised learning will provide a better approximation to the function to be learned for a given number of hidden units. This also means that a network with a fewer number of hidden units can be used for a given accuracy of approximation.

Wettschereck and Dietterich (1992) have found by experiments that Gaussian RBFs with unsupervised mode of fixing the centres and supervised mode of determining the weights in general does not perform as well as MLP with back-propagation, and that generalized RBFs with complete supervised mode perform better than MLP with back-propagation.

9.6.6 Selection of centres using the concept of generalized degrees of freedom

In a study by the author and his colleagues (Jayawardena et al., 2006), a simple algorithm in which the centres are neither 'too' distant from the input vector $X(t)$, nor 'too' close to each other, and the 'sets' (depends on the number of centres chosen) of centres nearest to the input vector $X(t)$ are included in the prediction process, is introduced. To do this, the prediction origin, t_0, of the univariate time series is first identified. The input vector will then be

$$X(t_0) = (x(t_0), x(t_0 - \tau),..., x(t_0 - (d_e - 1)\tau)) \tag{9.90}$$

where d_e is the embedding dimension and τ is the time delay (see Chapter 11 for the definitions and details of embedding dimension, time delay, and phase space reconstruction). The location of the first centre, $(C_i(t), i = 1)$, is chosen from the reconstructed phase space (which are constructed by the partial data from time $t = 1$ to $t = t_0$) arbitrarily. Locations of subsequent centres, $C_{i+1}(t)$, are chosen to satisfy the condition that if $\hat{X}(t)$ is a point in the reconstructed phase space, and if $\min_{1 \leq j \leq i} \| \hat{X}(t) - C_j(t) \| > \varepsilon$, a small positive quantity, then

$$C_{i+1}(t) = \hat{X}(t) \tag{9.91}$$

The small parameter ε ensures that there are enough centres in the local RBF model. However, a too small value of ε makes the distance between two centres too small to the extent that the two centres give similar functions in the local model. A value of $\varepsilon = 0.02 d_{max}$, where $d_{max} = \max\|X_1(t) - X_2(t)\| = $ 'diameter' of the space, X_1, X_2 being two points in the phase space, has been assumed in the above referenced study. Here X_1 and X_2 cover the entire phase space; that is, $X_1 = (x_i, x_{i+1}, x_{i+2}, ..., x_{i+\tau})$; $X_2 = (x_j, x_{j+1}, x_{j+2}, ..., x_{j+\tau})$; $1 \leq i \leq N - m$; $1 \leq j \leq N - m$; m is the embedding dimension. The centres chosen in this way will not be too close to each other. The first N_C centres that are nearest to the input vector $X(t_0)$ are collected, ensuring that they will not be too far away from $X(t_0)$ (a far away centre is not useful in the prediction process). Although this choice is purely empirical, the results justified that it is practical.

9.6.6.1 Training of RBF networks

Training of an RBF network involves the determination of the number of centres C_i, the widths of the receptive fields σ, and the connection weights to the output layer w_i. The first two parameters are usually determined in an unsupervised mode, whereas the weights can be determined in a supervised mode. In an RBF network, the output layer weights are linear, and can therefore be estimated by the least squares method for given values of σ and the number of centres, using the fact that a set of N_B nearest neighbours, $X^r(t)$; $r = 1, 2, 3,...,N_B$ of $X(t)$, with

$$X^r(t) = (x^r(t), x^r(t - \tau),..., x^r(t - (d_e - 1)\tau)) \tag{9.92}$$

will, at time level $t + \tau$, evolve to $X^r(t + \tau)$, which will be in the neighbourhood of $X(t + \tau)$. The weighted values w_i are then determined by minimizing

$$\sum_{r=1}^{N_B} \left[x^r(t + \tau) - \sum_{i=1}^{N_C} w_i \exp\left(-\frac{\|X^r(t) - C_i(t)\|^2}{2\sigma^2} \right) \right]^2 \tag{9.93}$$

where $x^r(t + \tau)$ is the evolved point at time level $t + \tau$ of $x^r(t)$ at time level t. This is known.

Optimization of the remaining two parameters, N_C and σ, is more difficult than optimizing a single parameter (Kantz and Schreiber, 1997, p. 212). The normal practice is to fix N_C, and choose the 'best' value (based on some error criterion) of σ from a given reference set. In the referenced study, an iterative method to obtain the best value of the width and the weights of the RBF network for a given number of centres has been introduced. The generalized degrees of freedom (GDF) method (Ye, 1998) is used to evaluate the models for different numbers of centres.

First, the details of determining the 'best' value of σ and the weights of the RBF network for a given number of centres (N_C) are described below.

The output, $x(t + \tau)$, of a Gaussian RBF, which is the predicted value of the input point $X(t)$, can be written as

$$x(t + \tau) = \sum_{i=1}^{N_C} w_i \exp\left(-\frac{\|X(t) - C_i(t)\|^2}{2\sigma^2} \right) \tag{9.94}$$

where $X(t)$ is the input vector that belongs to the reconstructed phase space, $C_i(t)$, $1 \le i \le N_C$, also points in the reconstructed phase space, are the centres, N_C is the number of centres, w_i, $1 \le i \le N_C$, are the weights of the RBF network output layer connections, σ is the receptive field width of the RBF, and $\|\cdot\|$ is the Euclidean norm (the centres and the receptive field width can be thought of as the mean and the standard deviation of the Gaussian distribution). The centres can be denoted as

$$C_i(t) = (c_i(t), c_i(t - \tau),..., c_i(t - (d_e - 1)\tau)) \tag{9.95}$$

For an assumed value of the width σ_n, Equation 9.94 can be written in matrix form as

$$\mathbf{V} = \mathbf{UW} + \mathbf{E} \tag{9.96}$$

where

$$\mathbf{V} = (x^1(t+\tau), x^2(t+\tau), \ldots, x^{N_B}(t+\tau))^T \tag{9.97}$$

are the evolved values at time level $t + \tau$, of the selected nearest neighbours $\mathbf{X}^r(t)$, $1 \le r \le N_B$, which are known

$$\mathbf{U} = \begin{bmatrix} \phi(\|\mathbf{X}^1 - \mathbf{C}_1\|) & \phi(\|\mathbf{X}^1 - \mathbf{C}_2\|) & \cdots & \phi(\|\mathbf{X}^1 - \mathbf{C}_{N_C}\|) \\ \phi(\|\mathbf{X}^2 - \mathbf{C}_1\|) & \phi(\|\mathbf{X}^2 - \mathbf{C}_2\|) & \cdots & \phi(\|\mathbf{X}^2 - \mathbf{C}_{N_C}\|) \\ \vdots & \vdots & \vdots & \vdots \\ \phi(\|\mathbf{X}^{N_B} - \mathbf{C}_1\|) & \phi(\|\mathbf{X}^{N_B} - \mathbf{C}_2\|) & \cdots & \phi(\|\mathbf{X}^{N_B} - \mathbf{C}_{N_C}\|) \end{bmatrix} \tag{9.98}$$

$$\phi(s) = \exp\left(-\frac{s^2}{2\sigma_n^2}\right) \tag{9.99}$$

$\mathbf{C}_1, \mathbf{C}_2, \ldots, \mathbf{C}_{N_C}$ are centres obtained from the algorithm introduced in Section 9.6.6, and arranged sequentially so that $\mathbf{C}_1$ is closest to $\mathbf{X}(t)$.

$\mathbf{W} = (w_1, w_2, \ldots, w_{N_C})^T$, are the connection weights of the RBF, which are functions of the assumed value of σ_n, and E is an error vector.

The corresponding estimation of the output layer weights can be obtained by the least squares method as (n here is an iteration counter)

$$\mathbf{W}^n = (\mathbf{U}'\mathbf{U})^{-1}\mathbf{U}'\mathbf{V} \tag{9.100}$$

The error term E is then transformed into a scalar function f of the assumed value of σ_n as

$$f(\sigma_n) = \|\mathbf{E}(\sigma_n)\|^2 = \sum_{r=1}^{N_B} \left[x^r(t+\tau) - \sum_{i=1}^{N_C} w_i^n \exp\left(-\frac{\|\mathbf{X}^r(t) - \mathbf{C}_i(t)\|^2}{2\sigma_n^2}\right) \right]^2 \tag{9.101}$$

where $\mathbf{W}^n = (w_1^n, w_2^n, \ldots, w_{N_C}^n)^T$.

To make the representation simpler, it is convenient to replace σ^2 by a new variable, say α. Then, $f(\sigma_n) = f(\alpha_n)$. The function f of σ (or α) is minimized using its derivatives f' and f''. The next better estimation of the width σ (or α) is

$$\alpha_{n+1} = \alpha_n - \frac{A_n}{B_n} \tag{9.102}$$

where

$$A_n = \frac{\partial f(\alpha_n)}{\partial \alpha_n}$$

$$= -\frac{1}{\alpha_n^2} \sum_{r=1}^{N_B} \left\{ \left[x^r(t+\tau) - \sum_{i=1}^{N_C} w_i^n \phi\left(\left\|\mathbf{X}^r - \mathbf{C}_i\right\|\right) \right] \right.$$

$$\left. \left[\sum_{i=1}^{N_C} w_i^n \phi\left(\left\|\mathbf{X}^r - \mathbf{C}_i\right\|\right)\left\|\mathbf{X}^r - \mathbf{C}_i\right\|^2 \right] \right\}$$

(9.103)

$$B_n = \frac{\partial^2 f(\alpha_n)}{\partial \alpha_n^2}$$

$$= -\frac{1}{\alpha_n^4} \sum_{r=1}^{N_B} \left\{ \left[x^r(t+\tau) - \sum_{i=1}^{N_C} w_i^n \phi\left(\left\|\mathbf{X}^r - \mathbf{C}_i\right\|\right) \right]\left[\sum_{i=1}^{N_C} w_i^n \phi\left(\left\|\mathbf{X}^r - \mathbf{C}_i\right\|\right)\left\|\mathbf{X}^r - \mathbf{C}_i\right\|^4 \right] \right\}$$

$$+\frac{1}{\alpha_n^4} \sum_{r=1}^{N_B} \left[\sum_{i=1}^{N_C} w_i^n \phi\left(\left\|\mathbf{X}^r - \mathbf{C}_i\right\|\right)\left\|\mathbf{X}^r - \mathbf{C}_i\right\|^2 \right]^2$$

(9.104)

$$+\frac{2}{\alpha_n^3} \sum_{r=1}^{N_B} \left\{ \left[x^r(t+\tau) - \sum_{i=1}^{N_C} w_i^n \phi\left(\left\|\mathbf{X}^r - \mathbf{C}_i\right\|\right) \right]\left[\sum_{i=1}^{N_C} w_i^n \phi\left(\left\|\mathbf{X}^r - \mathbf{C}_i\right\|\right)\left\|\mathbf{X}^r - \mathbf{C}_i\right\|^2 \right] \right\}$$

In fact, the above process to estimate a better value of α_{n+1} is just a standard Newton iteration procedure for the error term given in Equation 9.96 for given $\mathbf{W}^n$. Repeating the procedure, the following iteration sequence can be obtained (reverting to the original variable σ):

$$\sigma_0 \to \mathbf{W}^1 \to \sigma_1 \to \mathbf{W}^2 \to \sigma_2 \to \ldots \to \mathbf{W}^{n-1} \to \sigma_{n-1} \to \mathbf{W}^n \to \ldots$$

(9.105)

The procedure is terminated when σ_n converges giving better estimations of both the weighted values and widths for the local RBF network.

The method described above shows how to estimate the parameters σ and the weights w_i's for an assumed number of centres, which would certainly not be the only possible number of centres and may not necessarily be the best. A method of choosing a better number of centres using the GDF concept proposed by Ye (1998) is given next.

The GDF can be defined as the sum of the average sensitivities of each fitted value of the model to perturbations in the corresponding observed value. It can be applied to many modelling situations, such as artificial neural networks (ANN), and the prediction of chaotic data using the nearest-neighbourhood approach (Jayawardena et al., 2002). Given $V(t)$, the observed value, the objective in the present application is to choose locally the best set of centres in the network using the GDF. The procedure is as follows:

For a given (assumed) number of centres and the best receptive field σ determined from the method described above (Equation 9.105), the fitted mean vector $\hat{\mu}$ of the actual mean vector μ of V is estimated from Equation 9.96 as

$$\hat{\mu} = \mathbf{UW}$$

(9.106)

which, by the least squares method, is

$$\hat{\mu} = U(U'U)^{-1}U'V \tag{9.107}$$

For different values of the number of centres N_C, there will be different fitted vector functions $\hat{\mu}$, which in general will be different from the actual observed values V. A better model is one that has a smaller variance.

To estimate the value of the variance for a given number of centres, its GDF D is necessary. Since the assumed local relationship (Equation 9.107) is non-linear, the GDF is estimated using the Monte Carlo method introduced by Ye (1998). For a vector V of length n, it works as follows:

- Generate n perturbations $\Delta_t = (\delta_{1t}, \delta_{2t},..., \delta_{nt})'$ from the n-variate normal distribution function $N(0, \tau^2 I)$, with density:

$$g(x) = (2\pi)^{-\frac{n}{2}}(\tau^{2n})^{-\frac{1}{2}}\exp\left(-\frac{1}{2}x(\tau^2 I)^{-1}x\right) = \prod_{i=1}^{n}\frac{1}{\sqrt{2\pi}\tau}\exp\left(-\frac{1}{2}\frac{x_i^2}{\tau^2}\right)$$
$$= \prod_{i=1}^{n}\frac{1}{\tau}h\left(\frac{x_i}{\tau}\right) \tag{9.108}$$

where $x = (x_1,...,x_n)^T$; h is the standard normal density function; $t = 1,2,...,T$ refers to the number of perturbations with $T \geq n$; and τ is a tuning parameter, which can be thought of as the perturbation size. A reasonable size for τ is 0.6σ (Ye, 1998) where σ is the standard deviation of the deterministic time series, $X(t)$ in this case. The parameter n is equal to the sample size of the response function $V(t)$.
- Evaluate $\hat{\mu}_i(V + \Delta_t)$ – the i th component of the fitted vector of $(V + \Delta_t)$, based on the model corresponding to the assumed number of centres.
- Calculate $\hat{h}_i$ as the regression slope from the T equations, for $t = 1,2,...,T$:

$$\hat{\mu}_i(V + \Delta_t) = \alpha + \hat{h}_i\delta_{it}$$

Then an estimation of $D(N_C)$ is given by

$$\hat{D}(N_C) = \sum_i \hat{h}_i \tag{9.109}$$

The sum of squared residuals for a chosen number of centres, RSS, is then expressed as

$$RSS = (V - \hat{\mu})(V - \hat{\mu})^T \tag{9.110}$$

An unbiased estimation of the variance v^2 is given as

$$v^2(N_C) = \frac{RSS}{N_B - D} = \frac{(V - \hat{\mu})(V - \hat{\mu})^T}{N_B - D} \tag{9.111}$$

The model with the number of centres that has the least variance is chosen as the 'best' model for predictions.

9.6.6.2 Computational procedure

For a given prediction origin t_0, the partial univariate data from time $t = 1$ to $t = t_0$ is used to construct the time-delay phase space (see Chapter 11 for the determination of the time-delay τ and the embedding dimension of the phase space). To predict the signal at time $t_0 + \tau$ for a given point $\mathbf{X}(t_0)$ in the phase space, N_B nearest (to $\mathbf{X}(t_0)$) neighbours are employed to train the RBF network. The number of neighbours N_B is empirically set with the condition that it should be greater than the number of centres. Then, N_C nearest centres $\mathbf{C}_i(t)$, to $\mathbf{X}(t_0)$ (defined in Section 9.6.6.1), $N_C = N_C(\text{min}),\ldots, N_C(\text{max})$ are chosen using Equation 9.91. For a given N_C centres, the variance (Equation 9.111) is calculated via the GDF method, and N_C is

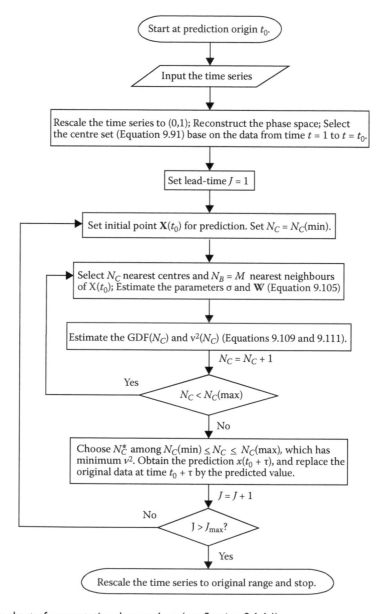

Figure 9.3 Flow chart of computational procedure (see Section 9.6.6.1).

chosen so that the variance is a minimum. Consequently, the weight $\mathbf{W} = (w_1, w_2, \ldots, w_{N_C})^T$, and the receptive fields σ of the RBF can be estimated (Section 9.6.6.1). With all these ingredients, the RBF network is then designed, and capable of predicting the univariate value at time $t_0 + \tau$. The sequence of computational steps is given in the flow chart depicted in Figure 9.3. In the flow chart, the minimum number of centres, $N_{c(min)}$; the maximum number of centres, $N_{c(max)}$; the maximum number of neighbours, M; and the maximum lead time for prediction, J_{max}, are user-defined parameters.

9.6.7 Other methods of learning

Chang and Chen (2003), instead of using the k-means clustering algorithm, have proposed a training algorithm that uses the fuzzy min–max clustering method (Simpson, 1993). The fuzzy min–max clustering algorithm involves three phases: expansion of a hyperbox, overlap test, and contraction of a hyperbox. During the network training process, a large number of n-dimensional hyperboxes that range from 0 to 1 along each dimension are generated. Each hyperbox is viewed as a hidden node, and the maximum and minimum are used to define the boundary of the hyperbox. The advantage of using fuzzy min–max clustering is that the number, centres, and the widths of the RBF can be determined systematically and automatically. There is only one parameter that needs to be user defined.

Song et al. (2005) introduced a new type of RBF, which they refer to as Shannon radial basis functions, which are obtained from the Shannon sampling theorem. The Shannon sampling theorem, which is also referred to as the Nyquist–Shannon sampling theorem, states that if a function $x(t)$ contains no frequencies higher than B cycles per second, it is completely determined by giving its ordinates at a series of points spaced $\dfrac{1}{2B}$ seconds apart. In their reformulated RBF network, the hidden nodes, which are smaller in number than traditional RBFs, are determined automatically and claimed to have faster learning speed and better accuracy than other methods.

Lin and Chen (2004) used a fully supervised learning algorithm that automatically determines the number of neurons in the hidden layer instead of the traditional hybrid technique. Their procedure starts with no hidden neurons. New hidden units are added as and when new training data are received. The location of the first centre is chosen from the training set and the width of the RBF of the j-th neuron

$$\rho = \sqrt{\frac{d_{max}^2}{j+1}} \tag{9.112}$$

where d_{max} is the maximum distance between the training data. Next, a training data point is chosen as the new hidden unit. The relative root mean square error (RRMSE) is computed after the output is obtained as a linear summation of the single hidden node output. In the same way, other data points are chosen sequentially, and the training data point that has the minimum RRMSE is added to the hidden layer. The procedure is continued until all data points have been used. A problem that may occur in this kind of training is the possibility that the best-performing parameter values so selected may end up overfitting the validation data. To overcome this problem, early stopping using cross validation is used. Early stopping is implemented when the RRMSEs of latter two consecutive neurons are larger than that of the former neuron. Use of the two consecutive neurons can avoid fluctuation of the RRMSE of the validation data.

The decision as to whether a particular input–output pair should give rise to a new hidden node is made using the following two criteria:

$$\text{num}_{\text{unit}} \leq \text{num}_{\text{max}}$$

$$\text{RRMSE} \leq e_{\text{min}}$$

where num_{unit} is the number of units in the hidden layer and num_{max} and e_{min} are thresholds to be appropriately chosen. If the above two conditions are satisfied, a new hidden node is added. The first condition implies that the number of hidden nodes is limited, while the second condition implies that the error between the network output and the desired output must be significant.

Nor et al. (2007) and Besaw et al. (2010) used a generalized regression neural network (GRNN) with added recurrent feedback loops that allow future predictions to be based on time-lagged predictions rather than time-lagged observations. This approach is well suited for ungauged catchments. The inputs to the model are the predicted flows and time-lagged measured rainfall and other influencing variables. The method is non-parametric and memory based.

GRNN consists of four layers: input, pattern, summation, and output. Each layer is fully connected to the adjacent layer via a set of weights. The pattern layer has one node for each pair of input data. Each node in the pattern layer is connected to two summation layer nodes, S_1 and S_2. The weights linking the pattern layer nodes with the summation node S_1 store the target output values for each input–output training pattern. The weights from the pattern layer to the summation node S_2 are set to unity.

The prediction y is computed as

$$y = \frac{S_1}{S_2} = \frac{\sum_{i=1}^{N} d_i \exp\left(-\frac{\|x - w_i\|^2}{2\sigma^2}\right)}{\sum_{i=1}^{N} \exp\left(-\frac{\|x - w_i\|^2}{2\sigma^2}\right)} \tag{9.113}$$

where S_1 and S_2 are the two summations and N is the number of training samples. More details of the GRNN method can be found in Specht (1991), Nor et al. (2007), and Besaw et al. (2010). GRNN is a kind of kernel regression, and can be designed quickly, has fast learning, and effectively uses historical data.

9.7 CURSE OF DIMENSIONALITY

The curse of dimensionality – the exponential increase of hidden units as a function of the input dimensionality – was first introduced by Bellman (1961) in the context of adaptive control. It is based on the notion that if the function is complex, more samples are needed to learn it. In high dimensions, it is difficult to find dense samples. It is the exponential growth in complexity with increasing dimensionality that leads to the deterioration of the space-filling properties for uniformly randomly distributed points in high-dimension space; that is, the number of parameters needed for the approximating function to attain a prescribed degree of accuracy increases exponentially with increasing input dimensionality. Although some approximate techniques that enable the rate of convergence to be independent of the input dimension have been proposed, it cannot be completely eliminated.

In general, the selection of units (nodes) in the hidden layer is a trial-and-error process. A guideline is that the number of nodes should be less than the number of training examples. Otherwise, the learning will not lead to a generalization. It should also be noted that a two hidden layer network might perform better than a single hidden layer network with too many units. This also applies to MLP networks.

9.8 PERFORMANCE CRITERIA

The performance of ANN models that use RBFs are normally measured by the same error indicators as for MLP-type networks. The most commonly used ones are the mean absolute error (MAE), root mean square error (RMSE), relative root mean square error with respect to the expected mean value (RRMSE), the Nash–Sutcliffe coefficient of efficiency (NSE), modified coefficient of efficiency (MNSE), coefficient of correlation r, and the coefficient of determination (CD) which is different from R^2 in statistics. These are defined in Equations 8.117 through 8.124 in Chapter 8.

9.9 COMPARISON OF MLP VERSUS RBF NETWORKS

The key similarities and differences between MLP- and RBF-type neural networks can be summarized as follows:

- Both are universal approximators.
- MLP can be used where RBF works and vice versa.
- RBF has a single hidden layer, whereas MLP can have more than one hidden layer.
- The hidden layer of an RBF is non-linear, whereas the output layer is linear. In an MLP network, hidden and output layers can be non-linear, particularly for pattern classification. For non-linear regression in MLP networks, a linear activation function in the output layer is preferred.
- The argument in the activation function of an RBF computes the Euclidean norm (distance) between the input vector and the centre of that unit, whereas in an MLP network, it computes the inner product[12] of the input vector and the synaptic weights of that unit.
- MLP is a global approximation non-linear input–output mapping technique. RBF does localized non-linear input–output mapping.
- RBFs do not have the problem of getting trapped in local minima in the optimization process.
- MLPs require a smaller number of parameters compared with RBFs for the same accuracy. The only fundamental difference is the way in which hidden units combine values.
- RBF networks do not have anything that is exactly the same as the bias term in an MLP.

---

[12] The inner product (or, dot product) between two vectors $x_i, x_j; i, j = 1, 2, ..., m$ is given by $x_i^T x_j = \sum_{k=1}^{m} x_{ik} x_{jk}$ where x_{ik} is the k-th element of x_i and x_{jk} is the k-th element of x_j. The inner product converts a vector space to a scalar quantity.

- Some types of RBFs have a 'width' associated with each hidden unit or with the entire hidden layer; instead of adding it in the combination function like a bias, the Euclidean distance is divided by the width.
- The output activation function in RBF networks is usually the identity.
- MLP networks cannot be trained by 'hybrid' methods, whereas RBF networks are usually trained by 'hybrid' methods.
- Nodes in an MLP network typically share a common neural model, whereas hidden and output nodes of an RBF network are functionally distinct.
- RBF networks train rapidly (usually orders of magnitude faster than MLP) while exhibiting none of back-propagation training pathologies such as paralysis (learning reaching a stage where the weights can no longer be adjusted) or local minima problems.
- Better generalization can be achieved with RBF networks. These models can also be interpreted as fuzzy connectionist models, as the RBFs can be considered as membership functions.
- Problem of finding the appropriate number of hidden nodes in an RBF is a disadvantage. Too many or too few hidden nodes will prevent RBF from properly approximating the data.
- After training, RBF networks are slower to use.
- RBFs suffer the curse of dimensionality – the exponential increase of hidden units as a function of the input dimensionality.

9.10 APPLICATIONS

RBF-type neural networks have not been used to the same extent as MLP-type networks for prediction of hydrological or environmental phenomena. Unlike MLPs with sigmoid or hyperbolic tangent activation functions, RBFs respond to relatively small region of the input space, thereby requiring more neurons in the hidden layer. With increasing input patterns, this requirement may lead to the 'curse of dimensionality'. However, RBFs also have advantages over MLPs such as linearity in the output layer and not being prone to getting trapped in local minima.

A comprehensive description of all applications that use RBFs to model hydrological or environmental systems is beyond the scope of this chapter. Instead, an attempt to give a brief review of the studies in which the author had a role to play would be made. The earliest study (Jayawardena et al., 1997) was to predict the water level at a gauging station in the Shan Qiao experimental basin (area = 110 km^2; latitude: 22°40′N; longitude: 112°30′E) in Guangdong Province of China. The reason for using water levels instead of discharges is because they are more practical indicators of the level of flooding. Hourly rainfalls at Ban Chun and Shang Qiao and hourly water levels at Shang Qiao hydrological station, for six storm periods extracted from hydrological data yearbooks compiled by the Guangdong Provincial General Hydrological Station, were used in the study. Two-hour-lead-time water level predictions were made by taking the present water level at Shang Qiao station, SWL, to be dependent on its past values, SWL_{t-2}, SWL_{t-3}, SWL_{t-4}, past rainfall values at Shang Qiao, SR_{t-2}, SR_{t-3}, SR_{t-4}, and Ban Chun, BR_{t-2}, BR_{t-3}, BR_{t-4}. The data for the first storm period were used for training, and the remainder was used for testing the performance of the models. The results of two models, one that uses an MLP network and the other that uses an RBF network, were found to be comparable in terms of the accuracy. A similar study has been made to predict the water level at Tai Bin Chong gauging station (latitude: 23°20′N; longitude: 113°30′E), which lies in the downstream part of the Liu Xie River, a tributary of the Pearl River in southern China. It is the last gauged station where the water levels are not affected by the tidal variations. The aim is to predict water level at this station using the upstream water levels and rainfall in the region. Data corresponding to five flood events that occurred in 1987 and 1988, extracted from the hydrological data books, were used. In both these examples, the centres of the RBFs were determined by using the k-means clustering algorithm.

A similar RBF-type model that uses the k-means clustering algorithm was applied to predict hourly runoff at the Yamaguchi gauging station (area = 3.12 km²; latitude: 36°16′N; longitude: 140°07′E) in a small experimental catchment located in Kamihonsha, Uratsukuba area of Japan (Jayawardena et al., 1998). The data used in this application consist of hourly rainfall measured at Kamihonsha and hourly runoff discharge measured at Yamaguchi station for 12 storm events. They were provided by the Hydrology Laboratory of the Public Works Research Institute of Japan. In this application, the present discharge Q_t is taken to be dependent on past discharge values Q_{t-1}, Q_{t-2}, and Q_{t-3} and past rainfall values R_{t-1}, R_{t-2}, and R_{t-3}, thereby limiting the number of input nodes to six. The results have been, as before, quite comparable with those from MLP-type models, but with the added advantage of faster computational speed.

The method of centre selection that uses the concept of GDF described in Section 9.6.6 above has been used first to predict two non-linear theoretical functions, the Henon map and the Lorenz map (more details of these two functions can be found in Chapter 10), which are known to become chaotic under certain parameter conditions. The hydrological data sets used in this study include the daily discharges of Mekong River at Nong Khai (17.87°N and 102.72°E, basin area, 302,000 km², GRDC #2969090) in Thailand, and at Pakse (15.12°N and 108.80°E, basin area, 545,000 km², GRDC #2469260) in Lao PDR for the period April 1980 to December 1991, the daily discharges of Chao Phraya River at Nakhon Sawan (15.67°N and 100.2°E, basin area, 110,569 km², GRDC #2964100) in Thailand for the period April 1978 to March 1994, and the monthly mean sea surface temperature (SST) anomaly over the region bounded approximately by 6°N–6°S and 180°–90°W for the period January 1872 to December 1986, which has been defined as the S-index by Wright (1984) and used to identify climatic anomalies attributed to El Niño and Southern Oscillation. The first three data sets were obtained from the Global Runoff Data Centre (GRDC) in Koblenz, Germany, and the last one from a table compiled by Wright (1989). A few missing records of the data sets were replaced by their long-term averages. All the real data sets used were noise reduced by methods described in a separate study (Jayawardena and Gurung, 2000) before using them in the modelling and prediction process. In the analysis, it is implicit that the time series have been treated as chaotic attractors (see Chapter 10). The basic information of the data used in the study is given in Table 9.1. Better predictions, as measured by the cumulative mean square errors (CMSE), can be obtained by the method in which the number of centres and the widths of the RBF are optimized on the basis of the GDF as illustrated by a

Table 9.1 Basic information of data used in the study

Type of data	No. of data	Embedding dimension	Time delay	Origin for prediction Set (a)	Set (b)
Henon	10,000	2	1	5000	5500
Lorenz	10,000	3	10	5000	5500
Mekong at Nong Khai	4292	3	1	3500	4000
Mekong at Pakse	4292	3	1	3500	4000
Chao Phraya	5844	3	1	5000	5500
S-index	1380	3	1	500	1000

Table 9.2 Cumulative mean square errors for 10 lead-time steps with Set (a) as the origin for prediction

Type of data	Radial basis function method with chosen number of centres	Radial basis function method with fixed number of centres	Linear model	ARMA model
Henon	7.60E − 08	2.40E − 07	3.64E − 06	NA
Lorenz	9.87E − 06	1.41E − 05	6.17E − 05	NA
Mekong at Nong Khai	8.83E + 02	1.55E + 03	7.04E + 03	2.05E + 06
Mekong at Pakse	5.49E + 03	7.80E + 03	4.55E + 04	7.17E + 06
Chao Phraya	3.25E + 01	3.41E + 01	1.51E + 02	4.88E + 03
S-index	1.59E + 02	4.38E + 02	1.86E + 01	1.54E + 01

typical set of results shown in Table 9.2, which also gives the predictions obtained by a linear model and a stochastic model (Jayawardena et al., 2002). In this context, CMSE refers to the summation of squared errors up to a particular lead-time step divided by the corresponding number of lead-time steps, which can increase or decrease with increasing number of lead-time steps. For the theoretical data sets, there is hardly any difference between the actual values and the predictions, at least up to the limit of predictability. It should be noted that long-term predictions cannot be made using method described above. For the real data sets, the errors are at least one order of magnitude lower than those obtained by using a fixed number of centres.

Applications using the centre selection methods described in Section 9.6.7 include runoff predictions using rainfall data in the Sungai Bekok Catchment in Johor, and Sungai Ketil catchment in Kedah, Malaysia (Nor et al., 2007), streamflow predictions using time-lagged values of rainfall and reservoir inflow rates in the Fei-Tsui reservoir watershed in northern Taiwan (Lin and Chen, 2004), water level prediction of the Tanshui River in Taiwan under tidal effect (Chang and Chen, 2003), and streamflow prediction in Dog River at Wrightsville, northwestern Vermont, United States (Besaw et al., 2010).

9.11 CONCLUDING REMARKS

RBFs are a type of curve-fitting technique that can be used to solve interpolation problems in multidimensional space. In the context of ANN, RBFs play an important role, although the application areas in environmental and hydrological fields are not as much as MLPs. This chapter describes the fundamentals of RBFs, their variations, learning strategies, selection of centres and widths using the concept of GDF as well as other methods, comparative features between RBFs and MLPs, and some application areas.

REFERENCES

Bellman, R. (1961): *Adaptive Control Processes: A Guided Tour*. Princeton University Press, Princeton, NJ.

Besaw, L.E., Rizzo, D.M., Bierman, P.R. and Hackett, W.R. (2010): Advances in ungauged streamflow prediction using artificial neural networks. *Journal of Hydrology*, 386, 27–37.

Bishop, C. (1991): Improving the generalization properties of radial basis function neural networks. *Neural Computation*, 3, 579–588.

Broomhead, D.S. and Lowe, D. (1998): Multivariable functional interpolation and adaptive networks. *Complex Systems*, 2, 321–355.

Bugmann, G. (1998a): Classification using networks of normalized radial basis functions. In: Singh, S. (Ed.), *Proceedings (Tutorials) of the International Conference on Advances in Pattern Recognition (ICAPR'98)*, Plymouth, Nov. 23–25, 1998.

Bugmann, G. (1998b): Normalized Gaussian radial basis function networks. *Neurocomputing*, 20(1–3), 97–100.

Cha, I. and Kassam S.A. (1995): Interference cancelation using radial basis function networks. *Signal Processing*, 47, 247–268.

Chang, F.-J. and Chen, Y.-C. (2003): Estuary water-stage forecasting by using radial basis function neural network. *Journal of Hydrology*, 270(1–2), 158–166.

Chen, S. (1995): Non-linear time series modeling and prediction using Gaussian RBF networks with enhanced clustering and RLS learning. *Electronic Letters*, 31(2), 117–118.

Chen, S. (2006): Local regularization assisted orthogonal least squares regression. *Neurocomputing*, 69, 559–585.

Chen, S., Billings, S.A., Cowan, C.F.N. and Grant, P.M. (1990a): Non-linear systems identification using radial basis functions. *International Journal of System Sciences*, 21(12), 2513–2539.

Chen, S., Billings, S.A., Cowan, C.F.N. and Grant, P.M. (1990b): Practical identification of NARMAX models using radial basis functions. *International Journal of Control*, 52(6), 1327–1350.

Chen, S., Billings, S.A. and Luo, W. (1989): Orthogonal least squares methods and their application to non-linear system identification. *International Journal of Control*, 50(5), 1873–1895.

Chen, S., Chng, E.S. and Alkadhimi, K. (1996): Regularized orthogonal least squares algorithm for constructing radial basis function networks. *International Journal of Control*, 64(5), 829–837.

Chen, S., Cowan, C.F.N. and Grant, P.M. (1991): Orthogonal least squares learning algorithm for radial basis function networks. *IEEE Transactions on Neural Networks*, 2(2), 302–309.

Chen, S., Hong, X., Harris, C.J. and Hanzo, L. (2008): Fully complex-valued radial basis function networks: Orthogonal least squares regression and classification. *Neurocomputing*, 71(16–18), 3421–3433.

Cover, T.M. (1965): Geometrical and statistical properties of systems of linear inequalities with applications in pattern recognition. *IEEE Transactions on Electronic Computers*, EC-14, 326–334.

Craven, P. and Wahba, G. (1979): Smoothing noisy data with spline functions: Estimating the correct degrees of smoothing by the method of generalized cross validation. *Numerische Mathematik*, 31, 377–403.

Duda, R.O. and Hart, P.E. (1973): *Pattern Classification and Scene Analysis*. Wiley, New York.

Fun, M.H. and Hagan, M.T. (1999): Recursive orthogonal least squares learning with automatic weight selection for Gaussian neural networks. In: *Proc. Int. Joint Conf. on Neural Networks (IJCNN'99)*, vol. 3, pp. 1496–1500. Available at http://www.hagan.ecen.ceat.okstate.edu/fun_hag_nn99.pdf.

Golub, G. (1965): Numerical methods for solving linear least squares problems. *Numersche Mathematik*, 7, 206–216.

Haykin, S. (1999): *Neural Networks: A Comprehensive Foundation*. Prentice-Hall, New Jersey, 842 pp.

Householder, A. S. (1958): Unitary triangularization of a nonsymmetric matrix. *Journal of the ACM*, 5(4), 339–342.

Hush, D.R. and Horne, B.G. (1993): Progress in supervised neural networks: What's new since Lippmann? *IEEE Signal Processing Magazine*, 10, 8–39.

Jang, J.S.R. and Sun, C.T. (1993): Functional equivalence between Radial Basis Function Networks and Fuzzy Inference Systems. *IEEE Transactions on Neural Networks*, 4, 156–159. http://www.cs.nthu.edu.tw/~jang/publication.htm.

Jayawardena, A.W., Fernando, D. and Achela, K. (1998): Use of radial basis function type artificial neural networks for runoff simulation. *Computer-Aided Civil and Infrastructure Engineering*, 13(2), 91–99.

Jayawardena, A.W., Fernando, D.A.K. and Zhou, M.C. (1997): Comparison of multilayer perceptron and radial basis function networks as tools for flood forecasting. In: *Destructive Water: Water-Caused Natural Disasters, their Abatement and Control (Proceedings of the Conference held at Anaheim, CA, June 1996)*, IAHS Publication no. 239, pp. 173–181.

Jayawardena, A.W. and Gurung, A.B. (2000): Noise reduction and prediction of hydrometeorological time series: Dynamical systems approach vs. stochastic approach. *Journal of Hydrology*, 228(3–4), 242–264.

Jayawardena, A.W., Li, W.K. and Xu, P. (2002): Neighbourhood selection for local modelling and prediction of hydrological time series. *Journal of Hydrology*, 258, 40–57.

Jayawardena, A.W., Xu, P.C., Tsang, F.L. and Li, W.K. (2006): Determining the structure of a radial basis function network for prediction of nonlinear hydrological time series. *Hydrological Sciences Journal*, 51(1), 21–44.

Kantz, H. and Schreiber, T. (1997): Nonlinear Time Series Analysis, Cambridge nonlinear science series 7. Cambridge University Press, Cambridge, UK.

Lin, G.-F. and Chen, L.-H. (2004): A non-linear rainfall-runoff model using radial basis function network. *Journal of Hydrology*, 289, 1–8.

Mackay, D.J.C. (1992): Bayesian interpolation. *Neural Computation*, 4, 415–447.

MacQueen, J.B. (1967): Some methods for classification and analysis of multivariate observations. *Proceedings of 5th Berkeley Symposium on Mathematical Statistics and Probability*, University of California Press, Berkeley, CA, vol. 1, pp. 281–297.

Moody, J. and Darken, C. (1989): Fast-learning in networks or locally tuned processing units. *Neural Computations*, 1(2), 281–294.

Nadaraya, E.A. (1964): On estimating regression. *Theory Probab. Appl.*, 9(1), 141–142.

Nor, N.I.A., Harun, S. and Kassim, A.H.M. (2007): Radial basis function modeling of hourly streamflow hydrograph. *Journal of Hydrologic Engineering, ASCE*, 12(1), 113–123.

Orr, M.J.L. (1996): Introduction to radial basis function networks. Research Report, Centre for Cognitive Science, University of Edinburgh, Edinburgh, UK, p. 67.

Poggio, T. and Girosi, F. (1990a): Networks for approximation and learning. *Proceedings of the IEEE*, 78(9), 1481–1497.

Poggio, T. and Girosi, F. (1990b): Regularization algorithms for learning that are equivalent to multilayer networks. *Science, New Series*, 247, 4945, 978–982.

Powell, M.J.D. (1985): Radial basis functions for multivariable interpolation: A review. *IMA Conference on Algorithms for Approximation of Functions and Data*, RMCS, Shrivenham, England, pp. 143–167.

Powell, M.J.D. (1987a): Radial basis function approximations to polynomials. *Proceedings, 12th Biennial Numerical Analysis Conference*, Dundee, pp. 223–241.

Powell, M.J.D. (1987b): Radial basis functions for multivariable interpolation: A review. In: Mason, J.C. and Cox, M.G. (Eds.), *Algorithms for Approximation*. Clarendon Press, Oxford, pp. 143–167.

Rao, A.V., Miller, D., Rose, K. and Gersho, A. (1997): Mixture of experts regression modeling by deterministic annealing. *IEEE Transactions on Signal Processing*, 45, 2811–2820.

Servin, M. and Cuevas, F.J. (1993): A new kind of neural network based on radial basis functions. *Revista Mexicana de Fisica*, 39(2), 235–249.

Simpson, P.K. (1993): Fuzzy min–max neural networks. Part 2: Clustering. *IEEE Transactions on Fuzzy Systems*, 1(1), 32–45.

Song, S., Yu, Z. and Chen, X. (2005): A novel radial basis function neural network for approximation. *International Journal of Information Technology*, 11(9), 46–53.

Specht, D.F. (1991): A general regression neural network. *IEEE Transactions on Neural Networks*, 2, 568–576.

Tikhonov, A.N. (1963): On solving incorrectly problems and method of regularization. *Doklady Akademii Nauk, USSR*, 151, 501–504.

Watson, G.S. (1964): Smooth regression analysis, Sankhya. *The Indian Journal of Statistics, Series A* 26(4), 359–372.

Wettschereck, D. and Dietterich, T. (1992): Improving the performance of radial basis function networks by learning centre locations. *Advances in Neural Information Processing Systems*, 4, San Mateo, CA, pp. 1133–1140.

Wright, P.B. (1984): Relationship between the indices of the southern oscillation. *Monthly Weather Review*, 112, 1913–1919.

Wright, P.B. (1989): Homogenized long-period southern oscillation indices. *International Journal of Climatology*, 9, 33–54.

Ye, J. (1998): On measuring and correcting the effects of data mining and model selection. *Journal of the American Statistical Association*, 93(441), 120–131.

Chapter 10

Fractals and chaos

10.1 INTRODUCTION

Fractals are geometric objects that can be subdivided into parts, each of which is a copy of the whole. This property is known as self-similarity, which may be exact, quasi, or statistical. Fractals that possess exact self-similarity appear identical at different scales. Fractals defined by iterated functions belong to this category. Quasi-self-similar fractals appear approximately identical at different scales and are generally defined by recurrence relations. In statistically self-similar fractals, the statistical measures are preserved at different scales. Random fractals are examples that are statistically self-similar, but neither exactly nor quasi self-similar. Fractals have fine structures at infinitely small scales. They can be generated by simple and well-defined recursive equations. Examples in nature include clouds, mountains, coastlines, trees, ferns, river networks, cauliflower, system of blood vessels, snowflakes, etc. An exception of a self-similar object that is not a fractal is a straight line. Historically, the first fractal perhaps is the Weierstrass function (Weierstrass, 1872), which in its original form is expressed as

$$f(x) = \sum_{n=0}^{\infty} a^n \cos(b^n \pi x) \tag{10.1}$$

with $0 < a < 1$; b is a positive odd integer; $ab > 1 + \frac{3}{2}\pi$. This function is continuous everywhere but differentiable nowhere and has self-similarity.

Among the well-known fractals are the Cantor set (after Georg Cantor, 1883), the Koch curve and snowflakes (after Helge von Koch, 1904), and the Sierpinski triangle (after Waclow Sierpinski, 1915) in the plane one- and two-dimensional spaces, and Mandelbrot (after Benoit Mandelbrot, 1960s; he coined the word 'fractal' to an object whose Hausdorff–Besicovitch dimension is greater than the topological dimension) and Julia (after Gaston Julia, 1918) sets in the complex plane. The Sierpinski triangle and Koch snowflake exhibit exact self-similarity. The Mandelbrot set is quasi self-similar, as the satellites are approximations of the entire set, but not exact copies. Fractals can be deterministic or stochastic. Trajectories of Brownian motion are generated by stochastic fractals. Random fractals can be used to describe many highly irregular objects. Other applications include fractal landscape, enzymology, generation of new music, signal and image compression, seismology, soil and rock mechanics, and video games, among many others.

Most dynamical systems, after a certain transient period, will eventually attain a steady (no further motion) or periodic state (limit cycle). Limit cycles cannot occur in linear systems. They occur in many physical/biological systems, for example, heartbeat, self-excited vibration of bridges or aircraft wings. In these systems, there is a standard preferred periodicity and when slightly perturbed will return to the preferred state. Chaotic motions are neither steady nor periodic. They appear to behave like random systems but are governed by well-defined physical laws and therefore deterministic. Both fractals and chaos are generated by recursive functions and have infinite details. They are both complex and have specific properties. For example, an attractor in the phase space of a dynamical system can be fractal with non-integer dimension. Strange attractors in the phase space are generated by the solution of deterministic dynamical systems, which are, in essence, initial value differential equations that exhibit chaos. The time series of many natural phenomena are also fractal and therefore their information at different scales can be linked.

In this chapter, basic concepts of fractals and chaos are introduced. Definitions and estimation methods of fractal dimension and invariant measures for chaotic systems, as well as some examples of well-known fractals and chaotic maps are also given.

10.2 FRACTAL DIMENSIONS

In general, the dimension of a system can be thought of as the number of degrees of freedom, or the number of independent variables needed to define the state of the system. In Euclidean geometry, a point has zero dimension, a line has a dimension of one, a plane a dimension of two, and a solid a dimension of three. These, which are non-negative integer numbers, may also be considered as their respective topological dimensions. However, not all geometrical shapes can be described within the framework of Euclidean geometry. In fractals, several dimensions that are essentially non-integers can be identified. Quantitatively, many measures of these dimensions end up with the same numerical value.

10.2.1 Topological dimension

Topological dimension is the normal dimension in the Euclidean sense. For example, the topological dimension of a point is zero, of a line is one, of a plane is two, and of a solid is three.

10.2.2 Fractal dimension

For some well-known fractals, different measures of dimensions do coincide; however, in general, they are not equivalent. For example, the topological dimension of the Koch snowflake is one and the length between any two points on it is infinite. It is neither a line nor a plane and could be considered as something in between. Its fractal dimension should lie somewhere between one and two.

Fractal dimension is a statistical quantity that gives a measure of how closely a fractal fills space as it is scaled down. It is a measure of the geometrical complexity of the object. Fractal dimension is a non-negative real number. The Hausdorff dimension, the packing dimension and, more generally, the Rényi dimension are the most important ones from a theoretical point of view; however, the box-counting dimension and the correlation dimension are widely used in practice. A general definition is of the form

$$D_{\text{Fractal}} = \underset{\text{scale} \to 0}{\text{Limit}} \left[\frac{\ln(\text{number of self-similar objects})}{\ln(\text{magnification factor})} \right] \qquad (10.2a)[1]$$

---

[1] The logarithms in dimension definitions can be to any base.

or

$$D_{\text{Fractal}} = \underset{\varepsilon \to 0}{\text{Limit}} \left[\frac{\ln N(\varepsilon)}{\ln \left(\dfrac{1}{\varepsilon} \right)} \right] \tag{10.2b}$$

where $N(\varepsilon)$ is the number of self-similar objects needed to cover the original object and ε is the linear size in Euclidean space. The fractal dimension is always greater than or equal to the corresponding topological dimension.

The concept of fractal dimension leads to an ambiguous interpretation of the concept of 'length'. For example, if a straight line of length 1 m is measured by a ruler 20 cm long, it would need five measurements; with a ruler 10 cm long, it will need 10 measurements, and so on. As the length of the ruler becomes smaller and smaller, the number of measurements needed to cover the length becomes larger and larger. In the limit as the length of the ruler tends to zero, the measured length will approach the true length. In the case of a curve, the accuracy will depend on the length of the ruler. The smaller the ruler, the more accurate the measurement will be because a shorter ruler can fit more closely to a curve than a longer ruler.

In a curved space such as a coastline or a national border, as the length of a measuring scale becomes smaller and smaller, meaning that the number of measurements needed to cover the length will become larger and larger, the length will also become larger and larger. For example, the border between Spain and Portugal has been quoted as 987 or 1214 km, while that between the Netherlands and Belgium as 380 or 449 km. Richardson (1961), while collecting data, observed that there was considerable variation in the gazetted lengths of international borders. He found regularity between the length of a national boundary and the scale of measurement, which has been expressed in the form

$$\log(L(\varepsilon)) = (1 - D)\log(\varepsilon) + \text{constant} \tag{10.3}$$

where $L(\varepsilon)$ is the length, ε is the size of the scale, and D is the fractal dimension. This relationship plots as a straight line in a log–log scale with a slope equal to $1 - D$. The results of Richardson's studies on the comparison of the lengths of national boundaries of some countries is shown in Table 10.1.

Although it can be assumed that the measured lengths would converge to the true length of the object, Richardson demonstrated that the measured length of coastlines and other natural features appears to increase without limit as the unit of measurement is made smaller. This is known as the Richardson effect.

Mandelbrot (1967) discussed the work of Richardson (1961) and conjectured that coastlines and other geographic borders can display statistical self-similarity. He identified D in Equation 10.3 as the Hausdorff dimension of the border.

Table 10.1 Fractal dimensions of some national boundaries

Country	Length scale relationship	Fractal dimension D
Australia	$\log(L(\varepsilon)) = -0.13\log(\varepsilon) + 4.4$	1.13
South Africa	$\log(L(\varepsilon)) = -0.04\log(\varepsilon) + 3.8$	1.04
Germany	$\log(L(\varepsilon)) = -0.12\log(\varepsilon) + 3.7$	1.12
Great Britain	$\log(L(\varepsilon)) = -0.24\log(\varepsilon) + 3.7$	1.24
Portugal	$\log(L(\varepsilon)) = -0.12\log(\varepsilon) + 3.1$	1.12

10.2.3 Hausdorff dimension

Hausdorff dimension, sometimes known as Hausdorff–Besicovitch dimension, was introduced by Felix Hausdorff in 1918 with computational details provided by Abram Samoilovitch Besicovitch. Suppose the linear size of an object residing in Euclidean dimension D is reduced by $\frac{1}{r}$ in each spatial direction, its measure (length, area, or volume) would increase to r^D times the original. For example, in a one-dimensional case of a line, and if $r = 1$ (there is no reduction of the length of line), the number of self-similar objects, $N = r^D$ remain unchanged at 1 for one, two, and three dimensions; if $r = 2$, $N = r^D$ becomes 2 for one dimension, 4 for two dimensions, and 8 for three dimensions; if $r = 3$, $N = r^D$ becomes 3 for one dimension, 9 for two dimensions, and 27 for three dimensions. When the logarithms of the relationship $N = r^D$ are taken, the topological dimension D can be expressed as

$$D = \frac{\ln N(r)}{\ln(r)} \tag{10.4a}$$

which means that the dimension need not be an integer as in Euclidean geometry. Using Equation 10.4a, the Hausdorff dimension D_H can be defined as

$$D_H = \underset{\varepsilon \to 0}{\text{Limit}} \left[\frac{\ln(N(\varepsilon))}{\ln\left[\dfrac{1}{\varepsilon}\right]} \right] \tag{10.4b}$$

where $N(\varepsilon)$ is the number of self-similar objects of linear size ε needed to cover the whole object. For many geometrical shapes, the Hausdorff dimension is an integer, but for fractals, it is a non-integer.

10.2.4 Box-counting dimension

The box-counting dimension is defined as

$$D_0 = \underset{\varepsilon \to 0}{\text{Limit}} \left\{ \frac{\ln(N(\varepsilon))}{\ln\left[\dfrac{1}{\varepsilon}\right]} \right\} \tag{10.5a}$$

where $N(\varepsilon)$ is the minimum number of boxes (cubes) of edge length ε needed to cover the space when it is divided into a grid of boxes of size ε. For a point $N(\varepsilon) = 1$ and therefore $D = 0$; for a limit cycle $N(\varepsilon) \approx \frac{l}{\varepsilon}$. In the case of a straight line subdivided into N self-similar segments of edge length ε, the box-counting dimension would be

$$D = \underset{\varepsilon \to 0}{\text{Limit}} \frac{\ln(N(\varepsilon))}{\ln\left(\dfrac{1}{\varepsilon}\right)} = \frac{\ln(N)}{\ln(N)} = 1 \tag{10.5b}$$

Here, the length of each self-similar segments, ε, is $\dfrac{1}{N}$. For a two-dimensional Euclidean space of a square divided into four self-similar smaller squares, $N = 4$, and $\varepsilon = 1/2$. Therefore, the box-counting dimension is

$$D = \underset{\varepsilon \to 0}{\text{Limit}} \frac{\ln(N(\varepsilon))}{\ln\left(\dfrac{1}{\varepsilon}\right)} = \frac{\ln(4)}{\ln(2)} = \frac{\ln(2^2)}{\ln(2)} = 2\frac{\ln(2)}{\ln(2)} = 2 \tag{10.5c}$$

For three-dimensional Euclidean space, division of each edge into three equal segments $\left(\varepsilon = \dfrac{1}{3}\right)$ will produce 27 self-similar cubes. Thus, the box-counting dimension would then be

$$D = \underset{\varepsilon \to 0}{\text{Limit}} \frac{\ln(N(\varepsilon))}{\ln\left(\dfrac{1}{\varepsilon}\right)} = \frac{\ln(27)}{\ln\left(\dfrac{1}{3}\right)} = \frac{\ln(3^3)}{\ln(3)} = 3\frac{\ln(3)}{\ln(3)} = 3 \tag{10.5d}$$

10.2.5 Similarity dimension

Fractals that show obvious self-similarity can be characterized by the similarity dimension that is the same as other measures of fractal dimension. It is defined as

$$D = \underset{r \to 0}{\text{Limit}} \frac{\ln(N(r))}{\ln\left(\dfrac{1}{r}\right)} \tag{10.5e}$$

where $N(r)$ is the number of self-similar parts of the whole and r is the ratio of similarity.

10.2.6 Packing dimension

Mathematically, the packing dimension is another concept that can be used to define the dimension of a subset of a metric space. In some sense, it is the dual of the Hausdorff dimension since the packing dimension is constructed by packing small open balls inside the given subset, whereas the Hausdorff dimension is constructed by covering the given subset by such small open balls.

10.2.7 Information dimension

Information dimension is defined as follows:

$$D_{\text{information}} = \underset{\varepsilon \to 0}{\text{Limit}} \left[\frac{\displaystyle\sum_{i=1}^{N} p_i(\varepsilon)\ln[p_i(\varepsilon)]}{\ln(\varepsilon)} \right] \tag{10.6}$$

where the numerator is the information function, $p_i(\varepsilon)$ is the natural measure, or the probability that the element i is populated, and that $\sum_i p_i(\varepsilon) = 1$. If every element has equal probability of being visited, then

$$\sum_{i=1}^{N} p_i(\varepsilon) = N[p_i(\varepsilon)] = 1 \Rightarrow p_i(\varepsilon) = \frac{1}{N}$$

and

$$D_{information} = \underset{\varepsilon \to 0}{\text{Limit}} \left[\frac{\sum_{i=1}^{N} \frac{1}{N} \ln\left(\frac{1}{N}\right)}{\ln(\varepsilon)} \right]$$

$$= \underset{\varepsilon \to 0}{\text{Limit}} \left[\frac{\ln(N^{-1})}{\ln(\varepsilon)} \right] = \underset{\varepsilon \to 0}{\text{Limit}} - \frac{\ln(N)}{\ln(\varepsilon)} = \underset{\varepsilon \to 0}{\text{Limit}} \frac{\ln(N)}{\ln\left(\frac{1}{\varepsilon}\right)} = D_{capacity}$$

(10.7)

10.2.8 Capacity dimension

Capacity dimension is defined as

$$D_{capacity} = \underset{\varepsilon \to 0}{\text{Limit}} \left[\frac{\ln(N(\varepsilon))}{\ln\left(\frac{1}{\varepsilon}\right)} \right]$$

(10.8)

where $N(\varepsilon)$ is the minimum number of open sets of diameter ε needed to cover a set. If each element of fractal has equal probability of being visited, then

$$D_{capacity} = D_{information}$$

(10.9)

10.2.9 Rényi dimension

The Rényi dimension is defined as

$$D_q = \underset{\varepsilon \to 0}{\text{Limit}} \left[\frac{\frac{1}{1-q} \ln\left(\sum_i p_i^q\right)}{\ln\left(\frac{1}{\varepsilon}\right)} \right]$$

(10.10)

where the numerator is the Rényi entropy of order q (Rényi, 1970) and p_i is the fraction of the measure contained in the partition $P_i(\varepsilon)$ of side length $\leq \varepsilon$. Rényi dimension considers the frequency with which the boxes are visited by weighting them according to their natural measure. The strength of this weighting is given by an exponent q. When $q = 0$, Rényi dimension can be numerically calculated in the same way as for the box-counting dimension. Rényi dimension is a non-increasing function of q, that is, $D_{q_1} \leq D_{q_2}$, if $q_1 > q_2$.

Rényi dimension is a general fractal dimension. For example, when $q = 1$, D_q is equal to the information dimension; when $q = 2$, D_q is equal to the correlation dimension. An attractor whose Rényi dimensions are not equal is said to be multifractal. In a multifractal set, the fractal dimension may vary across the attractor giving rise to a multifractal spectrum $f(\alpha)$ where α is the individual dimension and $f(\alpha)$ is their relative weight distribution. They occur in nature particularly in geophysics such as the earth's topography where each location has an altitude. Topographical contours define different fractals with different fractal dimensions. Other examples of multifractals include fully developed turbulence and heartbeat dynamics.

10.2.10 Correlation dimension

Grassberger and Procaccia (1983a,b) defined correlation dimension with respect to the correlation integral, or correlation sum in discrete form, which is

$$C_m(r) = \frac{1}{N(N-1)} \sum_{i,j=1;i\neq j}^{\infty} H\left(r - \|x_i - x_j\|\right); \quad i \neq j \tag{10.11}$$

where m is the embedding dimension, N is the number of data points, r is the radius of sphere centred on either of the point $\{x_i\}$ or $\{x_j\}$, and H is the Heaviside step function with $H(u) = 1$ for $u > 0$, and $H(u) = 0$ for $u \leq 0$. The norm $\|x_i - x_j\|$ may be any of the three usual norms, the maximum norm, the diamond norm, or the Euclidean norm. Correlation functions are calculated for a series of embedding dimensions. Euclidean norms are calculated as

$$r_{ij} = \|x_i - x_j\| = \sqrt{(x_i - x_j)^2} \text{ for one-dimensional phase space} \tag{10.12a}$$

$$r_{ij} = \sqrt{(x_i - x_j)^2 + (y_i - y_j)^2} \text{ for two-dimensional phase space, and} \tag{10.12b}$$

$$r_{ij} = \sqrt{(x_i - x_j)^2 + (y_i - y_j)^2 + (z_i - z_j)^2} \text{ for three-dimensional phase space} \tag{10.12c}$$

Equation 10.11 is equivalent to counting the number of points inside a circle of radius r drawn around a point and repeating the same procedure for different values of r and different points in the phase space. In higher dimensions, the circles become hyperspheres.

If an attractor for the system exists, then for small r, it can be shown that

$$C(r) \cong r^d \tag{10.13}$$

where d is the correlation exponent. It may be estimated by the slope of a straight line in the plot of log $(C(r))$ versus log (r) for each value of the embedding dimension m.

For random processes, d varies linearly with increasing m without reaching a saturation value, whereas for deterministic processes, the value of d levels off after a certain m. The saturation value of d is defined as the correlation dimension D_2 of the attractor or a time series. It can also be defined as

$$D_2 = \underset{r \to 0}{\text{Limit}} \underset{N \to \infty}{\text{Limit}} \frac{d\ln(C(r))}{d\ln(r)} \tag{10.14}$$

The nearest integer above the saturation value of d provides the minimum number of embedding dimensions of the phase space necessary to model the dynamics of the attractor.

Although it is easy to implement the algorithm of Grassberger and Proccacia (1983a,b) for noise-free data, it is not the case for noisy data sets. Noise blurs the lower region of the length and causes deviation of the trajectory in all lengths, increasing the interpoint distance in the plot of correlation integral versus length. The usual procedure for reconstructing the phase space also would not work in the presence of noise.

Other methods of determining the correlation dimension include that proposed by Theiler (1987) in which the whole attractor's extent is divided into several boxes (grids) and all points are assigned to the boxes, and the method proposed by Schouten et al. (1994) built into the software RRCHAOS (Schouten and van den Bleek, 1994) in which the increase in interpoint distance is assumed to be due to noise.

It should be noted that in the plots of D_2 versus log r, there are large statistical errors for small and large values of r. In between, however, there is a region in which D_2 remains reasonably constant. This region is called the scaling region. In general, $D_{\text{correlation}} \leq D_{\text{information}} \leq D_{\text{capacity}}$.

In all dimension definitions, it is implied that irregularities of size less than ε are ignored. The measurement at scale ε is common to all dimensions. The power law in which the exponent D can be a non-integer holds for most fractals, that is, $N(\varepsilon) \alpha \dfrac{1}{\varepsilon^D}$ where $N(\varepsilon)$ is the minimum number of self-similar objects needed to cover the original space.

10.3 EXAMPLES OF SOME WELL-KNOWN FRACTALS

10.3.1 Cantor set

The Cantor set (Figure 10.1) can be generated by repeated removal of the middle third of a straight line infinitely. For a line that originally has a length of 1, that is, end points (0, 1), the first removal of the middle third will leave two lines each of length $\dfrac{1}{3}$. After the second removal, there will be four lines each of length $\left(\dfrac{1}{3}\right)\left(\dfrac{1}{3}\right)$, and after the third removal, eight lines each of length $\left(\dfrac{1}{3}\right)\left(\dfrac{1}{3}\right)\left(\dfrac{1}{3}\right)$, and so on. After the first removal, the end points 0 and 1 will remain for all subsequent removals. After the second removal, the end points $\dfrac{1}{3}$ and $\dfrac{2}{3}$ will also remain; after the third removal, the end points $\dfrac{1}{9}, \dfrac{2}{9}, \dfrac{7}{9}$, and $\dfrac{8}{9}$ will

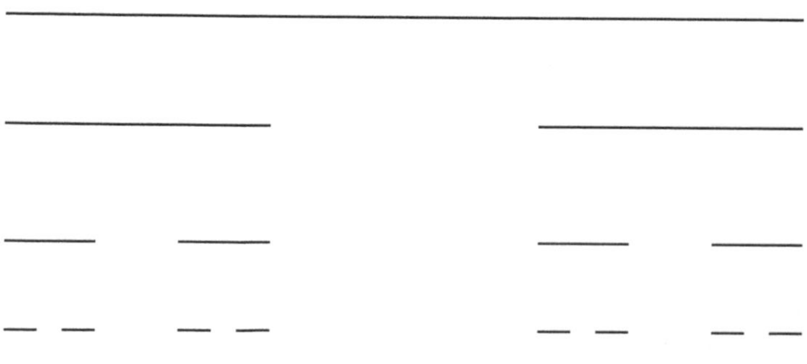

Figure 10.1 Cantor set.

also remain. As the number of removals tends to infinity, the points remaining will be $0, 1, \dfrac{1}{3}, \dfrac{2}{3}, \dfrac{1}{9}, \dfrac{2}{9}, \dfrac{7}{9}, \dfrac{8}{9},\dots$, which is uncountable.[2]

The total length removed is

$$\sum_{n=0}^{\infty} \frac{2^n}{3^{n+1}} = \frac{1}{3}+\left(\frac{1}{3}\right)\left(\frac{2}{3}\right)+\left(\frac{1}{3}\right)\left(\frac{2}{3}\right)^2+\left(\frac{1}{3}\right)\left(\frac{2}{3}\right)^3+\dots = \left(\frac{1}{3}\right)\left(\frac{1}{1-\dfrac{2}{3}}\right)=1 \tag{10.15}$$

Therefore, the remaining length should be zero! This may appear not surprising because the sum of the lengths of the removed intervals is equal to the original interval. However, a closer look at the process of removal of middle third reveals that there must be something remaining because the end points are not removed. For example, removing the segment $\left(\dfrac{1}{3}, \dfrac{2}{3}\right)$ from the original interval (0, 1) does not remove the points $\dfrac{1}{3}$ and $\dfrac{2}{3}$. Subsequent removals also do not remove other end points. It can also be seen that the points $\dfrac{1}{4}$ and $\dfrac{3}{10}$ are never in the middle third or as end points and thus never removed. The result is that the Cantor set is not empty but contains an infinite number of points, which is also sometimes referred to as a Cantor dust.

In the case of the Cantor set, the original line is divided into three equal segments. After removing the middle third, it is replaced by two segments each having lengths equal to that of the removed segment. Thus, three connecting segments of length $\dfrac{1}{3}$ becomes four connecting segments of length $\left(\dfrac{1}{3}\right)^2$. Therefore, the box-counting dimension becomes

$$D = \underset{\varepsilon \to 0}{\text{Limit}}\, \frac{\ln(N(\varepsilon))}{\ln\left(\dfrac{1}{\varepsilon}\right)} = \frac{\ln(2)}{\ln(3)} = 0.6309 \tag{10.16a}$$

In this case ε, the length of the self-similar segment is $\dfrac{1}{3}$.

Alternatively, since there are two identical straight lines each of which requires a magnification of three to become identical to the original line, the Hausdorff–Besicovitch dimension (see Section 10.2.3 for the definition) of the Cantor set is

$$D_H = \frac{\log(2)}{\log(3)} = \frac{\log(2^2)}{\log(3^2)} = \frac{2\log(2)}{2\log(3)} = \frac{\log(2)}{\log(3)} = 0.6309 \tag{10.16b}$$

By extending the same definition to 2^N self-similar objects with magnification factor 3^N, the same result can be obtained; that is,

$$D_H = \frac{\log(2^N)}{\log(3^N)} = \frac{N\log(2)}{N\log(3)} = \frac{\log(2)}{\log(3)} = 0.6309 \tag{10.16c}$$

The significance of the Cantor set in chaos is that certain functions when iterated many times lead to objects that look like a Cantor set.

---

[2] An uncountable set has its cardinal number larger than that of the set of all natural numbers. Cardinal numbers measure the size of sets. Infinite sets can have different cardinal numbers. For example, the set of all real numbers in the interval (0, 1) is uncountable, whereas the set of all integers is countable.

10.3.2 Sierpinski (gasket) triangle

The Sierpinski triangle (Figure 10.2), named after the Polish mathematician Waclaw Sierpinski (1915), can be constructed in a two-dimensional plane starting with an equilateral triangle with the base parallel to the horizontal axis, and successively removing the middle fourth of the triangle, thus leaving a triangular hole inside the original triangle. This leaves behind three equilateral triangles in each corner of the original triangle after each iteration. If the area of the original triangle is A, then the area after the first iteration is $\left(\dfrac{3}{4}\right)A$; after the second iteration, $\left(\dfrac{3}{4}\right)\left(\dfrac{3}{4}\right)A$; after the third iteration, $\left(\dfrac{3}{4}\right)\left(\dfrac{3}{4}\right)\left(\dfrac{3}{4}\right)A$, and so on. After the n-th iteration, the area would be

$$\text{Area} = \left(\frac{3}{4}\right)^{n} A \tag{10.17}$$

Table 10.2 illustrates that the final area tends to zero as n tends to infinity.

Since there are three identical triangles each of which require a magnification of 2 to become identical to the whole, the Hausdorff–Besicovitch dimension (same as the fractal dimension) of the Sierpinski triangle using Equation 10.4b is

$$D_{H} = \frac{\log(3)}{\log(2)} \approx 1.585 \tag{10.18a}$$

Figure 10.2 Sierpinski triangle. (From http://en.wikipedia.org/wiki/Sierpinski_triangle.)

Table 10.2 Sierpinski triangle area reduction

Iteration number	Area
0	$1 \times A$
1	$0.75 \times A$
2	$0.5625 \times A$
3	$0.42187 \times A$
10	$0.05631 \times A$
100	$3.207 \times 10^{-13} \times A$
200	$1.0286 \times 10^{-25} \times A$
500	$3.3934 \times 10^{-63} \times A$
1000	≈ 0

The Sierpinski triangle when subdivided into nine self-similar pieces with magnification factor 4, Equation 10.4a gives

$$D_H = \frac{\log(9)}{\log(4)} = \frac{\log(3^2)}{\log(2^2)} = \frac{2\log(3)}{2\log(2)} = \frac{\log(3)}{\log(2)} \approx 1.585 \tag{10.18b}$$

By extending the same definition to 3^N self-similar objects with magnification factor 2^N, the same result can be obtained; that is,

$$D_H = \frac{\log(3^N)}{\log(2^N)} = \frac{N\log(3)}{N\log(2)} = \frac{\log(3)}{\log(2)} \approx 1.585 \tag{10.18c}$$

Sierpinski triangle can also be constructed using an iterated function system. For example, starting with a point at the origin $(x_0 = 0; y_0 = 0)$, the successive points are generated by randomly iterating using a coordinate transformation such as

$$x_{n+1} = 0.5x_n; \quad y_{n+1} = 0.5y_n \tag{10.19}$$

$$x_{n+1} = 0.5x_n + 0.25; \quad y_{n+1} = 0.5y_n + 0.5\frac{\sqrt{3}}{2} \tag{10.20}$$

$$x_{n+1} = 0.5x_n + 0.5; \quad y_{n+1} = 0.5y_n \tag{10.21}$$

This is also called the 'chaos game'. The Sierpinski carpet is a similar set that can be constructed using the same procedure but starting with a square. At each iteration, each remaining square is divided into nine smaller squares. The middle one is discarded and the remaining eight are retained. The process when repeated a large number of times leads to a 'carpet'-like design that has a Hausdorff–Besicovitch dimension of

$$D_H = \frac{\log(8)}{\log(3)} = \frac{\log(8^2)}{\log(3^2)} = \frac{\log(8^3)}{\log(3^3)} = \cdots = \frac{\log(8^N)}{\log(3^N)} = \frac{\log(8)}{\log(3)} \approx 1.8928 \tag{10.22}$$

The three-dimensional equivalent of the Sierpinski carpet is the Menger sponge, which starts with a cube. The construction starts with a unit cube from which seven smaller cubes of length $\frac{1}{3}$ are removed. The remaining part consists of eight corner cubes joined by 12 bridging cubes. The process is repeated until a sponge-like structure of volume approaching zero remains. The Hausdorff–Besicovitch dimension of the Menger sponge is

$$D_H = \frac{\log(20)}{\log(3)} \approx 2.7268 \tag{10.23}$$

The Sierpinski triangle and Menger sponge have been used (Wong, 1988; Frieson and Mikula, 1987) in the description of the fractal geometry of porous media.

10.3.3 Koch curve

The Koch curve, named after the Swedish mathematician Helge von Koch (1904), is generated by starting with a straight line and dividing it into three equal segments (Figure 10.3). The middle third is then replaced by two sides of an equilateral triangle pointing outwards and with the unconnected base coinciding with the removed part. This procedure is repeated infinitely and the result is a continuous curve that is not differentiable anywhere. After each iteration, the length of the segment decreases by a third and the number of segments increases by four, thereby increasing the length by $\frac{4}{3}$. As the procedure is repeated, the length of the curve increases progressively, and after 'n' iterations, the total length of the curve would be $\left(\frac{4}{3}\right)^n$, which tends to infinity as n tends to infinity. However, the area under the curve remains finite at 1.8 if the area after the first iteration is taken to be unity. As there are four identical segments after each iteration, which require a magnification of 3 to become identical to the original, the Hausdorff–Besicovitch dimension of the Koch curve is

$$D_H = \frac{\log(4)}{\log(3)} = 1.26185 \tag{10.24}$$

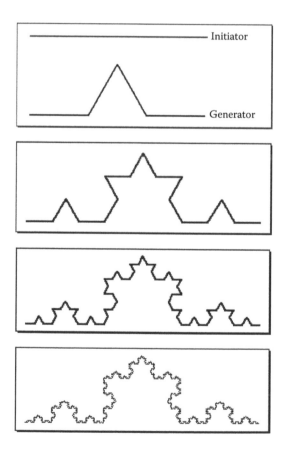

Figure 10.3 Koch curve.

10.3.4 Koch snowflake (or Koch star)

The Koch snowflake fractal is very similar to the Koch curve (Figure 10.4), except that it starts with an equilateral triangle instead of a straight line. It can also be generated by placing three copies of the Koch curve on the three sides of an equilateral triangle. As in the Koch curve, it can be shown that the length of the snowflake tends to infinity as the number of iterations tends to infinity, while the area enclosed remains finite at $1.6 = \left(\dfrac{8}{5}\right)$ times the area of the original triangle. The Hausdorff–Besicovitch dimension of the Koch snowflake is the same as that for the Koch curve at 1.26185.

10.3.5 Mandelbrot set

The Mandelbrot set (Figure 10.5), named after the French and American mathematician Benoit Mandelbrot (1983), is a quasi-self-similar set with the small-scale details not identical to the whole. It is a set of points in the complex plane the boundary of which is fractal, and is infinitely complex but generated using a simple quadratic equation involving complex numbers. It is one of the most widely known fractals, and is generated by the equation

$$z_{n+1} = f(z) = z_n^2 + c \tag{10.25}$$

where z is a complex number (having a real part and an imaginary part; $z = x + iy$; $i^2 = -1$) and c is a point on the plane (also a complex number). Depending on the value of c, the

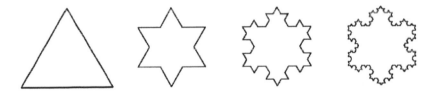

Figure 10.4 Koch snowflake. (From http://mathworld.wolfram.com/KochSnowflake.html.)

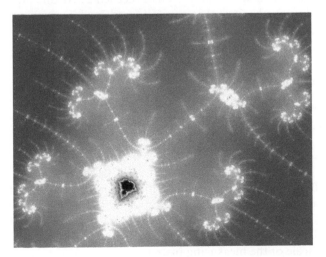

Figure 10.5 Mandelbrot set. (From http://www.olympus.net/personal/dewey/mandelbrot.html.)

numbers generated by the above equation will either be inside or outside of the Mandelbrot set. Starting with $z = 0$, and as the iteration continues, the magnitude of z will change for any value of c. The possible outcomes of this iteration are that the magnitude of z will either be less than 2 indefinitely, or that after a certain number of iterations, it will exceed 2 and continue to increase indefinitely. The former case where the points lie within 2 belongs to the Mandelbrot set, whereas in the latter case the points lie outside of the Mandelbrot set. The plot of a large number of points inside the Mandelbrot set leads to the well-known beautiful fractal that when blown up (increasing the number of iterations) will uncover unseen details. It is easy to see that values of z for $c = 0, -1, -2, i$ (i is the imaginary part of the complex number) lie within the Mandelbrot set (bounded), whereas $c = 1, 3, 2i, 1 + i$ are outside the Mandelbrot set (unbounded and tending to infinity). The area within the Mandelbrot set has been estimated to be 1.7274 using 240,000 terms of the Laurent series, which converges very slowly (Ewing and Schober, 1992). Slightly varying estimates have also been obtained by other researchers. Adding colours to points that are not inside the set according to how many iterations were needed before z surpassed 2 enhances the image aesthetically and also helps highlight parts of the set that are too small to see in the graph as shown in Figure 10.5. For example, if a point $(0.5, 0.5i)$ has to be coloured, iterations start with $c = (0.5, 0.5i)$ and $z_0 = 0$, and proceed as follows:

$$z_1 = z_0^2 + (0.5, 0.5i) = 0.5 + 0.5i = \sqrt{0.5} = 0.707 \tag{10.26a}$$

$$z_2 = z_1^2 + (0.5, 0.5i) = 0.5 + 1i = 1.118 \tag{10.26b}$$

$$z_3 = z_2^2 + (0.5, 0.5i) = -0.25 + 1.5i = 1.5207 \tag{10.26c}$$

$$z_4 = z_3^2 + (0.5, 0.5i) = -1.6875 - 0.25i = 1.7058 \tag{10.26d}$$

$$z_5 = z_4^2 + (0.5, 0.5i) = 3.2851 - 1.34375i = 3.549 > 2 \tag{10.26e}$$

At the fifth iteration, the magnitude of z has exceeded 2. At this point, the pixel at $(0.5, 0.5i)$ is colour coded on the basis of the required number of iterations (5). The procedure is repeated for all points (pixels).

The iterated function $f(z)$ can take other non-linear forms, such as

$$
\begin{aligned}
z_{n+1} &= c\sin(z_n) \\
z_{n+1} &= c\exp(z_n) \\
z_{n+1} &= ci\cos(z_n) \\
z_{n+1} &= cz_n(1 - z_n)
\end{aligned} \tag{10.27}
$$

The Hausdorff–Besicovitch dimension of the boundary of Mandelbrot set has been mathematically proven to be the integer 2 (Shishikura, 1998), which is unusual for a fractal. Mandelbrot (1967) is also known for his paper entitled 'How long is the coast of Britain? Statistical self-similarity and fractional dimension', which showed that the length of a coastline varies with the scale of the measuring instrument, has self-similarity at all scales, and is infinite in length for an infinitesimally small measuring device.

Figure 10.6 Julia set. (From http://en.wikipedia.org/wiki/Julia_set.)

10.3.6 Julia set

The Julia set (Figure 10.6) and the Fatou set named, respectively, after the French mathematicians Gaston Julia (1918) and Pierre Fatou (1917), are two complementary sets, the former chaotic and the latter regular, defined from a function. For generating the Julia set using the same function as for the Mandelbrot set, iterations are done by changing z_0 for different values of c resulting in an infinite number of Julia sets (there is only one Mandelbrot set for a given function). For example, starting with $z_0 = (-1, 0.5i)$, and a value of $c = (1, 0.25i)$, the iterations proceed as follows:

$$z_0 = -1 + 0.5i \tag{10.28a}$$

$$z_1 = z_0^2 + (1, 0.25i) = 2.25 - 0.75i = 2.371 > 2 \tag{10.28b}$$

In this case, the magnitude of z has exceeded 2 in the first iteration. At this point, the pixel at $(-1, 0.5i)$ is colour coded on the basis of the required number of iterations (1). The procedure is repeated for all points (pixels).

As a consequence of the definition of the Mandelbrot set, there is a close correspondence between the geometry of the Mandelbrot set at a given point and the structure of the corresponding Julia set. The Mandelbrot set can be thought of as an index for the Julia sets. This principle is exploited in virtually all results on the Mandelbrot set.

10.4 PERIMETER–AREA RELATIONSHIP OF FRACTALS

The perimeter–area (P–A) relationship for a Euclidean shape is

$$P \alpha A^{1/2} \tag{10.29}$$

For irregular shapes, the corresponding relationship is (Mandelbrot, 1982)

$$(P(\varepsilon))^{1/D} \alpha (A(\varepsilon))^{1/2} \tag{10.30}$$

where $P(\varepsilon)$ is the perimeter of the irregular shape measured with a measuring stick of length ε, and $A(\varepsilon)$ is the area measured in units of ε^2. This is the basic perimeter–area relationship for fractal sets.

The ratio of the perimeter to the area, defined as

$$\alpha_D(\varepsilon) = \frac{(P(\varepsilon))^{1/D}}{(A(\varepsilon))^{1/2}} \tag{10.31}$$

depends only on the scale of measurement ε and independent of the shape and size of the fractal set; in other words, scale independent.

Many irregular sets such as clouds and rain patches satisfy this relationship. The fractal dimension is an important parameter that characterizes the process of producing clouds and rains. A study by Lovejoy (1982) has shown that the radar pictures of tropical Atlantic rain areas at a resolution of 1 km × 1 km and the infrared pictures of the Indian Ocean clouds measured by the Geostationary Operational Environmental Satellite (GOES) sampled at a resolution of 4.8 km × 4.8 km follow the above perimeter–area relationship. The study shows that the area $A(\varepsilon)$, which covered a range of 1 km² to 10^6 km², and the perimeter $P(\varepsilon)$ plotted as two parallel straight lines. The separation of the two straight lines can be attributed to differences in the spatial resolution (satellite data resolution is coarser than radar data resolution). The radar data shows a longer perimeter than that by satellite. By multiplying the satellite-measured perimeters by a correction factor, the two parallel straight lines could be made to coincide and yield a least squares estimate of fractal dimension of 1.35 with a correlation coefficient of 0.994 (Lovejoy [1982], as cited in Xie [1993]). This result clearly illustrates the scale independence.

The perimeter–area relationship can also be extended to rivers. Hack (1957) has obtained the following perimeter–area relationship for rivers in Virginia and Maryland in the United States:

$$P = 1.4\sqrt{A^D} \tag{10.32}$$

where P is measured along the longest stream, A is measured from topographical maps, and $D = 1.2$. Other researchers (e.g., Rosso et al., 1991) have found slightly different values for D ranging from 1.036 to 1.29. Equation 10.32 establishes a link between fractals and Horton's law of stream numbers. It can be shown (Xie, 1993, p. 46) that

$$P = P_0 \left(\frac{A}{A_0} \right)^{\frac{D_u}{2}} \tag{10.33}$$

where

$$D_u = \frac{\ln(r_b)}{\ln(r_L)} \tag{10.34}$$

and r_b and r_L, respectively, are the bifurcation and length ratios of streams defined as

$$r_b = \frac{N_i}{N_{i+1}} \tag{10.35a}$$

$$r_L = \frac{P_i}{P_{i-1}} \tag{10.35b}$$

In Equations 10.35a, and 10.35b, N_i is the number of streams of order i, and N_{i+1} is the number of streams of order $i + 1$, P_i is the average length of streams of order i, P_{i-1} is the average length of streams of order $i - 1$, and P_0, A_0 are the average length and area of the smallest streams. The bifurcation and length ratios r_b and r_L can respectively be obtained as the gradients of the plots of N_i versus N_{i+1} and P_i versus P_{i-1}. Subscript u indicates the order of the streams. However, it is to be noted that rivers and streams are not exactly self-similar.

10.5 CHAOS

'Chaos' literally means 'disorder'. However, mathematical chaos is deterministic and thus 'orderly' in some sense, and has well-defined statistics. In this context, chaos is synonymous with dynamical instability, a condition discovered by physicist Henri Poincaré (1854–1912) in the early 20th century that refers to an inherent lack of predictability in some physical systems. In mathematics and physics, the chaos theory describes the behaviour of certain non-linear dynamical systems that may exhibit dynamics that are highly sensitive to initial conditions (popularly known as the 'butterfly effect'). As a result of this sensitivity, which manifests itself as an exponential growth of perturbations in the initial conditions, the behaviour of chaotic systems appears to be random. This happens even though the systems are deterministic, meaning that their future dynamics are fully defined by their initial conditions, with no random elements involved. Such systems are governed by physical laws and are very difficult to predict accurately. Chaos appears to show randomness and lack of order as results of the complex systems and interactions among systems. Weather forecasting, which is predictable in the short-term if enough information is available but unpredictable with certainty in the long term, is a commonly cited example. The principal characteristics of deterministic chaotic systems include the difficulty or impossibility to predict the long-term behaviour; sensitivity to initial conditions, meaning that two trajectories starting from very close initial conditions very rapidly moving to divergent states; topological mixing, meaning that a system will evolve over time so that any given region of its phase space will eventually overlap with any other given region; dense periodic orbits, meaning that every point in the space is approached arbitrarily closely by periodic orbits; and the exponential amplification of errors. Topologically mixing systems failing the dense periodic orbit condition may not display sensitivity to initial conditions, and hence may not be chaotic.

Newtonian laws of physics, which are based on complete determinism, assume that precise measurements are possible, and that the more precise the measurements are the more precise the predictions of future conditions, at least in theory. Thus, according to this assumption, almost perfect predictions of a physical system are possible if sufficient precise measurements are available, meaning that two or more identical sets of initial conditions would yield identical results. In the case of chaotic deterministic systems, Poincaré proved mathematically that even if the initial conditions are measured to the highest level of accuracy, the prediction uncertainties remain large.

In continuous systems, linear and two-dimensional (plane) space cannot exhibit chaos. Only three-dimensional, or non-Euclidean geometry, can exhibit chaotic behaviour. In discrete dynamical systems, even one- or two-dimensional phase space can show chaos.

10.5.1 Butterfly effect

The term 'butterfly effect' became known after Edward Lorenz (1972) gave a paper to the American Association for the Advancement of Science in Washington, DC, entitled 'Predictability: Does the flap of a butterfly's wings in Brazil set off a tornado in Texas?'. The message conveyed in this classic paper is the sensitivity to initial conditions and the long-term unpredictability. The flapping of wings represents a small change in the initial conditions that may grow exponentially leading to a large-scale phenomenon. In the absence of the flapping, the system may have ended in a completely different state. This concept is most relevant to weather forecasting, which has its limits of predictability. Lorenz, while studying numerical weather prediction starting with a set of 12 equations, proved with a computer of the time (1963), that a slight difference in the initial values (<0.1%) was sufficient to drive the system in a completely different direction. The story is that instead of using an initial value of 0.506127, which the computer gave at some point in the middle of computation, he recorded on paper a rounded-off value of 0.506 to save space, which he used to continue to run the model after a brief stop. For his work, Lorenz is credited with the honour of being cited as the 'father of chaos theory', although, in the late 19th century, the French mathematician Henri Poincaré showed that the three-body problem was complex and impossible to solve even though the governing equations seem simple. Poincaré is also credited with the same honour.

10.5.2 The *n*-body problem

The description of the motion of '*n*' bodies under the interaction of each other requires $6n$ variables: $3n$ space displacements and $3n$ velocities. This problem has attracted the attention of scientists for over a century, as it is important to the understanding of the motion of celestial bodies under the action of gravitational forces. However, there is no general solution to this problem at the present time. In recent years, more focussed attention has been given to the two-body and three-body problems in particular. Examples of the two-body problem include the orbiting of a planet around the sun and the orbiting of a satellite around a planet in celestial systems, and the orbiting of an electron around a nucleus in atomic physics. The two-body problem can be reformulated as 2 one-body problems, which in most cases can be solved exactly. The corresponding 2-body problems can also be solved using Newton's laws of motion. The basis for solving the two-body problem was laid by Johannes Kepler (1609) who proposed the laws of planetary motion in 1609. The earlier belief about celestial motion was based on the hypothesis conjectured by Claudius Ptolemy (circa 150 AD), which stated that Earth was the centre of the universe and all other celestial bodies, including the sun, orbited around it. This belief was superseded by the heliocentric theory proposed by Nicolaus Copernicus sometime between 1507 and 1515, which stated that there was no single centre of the universe and that the earth orbited around the sun in circles. Kepler's law, based on the analysis of empirical data, states that the orbits of the solar system are elliptical. Sir Isaac Newton in 1687 provided the missing principles through his three laws of motion and the inverse square law of gravitation. According to Newton's laws, the future state of any system is predictable if the initial state is known and if the laws can be translated to quantifiable mathematical equations. A problem that became a hurdle to achieving accurate predictions was the uncertainty associated with the measurement of the initial condition. In any real system, measurement to an infinite level of accuracy is impossible. It was then thought, and accepted, that future states can be predicted with the same level of uncertainty as that of the measurement of initial condition, until Henri Poincaré concluded that the state of any systems involving more than two interacting bodies cannot be predicted with any kind of accuracy under certain initial conditions. His findings, which contradict Newton's laws, were based on the dynamics of the three-body problem.

The three-body problem is quite different; it has no solution in the general case, although Newton's laws of gravity and motion assure predictability. Examples of the three-body problem include the motion of two planets around the sun, the motion of the sun-earth-moon system, and the motion of a satellite between the earth and the moon. An attempt to solve the three-body problem was made by Henri Poincaré who could not find a solution but made an important discovery of the existence of chaotic deterministic systems. The three-body problem involving the sun, earth, and moon are governed by deterministic laws, and given the initial conditions, their subsequent positions and velocities should be predictable. However, Poincaré found that the evolution of such a system is often chaotic, meaning that a small perturbation in the initial conditions would lead to results completely different from those of the unperturbed system. His conclusion was that if small perturbations in the initial conditions are not detectable by the measuring instrument, then the prediction of the final state is impossible. For his work, Poincaré was awarded a prestigious scientific prize by King Oscar II of Sweden in 1887. Poincaré is also recognized as the father of chaos theory.

10.6 SOME DEFINITIONS

10.6.1 Metric space

Metric space is a set where a distance (or metric) is defined. Euclidean metric defines the distance between two points as the length of the straight line connecting them. In a metric space,

$$d(x, y) \geq 0 \tag{10.36a}$$

$$d(x, y) = 0 \text{ if and only if } x = y \tag{10.36b}$$

$$d(x, y) = d(y, x) \tag{10.36c}$$

$$d(x, z) \leq d(x, y) + d(y, z) \tag{10.36d}$$

The distance function d is also called the metric. The Euclidean metric is defined as

$$d(x, y) = \|x - y\| = \sqrt{\sum_{i=1}^{n} (x_i - y_i)^2} \tag{10.37}$$

10.6.2 Manifold

A manifold is a mathematical space in which every point has a neighbourhood that resembles a Euclidean space. For example, in a one-dimensional manifold, every point has a neighbourhood that looks like a segment of a line. A two-dimensional manifold has a neighbourhood that looks like a disk. Manifolds need not be connected nor be closed nor be finite. Thus, segments of a line, a pair of circles, a parabola, and a hyperbola are all examples of manifolds. However, a pair of circles that touch each other is excluded.

10.6.3 Map

In physics and mathematics, a 'map' is any mathematical transformation that is applied over and over again in sequence. Sometimes, the term 'mapping' is used instead of 'map'. An example of a one-dimensional map is

$$x_{n+1} = x_n + 10 \tag{10.38}$$

with a given initial value for x_0.

A two-dimensional map that is uncoupled, meaning that the two variables are independent of each other, is of the form

$$x_{n+1} = x_n + 10 \tag{10.39a}$$

$$y_{n+1} = y_n + 5 \tag{10.39b}$$

On the other hand, a two-dimensional map that is coupled, meaning that one or both variables are dependent on each other, is of the form (with given initial values x_0 and y_0)

$$x_{n+1} = x_n + 10 \tag{10.40a}$$

$$y_{n+1} = y_n + x_n \tag{10.40b}$$

10.6.4 Attractor

Attractor is a subset of the phase space of a dynamical system. The region into which a large set of initial conditions converge is called an attractor. The points that are sufficiently close to the attractor will remain close even when slightly perturbed. Geometrically, an attractor can be a point, a curve, or a manifold. Fixed point and limit cycle are two simple attractors. A fixed point is one in which a system eventually evolves towards a point, such as a ball rolling inside a concave bowl. A limit cycle is a periodic orbit of an isolated system, such as the swing of a pendulum clock. Some dynamical systems are chaotic everywhere, while in many cases the chaotic behaviour is found only within the attractor.

10.6.4.1 Strange attractor

When the region is chaotic, it is called a strange attractor. It is also characterized by a non-integer dimension. The trajectory of a strange attractor cannot intersect with itself. Trajectories remain confined to a bounded region of the phase space and yet nearby trajectories diverge exponentially (butterfly effect!). Trajectories do not fill the phase space. This is possible due to repeated stretching and folding. (Imagine the making of Chinese noodle where the dough is flattened, stretched, and folded over and over again until the length becomes so large and the cross section becomes so small.) Strange attractors can occur in both continuous (e.g., the Lorenz map) and discrete systems (e.g., Hénon map). In continuous systems, strange attractors can occur only in three or more dimensions, whereas in discrete systems, they can occur in two dimensions and in some cases also in one dimension. Systems that lead to strange attractors never repeat (non-periodic).

10.6.5 Dynamical system

A dynamical system describes the evolution of a system from an initial state, according to a fixed rule. The fixed rule describes the time dependence of the position of a point in ambient space that may be multidimensional. If the fixed rule is deterministic, then the dynamical system is deterministic, whereas a system with possible outcomes described by a probability distribution is stochastic or random.

Dynamical systems are generally described by ordinary differential equations, but they may also be described by difference equations. Mathematically, they describe initial value problems. The future states of the system are determined by solving the governing equations iteratively, starting from an initial state. The locus of the change from the initial state to the final state is referred to as the trajectory or orbit of the system. A dynamical system is represented in an abstract phase space (or state space) whose coordinates describe its state at any instant of time.

The evolution of a dynamical system can be defined by an equation of the form

$$x_{n+1} = f(x_n) \tag{10.41}$$

10.6.6 Phase (or state) space

Phase space is an abstract multidimensional coordinate system in which every degree of freedom of the system is represented as an axis. A possible state of the system is represented by a point in the phase space. The phase space diagram may provide properties and details of the system that may not otherwise be seen.

10.7 INVARIANTS OF CHAOTIC SYSTEMS

10.7.1 Lyapunov exponent

The Lyapunov exponent, named after the Russian mathematician Aleksandr Mikhailovich Lyapunov (1857–1918), is an invariant whose sign indicates the existence or otherwise of chaos, and its magnitude measures the degree of chaoticity. A positive Lyapunov exponent in a dynamical system indicates that the system is chaotic, and the exponent itself gives a measure of the average rate at which the predictability is lost. An n-dimensional dynamical system has n Lyapunov exponents, orthogonal to one another. The global Lyapunov exponent gives the average exponential rate of divergence of two initially nearby orbits. The local (at x_0) and global Lyapunov exponents are respectively defined as

$$\lambda \approx \ln|f'(x_0)| \tag{10.42}$$

and

$$\lambda = \operatorname*{Limit}_{N \to \infty} \frac{1}{N} \sum_{n=0}^{N-1} \ln\left|f'(x_n)\right| \tag{10.43}$$

where $f'(x_n) = \dfrac{\mathrm{d}f}{\mathrm{d}x}$ and $x_{n+1} = f(x_n)$ is a one-dimensional map.

For establishing whether a system is chaotic or not, it is not necessary to calculate all Lyapunov exponents. Whether a system is chaotic or not is determined by the largest Lyapunov exponent. For determining the average predictability, it is necessary to calculate all positive Lyapunov exponents. It is relatively easy to calculate the Lyapunov exponents for low-dimensional systems; however, for high-dimensional systems, it is not so easy and is usually numerically computed. The numerical methods are also computationally intensive. To obtain reliable results, it is necessary to use double precision to minimize round-off errors. A general method that is applicable for any system in any dimension without having to calculate the matrix of partial derivatives has been described by Sprott (2003). A method claimed to be robust to changes in embedding dimension, size of data set, time delay, and noise level, and which has the added benefit of giving the correlation dimension, has been proposed by Rosenstein et al. (1992). Several algorithms are also available (e.g., Wolf et al., 1985).

The Lyapunov exponent can also be interpreted as the divergence of two initially nearby states. In quantitative terms, it may be expressed as

$$|\delta z(t)| \approx e^{\lambda t}|\delta z(0)| \tag{10.44}$$

where $\delta z(0)$ is the initial separation and $\delta z(t)$ is the corresponding separation after time t. The rate of separation may be different for different orientations of the initial separation vector, resulting in as many Lyapunov exponents as there are dimensions in the phase space.

10.7.2 Entropy of a dynamical system

There are several types of entropies defined in the literature. In the context of thermodynamics, entropy refers to the amount of 'disorder' in the system—the higher the entropy, the higher the amount of disorder. For a closed thermodynamic system, it is the measure of the amount of thermal energy not available to do mechanical work. In statistical mechanics, it refers to the amount of uncertainty in the system. In information theory, it is a measure of the uncertainty associated with a random variable. Shannon entropy (Shannon, 1948) refers to a measure of the average information content that is missing by not knowing the value of the random variable. In statistical thermodynamics, it measures the degree to which the probability of the system is spread out over all possible substates.

In a time series, entropy is an important entity as its inverse gives the time scale relevant for the predictability of the system. It also provides the topological information about the folding process. Several definitions of entropies can be found in the literature (e.g., Kantz and Schreiber, 2003), and some of them are briefly given here.

In information theory, the entropy of a random variable is defined as

$$K(Z) = -\sum_i p_i \ln p_i \tag{10.45}$$

where p_i is the probability of the random variable Z taking a specified value z_i (i.e., $\Pr(Z = z_i) = p_i$). In thermodynamics, a similar expression is

$$K(Z) = -k_B \sum_i p_i \ln p_i \tag{10.46}$$

where k_B is the Boltzmann constant.

The order-q Rényi entropy is defined as (Rényi, 1970)

$$K_q(P_\varepsilon) = \frac{1}{1-q} \ln \sum p_i^q \tag{10.47}$$

where p_i is the fraction of the measure contained in the partition $P_i(\varepsilon)$ of side length $\leq \varepsilon$. The Shannon entropy, evaluated by the l'Hospital rule for $q = 1$, is

$$K_1(P_\varepsilon) = -\sum_i p_i \ln p_i \tag{10.48}$$

Block entropies of block size n are defined as

$$K_q(n, P_\varepsilon) = \frac{1}{1-q} \ln \sum_{i_1, i_2, \ldots, i_m} p^q_{i_1, i_2, \ldots, i_n} \tag{10.49}$$

where $p_{i_1, i_2, \ldots, i_n}$ are the joint probabilities that at an arbitrary time n, the observable falls into the interval I_{i_1}, at time $n + 1$, to the interval I_{i_2}, and so on, where I_t is the interval. Then, the order-q entropies are

$$k_q = \sup_P \lim_{n \to \infty} \frac{1}{n} K_q(n, \varepsilon) = \sup_P \lim_{n \to \infty} (K_q(n+1, P_\varepsilon) - K_q(n, P_\varepsilon)) \tag{10.50}$$

where the supremum[3] $\sup_P$ indicates that one has to maximize over all possible partitions P_ε and usually implies the limit $\varepsilon \to 0$.

In Equation 10.50, k_1 (when $q = 1$) is the Kolmogorov–Sinai entropy (after Kolmogorov, 1958; Sinai, 1959), or K–S entropy, and k_0 (when $q = 0$) is the topological entropy. K–S entropy is also sometimes called the metric entropy, or simply the Kolmogorov entropy. K–S entropy has a numerical value of zero for non-chaotic systems and positive values for chaotic systems. It can also be thought of as the additional information obtained by observing the state of the system at a certain time given *a priori* knowledge of the entire past. Numerically, entropies are computed for a finite order, q, which in the limit as $q \to \infty$ converges to the K–S entropy. However, there are problems in estimating K–S entropy by using Equation 10.50, as it would require box counting, which would be difficult in a high-dimensional phase space. There are also difficulties arising from the limitations of the data set as it is non-trivial to obtain the limit process as $\varepsilon \to 0$.

10.7.2.1 Kolmogorov–Sinai (K–S) entropy

The sum of positive Lyapunov exponents is known as the Kolmogorov–Sinai (K–S) entropy, which is also another invariant in a dynamical system. It is also considered as a measure of the average rate at which predictability is lost. Therefore, its inverse can be thought of as a time for which a reasonable prediction can be expected. A pure random process has infinite entropy, whereas a periodic process has zero entropy. The K–S entropy is difficult to calculate, and hence the correlation entropy, which is a reasonable lower bound to the K–S entropy in the same way as correlation dimension is to capacity dimension. Correlation entropy is defined as

$$K_2 = \operatorname*{Limit}_{r \to 0} \operatorname*{Limit}_{m \to \infty} \operatorname*{Limit}_{N \to \infty} \log \frac{C(m,r)}{C(m+1,r)} \tag{10.51}$$

where $C(m,r)$ is the correlation sum for an embedding dimension of m. For practicable limits in this equation, a plateau in the plot of $\log \frac{C(m,r)}{C(m+1,r)}$ versus $\log r$ is sought.

For a time series of N values, approximately N^2 calculations are required for estimating the correlation dimension. Care should therefore be taken to ensure the algorithm is written in the most efficient way. There are various claims to the minimum number of points required for the estimation of invariants. In general, it is of the order of 10^D where D is the

---

[3] Supremum, or least upper bound, is the smallest real number greater than or equal to every number in the set, for example, Sup{1,2,3} = 3.

dimension of the attractor. Other values quoted in the literature include 42^m (Smith, 1988) where m is the embedding dimension, $10^{D/2}$ (Ding et al., 1993), $10^{2+0.4D}$ (Tsonis, 1992). These conditions are necessary but may not be sufficient.

10.7.2.2 Modified correlation entropy

Estimating the correlation entropy accurately is important for the simulation and forecasting of a chaotic time series. However, this is made difficult by the presence of noise that blurs the signal of the time series, thereby introducing errors in the calculation of the correlation sum. The presence of additive noise strongly affects the evolution of the correlation sum, which in turn affects the estimation of the correlation entropy. Methods to calculate various entropy measures for a chaotic time series have been proposed by several authors (e.g., Grassberger and Procaccia, 1983c; Kantz and Schürmann, 1996; Schürmann and Grassberger, 2002; Bonachela et al., 2008). Most of the methods use the correlation integral equations that depend non-linearly on the K–S entropy. Hence, it is difficult to estimate the K–S entropy by these correlation integral equations.

To overcome this difficulty, a new quantity, referred to as the modified correlation entropy as an approximation of the K–S entropy, has been introduced by Jayawardena et al. (2010). For noise-free data, the modified correlation entropy is equivalent to the general correlation entropy. For noisy chaotic data, the modified correlation entropy is closer to the K–S entropy than the general correlation entropy. The essence of the method, which employs the correlation integral equation obtained by Diks (1999, p. 123), and Oltmans and Verheijen (1997) for a chaotic time series with additive Gaussian noise is described below.

For a noise-free chaotic time series, Frank et al. (1993) proved that the correlation sum behaves as

$$C_m(r) = \phi[\exp(-m\tau K)]\left(\frac{r}{\sqrt{m}}\right)^D \text{ for } r \to 0, \ m \to \infty \tag{10.52}$$

where ϕ is a constant, D is the correlation dimension, K is the K–S entropy (order 2 entropy), m is the embedding dimension, and τ is the time delay.

By comparing the correlation sum (Equation 10.52) for two different values of the embedding dimension, m and $m + 2$, we have when $r \approx 0$

$$\frac{C_m(r)}{C_{m+2}(r)} = \exp(2\tau K)\left(1 + \frac{2}{m}\right)^{D/2} \tag{10.53}$$

which can be written as

$$\ln\frac{C_m(r)}{C_{m+2}(r)} = 2\tau K + \ln\left(1 + \frac{2}{m}\right)^{D/2} \tag{10.54}$$

The correlation entropy K_2 as defined by Diks (1999, p. 111) is

$$K_2 = \frac{1}{2\tau}\ln\left[\frac{C_m(r)}{C_{m+2}(r)}\right] \tag{10.55}$$

When the time series is noise free, Equation 10.53 holds, and Equations 10.54 and 10.55 then simplify to a relationship between the correlation entropy K_2 and the K–S entropy K as

$$K_2 = K + \frac{1}{2\tau}\ln\left[\left(1 + \frac{2}{m}\right)^{D/2}\right] \tag{10.56}$$

which converges to K as $m \to \infty$; that is,

$$\lim_{m \to \infty} K_2 = K \tag{10.57}$$

The correlation sum for a noisy chaotic time series, as given by Diks (1999, p. 123), and Oltmans and Verheijen (1997) is

$$C_m(r) = \frac{\phi e^{-m\tau K} m^{-D/2} \sigma^{D-m} 2^{-m} r^m}{\Gamma(m/2+1)} M\left(\frac{m-D}{2}, \frac{m}{2}+1, -\frac{r^2}{4\sigma^2}\right) \tag{10.58}$$

where σ is the Gaussian noise level and $M(a,b,z)$ is the Kummer's confluent hypergeometric function, which has the following integral representation

$$M(a,b,z) = \frac{\Gamma(b)}{\Gamma(a)\Gamma(b-a)} \int_0^1 e^{zt} t^{a-1} (1-t)^{b-a-1}\, dt \tag{10.59}$$

with

$$\begin{aligned} a &= \frac{m-D}{2} \\ b &= \frac{m}{2}+1 \\ z &= -\frac{r^2}{4\sigma^2} \end{aligned} \tag{10.60}$$

Considering two different values of the embedding dimension, m and $m + 2$,

$$\begin{aligned} \frac{C_m(r)}{C_{m+2}(r)} &= \frac{m^{-D/2}}{e^{-2\tau K}(m+2)^{-D/2} 2^{-2}\sigma^{-2}r^2} \frac{\Gamma(m/2+2)}{\Gamma(m/2+1)} \frac{M(a,b,z)}{M(a+1,b+1,z)} \\ &= \exp(2\tau K)\left(\frac{m+2}{m}\right)^{D/2} \frac{2\sigma^2(m-D)M(a,b,z)}{r^2 dM(a,b,z)/dz} \\ &= \exp(2\tau K)\left(\frac{m+2}{m}\right)^{D/2} \frac{2\sigma^2(m-D)}{r^2 d\ln[M(a,b,z)]/dz} \end{aligned} \tag{10.61}$$

Hence,

$$\exp(2\tau K)\left(\frac{m+2}{m}\right)^{D/2}(m-D) = \frac{C_m(r)}{C_{m+2}(r)} \frac{r^2}{2\sigma^2} \frac{d\ln[M(a,b,z)]}{dz} \tag{10.62}$$

Taking natural logarithms for both sides of Equation 10.58 gives

$$\ln(C_m(r)) = \ln(\phi e^{-m\tau K} m^{-D/2} 2^{-m} \sigma^{D-m}) + m\ln(r)$$

$$- \ln(\Gamma(m/2+1)) + \ln\left[M\left(\frac{m-D}{2}, \frac{m}{2}+1, -\frac{r^2}{4\sigma^2} \right) \right] \tag{10.63}$$

Differentiating Equation 10.63 with respect to r,

$$\frac{d}{dr}\ln[C_m(r)] = \frac{m}{r} + \frac{d}{dz}\ln\left[M\left(\frac{m-D}{2}, \frac{m}{2}+1, -\frac{r^2}{4\sigma^2} \right) \right]\left(-\frac{r}{2\sigma^2} \right)$$

from which,

$$\frac{r^2}{2\sigma^2} \frac{d}{dz}\ln[M(a,b,z)] = m - r\frac{d}{dr}\ln[C_m(r)] \tag{10.64}$$

Substituting Equation 10.64 into Equation 10.62 gives

$$\exp(2\tau K)\left(\frac{m+2}{m} \right)^{D/2} (m-D) = \frac{C_m(r)}{C_{m+2}(r)}\left(m - r\frac{d}{dr}\ln[C_m(r)] \right) \tag{10.65}$$

Taking natural logarithms for both sides of Equation 10.65 gives

$$2\tau K + \ln\left(\frac{m+2}{m} \right)^{D/2} = \ln\left(\frac{C_m(r)}{C_{m+2}(r)}\left(m - r\frac{d}{dr}\ln[C_m(r)] \right) \right) - \ln(m-D) \tag{10.66}$$

Note that the left-hand side of Equation 10.66 is just the right-hand side of Equation 10.54. Since the right-hand side of Equation 10.66 does not depend on K, we can define a new correlation entropy for the noisy time series by

$$\bar{K}_2 = \frac{1}{2\tau}\ln\left[\frac{C_m(r)}{C_{m+2}(r)} \right] + \frac{1}{2\tau}\ln\left[\left(m - \frac{d\ln(C_m(r))}{d\ln r} \right)/(m-D) \right] \tag{10.67}$$

The correlation entropy estimated by Equation 10.67 is referred to as the 'modified correlation entropy' for a chaotic time series (Jayawardena et al., 2010). When the chaotic time series is noise free, by using Equation 10.52, or using Equations 10.58 and 10.59 (Jayawardena et al., 2008), it can be shown that

$$D = \frac{d\ln[C_m(r)]}{d\ln(r)} \tag{10.68}$$

Equation 10.67 ensures that when the chaotic time series is noise free, the modified correlation entropy and the correlation entropy (Equation 10.55) are equivalent; that is,

$$\bar{K}_2 = K_2 \tag{10.69}$$

The estimation of the modified correlation entropy by Equation 10.67 needs the correlation dimension of the chaotic time series, which can be given *a priori*, or be estimated using the correlation sum. In the former case, it can be given approximately as

$$D \approx m_d - 1 \qquad (10.70)$$

where m_d is the minimum embedding dimension of the chaotic time series, which can be obtained by the false nearest neighbour (FNN) method (Kantz and Schreiber, 2003). In the latter case, it can be estimated by employing the correlation sum using Equation 10.11. Given several values $r_1, r_2, ..., r_L$, the corresponding correlation sums are then obtained as $C_m(r_1), C_m(r_2), ..., C_m(r_L)$. Then, the correlation dimension is approximated by

$$D \approx \frac{1}{L} \sum_{i=1}^{L} \frac{d \ln[C_m(r_i)]}{d \ln(r_i)} \qquad (10.71)$$

The estimation process for the modified correlation entropy does not involve the noise level of the noisy chaotic time series. Hence, the above estimation method would be easier for a noisy chaotic time series than the estimation method based on the method given by Diks (1999), which is of the form

$$K = \frac{1}{\tau} \ln\left[\frac{C_m(r)}{C_{m+1}(r)}\right] - \frac{1}{\tau} \ln\left[\frac{r}{2\sqrt{m}\sigma}\right] \qquad (10.72)$$

for large embedding dimensions. Furthermore, the modified correlation entropy is more robust to noise than the general correlation entropy obtained by Equation 10.56.

The application of this method has been demonstrated with four chaotic time series: two artificial and two real-world. The numerical results show that the modified correlation entropy is a robust estimation for the K–S entropy. The K–S entropy is also estimated by using the Lyapunov spectrum, and numerical results for the four data sets used show that the modified correlation entropy is closer to the K–S entropy than the correlation entropy.

10.7.2.3 K–S entropy and the Lyapunov spectrum

In a dynamical system, the Lyapunov exponent characterizes the rate of separation of initially nearby trajectories. The rate of separation can be different for different initial separations, thus leading to a whole spectrum of Lyapunov exponents with the number of such exponents being equal to the dimension of the phase space. The largest of them determines the predictability of the system and a positive largest exponent gives an indication that the system is chaotic. For an evolutionary system, the spectrum of Lyapunov exponents, $\{\lambda_1, \lambda_2, ..., \lambda_n\}$, depends on the starting point x_0 and are defined from the Jacobian matrix as follows:

$$\lambda_j = \lim_{N \to \infty} \frac{1}{N} \left\| J[F^{(N)}(x_0)]e_j \right\| \qquad (10.73)$$

$$J[F^{(N)}(x_0)] = \frac{dF^{(N)}(x)}{dx}\bigg|_{x_0} \qquad (10.74)$$

where $J(F)$ is the Jacobian matrix of the non-linear function F, $F^{(n)}$ is the n-th iteration of the non-linear evolution function, and $\{e_j\}, j = 1, 2, ..., m$ is a set of orthogonal basis of the tangent space at $\mathbf{x}_0$. The evolution function relates the future value to the present value according to the equation

$$\mathbf{x}_{n+1} = F(\mathbf{x}_n) \tag{10.75}$$

where $\{\mathbf{x}_n\}_{n=1}^{N}$ is the trajectory in the reconstructed phase space of the chaotic time series.

For a conservative system, the sum of all Lyapunov exponents is zero. For a dissipative system, it is negative. According to Pesin's (1977) theorem, the sum of all positive Lyapunov exponents gives an estimate of the K–S entropy. The Lyapunov spectrum estimated as described below can be used to estimate the K–S entropy to compare with the modified correlation entropy.

The Jacobian of Equations 10.73 and 10.74, by the chain rule, can be written as

$$
\begin{aligned}
JF^{(N)}(x_0) &= [JF(x_{N-1})][JF^{(N-1)}(x_0)] \\
&= [JF(x_{N-1})][JF(x_{N-2})][JF^{(N-2)}(x_0)] \\
&= \cdots \\
&= [JF(x_{N-1})][JF(x_{N-2})]\ldots[JF(x_1)][JF(x_0)] \\
&= A_n A_{n-1} \ldots A_2 A_1
\end{aligned} \tag{10.76}
$$

where $A_n = JF(x_{n-1})$, $1 \le n \le N$. Using the QR decomposition, matrix $\mathbf{A}$ can be decomposed into two matrixes: an orthogonal matrix $\mathbf{Q}$ and an upper triangular matrix $\mathbf{R}$, which can be defined via

$$\mathbf{A}_{n+1}\mathbf{Q}_n = \mathbf{Q}_{n+1}\mathbf{R}_{n+1} \tag{10.77}$$

where $\mathbf{Q}_0 = I$, the identity matrix. Then the Lyapunov spectrum of the system is given by

$$\lambda_j = \lim_{N \to \infty} \frac{1}{N} \sum_{n=1}^{N} \ln\left(\left\|\mathbf{R}_n(j,j)\right\|\right) \tag{10.78}$$

where $\mathbf{R}_n(j,j)$ is the element of matrix $\mathbf{R}_n$ at j-th row and j-th column. The Jacobian matrix at each time point in the trajectory of the non-linear system is evaluated by employing the least squares method to some nearest neighbours of the corresponding points in the reconstructed phase space. The above procedure follows that developed by Brown et al. (1991).

10.8 EXAMPLES OF KNOWN CHAOTIC ATTRACTORS

10.8.1 Logistic map

A description of the logistic map (see Equation 3.19 in Chapter 3), including some of its properties, is given in Chapter 3. The logistic map is non-invertible, implying that given x_{n+1}, it is not possible to solve for x_n uniquely. Some dimensions of the logistic map are as follows:

Correlation dimension = 0.5 ± 0.005 (Grassberger, 1983)
Hausdorff dimension ≈ 0.538 (Grassberger, 1981)
Information dimension = 0.5170976 (Grassberger, 1983)
Lyapunov exponent = $\ln(2) = 0.6931$

Figures 3.4 and 3.5 in Chapter 3, reproduced here as Figures 10.7a and 10.7b for convenience, respectively illustrate the behaviour x_t and with increasing time for r in the ranges (0, 1), and (1, 2). This behaviour is independent of the initial value. In the former case, x_t decays to zero, whereas in the latter case it attains a steady state value of $\dfrac{r-1}{r}$, which for $r = 1.5$ is 0.333. When r lies within (2, 3) also, x_t attains a steady state value of $\dfrac{r-1}{r}$ with initial oscillations around the limiting value as shown in Figure 10.7c. When r lies within (3, 3.45), x_t oscillates between two values indefinitely as shown in the same figure. When r lies within

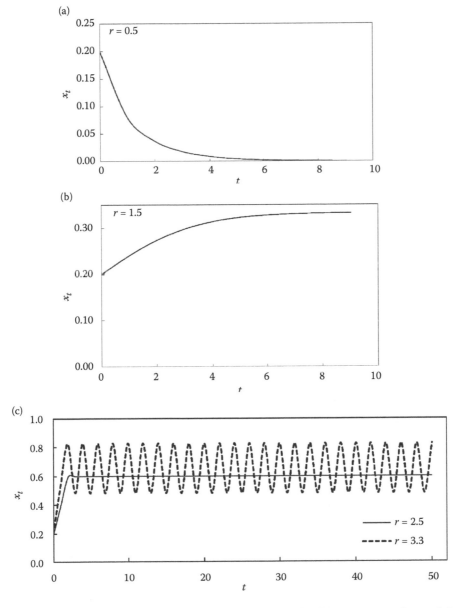

Figure 10.7 (a) Logistic curve for $r = 0.5$. (b) Logistic curve for $r = 1.5$. (c) Logistic curve for $r = 2.5$ and 3.3. (d) Logistic curve for $r = 3.45$ and 3.54. (e) Logistic curve for $r = 3.57$ and 4.0. (f) Logistic curve for $r > 4.0$.

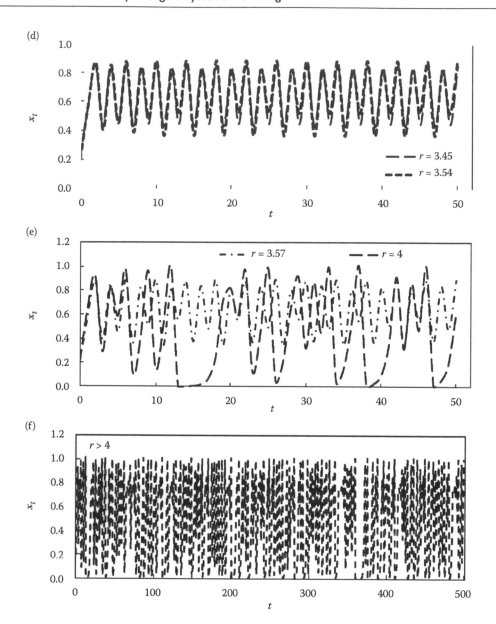

Figure 10.7 (Continued) (a) Logistic curve for $r = 0.5$. (b) Logistic curve for $r = 1.5$. (c) Logistic curve for $r = 2.5$ and 3.3. (d) Logistic curve for $r = 3.45$ and 3.54. (e) Logistic curve for $r = 3.57$ and 4.0. (f) Logistic curve for $r > 4.0$.

the range (3.45, 3.54), x_t oscillates between four values indefinitely (Figure 10.7d). When $r > 3.54$, x_t oscillates between 8, 16, 32, etc., values indefinitely (Figure 10.7e). When $r = 3.57$, the onset of chaos begins to appear and there are no oscillations. Slight changes in the initial conditions would lead to completely different results. Within the chaotic region ($r \geq 3.57$), there exist some isolated values of $r(1 + \sqrt{8} \approx 3.83)$ where there are some islands of stability, for example, oscillations between 3 values, 6 values, and 12 values, etc. When $r > 4$, x_t goes outside (0, 1) and diverges for all initial values (Figure 10.7f).

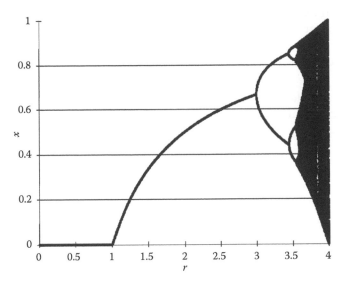

Figure 10.8 Bifurcation diagram for the logistic map.

10.8.1.1 Bifurcation

Bifurcation generally means separation of a structure into two branches; however, in the context of dynamical systems, it refers to period doubling, quadrupling, etc., that lead to the onset of chaos. The bifurcation diagram for the logistic map, which is a fractal, is a plot of the values of the parameter r along the horizontal axis and the possible long-term values of x along the vertical axis (Figure 10.8). The bifurcation diagram shows that period doubling bifurcation leads to a new behaviour with twice the period of the original system and period halving bifurcation leads to a new behaviour with half the period of the original system. Period halving leads from chaos to order, whereas period doubling leads from order to chaos. Logistic map bifurcation diagram is a clear illustration of the link between fractals and chaos. The points of separation of the bifurcation diagram of the logistic map looks like a Cantor set with gaps of varying sizes.

The period doubling phenomenon is not limited to the logistic map only. In fact, it can occur in any smooth, one-dimensional, non-monotonic non-linear map, which is unimodal in shape and has a local maximum. Examples include the circle, ellipse, and the sine maps. The map here refers to the plot of x_{n+1} versus x_n. An important property that has been discovered by Feigenbaum (1978, 1979, 1980) is that the distance between successive bifurcations ($r_n - r_{n-1}$ in the plot of x vs. r), or between diameters of successive circles on the real axis of a Mandelbrot set, decreases geometrically such that

$$\underset{n \to \infty}{\text{Limit}} \frac{r_n - r_{n-1}}{r_{n+1} - r_n} \to \text{constant}, \ \delta = 4.669\ldots \tag{10.79a}$$

The constant δ, which is universal, is known as Feigenbaum number, a new mathematical constant just as π is to circles. A second constant, α, which is also universal and known as the scaling factor, is the ratio between the width of a tine[4] d_n and the width of one of its two

---

[4] A tine is a prong, or fork, that takes off from a bifurcation point.

subtines, d_{n+1} at bifurcation. Since the width of the tine varies with varying distance, they should be measured at corresponding points. Quantitatively, it can be expressed as

$$\underset{n\to\infty}{\text{Limit}}\ \frac{d_n}{d_{n+1}} \to \text{constant},\ \alpha = 2.5029\ldots \tag{10.79b}$$

The sign of α can alternate between positive and negative.

Examples of iterative maps that fall into this category include

$$x_{n+1} = x_n^2 + c \tag{10.80a}$$

$$x_{n+1} = c(\sin x_n) \tag{10.80b}$$

$$x_{n+1} = \sin(\pi x_n) + c \tag{10.80c}$$

where c is a real valued constant. The Feigenbaum constants have been used in mathematical models of animal population, hydrodynamics, and turbulence analysis. They ensure that every chaotic system that corresponds to the above description will bifurcate at the same rate.

10.8.2 Hénon map

The Hénon map (Hénon, 1976) is the most general two-dimensional map and given as

$$x_{n+1} = 1 - a x_n^2 + y_n \tag{10.81a}$$

$$x_{n+1} = b x_n \tag{10.81b}$$

or, combined into one equation, as

$$x_{n+1} = 1 - a x_n^2 + b x_{n-1} \tag{10.81c}$$

where a and b are parameters. It has a chaotic attractor when $a = 1.4$ and $b = 0.3$. For other values of a and b, it may be chaotic, intermittent, or periodic. The trajectory of x_n values of the Hénon map, generated with $x_0 = 0.3$, $x_1 = 1.2$ is shown in Figure 10.9.

An important property of the Hénon map is a set of self-similar strands, which appears to look like a smooth curve. As it is zoomed in further, even more strands can be seen where self-similarity continues to an arbitrary small scale. It is a fractal that itself could generate fractals endlessly. It is invertible; that is, each point has a unique past. It is dissipative and contracts the area at the same rate everywhere in the phase space. For some parameter values, there is a trapping region that contains the strange attractor in which it gets mapped into itself. The correlation dimension of the Hénon map has been estimated to be 1.42 ± 0.02 (Grassberger and Proccasia, 1983), and the Hausdorff–Besicovitch dimension as 1.261 ± 0.003 (Russell et al., 1980). It has a Lyapunov exponent of 0.426. The Hénon map as well as some other fractals have been used to generate electronic music (e.g., YouTube).

10.8.3 Lorenz map

The Lorenz map (Lorenz, 1963) is a simplified set of non-linear ordinary differential equations that arise from a system of equations (Navier–Stokes) that describe the two-dimensional flow of a fluid with an imposed temperature difference, under the influence of gravity, buoyancy, thermal diffusivity, and kinematic viscosity. The equations attempt to model how the

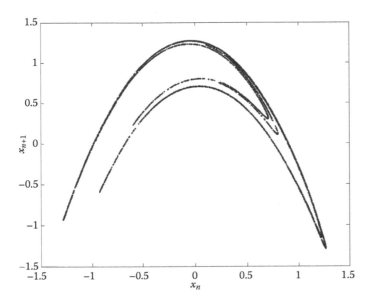

Figure 10.9 Hénon map.

air in the atmosphere is heated from bottom, rises by convection, and cooled at the top and descends. The simplified equations are defined as

$$\frac{dx}{dt} = \sigma(y - x) \tag{10.82a}$$

$$\frac{dy}{dt} = -xz + rx - y \tag{10.82b}$$

$$\frac{dz}{dt} = xy - bz \tag{10.82c}$$

where the parameters σ, r, and b are all non-negative, and in particular, $x = x(t)$ is a measure of the velocity of fluid flow in convection, and the circulation is clockwise if $x > 0$ and counterclockwise if $x < 0$; $y = y(t)$ is a measure of the temperature difference between ascending and descending fluid; $z = z(t)$ is a measure of the deviation of the temperature gradient from linearity; $\sigma = \dfrac{\nu}{\alpha}$ is the Prandtl number (dimensionless); ν is the kinematic viscosity of the fluid and α is the thermal diffusivity of the fluid; $r = \dfrac{Ra}{Ra_c}$, Ra is the Rayleigh number defined as $Ra = \dfrac{g\beta(T_s - T_\infty)d^3}{\nu\alpha}$, and Ra_c is the critical Rayleigh number (beginning of turbulent flow) defined as $Ra_c = \dfrac{\pi^4(1 + a^2)^3}{a^2}$, with g, the acceleration due to gravity; β, the thermal expansion coefficient; T_s, the surface temperature; T_∞, the quiescent temperature; d, a characteristic length (depth of fluid layer); and a, the length scale. $b = \dfrac{4}{1 + a^2}$ is a geometric parameter representing the size of the area being modelled.

In a three-dimensional phase space, x, y, z can be thought of as the coordinates of a point that changes with time. As time passes, the trajectory of this point from some initial state will give the evolution of the system. The system becomes chaotic for $\sigma = 10$, $r = 28$, and

$b = \dfrac{8}{3}$. In other words, a system that is fully deterministic leads to an outcome that looks like stochastic or random. Trajectories of initially nearby points will diverge, which is a characteristic hallmark of chaotic systems.

Lorenz's equations, though simplified, and autonomous (independent variable t does not appear explicitly on the right-hand sides of the equations) cannot be solved analytically. Approximate solutions to this initial value problem (or any other initial value problem) can, however, be obtained by numerical integration. Euler's method and the Runge–Kutta method, which are both based on Taylor series approximation of transforming the continuous derivatives to discrete differences, are widely used. The accuracy of the approximate solution depends on the time step used to march the solution forward from a known initial solution. Euler's first-order approximation works as follows:

For the x, y, and z components:

$$\frac{dx}{dt} = \sigma(y_t - x_t), \text{ approximated as } \frac{\Delta x}{\Delta t} = \sigma(y_t - x_t) \Rightarrow \Delta x = \Delta t \sigma(y_t - x_t) \tag{10.83a}$$

$$\frac{dy}{dt} = -x_t z_t + r x_t - y_t, \text{approximated as } \frac{\Delta y}{\Delta t} - x_t z_t + r x_t - y_t \Rightarrow \Delta y = \Delta t(-x_t z_t + r x_t - y_t) \tag{10.83b}$$

$$\frac{dz}{dt} = x_t y_t - b z_t, \text{ approximated as } \frac{\Delta z}{\Delta t} = x_t y_t - b z_t \Rightarrow \Delta z = \Delta t(x_t y_t - b z_t) \tag{10.83c}$$

from which

$$x_{t+\Delta t} = x_t + (\Delta x)_t \tag{10.84a}$$

$$y_{t+\Delta t} = y_t + (\Delta y)_t \tag{10.84b}$$

and

$$z_{t+\Delta t} = z_t + (\Delta z)_t \tag{10.84c}$$

Therefore, given the initial values of x, y, and z, and a choice of the time step Δt, all subsequent values can be computed recursively. The Euler method marches the solution forward in time by the time step Δt using the derivative at the beginning of the time step only and the resulting total accumulated error is of $O(\Delta t)^2$. A better approximate method is what is referred to as the mid-point method or the second-order Runge–Kutta method, which has errors of $O(\Delta t)^3$. In this method, the function value calculated at the mid-point of the interval is used to compute the full step. The method requires two evaluations for each time step. A more widely used method that is relatively stable and easy to implement is the fourth-order Runge–Kutta method in which the total accumulated error is of $O(\Delta t)^5$. The method that requires four evaluations of each function at each time step works as follows:

$$x_{t+\Delta t} = x_t + \frac{\Delta t}{6}\left(k_1^x + 2k_2^x + 2k_3^x + k_4^x\right)$$

$$y_{t+\Delta t} = y_t + \frac{\Delta t}{6}\left(k_1^y + 2k_2^y + 2k_3^y + k_4^y\right) \tag{10.85}$$

$$z_{t+\Delta t} = z_t + \frac{\Delta t}{6}\left(k_1^z + 2k_2^z + 2k_3^z + k_4^z\right)$$

where

$$k_1^x = \sigma(y_t - x_t) \tag{10.86a}$$

$$k_2^x = \sigma\left(y_t + \frac{\Delta t}{2}k_1^y - \left(x_t + \frac{\Delta t}{2}k_1^x\right)\right) \tag{10.86b}$$

$$k_3^x = \sigma\left(y_t + \frac{\Delta t}{2}k_2^y - \left(x_t + \frac{\Delta t}{2}k_2^x\right)\right) \tag{10.86c}$$

$$k_4^x = \sigma\left(y_t + \Delta t k_3^y - \left(x_t + \Delta t k_3^x\right)\right) \tag{10.86d}$$

$$k_1^y = -x_t z_t + r x_t - y_t \tag{10.87a}$$

$$k_2^y = -\left(x_t + \frac{\Delta t}{2}k_1^x\right)\left(z_t + \frac{\Delta t}{2}k_1^z\right) + r\left(x_t + \frac{\Delta t}{2}k_1^x\right) - \left(y_t + \frac{\Delta t}{2}k_1^y\right) \tag{10.87b}$$

$$k_3^y = -\left(x_t + \frac{\Delta t}{2}k_2^x\right)\left(z_t + \frac{\Delta t}{2}k_2^z\right) + r\left(x_t + \frac{\Delta t}{2}k_2^x\right) - \left(y_t + \frac{\Delta t}{2}k_2^y\right) \tag{10.87c}$$

$$k_4^y = -\left(x_t + \Delta t k_3^x\right)\left(z_t + \Delta t k_3^z\right) + r\left(x_t + \Delta t k_3^x\right) - \left(y_t + \Delta t k_3^y\right) \tag{10.87d}$$

$$k_1^z = x_t y_t - b z_t \tag{10.88a}$$

$$k_2^z = \left(x_t + \frac{\Delta t}{2}k_1^x\right)\left(y_t + \frac{\Delta t}{2}k_1^y\right) - b\left(z_t + \frac{\Delta t}{2}k_1^z\right) \tag{10.88b}$$

$$k_3^z = \left(x_t + \frac{\Delta t}{2}k_2^x\right)\left(y_t + \frac{\Delta t}{2}k_2^y\right) - b\left(z_t + \frac{\Delta t}{2}k_2^z\right) \tag{10.88c}$$

$$k_4^z = \left(x_t + \Delta t k_3^x\right)\left(y_t + \Delta t k_3^y\right) - b\left(z_t + \Delta t k_3^z\right) \tag{10.88d}$$

In the above notation for the superscripts and subscripts of k, the superscripts refer to the x, y, and z components, and the subscripts refer to the first, two middle, and end points where the derivatives are taken. The time step Δt used in the above formulations can either be fixed or varied in an adaptive way to cater for different rates of changes of the variables involved. The former is convenient for formulation but computationally not efficient, whereas the latter has the opposite effects. A point to note is that a smaller Δt does not necessarily mean better accuracy because a smaller Δt requires more iterations and therefore more round-off errors.

With the above formulation, for given initial values of x_t, y_t, and z_t, x_{t+1}, y_{t+1}, and z_{t+1} could be evaluated. Hence, the individual time series of x_t, y_t, and z_t could be obtained recursively from which their trajectories can be shown in the phase space diagram. Both the Euler method and the Runge–Kutta method described above are explicit numerical methods, and therefore have limitations resulting from stability. The time series plots of x_t, y_t, and z_t starting with nearby initial values will tend to diverge after a large number of iterations. Such divergence is a typical characteristic of chaotic systems. The Lorenz system solved numerically by the fourth-order Runge–Kutta method with a time step of 0.02 and initial values of $x_0 = 12.5$, $y_0 = 2.5$, and $z_0 = 1.5$ is shown in Figure 10.10. The values of (x, y, z) are calculated for time interval $t \in (0, 1000)$; however, to ensure that all points lie within the strange attractor, the trajectory is drawn for the interval $t \in (500, 1000)$. It has a correlation dimension of 2.068 ± 0.086 and a Lyapunov exponent of 0.9056.

10.8.4 Duffing equation

Although the Duffing equation (Duffing, 1918) was first published for studying classic oscillators in electronics, it has later found applications in biology, particularly in the study of modelling the brain. The equation displays catastrophic jumps of amplitude and phase for gradual changes of the forcing frequency term. It is also considered as the simplest non-linear forced damped oscillator. In a mechanical system, the equation models a spring pendulum whose spring constant (stiffness) does not strictly obey Hooke's law. The most general form of the Duffing equation, which is a second-order non-linear ordinary differential equation, is

$$\frac{d^2x}{dt^2} + \alpha\frac{dx}{dt} + f(x) = e(t) \tag{10.89}$$

where $e(t)$ is a periodic function of period 2π. More specifically, it takes the form

$$\frac{d^2x}{dt^2} + \alpha\frac{dx}{dt} \pm \omega_0^2 x + \beta x^3 = b\cos(\omega t + \phi) \tag{10.90}$$

which describes various physical problems depending on the magnitudes of the parameters α, which represents a damping constant; β, a non-linearity parameter; ω, a forcing frequency; ϕ, a forcing phase angle; ω_0^2, the ratio of the spring constant to the mass; and b, a

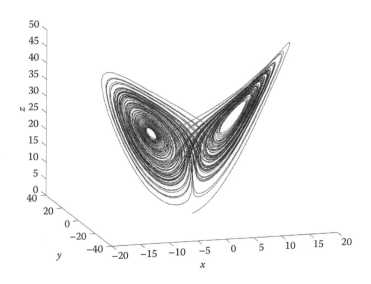

Figure 10.10 Lorenz map.

forcing amplitude. For example, with $\alpha = 0$, $\beta = 0$, $b = 0$, $\omega_0^2 = 1$, and taking the positive sign for the third term, the equation simplifies to that of a simple harmonic motion (no damping, no forcing, no non-linearity):

$$\frac{d^2x}{dt^2} + x = 0 \tag{10.91a}$$

When $\beta = 0$ and $b = 0$, the equation reduces to that of an unforced damped oscillation:

$$\frac{d^2x}{dt^2} + \alpha\frac{dx}{dt} + \omega_0^2 x = 0 \tag{10.91b}$$

When $\alpha = 0$, $\beta = 0$, and $\omega_0^2 = 0$, the equation reduces to that of an undamped forced oscillation:

$$\frac{d^2x}{dt^2} = b\cos(\omega t + \phi) \tag{10.91c}$$

When $\beta = 0$, the equation reduces to that of a damped forced oscillation:

$$\frac{d^2x}{dt^2} + \alpha\frac{dx}{dt} \pm \omega_0^2 x = b\cos(\omega t + \phi) \tag{10.91d}$$

Equations 10.91a and 10.91b are autonomous, whereas Equations 10.91c and 10.91d are non-autonomous. All these four equations are linear. Adding the non-linear term with $\beta = 1$ and $\phi = 0$ gives the non-linear forced damped oscillation, which is generally referred to as the Duffing equation.

$$\frac{d^2x}{dt^2} + \alpha\frac{dx}{dt} \pm \omega_0^2 x + x^3 = b\cos(\omega t) \tag{10.91e}$$

When $\alpha = 0$ and $\phi = 0$, Equation 10.90 reduces to that of undamped forced oscillation:

$$\frac{d^2x}{dt^2} \pm \omega_0^2 x + \beta x^3 = b\cos(\omega t) \tag{10.91f}$$

Equation 10.90 can also be written as two first-order ordinary differential equations:

$$\frac{dx}{dt} = y \tag{10.91g}$$

$$\frac{dy}{dt} = -x - x^3 - \alpha y + b\cos(\omega t) \quad \left(\text{with } \omega_0^2 = 1;\ \beta = 1;\ \phi = 0\right) \tag{10.91h}$$

which, in the unforced case ($b = 0$), using the original single variable form, is

$$\frac{d^2x}{dt^2} + \alpha\frac{dx}{dt} + x + x^3 = 0 \tag{10.91i}$$

The equation subsequently has been written in several different forms. One such form is that proposed by Ueda (1979, 1980), which takes the form (forced, damped, and non-linear)

$$\frac{d^2x}{dt^2} + \alpha\frac{dx}{dt} + x^3 = b\cos(t) \tag{10.91j}$$

and becomes chaotic for $\alpha = 0.1$, $b = 12.0$. He also experimented with other values of these two parameters and found that the system exhibits chaotic behaviour in the ranges of $\alpha = (0, 0.3)$ and $b = (6–13)$.

Another variation of the Duffing equation is (forced, undamped, non-linear)

$$\frac{d^2x}{dt^2} + \omega_0^2 x + x^3 = b\cos(t) \tag{10.91k}$$

The phase space diagram changes from the shape of a limit cycle for low values of b ($= 0.5$) to more complex portraits as the value of b increases. For low values of b, the corresponding time series is almost sinusoidal. Figure 10.11 shows the attractor obtained by solving Equation 10.91k with parameter values $\omega_0^2 = 0.3$ and $b = 34$, by the fourth-order Runge–Kutta method with initial conditions $x_0 = 0.1$, $y_0 = 0.1$, and a time step of 0.02. The Duffing system has a correlation dimension of 2.334 ± 0.114 and a Lyapunov exponent of 0.1572.

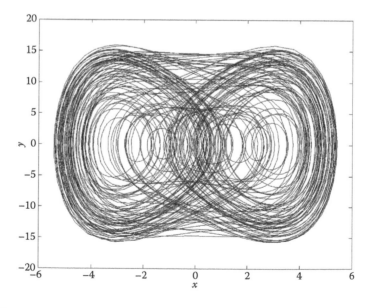

Figure 10.11 Duffing attractor.

10.8.5 Rössler equations

The Rössler equations (Rössler, 1976), which are a set of three-dimensional non-linear ordinary differential equations, is defined as

$$\frac{dx}{dt} = -y - z \tag{10.92a}$$

$$\frac{dy}{dt} = x + ay \tag{10.92b}$$

$$\frac{dz}{dt} = b + z(x - c) \tag{10.92c}$$

where the parameters a, b, and c are adjustable. Depending on the values of the parameters, the system can have a periodic behaviour, converge to a fixed point, or escape towards infinity. For example, when the value of $a \leq 0$, the system converges to a fixed point. When $a = 0.1$, it has a unit period. As the value of a increases from 0.2 (within a narrow range from 0.2 to 0.38), the system becomes increasingly chaotic. As the value of b approaches zero, the attractor approaches infinity. In the case of the parameter c, the bifurcation diagram indicates that the system is periodic for low values of c (period 1 for values up to about 4; period 3 for values up to about 12) but quickly becomes chaotic as c increases. The values that Rössler used are $a = 0.2$, $b = 0.2$, and $c = 5.7$. Applications of Rössler equations have been found in Lotka–Volterra predator–prey models (Gilpin, 1979), in food chain models (Paine, 1966), and in chemical oscillator models (Roux et al., 1983).

The Hausdorf–Besicovitch dimension of the Rössler attractor is estimated to be between 2.01 and 2.02. That means it is more than a collection of lines and slightly more than a collection of planes. It is a fractal (having a non-integer Hausdorf–Besicovitch dimension), which is a strange attractor, implying that the attractor never retraces a previously traversed path. It has a correlation dimension of 1.991 ± 0.065 and a Lyapunov exponent of 0.0714.

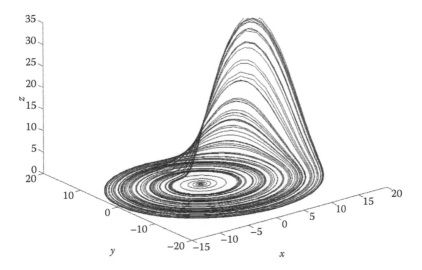

Figure 10.12 Rössler attractor.

The Rössler equations can be solved by the same methods described for the solution of the Lorenz attractor. A solution obtained by Jayawardena et al. (2008) for $a = 0.15$, $b = 0.2$, and $c = 10$, and initial values $x_0 = 0.1$, $y_0 = 0.1$, and $z_0 = 0.1$ is shown in Figure 10.12.

10.8.6 Chua's equation

Chua's circuit (Chua and Lin, 1990) is the simplest electronic circuit that exhibit chaos. It is a parallel LRC[5] circuit consisting of two linear capacitors, one linear inductor, two linear resistors, and a non-linear resistor (referred to as Chua's diode). The set of equations that relates the voltages across the two capacitors, according to Kirchoff's laws, take the form

$$C_1 \frac{dV_1}{dt} = \frac{1}{R}(V_2 - V_1) - f(V_1) \tag{10.93a}$$

$$C_2 \frac{dV_2}{dt} = -\frac{1}{R}(V_2 - V_1) + I \tag{10.93b}$$

$$L \frac{dI}{dt} = -rI - V_2 \tag{10.93c}$$

where V_1 is the voltage across capacitor C_1 (and the non-linear resistance), V_2 is the voltage across capacitor C_2 (and the voltage across the inductor), I is the current through the inductor, L is the inductance of the inductor, R is the resistance of the resistor, and $f(V_1)$ is a current–voltage characteristic. Normalizing these equations, we get

$$\frac{dx}{dt} = a(y - x) - g(x) \tag{10.94a}$$

$$\frac{dy}{dt} = \sigma[-a(y - x) + z] \tag{10.94b}$$

$$\frac{dz}{dt} = -c(y + \gamma z) \tag{10.94c}$$

where $x = \dfrac{V_1}{V_c}$, $y = \dfrac{V_2}{V_c}$, $z = \dfrac{R_1 I}{V_c}$; $a = \dfrac{R_1}{R}$, $b = 1 - \dfrac{R_1}{R_2}$, $c = \dfrac{C_1 R_1^2}{L}$, $\sigma = \dfrac{C_1}{C_2}$, $\gamma = \dfrac{r}{R_1}$, and V_c and R_1, respectively, are reference voltages and resistances. The normalized form of the current–voltage characteristic $g(x)$ is given as

$g(x) = -x$ for $|x| < 1$, for which the slope is -1

$g(x) = -1 + b \{|x| - 1\} \operatorname{sgn}(x)$, for $1 < |x| < 10$, for which the slope is $-b$

$g(x) = 10\{(|x| - 10) + (9b - 1)\} \operatorname{sgn}(x)$, for $|x| > 10$, for which the slope is 10

---

[5] LRC circuit (or sometimes RLC circuit) refers to an electrical circuit consisting of an inductor, a resistor, and a capacitor connected in series or in parallel. The letters LRC are the symbols used to represent inductance, resistance and capacitance, respectively, in electrical engineering.

There exist several versions of Chua's equations. For example, the third equation is sometimes written as a function of y only. The current–voltage characteristic, which is normally taken as a three-segment piecewise linear function is sometimes replaced by a smooth cubic polynomial. A version that has been used by Elwakil and Kennedy (2000), in dimensionless form, and that gives a double scroll (Suykens et al. [1997] have presented an 'n' scroll attractor) is as follows:

$$\frac{dx}{dt} = (1 - K - \varepsilon_r K)y - (1 + \varepsilon_r)x + \varepsilon_r z \tag{10.95a}$$

$$\frac{dy}{dt} = x + (K - 2)y \tag{10.95b}$$

$$\varepsilon_c \frac{dz}{dt} = \varepsilon_r(x + Ky - z) - f(z) \tag{10.95c}$$

with

$$f(z) = \alpha_1 z + \alpha_2(|z + 1| - |z - 1|),$$

and

$$K = 3.25,\ \varepsilon_r = 1/6,\ \varepsilon_c = 0.06,\ \alpha_1 = 0.8,\ \alpha_2 = -0.5.$$

Chua's system has only one non-linearity compared with two for the Lorenz equations. In the laboratory, it can be used to generate pseudo-random signals. It is also used to simulate brain dynamics and music composition. The trajectory obtained by solving these equations using the fourth-order Runge–Kutta method as described above, with a time step of 0.02 and an initial set of $x_0 = 0.1$, $y_0 = 0.1$, and $z_0 = 0.1$, is illustrated in Figure 10.13, which

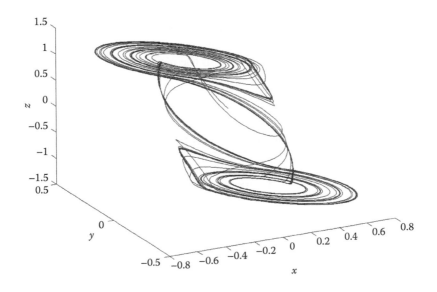

Figure 10.13 Chua's circuit attractor.

clearly demonstrates the chaotic behaviour of Chua's system. Chua's system has a correlation dimension of 2.125 ± 0.098 and a Lyapunov exponent of 0.3271.

10.9 APPLICATIONS AREAS OF CHAOS

The chaos theory has been applied to many disciplines. Examples can be found in mathematics, meteorology, computer science, various fields in engineering, finance, physics, biology, and population dynamics, just to name a few. In the laboratory, chaotic behaviour has been observed in electrical circuits, lasers, oscillating chemical reactions, fluid dynamics, magnetometry, etc. In nature, chaotic behaviour has been observed in weather, dynamics of satellites in the solar system, motion of celestial bodies, and ecology where population growth under density dependence as shown by Ricker (1954) model (see also Chapter 3). It is also now being applied in the medical field. The n-body problem is a classic example of chaotic dynamics.

The names chaos and chaos theory have also been used recently by the entertainment industry in several movies (e.g., Jurassic Park [1993], π [1998], The Butterfly Effect [2004], The Science of Sleep [2006], and Chaos [2006]) and music composed by using fractals (e.g., Brown2, Brownian, Drums etc., as well as others available from the website third.apex.to.fractovia [http://www.fractovia.org/art/fmusic/index.html] [4]).

10.10 CONCLUDING REMARKS

The objective of this chapter was to give an introduction to fractals and chaos. Various measures of fractal dimension have been introduced, including their estimation methods and relationships to one another. Examples of well-known fractals such as the Cantor set, Sierpinski triangle, Koch curve, Koch snowflakes, and Mandelbrot and Julia sets have been described including their methods of generation. The perimeter–area relationship of fractals, including their relationship to natural landscapes, is also presented. On the 'chaos' part, a brief history including the 'butterfly effect', the 'n'-body problem, and their relevance to dynamical systems is given. Finally, the mathematical details of the estimation methods of invariants in chaotic systems including some recent developments, together with few examples of known chaotic systems and their properties, are presented. More details of chaos in dynamical systems will be presented in the next chapter.

REFERENCES

Bonachela, J.A., Hinrichsen, H. and Munoz, M. (2008): Entropy estimations of small data set. *Journal of Physics A: Mathematical and Theoretical*, 41, 202001.

Brown, R., Bryant, P. and Abarbanel, H.D. (1991): Computing the Lyapunov spectrum of a dynamical system from an observed time series. *Physical Review A*, 43, 2787–2806.

Chua, L.O. and Lin, G.N. (1990): Canonical realization of Chua's circuit family. *IEEE Transactions on Circuits and Systems*, 37(7), 885–902.

Diks, C. (1999): *Nonlinear Time Series Analysis, Methods and Applications*. World Scientific Publishing, Singapore.

Ding, M., Grebogi, C., Ott, E., Sauer, T. and Yorke, J.A. (1993): Plateau onset for correlation dimension: When does it occur? *Physical Review Letters*, 70, 3872–3875.

Duffing, G. (1918): *Erzwungene Schwingungen bei Veränderlicher Eigenfrequenz*. F. Vieweg u. Sohn, Braunschweig.

Elwakil, A.S. and Kennedy, M.P. (2000): Chua's circuit decomposition: A systematic design approach for chaotic oscillators. *Journal of the Franklin Institute*, 337, 251–265.

Ewing, J.H. and Schober, G. (1992): The area of the Mandelbrot set. *Numerische Mathematik*, 61, 59–72.

Feigenbaum, M.J. (1978): Quantitative universality for a class of non-linear transformations. *Journal of Statistical Physics*, 19, 25–52.

Feigenbaum, M.J. (1979): The universal metric properties of nonlinear transformations. *Journal of Statistical Physics*, 21, 669–706.

Feigenbaum, M.J. (1980): The metric universal properties of period doubling bifurcations and the spectrum for a route to turbulence. *Annals of the New York Academy of Science*, 357, 330–336, 1980.

Frank, M., Blank, H.-R., Heindl, J., Kaltenhauser, M., Kochner, H., Kreissche, W., Muller, N., Pocher, S., Sporer, R. and Wagner, T. (1993): Improvement of K_2 entropy calculations by means of dimension scaled distances. *Physica D*, 65, 359–364.

Frieson, W.I. and Mikula, R.J. (1987): Fractal dimensions of coal particles. *Journal of Colloid and Interface Science*, 20(1), 263–271.

Gilpin, M.E. (1979): Spiral chaos in a predator prey model. *American Naturalist*, 113, 306–308.

Grassberger, P. (1981): On the Hausdorff dimension of fractal attractors. *Journal of Statistical Physics*, 26, 173–179, doi:10.1007/BF01106792.

Grassberger, P. and Procaccia, I. (1983a): Characterization of strange attractors. *Physical Review Letters*, 50(5), 346–349.

Grassberger, P. and Procaccia, I. (1983b): Measuring the strangeness of strange attractors. *Physica D*, 9, 189 208.

Grassberger, P. and Procaccia, I. (1983c): Estimation of the Kolmogorov entropy from a chaotic signal. *Physical Review A*, 28, 2591–2593.

Hack, J.T. (1957): Studies of longitudinal stream profile in Virginia and Maryland, US. *Geological Survey Professional Paper*, 294-B, 45–97.

Hénon, M. (1976): A two-dimensional mapping with a strange attractor. *Communications in Mathematical Physics*, 50, 69–77, doi:10.1007/BF01608556.

Jayawardena, A.W., Xu, P.C. and Li, W.K. (2008): A method of estimating the noise level in a chaotic time series. *Chaos, American Institute of Physics*, 18(2), 023115–023115-11, doi:10.1063/1.2903757.

Jayawardena, A.W., Xu, P.C. and Li, W.K. (2010): Modified correlation entropy estimation for a noisy chaotic time series. *Chaos, American Institute of Physics*, 20, 023104, doi:10.1063/1.3382013, 023104-1–023104-11.

Kantz, H. and Schreiber, T. (2003): *Nonlinear Time Series Analysis*. Cambridge University Press, Cambridge.

Kantz, H. and Schürmann, T. (1996): Enlarger scaling ranges for the KS-Entropy and information dimension. *Chaos*, 6, 167–171.

Kolmogorov, A.N. (1958): A new invariant of transitive dynamical systems. *Doklady Akademii Nauk, USSR*, 119, 861.

Lorenz, E.N. (1963): Deterministic non-periodic flow. *Journal of the Atmospheric Sciences*, 20, 130–141.

Lorenz, E.N. (1972): Predictability: Does the flap of a butterfly's wings in Brazil set off a tornado in Texas? *139th Annual Meeting of the American Association for the Advancement of Science*, December 29, In: Lorenz, E.N., *The Essence of Chaos (1995)*, University of Washington Press, Seattle.

Lovejoy, S. (1982): Area–perimeter relation for rain and cloud areas. *Science*, 216, 185–187.

Mandelbrot, B.B. (1967): How long is the coast of Britain? Statistical self-similarity and fractional dimension. *Science, New Series*, 156(3775), 636–638.

Mandelbrot, B.B. (1982): *The Fractal Geometry of Nature*. W.H. Freeman, New York, pp. 188–189.

Oltmans, H. and Verheijen, P.J.T. (1997): Influence of noise on power-law scaling functions and an algorithm for dimension estimations. *Physical Review E*, 56, 1160–1170.

Paine, R.T. (1966): Food web complexity and species diversity. *American Nature*, 100, 65–75.

Pesin, Y.B. (1977): Characteristic Lyapunov exponents and smooth ergodic theory. *Russian Mathematical Surveys*, 32, 55.

Rényi, A. (1970): *Probability Theory*. American Elsevier Publishing Company, Inc., New York, 666 pp.

Richardson, L.F. (1961): The problem of contiguity: An appendix to Statistic of Deadly Quarrels. *General Systems: Yearbook of the Society for the Advancement of General Systems Theory*, 6(139), 139–187. Ann Arbor, MI: The Society, 1956: Society for General Systems Research. OCLC 1429672.

Ricker, W.E. (1954): Stock and recruitment. *Journal of Fisheries Board of Canada*, 11(5), 559–623.

Rosenstein, M.T., Collins, J.J. and De Luca, C.J. (1992): A practical method for calculating largest Lyapunov exponents from small data sets. http://www.physionet.org/physiotools/lyapunov/RosensteinM93.pdf.

Rössler, O.E. (1976): An equation for continuous chaos. *Physics Letters A*, A57(5), 397–398.

Rosso, R., Bacchi, B. and La Barbera, P. (1991): Fractal relation of mainstream length to catchment area in river networks. *Water Resources Research*, 27(3), 381–388.

Roux, J.-C., Simoyi, R.H. and Swinney, H.L. (1983): Observation of a strange attractor. *Physica*, 8D, 257–266.

Russell, D.A., Hanson, J.D. and Ott, E. (1980): Dimension of strange attractors. *Physical Review Letters*, 45, 1175–1178.

Schouten, J.C. and van den Bleek, C.M. (1994): *RRCHAOS: A Menu-Driven Software Package for Chaotic Time Series Analysis*. Reactor Research Foundation, Delft, The Netherlands.

Schouten, J.C., Takens, F. and van den Bleek, C.M. (1994): Estimation of the dimension of a noisy attractor. *Physical Review E*, 50(3), 1851–1861.

Schürmann, T. and Grassberger, P. (2002): Entropy estimation of symbol sequences. *Chaos*, 6, 414–427.

Shannon, C.E. (1948): A mathematical theory of communication. *Bell System Technology Journal*, 27, 379–423, 623–656.

Shishikura, M. (1998): The Hausdorff dimension of the boundary of the Mandelbrot set and Julia sets. *Annals of Mathematics*, 147, 225–267. (First appeared in 1991 as a Stony Brook IMS Preprint, available as arXiv:math.DS/9201282.)

Sinai, A.G. (1959): On the concept of entropy of a dynamical system. *Doklady Akademii Nauk, USSR*, 142, 768.

Smith, L.A. (1988): Intrinsic limits on dimension calculations. *Physics Letters A*, 133(6), 283–288.

Sprott, J.C. (2003): *Chaos and Time-Series Analysis*. Oxford University Press, Oxford.

Suykens, J., Huang, A. and Chua, L. (1997): A family of n-scroll attractors from a generalized Chua's circuit. *Archiv fur Elektronik und Ubertragungstechnik (International Journal of Electronics and Communications)*, 51(3), 131–138.

Theiler, J. (1987): Efficient algorithm for estimating the correlation dimension from a set of discrete points. *Physical Review A*, 36(9), 4456–4462.

Tsonis, A.A. (1992): *Chaos from Theory to Application*. Plenum Press, New York.

Ueda, Y. (1979): Randomly transitional phenomena in the system governed by Duffing's equation. *Journal of Statistical Physics*, 20, pp. 181–196.

Ueda, Y. (1980): Steady motions exhibited by Duffing's equation: A picture book of regular and chaotic motions. In: P.J. Holmes (Ed.), *New Approaches to Nonlinear Problems in Dynamics*. SIAM, Philadelphia, pp. 311–322.

Weierstrass, K. (1872): Über continuirliche Functionen eines reellen Arguments, die für keinen Werth des letzeren einen bestimmten Differentialquotienten besitzen, collected works. English translation: On continuous functions of a real argument that do not have a well-defined differential quotient. In: G.A. Edgar, *Classics on Fractals*. Addison-Wesley Publishing Company, Reading, MA, 1993, pp. 3–9.

Wolf, A., Swift, J.B., Swinney, H.L. and Vastano, J.A. (1985): Determining Lyapunov exponents from a time series. *Physica D*, 16, 285–317.

Wong, P.-Z. (1988): The statistical physics of sedimentary rock. *Physics Today*, 41, 24–32.

Xie, H. (1993): *Fractals in Rock Mechanics*. Geomechanics Research Series 1. A.A. Balkema, Rotterdam, The Netherlands.

Chapter 11

Dynamical systems approach of modelling

11.1 INTRODUCTION

Time series analysis is a hot topic in many areas of science and engineering. Many experimental and field observations of physical, biological, chemical, as well as economic and social processes are made as functions of time that constitute time series. The basic objective of time series analysis is to understand the characteristics of the processes that generate the time series of data and to make future predictions and simulations under different scenarios. One of the underlying assumptions in such time series analysis in the past has been that they are stationary and therefore their statistical behaviour in the future will be the same as in the past, thereby imposing limitations in the applicability.

Linear time series analysis has reached a stage of maturity some years ago. Well-established methods of analysis such as the autoregressive moving average modelling (ARMA) and Fourier analysis have been in existence for a number of years, and several types of tools and software based on such methods are readily available nowadays. There are many advantages of the linear methods of analysis, such as their generality, ease of extrapolation, relatively easy mathematics and formulation, and versatility. Solutions obtained for a problem in a particular field can be directly applied to a mathematically equivalent problem in a completely different field.

On the other hand, non-linear time series analysis is still in a stage of development, and at the moment mainly confined to the mathematics and physics disciplines. The techniques are not yet well developed and well established to a level whereby they are directly suitable for engineering practitioners. The techniques available are problem specific and lack generality. The basic steps in non-linear model building in the context of time series analysis consist of testing for determinism of the system, making sure that linear methods are inadequate, and characterizing the dynamics of the system in terms of some statistical parameters.

In recent years, it has been realized that certain types of processes that are seemingly stochastic are in fact evolving from deterministic non-linear dynamical systems. By treating the systems that generate such processes as deterministic, it is possible to uncover the complicated dynamics and make realistic short-term predictions. Such systems exhibit stable properties, which are predictable at times but become 'chaotic' under certain initial conditions. The study of 'chaotic' systems, including the scaling behaviour as well as their application to model hydrological processes, has, in the recent past, drawn the attention of many researchers in several disciplines (e.g., Farmer and Sidorowich, 1987; Abarbanel et al., 1990; Sugihara and May, 1990; Elsner and Tsonis, 1992; Smith, 1992; Ott, 1993; Jayawardena and Lai, 1994; Sivakumar et al., 1999a; Jayawardena et al., 2002; Sivakumar and Jayawardena, 2002, 2003; Sivakumar et al., 2002; Regonda et al., 2004; Sivakumar, 2005). A system may have several independent variables and several dependent variables. The usual independent variables are space and time. In the context of this chapter, time is taken as the independent variable and therefore all variables used unless otherwise stated are treated as functions of time and the sequences of data constitute time series.

In a deterministic system, predictions can generally be made using an evolutionary equation in which the future state is considered to be dependent on present and past states. The prediction process therefore involves an accurate estimation of the mapping function (or evolutionary equation), which transforms present and past states into future states. In a chaotic system, the errors grow exponentially and the predictive power is lost very quickly because of sensitivity to initial conditions. The mapping function can be estimated using local models in which the function approximation at each time step is done from data sets of the local neighbourhood only in a piece-wise manner, or global models in which the function approximation is done for the whole domain. Local models include linear or polynomial function approximations in the local neighbourhoods, as well as radial basis functions (RBFs), whereas global models are generally of the polynomial type, although RBFs have also been used (Casdagli, 1989; Brown, 1993). In general, local models can be more accurate locally than global models; however, the penalty for better accuracy is that they have a large number of parameters that need adjustment. Global models can be relatively simpler, if suitable models describing the whole data set can be found. This is not easy.

Before an attempt to determine a mapping function, it is first necessary to establish whether a time series is in fact generated from a chaotic deterministic system. This is done by estimating certain invariant measures that characterize the time series, the most important of which are the correlation dimension, Lyapunov exponent, and Kolmogorov–Sinai (KS) entropy. Other invariant measures include the capacity dimension, topological dimension, fractal dimension, and Hausdorff dimension, all of which have non-integer values. The methods of estimation are varied and with limitations.

This chapter first gives a brief comparison between random and chaotic deterministic systems followed by introductions to the basic mathematical setting of a time series as a dynamical system, the concept of embedding including the embedding theorem and methods of estimating the embedding dimension, delay time, phase space reconstruction, and prediction. Next, the inverse problem and how it can be used to distinguish chaos from random processes, tests for detecting non-linearity and determinism, and the problem of noise in real data as well as methods of noise reduction and noise level estimation are addressed. Lastly, a brief review of some examples of applications of the dynamical systems approach to hydrological systems is given.

11.2 RANDOM VERSUS CHAOTIC DETERMINISTIC SYSTEMS

A random system has an infinite number of degrees of freedom and can only be described by a probability distribution function. A random variable takes real values and the properties are described by the statistics such as the mean and the variance. A probability distribution function does not give any information on the time or the frequency component of the process. A random process is sometimes referred to as a stochastic process. Examples of random processes include Brownian motion, random walk, stock market fluctuation, currency fluctuation, electrocardiogram (ECG), blood pressure, among many others. Most geophysical and meteorological processes have a deterministic and a stochastic component. The deterministic component generally governs the long-term behaviour, whereas the stochastic component the short-term behaviour. More details of stochastic processes and their modelling paradigms are given in Chapter 7.

A deterministic dynamical system, on the other hand, has a finite number of degrees of freedom, usually low, and is governed by certain laws or rules. With sufficient information of the current state, it is possible to predict the future state using the laws or rules that govern the system. A deterministic dynamical system can be discrete or continuous; however, only discrete systems are considered here as they are the only ones that can be processed and modelled by a digital computer. Extension to continuous systems is straightforward, and in fact, a continuous system can be thought of as a special case of discrete systems in which the sampling interval is made infinitely small.

Not all deterministic systems are chaotic, but all chaotic systems are deterministic. 'Chaos' literally means 'disorder'. However, mathematical chaos are deterministic and thus 'orderly' in some sense, and have well defined statistics. Chaos theory became formalized in the mid-19th century when it became evident that the linear theory could not explain certain phenomena. The inherent 'noise' (measurement error) present in any observation was then considered as a component of system dynamics. The main catalyst for the development of chaos theory was the computer. Much of the mathematics of chaos involves the repeated iteration of simple mathematical functions, which would be impractical to do manually. Computers made these repeated calculations possible and practical.

In mathematics and physics, the chaos theory describes the behaviour of certain non-linear dynamical systems that may exhibit dynamics that are highly sensitive to initial conditions (popularly referred to as the 'butterfly effect'[1]). As a result of this sensitivity, which manifests itself as an exponential growth of perturbations in the initial conditions, the behaviour of chaotic systems appears to be random. This happens even though the systems are deterministic, meaning that their future dynamics are fully defined by their initial conditions, with no random elements involved. Such systems are governed by physical laws and are very difficult to predict accurately. Weather forecasting, which is predictable in the short term if enough information is available but unpredictable with certainty in the long term, is a commonly cited example. The principal characteristics of deterministic chaotic systems include the difficulty or impossibility to predict the long-term behaviour; sensitivity to initial conditions, meaning that two trajectories starting from very close initial conditions very rapidly moving to divergent states; and the exponential amplification of errors.

All methods for distinguishing deterministic and stochastic processes rely on the fact that a deterministic system always evolves in the same way from a given starting point. Thus, given a time series to test for determinism, one can pick a test state and search the time series for a similar or 'nearby' state and compare their respective time evolutions. A deterministic system will have an error that either remains small (stable, regular solution) or increases exponentially with time (chaos). A stochastic system will have randomly distributed errors. Certain seemingly stochastic time series can in fact be evolving from deterministic non-linear dynamical systems.

11.3 TIME SERIES AS A DYNAMICAL SYSTEM

A continuous-time dynamical system can be described by a set of ordinary differential equations (ODEs) as follows:

$$\frac{dX(t)}{dt} = F(X(t)) \tag{11.1}$$

where X is an N-dimensional vector. This is a set of first-order autonomous ODEs, and the function $F(X(t))$ is continuously differentiable as many times as desirable.

---

[1] See Section 10.5.1 for a historical introduction.

In principle, for any given initial state of the system $X(0)$, future states of the system can be obtained by solving Equation 11.1. For example, when $N = 3$, $X = \{x^{(1)}, x^{(2)}, x^{(3)}\}$, and the graphical representation of the three components in a set of orthogonal axes is known as the phase space. The path followed by the system as it evolves with time is known as the trajectory (or orbit). A continuous-time dynamical system is sometimes referred to as a flow, analogous to the paths followed by the particles of a flowing fluid (Lagrangian description of fluid flow). In discrete systems, the corresponding description is called a map, which in vector form is

$$X_{n+1} = \phi(X_n) \tag{11.2}$$

where X_n and X_{n+1} are the magnitudes of the input function at times n and $n + 1$, respectively, and ϕ is a function and/or vector/matrix that maps the input function to a corresponding output function. The system is said to be deterministic if ϕ is deterministic. Furthermore, if the function ϕ is time invariant, the system is said to be stationary. Non-stationary dynamical systems are extremely difficult to analyse, and for all practical purposes, the types of systems that we normally treat are assumed to be stationary, although they may not really be so.

In the same way as in a continuous-time system, the future states of a discrete system can also be obtained recursively, given the initial state. It is possible to reduce the dimension N of a continuous-time system via Poincaré sections[2] to a discrete map of dimension $N - 1$. In continuous-time systems, chaos can occur for $N \geq 3$. Therefore, for one-dimensional (1D) and 2D continuous-time systems, chaotic behaviour can be ruled out. In the case of discrete maps, the existence of chaos depends upon the type of map. For invertible maps, meaning that ϕ^{-1} exists, and that

$$X_n = \phi^{-1}(X_{n+1}) \tag{11.3}$$

chaos can occur only for $N \geq 2$. On the other hand, for non-invertible maps, there can be chaos even for 1D systems.

11.3.1 Dynamical system

A dynamical system consists of an abstract phase (or state) space whose coordinates describe the state of the system at any instant. The system is governed by a dynamical rule that specifies the immediate future of all state variables, given their initial values. Mathematically, a dynamical system is described by an initial value problem.

A dynamical system can be represented by a set of ODEs of the form[3]

$$\frac{dY}{dt} = F(Y) \tag{11.4}$$

where $Y = \{y_1, y_2, \ldots\}$ is a multidimensional point S in the phase space that represents a state of the system. The dimension of the phase space is associated with the number of degrees of freedom of the system. In Equation 11.4, $F(Y)$ is a non-linear operator, or mapping function. Given an initial state $Y(0)$, the state or solution at any time t may be written as $Y(t) = \phi_t Y(0)$. The solutions to all possible initial values, $\phi_t S$, may be considered as the flow of points in the phase space.

---

[2] The Poincaré surface of section is a section transverse to the flow. It can be normal to the flow but not necessarily as long as it is not parallel to the flow. The surface of section that pierces through the trajectory reduces an N-dimensional continuous-time dynamical system to a $(N - 1)$-dimensional map.

[3] The notation Y is used to imply the description in the phase space as opposed to X in the original space.

Initially, the dimension of the set $\phi_t S$ is the same as the dimension of S itself. However, in dissipative systems, as the system evolves (asymptotic dynamics), the flow contracts into sets of lower dimensions, which are referred to as attractors. Mathematically, an attractor is considered as a set in the phase space that has a neighbourhood in which every point stays nearby and approaches the attractor as time goes to infinity. Examples include a ball rolling inside a bowl, which settles down to the bottom and the limit cycle, which is a periodic orbit. In the former case, the attractor is a point and the neighbourhood of the equilibrium point is known as the basin of attraction, whereas in the latter case it is a periodic cycle. An attractor exists within a smooth manifold[4] M, which will have a dimension less than that of S. The lowered dimensionality may be explained physically as due to the self-organization of the system.

A strange attractor is an attractor that has a fractal dimension. It is often associated with chaotic dynamics, and sometimes referred to as a chaotic attractor, but there are also non-chaotic strange attractors (Grebogi et al., 1984).

The complete solution of Equation 11.4 is equivalent to knowing the family of ϕ_t. This, however, is practically not possible. Instead, solutions are sought in an equivalent system via a process called embedding.

11.3.2 Sensitivity to initial conditions

An important characteristic of chaotic dynamical systems is the sensitivity to initial conditions. Orbits that are nearby initially diverge exponentially as they evolve with time; that is, small errors due to noise, measurement, and round-off magnify rapidly, causing the solution to change from what it should be if there were no errors. In the generation of the orbits of chaotic maps, several thousands of iterations are needed and if a round-off error of the order of 10^{-14} (which is typical for single precision computations) is propagated through thousands of iterations, the errors may become larger than the signal. This means that long-term prediction of chaotic systems is not possible. This phenomenon is well illustrated in the Lorenz system (see Chapter 10). Non-chaotic systems do not lead to long-term divergence of solutions. Small initial errors lead to small errors for all times.

The rate at which nearby orbits converge or diverge can be measured by the Lyapunov exponent. There are as many Lyapunov exponents as there are dimensions in the state space of a system; however, the largest is usually the most important. If the largest Lyapunov exponent is negative, the orbits converge in time and the dynamical system is insensitive to initial conditions. On the other hand, if the largest Lyapunov exponent is positive, the distance between initially nearby orbits grow exponentially in time and the system is sensitive to initial conditions.

11.4 EMBEDDING

Dynamical systems in phase space of different dimensions may have equivalence if their asymptotic behaviour is confined to attracting manifolds of the same dimension. This leads to the possibility that systems that are physically quite different may lead to qualitatively equivalent dynamics. The equivalent system may be arrived at by embedding, which is the application of dynamical systems techniques to a data series that involves the use of a single observable as a function of time with delay coordinates. It creates a phase space portrait from a single data series that is topologically equivalent to that of the full attractor. It is based on the idea of using delay coordinates to approximate the derivatives in ODEs. Such representations have been in

---

[4] Manifold is an abstract topological space that is locally Euclidean and in which every point has a neighbourhood.

use in time series modelling (e.g., linear autoregressive modelling; Yule, 1927), as well as in Newtonian dynamics (e.g., n-body problem, see Chapter 10).

Mathematically, embedding is a smooth map, Φ, from a manifold M of phase space S to a Euclidean space U such that its image $\Phi(M)$ is a smooth submanifold of U and that Φ is a diffeomorphism[5] between M and $\Phi(M)$. The manifold M lies within an attractor of dimension d of the Euclidean space U of dimension k $(k > d)$. If $\Phi(M)$ is one-to-one equivalent, then the inverse of $\Phi(M)$ can be defined. Embedding can also be interpreted as a mapping from a higher-dimensional full system of dimension k $(>d, m)$ to a lower-dimensional equivalent system of dimension m that has one-to-one correspondence.

11.4.1 Embedding theorem

The embedding theorem for Euclidean spaces was first given by Whitney (1936) who proved that a smooth manifold of dimension d within the k-dimensional space $(k > d)$ can be topologically embedded into an m-dimensional space provided that $m > 2d$. He stated that a generic[6] map from a d-dimensional manifold to a $(2d + 1)$-dimensional Euclidean space is an embedding. The image of the d-dimensional manifold is completely unfolded in the larger m-dimensional space. There is a unique one-to-one mapping from d-manifold to m-manifold. If $m > 2d$, then $\Phi(M)$ does not intersect itself.

As $2d + 1$ independent signals measured from a system can be considered as a map from the set of states to $2d + 1$ dimensional space, Whitney's theorem implies that each state can be identified uniquely by a vector of $2d + 1$ measurements, thereby reconstructing the phase space.

In reality, however, it is practically impossible to measure all the $(2d + 1)$ components of the signal vector. Takens (1981), by his delay embedding theorem, proved that the same goal could be achieved with a time series of measurements of a single component. He proved that instead of $2d + 1$ generic signals, the time-delayed measurements of one generic signal would be sufficient to embed the d-dimensional manifold.

Mathematically, Takens' theorem is stated as follows (Ott et al., 1994, p. 75):

Consider a d-dimensional compact manifold M. For pairs (F, f), where F is a smooth (C^2) vector field and f is a smooth function on M, it is a generic property that

$$\Phi_{F, f}(Y): M \rightarrow R^{m \geq 2d+1}$$

defined by

$$\Phi_{F, f}(Y) = \{f(y), f(\phi_1(y)), \ldots f(\phi_{2d}(y))\} \tag{11.5}$$

is an embedding where ϕ_t is the flow of F. The function $f(y)$ corresponds to the measurement made on the system in a state given by $y \in M$.

The delay embedding theorem provides a framework within which a chaotic dynamical system can be reconstructed from a single sequence of observations of the state. The reconstructed phase space preserves the properties of the dynamical system that do not change

---

[5] Two manifolds (or vectors) F and G are said to be C^k equivalent if there exists a C^k diffeomorphism, Φ, which takes orbits $\phi_t(Y)$ of F to orbits $\psi_{t'}(\Phi(Y))$ of G while preserving their orientation. Φ is called C^k diffeomorphism if the functions Φ and Φ^{-1} are k-times continuously differentiable. In other words, it means that Φ is invertible, non-linear, and smooth. $k = 0$ corresponds to homeomorphism, meaning continuous one-to-one correspondence in both directions, and is also known as topological equivalence. $k \geq 1$ corresponds to differentiable equivalence (Ott et al., 1994, p. 74).

[6] A property is generic over a class if any system in the class can be perturbed into one with that property. It is a precise way of defining a set, which is near every system in the collection (dense), which is not too big.

under smooth coordinate changes but does not preserve the shape of the phase space (dissipative characteristic). The single sequence may be written as

$$X(t) = \{x(t), x(t - \tau), x(t - 2\tau),..., x(t - (m - 1)\tau)\} \tag{11.6}$$

In Equation 11.6, m denotes the embedding dimension and τ the delay time. For example, for a time series of length $N = 10$, with a window length n of 5, $m = N - (n - 1) = 6$. If $\tau = 1$, then,

$$X(t) = \begin{bmatrix} x_1, x_2, x_3, x_4, x_5 \\ x_2, x_3, x_4, x_5, x_6 \\ x_3, x_4, x_5, x_6, x_7 \\ x_4, x_5, x_6, x_7, x_8 \\ x_5, x_6, x_7, x_8, x_9 \\ x_6, x_7, x_8, x_9, x_{10} \end{bmatrix} \tag{11.7}$$

The embedding theorem further states that there is diffeomorphism between the reconstructed and original phase spaces provided

$$m \geq 2d + 1$$

This implies that the invariants such as the correlation dimension, Lyapunov exponent, entropy, etc., of the reconstructed phase space are the same as those of the original attractor. It also means that there is sufficient information contained in the vector $X(t)$ (Equation 11.6) about the dynamical system when the reconstructed trajectory is disentangled. The rationale behind the embedding theorem can be interpreted as follows.

Suppose there is only one measurable variable in a dynamical system, it alone is not sufficient to describe the underlying dynamics. However, in addition to the scalar value $x(t)$, if the derivatives (including higher-order ones) such as $\frac{dx(t)}{dt}$, $\frac{d^2x(t)}{dx^2(t)}$, $\frac{d^3x(t)}{dx^3(t)}$, $\frac{d^4x(t)}{dx^4(t)}$,..., $\frac{d^{d_e}x(t)}{dx^{d_e}(t)}$ up to a certain finite level d_e of the measurable variable are also available, and if the dimension of the system is less than d_e, then there is sufficient information to describe the system completely. When the sampling interval is small, the derivatives are equivalent to the values of the variable at d_e different times. In other words,

$$X_n \to \left(x_n, x_{n-1}, x_{n-2},..., x_{n-d_e}\right)$$

contains the same information as the original system provided that 'd_e',[7] referred to as the embedding dimension of the dynamical system, is sufficiently large, data record is long enough, sampling interval is short enough, and the measurements are 'noise free'. These conditions are rarely satisfied in reality, but the embedded points behave approximately the same as the original system. To proceed further, we need a good estimate of the embedding dimension 'd_e'.

The new (embedded) system contains the same information as the original system. This interpretation is not exact from a theoretical point of view, but good enough for practical

---

[7] The variables 'd_e' and m both refer to the embedding dimension.

purposes. The embedding theorem is more prone to error in the presence of noise, a topic that will be dealt with later in the chapter. The reconstruction of the phase space diagram using Takens' theorem requires two parameters, the delay time τ and the embedding dimension m, which are generally not known *a priori*.

Delay time embedding is not the only method of embedding scalar data into multidimensional phase space, but it is the most widely used.

11.4.2 Embedding dimension

Embedding dimension can be thought of as the minimum number of dimensions (sometimes referred to as the degrees of freedom) in the state space needed to describe the system adequately. From a mathematical point of view, any dimension that satisfies Takens' theorem would work. However, if a dimension greater than the optimum one is used, that would add extra computational burden unnecessarily. More computations will also add more round-off errors.

The optimum (minimum) embedding dimension can be determined by a trial-and-error approach in which the performance of models with assumed different embedding dimensions, starting from a low value and gradually increasing, are compared and the one corresponding to the best model is selected. In this method, *a priori* knowledge of the model that can adequately describe the system is necessary. Another approach is to increase the embedding dimension gradually until trajectories no longer appear to intersect. This approach, however, does not work for high dimensions and in the presence of noise. It also has the problem of creating artificial symmetry of the phase space as the embedding dimension is increased. It is also possible to estimate the optimum embedding dimension as that corresponding to the saturation value of some dynamical invariant such as the correlation integral (see Section 10.2.10) for varying embedding dimensions (Grassberger and Procaccia, 1983a,b). The rationale for this approach is that any dynamic invariant should be independent of the embedding dimension provided it is large. Other methods of estimating the embedding dimension include the false nearest neighbour (FNN) method (Abarbanel, 1996), singular value decomposition (SVD) and the method of true vector fields.

11.4.2.1 False nearest neighbour (FNN) method

FNNs are points that appear to be nearest neighbours because the embedding space is too small, or viewed in a smaller embedding space; that is, $d < d_e$. In a small embedding dimension phase space, some neighbours that are far away will appear to be nearby because they are projected onto a lower-dimensional space. When the number of FNNs decreases to zero, the attractor is unfolded or embedded in m-dimensional Euclidean space. A criterion to measure the embedding error is the increase in distance between two neighbours when moving from dimension d to $d + 1$.

When the embedding dimension increases from d to $d + 1$, it is possible to differentiate between points on the trajectory that are true neighbours and those on the trajectory that are false neighbours. When the embedding dimension has exceeded the minimum embedding dimension d_e, all neighbours of every point in the phase space will be true neighbours.

The FNN method enables the determination of the optimum (minimum) embedding dimension needed to unfold the attractor from the data. A lower embedding dimension will cause some parts of the attractor that are far apart in the original phase space to overlap in the reconstructed phase space. The method detects the fraction of the nearest neighbours in the data set viewed in dimension d that fails to remain in the neighbourhood in dimension $d + 1$. The dimension d attains the optimum dimension d_e when all nearest neighbours are true; that is, they do not significantly move when the dimension is increased from d to $d + 1$.

The FNN method (Kennel et al., 1992; Abarbanel, 1996) works as follows:
For a point $Y(t)$ at time level t, defined as

$$Y(t) = (y(t), y(t - \tau), y(t - 2\tau), ..., y(t - (d_e - 1)\tau)) \tag{11.8}$$

in the reconstructed phase space of embedding dimension d_e, and delay time τ, there must exist another point $Y^{NN}(s)$

$$Y^{NN}(s) = (y(s), y(s - \tau), y(s - 2\tau), ..., y(s - (d_e - 1)\tau)) \tag{11.9}$$

at time level $s \neq t$, such that for every point in reconstructed phase space $Z(u)$, defined as

$$Z(u) = (y(u), y(u - \tau), y(u - 2\tau), ..., y(u - (d_e - 1)\tau)) \tag{11.10}$$

at time level $u \neq t$

$$\left\| Y^{NN}(s) - Y(t) \right\| \leq \left\| Z(u) - Y(t) \right\| \tag{11.11}$$

The norm here is usually the Euclidean norm and the nearest neighbour $Y^{NN}(t)$ of $Y(t)$ is $Y^{NN}(s)$, and is given by

$$Y^{NN}(t) = (y^{NN}(t), y^{NN}(t - \tau), y^{NN}(t - 2\tau), ..., y^{NN}(t - (d_e - 1)\tau)) \tag{11.12}$$

The time level t of $Y^{NN}(t)$ has very little relation to the time level at which $Y(t)$ appears.

The point $Y^{NN}(t)$ is called an FNN of $Y(t)$ if it arrives in its neighbourhood by projection from a higher dimension. If they are genuine neighbours, they became close due to the system dynamics and not by projection from a higher dimension, and should separate slowly. If the neighbours for a certain embedding dimension came by projection from a higher dimension, then that embedding dimension cannot unfold the attractor. If most points in the phase space have FNNs, then the number d_e would not be the embedding dimension for the chaotic data. By comparing the distance between the vectors $Y(t)$ and $Y^{NN}(t)$ in dimension d_e with that in dimension $d_e + 1$, it is possible to establish whether a nearest neighbour is true or false.

This is checked using the approximate condition that if $Y^{NN}(t)$ is the nearest neighbour of $Y(t)$, and if

$$\frac{\left| y(t + d_e\tau) - y^{NN}(t + d_e\tau) \right|}{\left\| Y(t) - Y^{NN}(t) \right\|} > R_T \tag{11.13}$$

then $Y^{NN}(t)$ is an FNN of $Y(t)$ (Kennel et al., 1992; Abarbanel, 1996). In Equation 11.13, R_T depends on the spatial distribution of the embedded data and varies in the range $10 \leq R_T \leq 50$ (Abarbanel et al., 1993) although a range of $10 \leq R_T \leq 30$ has also been reported (Small, 2005). If R_T is too small, true nearby neighbours will be counted as false and if it is too large, some false neighbours will not be counted. A value of $R_T = 15$ seems a good compromise. By checking for every point in the phase space whether it has an FNN, the percentage of FNN points could be obtained. If for a certain d_e, the percentage of FNN points is less than 5, it is accepted as the embedding dimension for the data set. For clean data from a chaotic system, it is expected that the percentage of FNNs drops from nearly 100 in dimension 1 to zero when d_e is reached. Illustration of this are given in Figure 11.1a through c for the logistic (see Equation 3.19), Hènon (see Equation 10.81), and Lorenz data sets (see

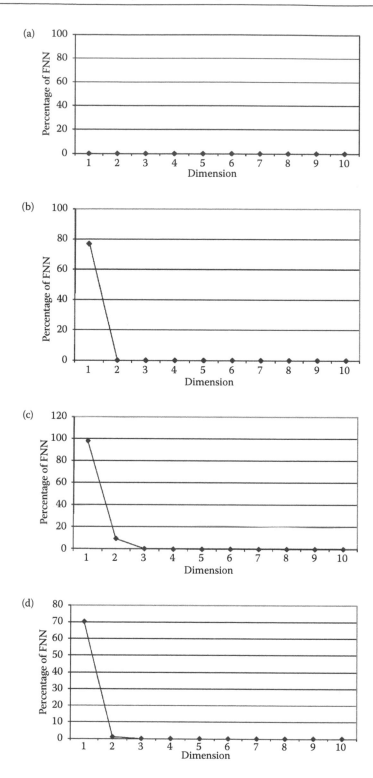

Figure 11.1 (a) Percentage of global false nearest neighbours for logistic data, (b) percentage of global false nearest neighbours for Hénon data, (c) percentage of global false nearest neighbours for Lorenz data, (d) percentage of global false nearest neighbours for Mekong data at Pakse.

Equation 10.82), which are known to be chaotic under certain parameter conditions, and which have embedding dimensions of 1, 2, and 3, respectively. Figure 11.1d shows a similar plot for daily discharges of the Mekong River measured at the Pakse gauging station.

Certain maps such as the logistic map, the Hénon map, and the Lorenz map are known to have the optimal or exact embedding dimensions. When the optimal embedding dimension is not known, such as in an observed time series, the correlation dimension is calculated for increasing embedding dimensions until it reaches a saturation value.

11.4.2.2 Singular value decomposition (SVD)

SVD refers to the decomposition (factorization) of any rectangular matrix into three matrices, an orthogonal[8] matrix, a diagonal[9] matrix, and the transpose of another orthogonal matrix as follows:

$$A = USV^T \tag{11.14}$$

where A is an $M \times N$ matrix, U is an $N \times N$ orthogonal matrix, S is an $N \times M$ diagonal matrix, and V is an $N \times N$ orthogonal matrix. It can be thought of as a method of transforming a set of correlated data into a set of uncorrelated data. It can also be thought of as a method of identifying and ordering data along dimensions that have most variations with the objective of reducing the dimensionality of the problem. In this sense, SVD has the same effect as the principal components analysis (PCA). The diagonal terms of S, $s_{i,j}$, $i = j$ are the singular values of A ($s_{i,j} = 0$ for $i \neq j$).

A great deal of information about the system could be derived from singular spectrum analysis without a prior knowledge of the underlying dynamics of the system. It is based on PCA of the univariate time series. The analysis begins with the construction of the trajectory matrix, A.

Consider an n-window of a time series of N data points, which will result in $m = (N - (n - 1))$ vectors as follows:

$$X_1 = \{x(1), x(1 + \tau), x(1 + 2\tau), x(1 + 3\tau),..., x(1 + (n - 1)\tau)\} \tag{11.15a}$$

$$X_2 = \{x(2), x(2 + \tau), x(2 + 2\tau), x(2 + 3\tau),..., x(2 + (n - 1)\tau)\} \tag{11.15b}$$

$$X_3 = \{x(3), x(3 + \tau), x(3 + 2\tau), x(3 + 3\tau),..., x(3 + (n - 1)\tau)\} \tag{11.15c}$$

...

...

$$X_m = \{x(m), x(m + \tau), x(m + 2\tau), x(m + 3\tau),..., x(m + (n - 1)\tau)\} \tag{11.15m}$$

where τ is the time interval between successive observation of x. In Equation 11.15m, m is the embedding dimension of the series. Without loss of generality, τ may be assumed to have a value of unity.

---

[8] The inverse of an orthogonal matrix M is its own transpose; that is, $M^{-1} = M^T$, implying $M^T M = M M^T = I$ (identity matrix).
[9] A diagonal matrix M (which need not be square) has $m_{ij} = 0$ for $i \neq j$.

For example, if $n = 5$ and $N = 10$, the data series will be $x(1)$, $x(2)$, $x(3),..., x(10)$. Then,

$$X_1 = \{x(1), x(1 + \tau), x(1 + 2\tau), x(1 + 3\tau), x(1 + 4\tau)\} \tag{11.16a}$$

$$X_2 = \{x(2), x(2 + \tau), x(2 + 2\tau), x(2 + 3\tau), x(2 + 4\tau)\} \tag{11.16b}$$

$$X_3 = \{x(3), x(3 + \tau), x(3 + 2\tau), x(3 + 3\tau), x(3 + 4\tau)\} \tag{11.16c}$$

$$X_4 = \{x(4), x(4 + \tau), x(4 + 2\tau), x(4 + 3\tau), x(4 + 4\tau)\} \tag{11.16d}$$

$$X_5 = \{x(5), x(5 + \tau), x(5 + 2\tau), x(5 + 3\tau), x(5 + 4\tau)\} \tag{11.16e}$$

$$X_6 = \{x(6), x(6 + \tau), x(6 + 2\tau), x(6 + 3\tau), x(6 + 4\tau)\} \tag{11.16f}$$

and, if $\tau = 1$,

$$X_1 = \{x(1), x(2), x(3), x(4), x(5)\} \tag{11.17a}$$

$$X_2 = \{x(2), x(3), x(4), x(5), x(6)\} \tag{11.17b}$$

$$X_3 = \{x(3), x(4), x(5), x(6), x(7)\} \tag{11.17c}$$

$$X_4 = \{x(4), x(5), x(6), x(7), x(8)\} \tag{11.17d}$$

$$X_5 = \{x(5), x(6), x(7), x(8), x(9)\} \tag{11.17e}$$

$$X_6 = \{x(6), x(7), x(8), x(9), x(10)\} \tag{11.17f}$$

These vectors are then used to construct the trajectory matrix A, which contains all the records of patterns that have occurred within that window:

$$A = \frac{1}{\sqrt{N}} \begin{bmatrix} X_1 \\ X_2 \\ X_3 \\ ... \\ ... \\ X_m \end{bmatrix} \tag{11.18}$$

The division by $\sqrt{N}$ is to normalize the matrix. In expanded form it is

$$A = \frac{1}{\sqrt{N}} \begin{bmatrix} x(1) & x(1+\tau) & x(1+2\tau) & ... & x(1+(n-1)\tau) \\ x(2) & x(2+\tau) & x(2+2\tau) & ... & x(2+(n-2)\tau) \\ x(3) & x(3+\tau) & x(3+2\tau) & ... & x(3+(n-3)\tau) \\ ... & ... & ... & ... & ... \\ ... & ... & ... & ... & ... \\ x(N-(n-1)\tau) & x(N-(n-2)\tau) & x(N-(n-3)\tau) & ... & x(N-(n-1)+(n-1)\tau) \end{bmatrix} \tag{11.19}$$

Since a continuous-time multivariate system can be described by the time-lagged values of a univariate system, the singular spectral analysis takes a univariate time series and makes it a multivariate (or multidimensional) time series (Elsner and Tsonis, 1997). The trajectory matrix defines the space of lagged coordinates as shown in Equation 11.19.

Computation of SVD is done by first finding the eigenvalues and eigenvectors[10] of AA^T and A^TA. The eigenvectors of AA^T and A^TA constitute the columns of U (left eigenvector) and V (right eigenvector), respectively. Non-zero singular values in S (diagonal terms) are the square roots of eigenvalues from AA^T or A^TA. They are arranged in descending order. The singular values are always real numbers. If the matrix A is real, then U and V are also real. The product matrix A^TA can be written as

$$A^TA = \frac{1}{N} \sum_{i=1}^{N} A_i A_i^T \tag{11.20a}$$

which, when expanded takes the form

$$\frac{1}{N} \begin{bmatrix} \sum_{i=1}^{N} x(i)x(i) & \sum_{i=1}^{N} x(i)x(i+1) & \cdots & \sum_{i=1}^{N} x(i)x(i+n-1) \\ \sum_{i=1}^{N} x(i+1)x(i) & \sum_{i=1}^{N} x(i+1)x(i+1) & \cdots & \sum_{i=1}^{N} x(i+1)x(i+n-1) \\ \cdots & \cdots & \cdots \cdots \\ \cdots & \cdots & \cdots \cdots \\ \sum_{i=1}^{N} x(i+n-1)x(i) & \sum_{i=1}^{N} x(i+n-1)x(i+1) & \cdots & \sum_{i=1}^{N} x(i+n-1)x(i+n-1) \end{bmatrix} \tag{11.20b}$$

Equation 11.20 is the covariance matrix of the components of the time series. Division by N is to normalize the matrix. SVD is then applied to the covariance matrix of the components of the time series. SVD transforms the trajectory matrix into two orthogonal and one diagonal matrix as shown in Equation 11.14.

The diagonal terms in the diagonal matrix give the singular values. The projections of the trajectory onto the columns of the empirical orthogonal matrix give the principal components. The rank[11] of the covariance matrix, n' ($n' \leq n$), which is equal to the number of non-zero eigenvalues, gives the minimum dimension of the subspace of the embedding space. The interpretation of the singular value analysis, however, should be done with caution as it does not recognize the difference between two processes that have almost equal Fourier spectra. This analysis assumes that the time series is noise free. Examples of numerical computations of SVD can be found in several websites (e.g., MATLAB® ver 1.1 (March 2004) PROPACK 1.1, Matlab.tar.gz). Although univariate time series are generically sufficient for phase space reconstruction, there is no guarantee that it is so. It has been shown (Ataei et al., 2003) that the SVD method of computing the optimum embedding dimension is not efficient for certain types of time series and that the problem can be resolved by using the interaction between multiple time series.

---

[10]An eigenvector is a non-zero vector that satisfies the equation $Ax = \lambda x$ where A is a square matrix, x is the eigenvector, and λ is the eigenvalue. λ can be determined by setting the condition that the determinant of the coefficient matrix is zero.

[11]The rank of a matrix is the maximum number of independent rows (or columns).

11.4.3 Delay time

Normally, the solution of Equation 11.1 or 11.2 requires measurement of all components of the vector X. However, this may not be practical in a complex system with many interacting components. In such situations, it is possible to reconstruct the phase space from a single scalar function of the state vector. This is done using the delay coordinate vector (Takens, 1981), which is given as

$$y(t) = \{x(t),\, x(t - \tau),\, x(t - 2\tau),\ldots,\, x(t - (m - 1)\tau)\} \tag{11.21}$$

where τ is known as the delay time. It is possible to express $y(t)$ as a function of $x(t)$ because the measurement $y(t)$ depends on the system state $x(t)$; that is, $y(t) = f(x(t))$.

Delay time is important for the phase space reconstruction. If it is too small, data may not be independent, and if it is too large, the link between two successive values may become random with respect to each other, thereby missing out some important information. Therefore, a compromise should be made. The delay time τ may be chosen as the lag time at which the autocorrelation falls below a threshold value that is commonly defined as $1/e$ (decorrelation time), especially if the autocorrelation function is approximately exponential (Tsonis and Elsner, 1988). Another method is to use the lag time at which the autocorrelation first becomes zero if it crosses the zero line (Holzfuss and Mayer-Kress, 1986; Mpitsos et al., 1987). Choosing delay time according to this criterion would make, on average over observations, $x(t)$ and $x(t + \tau)$ linearly independent. This does not, however, say anything about their non-linear dependence. It is therefore necessary to look for other criteria whereby the non-linear independence can be established. One such approach is to look at it from the point of view of information theory (Shaw, 1984; Fraser and Swinney, 1986; Fraser, 1989a,b), which leads to the mutual information criterion.

11.4.3.1 Average mutual information

The idea behind mutual information is about how much one can learn about the measurement of a system at a particular time from the measurements of the system made at a different time. This idea can also be extended to two different systems, for example, how much one can learn from a measurement a_i made at time i of system A about the measurement b_j made at time j of system B. Using probability distributions of the two outcomes from the two systems, the amount one can learn in bits[12] from the measurements made is described by (Gallager, 1968)

$$I_{AB}(a_i, b_j) = \log_2 \left[\frac{P_{AB}(a_i, b_j)}{P_A(a_i)P_B(b_j)} \right] \tag{11.22}$$

where I_{AB} is the mutual information of the two measurements a_i and b_j, P_A is the probability of observing a out of the set A, P_B is the probability of observing b in a measurement of B, and P_{AB} is the joint probability of measurement of a and b. The mutual information is symmetric in how much can be learned about b_j from measurements of a_i. Average mutual information is the average of all possible measurements from systems A and B and is given as

$$I_{AB}(T) = \sum_{a_i, b_j} P_{AB}(a_i, b_j) I_{AB}(a_i, b_j) \tag{11.23}$$

---

[12] When logarithms to the base 2 are taken in Equation 11.22, the units of I_{AB} will be in bits. In information theory, one bit is typically defined as the uncertainty of a binary random variable that is 0 or 1 with equal probability, or the information that is gained when the value of such a variable becomes known.

In the context of a time series, the set A can be thought of as the measurement $x(t)$ and the measurement $x(t + T)$ can be thought of as set B. Then, the average mutual information $I(T)$ between the observations at time t and $t + T$ is given as (Gallager, 1968)

$$I(T) = \sum_{t=1}^{N} P(x(t), x(t+T)) \log_2 \left[\frac{P(x(t), x(t+T))}{P(x(t))P(x(t+T))} \right] \tag{11.24}$$

with $I(T) \geq 0$.

The average mutual information that is defined for non-linear systems is analogous to the correlation function for a linear system. When the measurements of systems A and B are independent,

$$P_{AB}(a, b) = P_a(a)P_b(b) \tag{11.25}$$

and

$$I_{AB}(a_i, b_j) = 0 \tag{11.26}$$

The computation of $I(T)$ can be done using frequency diagrams (or histograms) of the time series $x(t)$, which in the case of long stationary time series would be the same as the frequency diagram of $x(t + T)$. The frequency diagram is equivalent to the probability distribution. The joint distribution is obtained by counting the number of times a box in the $x(t)$ versus $x(t + T)$ plane is occupied, which is then normalized.

Fraser and Swinney (1986) have suggested that the time T_m, where the first minimum of $I(T)$ occurs, should be taken as the delay time for phase space reconstruction. It has also been suggested (Fraser and Swinney, 1986; Fraser, 1989a,b) that the average mutual information function is a kind of non-linear autocorrelation function. When the data are periodic, a value equal to a quarter of the average period also has been suggested. This is approximately the same as the time at which the autocorrelation first becomes zero. The chosen delay time should be large enough so that the new coordinates contain as much new information as possible and should be small enough so that the coordinates are not entirely independent.

In some situations, there may not exist a minimum in the mutual information. In such situations, the value of T_m should be chosen intuitively. When a minimum of the average mutual information occurs, it is usually of the same order of magnitude as the first zero crossing of the autocorrelation.

For most analysis, the delay time is assumed a constant. However, non-uniform delay times can also be used (see, e.g., Judd and Mees, 1995). From a mathematical point of view, τ is arbitrary (Kantz and Schreiber, 1997, p. 130). Schouten et al. (1994) used a value of unity for convenience. Despite numerous suggestions, there is no rigorous method of determining an optimal value for τ. Since both parameters are needed for the phase space reconstruction, a question arises as to which one should be determined first. A common practice is to estimate the delay time first, and then choose the embedding dimension. For low-dimensional systems, starting with a delay time of unity seems to be appropriate.

11.4.4 Irregular embeddings

The usual method of estimating the embedding dimension and the delay time assumes that a single embedding dimension is sufficient to reconstruct the phase space diagram. Although this may be sufficient to estimate the invariant measures of the dynamical system, it may not be adequate to uncover the underlying dynamics in a non-linear system. In such situations, irregular embedding would be useful. Instead of the uniform embedding given in Equation 11.6, Judd and Mees (1995) proposed a non-uniform embedding described by the following equation:

$$Y(t) = (y(t - \tau_1), y(t - \tau_2), ..., y(t - \tau_{d_e})) \tag{11.27}$$

in which the parameters $0 \leq \tau_1 \leq \tau_2 \leq \tau_3 \leq, ..., \leq \tau_{d_e}$ are the delay times. The problem then is to find a set of delay times rather than a single delay time. They have, however, pointed out that this may also be not sufficient because some dynamical systems may have important variables that are different in different parts of the phase space. Such situations that lead to non-constant non-uniform embeddings are called variable embeddings.

The problem in irregular embedding is equivalent to finding the parameters $\tau_1, \tau_2, \tau_3, ..., \tau_k$ and the embedding window d_w, where $1 \leq \tau_1 \leq \tau_2 < ... < \tau_i < \tau_{i+1} \leq ... < \tau_k \leq d_w$, such that the embedding

$$Y(t) = (y(t - \tau_1), y(t - \tau_2), ..., y(t - \tau_k)) \tag{11.28}$$

is the 'optimal'. It has also been reported (Small, 2005) that superior results could be achieved by employing an 'optimal embedding strategy' by seeking the embedding that gives the minimum description length. The description length of a time series is approximately given as

$$DL\{x(t)\} \approx \frac{N}{2}(1 + \ln(2\pi)) + \frac{d}{2}\ln\left(\frac{1}{d}\sum_{i=1}^{d}(x_i - \bar{x})^2\right) + \frac{N-d}{2}\ln\left(\frac{1}{N-d}\sum_{i=d+1}^{N}e_i^2\right)$$
$$+ d + DL(d) + DL(\bar{x}) + DL(P) \tag{11.29}$$

where $d = \max_i(\tau_i) = \tau_{d_e}$, $\bar{x}$ is the mean of data, e_i is the model prediction error, and $DL(P)$ is the description length of the model parameters. The description length of an integer is given as

$$DL(d) = \log(d) + \log\{\log(d)\} + ... \tag{11.30}$$

where each term on the right-hand side is an integer and the last term is zero. Also, the two terms $\frac{N}{2}(1 + \ln(2\pi)) + DL(\bar{x})$ are independent of the embedding strategy. Therefore, the optimal embedding strategy is that which minimizes

$$\frac{d}{2}\ln\left(\frac{1}{d}\sum_{i=1}^{d}(x_i - \bar{x})^2\right) + \frac{N-d}{2}\ln\left(\frac{1}{N-d}\sum_{i=d+1}^{N}e_i^2\right) + d + DL(d) + DL(P) \tag{11.31}$$

11.5 PHASE (OR STATE) SPACE RECONSTRUCTION

A dynamical system can be described by a phase space diagram in which the evolution from some known initial state is described by its trajectories. A phase space is a set of coordinates within which the dynamics of the system can be examined. It is a collection of all possible states of the system. It can be finite, countably infinite,[13] or uncountably[14] infinite. A multidimensional point in the phase space specifies the system completely and it is all that is needed to predict its immediate future. The term 'state space' is more widely used in engineering literature, whereas 'phase space' is more common in physics literature. A trajectory would be a curve in a continuous system, whereas it would be a sequence of points in a discrete system. After the transients have died down, the trajectories of most dynamical systems converge to a single subspace called an *attractor*, regardless of the initial conditions. The attractor can be multidimensional and lies in the original phase space of dimension k but has dimension less than k. Reconstruction of the phase space from a univariate time series is carried out using the embedding theorem (Takens, 1981), and the important invariant measures such as the Lyapunov exponent, KS entropy, and correlation dimension, which give useful information for diagnosing, modelling, and predicting the system, are computed in this space.

The thinking behind the phase space reconstruction is that a system with a large number of degrees of freedom can be modelled in a much lower-dimensional space in which the attractor resides. Rather than attempting to study the behaviour of the original system in the higher-dimensional space, investigation of the evolution of the attractor in a lower-dimensional space would provide sufficient information about the original system. It has been argued (Mañé, 1981; Takens, 1981) that if the embedding dimension is large enough, the physical properties of the attractor will be the same when computed in the lagged coordinates and in the physical coordinates.

In autonomous systems (no time-dependent terms in the equations), it may be possible to reduce the dimensionality of the phase space by Poincaré sections. A Poincaré map represents a reduction of an N-dimensional continuous-time dynamical system (sometimes called a flow) to an $(N - 1)$-dimensional map. It is done via a technique called Poincaré surface of sections. Consider, for example, a 3D flow. The surface of section will then be a 2D plane in the 3D phase space. If a plane $x^{(3)} = $ constant is considered (this is not a requirement; a plane of any orientation that is appropriate to the problem would be equally satisfactory), it will intersect the orbit at two points in two successive crossings. Using one of the points as the initial condition, the other point can be uniquely determined in either direction. In this case, the Poincaré map is an invertible 2D map that transforms the coordinates $\{x^{(1)}(t),$ $x^{(2)}(t)\}$ at the t-th intersection with the orbit to $\{x^{(1)}(t + 1), x^{(2)}(t + 1)\}$ at the $(t + 1)$-th intersection. Thus, the condition to have a Poincaré surface of section, or an equivalence of an N-dimensional flow to an $(N - 1)$-dimensional invertible map, is $N \geq 3$.

In a dynamical system, the future state can be determined if the current state and the time derivatives of the system variable are known. Using Taylor series expansion, the first derivative of a scalar measurement of the system can be expressed as

$$\frac{ds}{dt} = \frac{s(t_0 + (n+1)\tau_s) - s(t_0 + n\tau_s)}{\tau_s} \tag{11.32}$$

With a finite value for τ_s, this is an approximation and therefore introduces some error. If higher-order derivatives are estimated in the same way, the errors tend to accumulate as the order of differentiation increases. Instead of estimating the time derivatives by the above approximation, the idea of using time-lagged coordinates for reconstructing the phase space has been suggested by several researchers (Packard et al., 1980; Eckmann and Ruelle, 1985;

---

[13] A set of integers is countably infinite.
[14] A set of all real numbers is uncountably infinite.

Mañé, 1981; Takens, 1981; Sauer et al., 1991). The main idea is that the phase space could be reconstructed by using a new vector consisting of time-lagged values of the observed variable; that is,

$$Y(n) = \{s(n), s(n + T), s(n + 2T), s(n + 3T),..., s(n + (d - 1)T)\} \tag{11.33}$$

where the value of T is yet to be established, and $s(n + jT)$ are some unknown non-linear combinations of the actual physical variables of the system. Although the relationship between $s(n + jT)$ and $Y(n)$ is unknown, it does not matter for the purpose of reconstructing a phase space. Guidelines for the choice of T and d are given in Section 11.4 (delay time and embedding). In phase space reconstruction, the objective is not to reconstruct the infinite dimensional original phase space but to model the evolution of the dynamics in the attractor, which will usually be of a much lower dimension. In other words, the focus is to seek a description of the time-asymptotic behaviour of the system.

11.6 PHASE SPACE PREDICTION

Forecasting in any field is a difficult task. In the classical approach, building a model from first principles and with initial conditions is not always possible, first because many systems are not well understood and the 'first principles' are not complete. In situations where the 'first principles' are clear enough, getting the measurements of initial conditions poses another problem. Therefore, it is necessary to look for alternative methods of modelling and forecasting. One approach is to build models by considering the time series of observations as a realization of a random process. An alternative approach is to consider the time series of observations as a realization of a low-dimensional deterministic system that appear to have random-like behaviour. The latter type of apparent randomness is what is referred to as chaos.

Traditional linear models such as the ARMA type will not work well with chaotic signals. If the parameters of such an ARMA model are estimated by least squares fitting, the error in fitting would be of the same order as the signal itself. Instead, if local piece-wise linear models are fitted in the neighbourhoods of the phase space, the errors can be made significantly smaller depending on the quantity of data and the size of the neighbourhood. Local linear models are the simplest non-linear models and therefore may not work in all situations. It is, however, worth attempting local linear models before embarking on more complicated non-linear models. The evolution of the function $x(t)$ is an initial value problem. Some changes in the evolution process are small and smooth, whereas some are large and abrupt.

The procedure for modelling a chaotic time series starts with embedding the data in a phase space. The details of the embedding processes introduced by Packard et al. (1980) and Takens (1981) have already been described in Sections 11.4 and 11.5. A requirement for embedding is that the embedded vector should have a dimension greater than twice the dimension of the attractor. The next step is to assume a forecasting function that relates the present value to the future value in the form

$$y(t + T) = f_T(y(t)) \tag{11.34}$$

where f_T is the predictor that approximates F of Equation 11.1. For chaotic systems, f_T is necessarily non-linear. Two approaches can be followed for finding f_T: global approximation and local approximation. In global approximation methods, polynomials can be fitted by estimating the parameters by least squares fitting. Other approaches include fitting a new function f_T for each time and recasting Equation 11.34 as a differential equation and express $y(t + T)$ as its integral. In all these approaches, the number of free parameters for a general polynomial is of the order of d_e^p, which becomes intractable for large d_e^p (d_e^p is the embedding dimension; p is the order of the polynomial).

In the local approximation approach, predictions are made using only nearby states. The basic idea in the local approximation method is to break up the domain into local neighbourhoods and fit parameters in each neighbourhood separately. The nearest neighbour is defined using a norm, which is a measure of the distance between two points. For example, $y(t')$ is the nearest neighbour of $y(t)$ for t', which minimizes $\|y(t) - y(t')\|$. The norm can be any one of the three commonly used ones – Euclidean, diagonal, or maximum. Possible local approximation methods include the following (Jayawardena and Lai, 1994):

- 'Zero-order' approximation in which the future value is assumed to be equal to the value of its nearest neighbour as

$$y(t, T) = y(t', T) \tag{11.35}$$

  where $y(t', T)$ is the point in the domain that is nearest to $y(t, T)$. This is the simplest local approximation model.
- Average of several neighbours as

$$y(t, T) = \frac{1}{k} \sum_{i=1}^{k} y(t_i' + T) \tag{11.36}$$

  where the number of neighbours k is determined by trial and error, or using a neighbour searching algorithm.
- Weighted average of several neighbours as

$$y(t, T) = \frac{\displaystyle\sum_{i=1}^{k} w_i y(t_i' + T)}{\displaystyle\sum_{i=1}^{k} w_i} \tag{11.37}$$

  where the weights w_i are dependent on the distances from each neighbour to the current state.
- Same as above but with exponential weightings. This method has been adopted by Sugihara and May (1990) and is of the form

$$y(t, T) = \frac{\displaystyle\sum_{i=1}^{k} \exp(w_i) y(t_i' + T)}{\displaystyle\sum_{i=1}^{k} \exp(w_i)} \tag{11.38}$$

- Linear approximation in which the prediction is obtained by fitting linear polynomials to k pairs $\{y(t_i'), y(t_i' + T)\}$. The predicted state is given as

$$y(t, T) = f_t y_t = A y_t + B \tag{11.39}$$

where the $m \times m$ matrix A is the Jacobian of f_t at y_t, B is an m-vector, and m is the embedding dimension.

A better approach would be to go for 'higher-order' approximations using k nearest neighbours. In the first order, or linear approximation, linear polynomials are fitted to $(y(t')$, $y(t' + T))$. Usually, the number of neighbours is selected to satisfy the condition $k > m + 1$. Although higher-order polynomials can be used, the results have not been better than with first order (Farmer and Sidorowich, 1987).

For non-autonomous systems, where the map or vector field depends explicitly on time, the phase space must include time as a phase space coordinate since the specific time must be specified to know the future state. Thus,

$$\frac{dz}{dt} = F(z, t) \tag{11.40}$$

is a dynamical system on the phase space consisting of (z, t), with the addition of the new dynamics $\dfrac{dt}{dt} = 1$.

11.7 INVERSE PROBLEM

The standard problem in dynamical systems is to describe the asymptotic behaviour from a non-linear map. The inverse problem is to construct the non-linear map given the asymptotic behaviour, which could be the predictive model. Solution of inverse problems can be done by a number of numerical procedures that involve interpolation or approximation of unknown functions from scattered data. The inverse approach can also be used to distinguish low-dimensional chaos from randomness. Casdagli (1989) has given a detailed description of the mathematical formulation of the inverse problem. The following summary is based on Casdagli's work.

In a continuous-time domain, an inverse problem is to construct a smooth map f_∞ in terms of the x_n, $1 \le n < \infty$ such that

$$x_{n+1} = f_\infty(x_n) \tag{11.41}$$

In a discrete-time domain, the problem is to construct a smooth map f_N in terms of x_n, $1 \le n < N$ such that

$$x_{n+1} = f_N(x_n), \ 1 \le n < N - 1 \tag{11.42}$$

Using the scalar variables $s(t)$ of a chaotic time series sampled at discrete intervals of time $t = n\tau$, $1 \le n < \infty$ of which the dynamics is that of a strange attractor lying on an l-dimensional invariant manifold of a dynamical system, the corresponding smooth mapping function f, according to Takens' theorem, satisfies the equation

$$s((n + m)\tau) = f(s(n + m - 1)\tau), s(n + m - 2)\tau, ..., s(n\tau)) \tag{11.43}$$

for $m \le 2l + 1$.

In Equation 11.43, m is referred to as the trial value of the embedding dimension, which has an optimal value equal to m^*. The inverse problem is to determine the smooth function f that satisfies Equation 11.43 for $1 \leq n < N - m$, given the time series $s(n\tau)$ for $1 \leq n < N$. In this formulation, the optimal embedding dimension m^* is not known *a priori*. For example, if $m = 5$, then by the local linear approximation method of prediction, neighbours of $s(N - 4)\tau$, $s(N - 3)\tau,..., s(N\tau)$ of the form $s(n - 4)\tau$, $s(n - 3)\tau,..., s(n\tau)$ would be needed. The predicted value of $s((N + 1)\tau)$ would be some weighted average of the neighbours $s((n + 1)\tau)$.

Interpolation functions can be global, local, radial basis, and artificial neural networks. In global interpolation functions, the parameters are chosen using the linear least squares method. This type is relatively easy to visualize and is guaranteed to converge to the true function as N and the degree of the polynomial increase. The disadvantage is that there will be a very large number of parameters to be determined for large embedding dimensions and higher-degree polynomials. It has also been proposed (Bayly et al., 1987) that rational predictors of the form $\dfrac{p}{q}$, where p and q are polynomials, can be used instead of just a single polynomial. Such predictors also lead to a linear least squares problem under certain parameter conditions.

In the local prediction approach, functions are fitted in a piece-wise manner to all the data points. To predict the value at a point x, k nearest neighbours of $x_1, x_2,..., x_{N-1}$ are found and a polynomial is fitted through the corresponding data points. The degree of the polynomial (assumed to be known) would normally be less than the embedding dimension. The parameters of the polynomials[15] are determined by the linear least squares method. The nearest neighbours can be found systematically using the concept of the generalized degrees of freedom (Jayawardena et al., 2002), or arbitrarily set to $2\,\dfrac{(m+p)!}{m!\,p!}$ (Casdagli, 1989). The predictor is a discontinuous function that may give rise to problems for long-time prediction. RBFs are global approximators with good local properties. Details of artificial neural networks and RBFs can be found in Chapters 8 and 9, respectively.

11.7.1 Prediction error

To test if the inverse approach has been successful, the variance of the prediction error using a time series of length $N + M$ has been obtained as follows (Casdagli, 1989):

$$\sigma^2(f_N) = \frac{1}{M} \sum_{n=N}^{N+M-1} \frac{(s((n+1)\tau) - f_N(s(n\tau),...,s((n-m+1)\tau)))^2}{\text{Var}} \qquad (11.44)$$

where the Var in Equation 11.44 is the variance of the time series.

Equation 11.44 above involves m, which is unknown. Its optimal value, m^* is also unknown *a priori*. The standard procedure is to start with $m = 1$, and repeat the estimation of the prediction error variance by increasing the value of m incrementally until a specified acceptable value is attained.

Through numerical experiments, it has been found (Casdagli, 1989) that for small N, RBFs are superior to local predictors. For large N, ignoring 10% of the worst predictions, piece-wise quadratic predictors are better than localized RBF predictors, which, in turn, are better than piece-wise linear predictors. For very low-dimensional systems, polynomial interpolations give very good results.

---

[15] A polynomial in m variables of degree p will have $^{(m+p)}C_m = \dfrac{(m+p)!}{m!\,p!}$ free parameters.

The non-linear prediction method can also be used to detect the presence (or absence) of chaos in a dynamical system. It is done by checking the prediction accuracy as functions of the embedding dimension and the lead time of prediction. If the system is chaotic, the accuracy, measured by some error criterion, would increase with increasing embedding dimensions up to the optimum embedding dimension and remain close to that value thereafter. In the case of stochastic systems, there would be no such increase of prediction accuracy with increasing embedding dimensions. It is also a well-known property that the prediction accuracy of a dynamical system deteriorates as the lead time increases much more rapidly than in the case of stochastic systems.

For a chaotic time series, the forecasting errors increase with increasing lead time at a rate that gives an estimate of the Lyapunov exponent. For uncorrelated noise, the forecasting errors are independent of the lead time. For time series with relatively fewer number of data points, a method of forecasting for short lead times has been proposed by Sugihara and May (1990). They claim that the method is capable of distinguishing chaos from measurement errors even with short runs of real data. In their approach, which has been tested with artificially generated chaotic data and applied to data on measles and chicken pox in humans, they first differenced the time series to reduce the effects of short-term linear autocorrelations. Using a delay time of $\tau = 1$, and an embedding dimension m, predictions have been made for an m-dimensional point with $m - 1$ past values with $m + 1$ nearest neighbours.

11.8 NON-LINEARITY AND DETERMINISM

11.8.1 Test for non-linearity

Given a time series of data, it is not known *a priori* whether it is of any practical use to go beyond the linear approximation. Even in a system that is known to be non-linear, a particular signal originating from that system may be linear. Therefore, it is necessary to justify the cause by establishing whether a time series is non-linear or not before attempting to carry out non-linear analysis. The procedure for establishing non-linearity involves the formulation of a suitable null hypothesis covering all Gaussian linear processes and then attempting to reject the null hypothesis by comparing the value of a non-linear test statistic estimated from the data with the probability distribution for the null hypothesis. This procedure is carried out using the method of surrogate data.

In testing for weak non-linearity, the test statistics should be powerful enough to distinguish between linear dynamics and weakly non-linear dynamics. It is relatively easy to detect strong non-linearity. An important criterion that can be used in selecting the test statistic is the discriminating power of the resulting test, which is defined as the probability that the null hypothesis is rejected when in fact it is false. This depends on the degree of deviation of the data from the null hypothesis.

11.8.1.1 Significance

Theiler et al. (1992a,b) defined a measure of 'significance', S as follows:

$$S = \frac{|Q_D - \mu_H|}{\sigma_H} \tag{11.45}$$

where Q_D is the statistic computed from the original data, Q_H is the corresponding statistic for the surrogate data generated under the null hypothesis, and μ_H and σ_H the mean and standard deviation of the distribution of Q_H. It gives an indication of the strength of discrimination between the original and surrogate data. The statistic S is a dimensionless quantity. However, it is usually referred to as sigmas (i.e., sigma is assumed as its unit). If the distribution of Q_H is Gaussian, a two-tailed test suggests rejection of the null hypothesis at 5% significance level if $|S| > 1.96$.

The significance with which non-linearity in a chaotic time series can be detected increases with increasing number of points in the series. However, increasing the complexity of the chaotic time series decreases the ability to distinguish from linearity (Theiler et al., 1992a,b).

11.8.1.2 Test statistics

Measures such as the autocovariance function, the power spectrum, correlation dimension, Lyapunov exponent, entropy, prediction error, etc., can be used to test non-linearity. In general, however, measures of non-linearity are derived from the autocovariance function or the power spectrum. Higher-order quantities such as the third-order quantity, defined as follows in Equation 11.46, which measures the asymmetry of a series under time reversal, is a preferred choice in the presence of time reversal asymmetry (Schreiber, 1999):

$$\phi^{\text{rev}}(\tau) = \frac{\displaystyle\sum_{n=\tau+1}^{N}(s_n - s_{n-\tau})^3}{\left[\displaystyle\sum_{n=\tau+1}^{N}(s_n - s_{n-\tau})^2\right]^{3/2}} \tag{11.46}$$

In Equation 11.46, s_n represents the time series of data and τ the lag time. It is to be noted that the statistics of linear stochastic processes are always symmetrical under time reversal. When a non-linearity test is performed, it is desirable to choose a test statistic that is useful for non-linear deterministic modelling, although a positive test for non-linearity does not guarantee the same result for determinism. When there is little or no time reversal asymmetry, the prediction error as the test statistic is reported to perform best (Schreiber, 1999).

11.8.1.3 Method of surrogate data

'Surrogate data' refers to a data set similar to the observed one and consistent with the null hypothesis. Surrogate data have the same properties as the original data but the non-linear phase relations are destroyed. Application of the method of surrogate data involves a null hypothesis against which observations are tested, and a discriminating statistic. If the discriminating statistic for the observed data is different from what is expected under the null hypothesis, then the null hypothesis can be rejected. In the method of surrogate data, the distribution of the discriminating statistic is estimated by generating (Monte Carlo simulation) an ensemble of surrogate data sets that will have the same statistical properties (e.g., mean, variance, autocorrelation, etc.) as those of the observed data. For each surrogate data set, the discriminating statistic from which the distribution can be approximated is computed. The method of surrogate data is an application of the 'bootstrap' method.

Methods of generating surrogate data include the non-windowed Fourier transform for linear Gaussian processes, windowed Fourier transform, and amplitude-adjusted Fourier transform (AAFT). In the non-windowed Fourier transform method, the surrogate data are generated to have the same Fourier spectra as the original data. In the windowed Fourier transform, the time series is multiplied by a window function of the form $w(t) = \sin\left(\dfrac{\pi t}{N}\right)$.

The AAFT method is associated with the null hypothesis that the observed time series is a monotonic non-linear transformation of a Gaussian process as described in Equation 11.51 below.

11.8.1.4 Null hypotheses

The null hypothesis specifies that certain statistical properties (e.g., the mean and variance) of the original data are preserved. The surrogate data are then generated to preserve these characteristics but to otherwise be random. Other properties such as the autocorrelation, Fourier spectrum, etc., can also be used as properties to be preserved. Possible choices of null hypothesis include

- Observed data are independently and identically (IID) distributed. If it is further assumed that the data come from a Gaussian distribution, the surrogate data can be easily generated using one of many pseudo-random number generators available, and then normalized to the mean and variance of the original data. IID surrogate data can also be generated by randomly shuffling the original data, thereby destroying the correlation structure but preserving the same amplitude structure. If the serial correlations in the original data and the shuffled data are significantly different, the hypothesis for independence is rejected.
- Observed non-IID data can be generated by a model of the form

$$x_t = a_0 + a_1 x_{t-1} + \sigma \eta_i \tag{11.47}$$

where the coefficients a_0, a_1, and σ are related to the mean, variance, and autocorrelation function of the original time series as follows:

$$a_1 = \rho_1; \quad a_0 = \mu(1 - a_1); \quad \sigma^2 = \upsilon\left(1 - a_1^2\right) \tag{11.48}$$

where μ, σ, and υ, respectively, are related to the mean of the original data, variance of the surrogate data, and variance of the original data, and, ρ_1 is the first autocorrelation coefficient of the original data, and η_i is the uncorrelated Gaussian noise of unit variance, $N(0,1)$.
- A general non-IID Gaussian process can be generated by extending Equation 11.47 to an arbitrary order that leads to an autoregressive model of the form

$$x_t = a_0 + \sum_{k=1}^{p} a_k x_{t-k} + \sigma \eta_i \tag{11.49}$$

The null hypothesis in this case would be that the time series is described by the autocorrelation function. The coefficients of Equation 11.49 can be estimated by any of the methods of parameter estimation of stochastic models, for example, the Yule–Walker equations. Surrogate data are generated to preserve the mean, variance, and the autocorrelation of the original series.

- A general linear stochastic process can be generated from

$$x_t = \sum_{i=1}^{P} a_i x_{t-i} + \sum_{i=0}^{Q} b_i \eta_{t-i} \qquad (11.50)$$

Here, two-point autocorrelations can be used as the test statistic. A problem in this case (and in the previous case) is that the test should be done not only for one particular set of values for a_i and b_i but for a whole class of processes. There is also the problem of the choice of P and Q.

- Another null hypothesis considers the process as essentially linear since the only non-linearity is contained in the measurement function $s(x_t)$. The data are generated by a simple generalization of the null hypothesis that explains deviations from the normal distribution by the action of a monotone static measurement function of the form

$$s_t = s(x_t) \qquad (11.51a)$$

$$x_t = \sum_{i=1}^{P} a_i x_{t-i} + \sum_{i=0}^{Q} b_i \eta_{t-i} \qquad (11.51b)$$

where x_t is a linear stochastic process (ARMA) and the non-linearity exists in the function that transforms x_t to s_t. Surrogate data for this hypothesis are generated by inverting $s(x_t)$ by rescaling the time series s_t to conform to a Gaussian distribution. The rescaled version is then phase randomized and the result is rescaled to the empirical distribution of s_t. It is reported that such AAFTs yield a correct test when N is large, correlation in the data is not too strong and the transfer function $s(x_t)$ is close to the identity (Schreiber, 1999).

In the case of chaotic time series, the correlation dimension, Lyapunov exponent, and forecasting error could be used as the discriminating statistic. They respectively represent the degrees of freedom, sensitivity to initial conditions, and determinism. Details of the methods of estimating the correlation dimension and Lyapunov exponent can be found in Chapter 10 (Sections 10.2.10 and 10.7.1).

11.8.2 Test for determinism

The usual procedure for the test of determinism involves the computation of different invariant measures. If a low-dimensional attractor or a positive finite value of Lyapunov exponent exists, it is claimed that the system is deterministic. However, with a finite amount of data and especially if they are noisy, the dimension estimation will at best be approximate and at worst, outright wrong. The algorithms can also, quite often, be easily fooled by linear correlation in the data (Theiler et al., 1992a,b). Although recent algorithms for computation of the invariant measures have been used without running into problems, in some situations the application of the dynamical systems approach may yield unexpected results. For example, finite positive values of the Lyapunov exponent have been obtained from data sets of stochastic processes by the algorithm of Wolf et al. (1985) (Abarbanel et al., 1990; Jayawardena and Lai, 1994; Kantz, 1994), posing a problem when solely dependent on this result. Any finite-length time series can be a particular realization of a random process, just as it is possible for a random-looking time series to come from a low-dimensional deterministic process (Theiler et al., 1992a,b). Furthermore, chaotic time series are always corrupted by noise, and it is difficult to delineate between chaos and noise.

Therefore, it is necessary to carry out some tests to avoid misinterpretation of the results. Here too, using the method of surrogate data, the distribution of the quantities, such as the correlation dimension, is computed for an ensemble of data that are just different realizations of a particular process. Discriminating statistical parameters (e.g., the correlation dimension) of different surrogate data sets are computed and a null hypothesis is then tested. Depending on the outcome of the hypothesis testing, a statistical conclusion about the original data could be made. Once the system characteristics have been identified, different modelling techniques can be used to predict the time series generated by them.

A test of determinism can also be carried out using the forecasting error because a system is deterministic if its future can be predicted. A discriminating statistic in this case would be an average error of forecasting obtained from non-linear modelling. The procedure involves splitting the time series into a calibrating (fitting) set of length N_1 and a testing set of length N_2 with $N = N_1 + N_2$, fitting a local linear model to the fitting part, using the fitted model to forecast values in the testing set and comparing with actual values. The difference between the predicted and actual values would be the forecasting error. Theiler et al. (1992a,b) have used the log median absolute prediction error as the discriminating statistic.

Kaplan and Glass (1992) have proposed a test of determinism based on the concept that that tangent to the trajectory generated by a deterministic system is a function of position in phase space, and therefore all the tangents to the trajectory in a given region of phase space will have a similar orientation.

II.9 NOISE AND NOISE REDUCTION

II.9.1 Noise in data

The old way of thinking has been to consider chaotic signals as noise and discard them. However, the new thinking is that chaotic signals contain valuable information. A difficulty in distinguishing chaotic signals from noise is that they both show broadband Fourier spectra. If there are no positive Lyapunov exponents, or zero KS entropy (KS entropy is the sum of positive Lyapunov exponents), then chaos cannot exist. Linear systems have zero or negative Lyapunov exponents and zero KS entropy. By reconstructing the phase space from a single scalar measurement, the evolution of the attractor is sought, but not the full system. Invariants are not sensitive to small perturbations of an orbit or initial condition, whereas the individual orbits are exponentially sensitive to such perturbations. Invariants can be topological or metric. The former is geometric and is robust to noise, and the latter depends on distances.

Noise can arise as a result of instrumental errors, observational errors, round-off errors in computations, and many other sources. Noise corrupts the signal and some means of separating the noise from the signal is a prerequisite to rigorous time series analysis. It is a problem common to both linear and non-linear analysis. Fourier techniques are not useful in this context. For many types of time series analysis, it is ideal, and sometimes necessary, to have data that are noise free. This is particularly so in the case of non-linear time series, which have signatures of chaotic dynamics, because the techniques of analysis and prediction have been developed under the assumption that the series are noise free. The presence of noise may limit the performance of the techniques of identification, estimation of invariant measures, model selection, and prediction of deterministic dynamical systems. For example, the presence of noise in the time series can destroy the self-similarity of the attractor, may distort the phase space reconstruction, and result in prediction errors being bounded from below regardless of the prediction method used. Noise causes adverse effects, such as blurring in the lower region of the plot of correlation integral (sum) versus distance and deviation of the trajectory causing increase in interpoint distance in the phase space.

Because of the potential problems that could be encountered in the analysis and prediction processes, and the fact that noise is inherently present in almost all experimental and observational time series, the question of dealing with noise has attracted the attention of many investigators from different disciplines (e.g., Schreiber, 1993a,b; Diks, 1999; Kantz and Schreiber, 1997; Oltmans and Verheijen, 1997; Jayawardena and Gurung, 2000; Jayawardena et al., 2008). Attempts have been made to deal with the problem by noise reduction methods (e.g., Schreiber, 1993b; Schreiber and Grassberger, 1991; Grassberger et al., 1993), as well as by modifying the scaling law (Schouten et al., 1994). The verification of such methods can be carried out only for series in which the clean signal is known *a priori*. For practical time series, such methods are therefore not very useful. Alternatively, methods for determining the noise level have been proposed by Schreiber (1993a), Oltmans and Verheijen (1997), and Diks (1999); however, their application is extremely difficult except in limiting situations.

Noise can be classified as either additive or dynamical. Additive noise is considered as due to measurement errors, whereas dynamical noise is due to perturbations of the system. In the latter case, there does not exist a rigid relationship between the present and future states of the system. The description and analysis that follow are restricted to additive noise. A data set corrupted by additive noise may be expressed as

$$x_n = y_n + v_n \tag{11.52}$$

where x_n is the noise corrupted data, y_n is the clean deterministic component, and v_n is the noise component. The deterministic component y_n follows the evolutionary law defined as

$$y_{n+1} = f(y_n) \tag{11.53}$$

where y_n and y_{n+1} are the states at time levels n and $n + 1$, respectively, and f is a function relating such states of a deterministic system. The term $\sqrt{\langle v^2 \rangle}$, known as the absolute noise level, is defined as

$$\langle v^2 \rangle = \frac{1}{N} \sum_{n=1}^{N} v_n^2 \tag{11.54}$$

and the relative noise level β is defined as

$$\beta = \sqrt{\frac{\langle v^2 \rangle}{\langle y^2 \rangle}} \tag{11.55}$$

Noise in a data set is also expressed as the signal-to-noise ratio, or SNR, in decibel units (dB) as

$$SNR = -20\log(\beta) \tag{11.56}$$

11.9.2 Noise reduction

Linear noise reduction techniques such as fast Fourier transform (FFT) remove important parts of the non-linear signal, and therefore the results of invariant measures would be incorrect if they are applied to non-linear systems. Recently, several methods of non-linear noise reduction have been proposed (e.g., Schreiber and Grassberger, 1991; Cawley and Hsu, 1992; Sauer, 1992; Schreiber, 1993b; Grassberger et al., 1993). Some reviews can also be found in Abarbanel et al. (1993), Jayawardena and Gurung (2000), and Ott et al. (1994), among others.

All noise reduction algorithms assume that the time series can be separated objectively into 'signal' and 'noise'. In linear systems, in the time domain, this may be on the basis of high- and low-frequency components – the high-frequency components considered as noise. In the frequency domain, sharp peaks are assumed to be the signal, whereas broadband spectra are considered as reflecting noise.

All non-linear noise reduction techniques accomplish their objectives in three steps: reconstruction of the attractor, approximation of the dynamics, and adjusting the approximations to better match with the observations (Kostelich and Schreiber, 1993). After noise removal, the clean series is written as

$$x_n^{correct} = y_n + v_n^{red} \tag{11.57}$$

where $x_n^{correct}$ is the noise reduced data and v_n^{red} is the correction made. These quantities facilitate in computing the improvement in the SNR and decrease in the violation of determinism (Grassberger et al., 1993), which are indicators of the performance of a noise reduction algorithm. The improvement in SNR is given by

$$\delta(SNR) = 20\log_{10}r_0 \tag{11.58}$$

where

$$r_0 = \left[\frac{\sum_n (v_n)^2}{\sum_n \left(v_n^{red}\right)^2} \right]^{1/2} \tag{11.59}$$

and the decrease in violation of determinism is given by

$$r_{dyn} = \left[\frac{\sum_n (x_n - f(x_{n-1}, x_{n-2}, ...))^2}{\sum_n \left(x_n^{corr} - f\left(x_{n-1}^{corr}, x_{n-2}^{corr}, ...\right)\right)^2} \right]^{1/2} \tag{11.60}$$

where f is the mapping function. These performance tests cannot, however, be applied to real data sets.

Given a time series contaminated with noise, the question is how to clean up the series so that the estimation of invariant measures such as the correlation dimension, Lyapunov exponent, KS entropy, etc., and model building can be carried out. In a linear system, this is quite a standard problem; that is, to separate sharp, narrowband signals from broadband noise. This can be achieved by Fourier spectral analysis in the frequency domain. In non-linear analysis, Fourier spectral analysis will not work and the signal–noise separation should be carried out in the time domain. To do this, some distinguishing characteristic of the non-linear time series must be first identified. If the corrupted time series $s(n)$ consists of the signal component $s_1(n)$ and other components $s_2(n)$, $s_3(n)$,..., such that

$$s(n) = s_1(n) + s_2(n) + s_3(n) + \ldots \tag{11.61}$$

then some characteristics of $s_1(n)$ that other $s_i(n)$, $i > 1$ do not possess must be identified. The best distinguishing characteristic would be the dynamical rule

$$y_1(k + 1) = F_1(y_1(k)) \tag{11.62}$$

in the reconstructed phase space of $s_1(n)$, which would be different from any other dynamical rule for the reconstructed phase space of other $s_i(n)$, $i > 1$. $F_1(.)$ is the mapping function for $s_1(n)$. The reconstructed phase space of $s_1(n)$ is given by

$$y_1(n) = [s_1(n), s_1(n + \tau_1), s_1(n + 2\tau_1),\ldots, s_1(n + (d_1 - 1)\tau_1)] \tag{11.63}$$

where τ_1 and d_1 are the delay time and the embedding dimension for $s_1(n)$, respectively. In the absence of any knowledge about the clean signal from the dynamics, or the dynamics itself, the separation of the signal from noise in which dynamics of either of the two are unknown can be accomplished by a method described by Abarbanel et al. (1993) in which local linear filters have been used and declaring the filtered data as 'clean'. The essence of the method, which has also been the basis of noise reduction by other researchers (e.g., Sauer, 1992; Schreiber and Grassberger, 1991; Cawley and Hsu, 1992), is summarized below.

Select N_B data points $y^r(n)$, $r = 1, 2, 3,\ldots, N_B$ in the neighbourhoods of every data point $y(n)$. Using $y^r(n)$ and $y(r,n + 1)$ into which they map, construct a local polynomial

$$y(r,n+1) = A + By^{(r)}(n) + Cy^{(r)}(n)y^{(r)}(n) + Dy^{(r)}(n)y^{(r)}(n)y^{(r)}(n) + \ldots = g(y^r(n),a) \tag{11.64}$$

Estimate the coefficients of the polynomial using local least squares minimization of the residuals in the local map $y^r(n) \to y(r, n + 1)$, for all the points. Using successive points along the orbits $y(n)$ and $y(n + 1)$ find small adjustments $\delta y(n)$ and $\delta y(n + 1)$, which minimize $\left\| y(n+1) + \delta y(n+1) - g(y(n) + \delta y(n)) \right\|^2$ at each point in the orbit.

The adjusted points $y(k) + \delta y(k)$ are taken to be the 'cleaned' signal.

Schreiber and Grassberger (1991) gave the correction as

$$x'_n = x_n - \alpha \frac{F(x_{n-r})}{\partial F(x_{n-r})/\partial x_n} \tag{11.65}$$

where the dynamics are given by $F(x_n, x_{n+1},\ldots, x_{n+m}) = 0^{16}$ and $m = 2r$. The method by Cawley and Hsu (1992) and Sauer (1992), sometimes referred to as Cawley–Hsu–Sauer method, consists of a compromise between orthogonal projections onto the subspaces defined by

$$F_q(x_n, x_{n+1},\ldots, x_{n+m}) = 0; \quad q = 1, 2,\ldots, Q \leq m \tag{11.66}$$

---

[16]This is the same as $x_{n+m} = f(x_n, x_{n+1},\ldots, x_{n+m-1})$.

where $m - Q + 1$ is the dimension of the manifold and m is the embedding dimension. By this method, expressions for normalized errors for 1D and 2D maps have been derived, and they conclude that their method is very efficient for oversampled time series despite the theoretical limitations. It is reported (Grassberger et al., 1993) that very good results have been obtained for higher dimensions (Lorenz and Rössler maps), but not as good for Hénon map.

Kostelich and Yorke (1988) have taken many nearby points in phase space to find a local approximation of the dynamics. Such approximations have then been used to produce a new time series whose dynamics would be more consistent with the dynamics of the phase space. The reconstructed attractor in their analysis considers the non-linear map f as approximated by a local linear map in the neighbourhood of x_i:

$$x_{i+1} = f(x_i) \approx A_i x_i + b_i \equiv L(x_i) \tag{11.67}$$

where A_i is the Jacobian matrix of f at x_i and b_i is a vector, which can be determined by least squares procedure, and L is meant to be linear operator. The equation for least squares solution may be written as

$$y_j^{(k)} = \sum a_i^{(k)} x_j + b_i^{(k)} \tag{11.68}$$

where $y_j^{(k)}$ is the k-th component of y_j and $y_i = f(x_j)$.

Better estimates of the errors can be obtained by choosing a window of consecutive points in the time series, finding a neighbourhood that also best satisfy the corresponding linear maps as ($\hat{x}_i$ is a neighbour of x_i)

$$[x_i, x_{i+1},..., x_{i+p}]$$

and

$$[\hat{x}_i, \hat{x}_{i+1},..., \hat{x}_{i+p}]$$

and estimating A_i and b_i by minimizing

$$\sum_{k=0}^{p} \left\| \hat{x}_{i+k} - L(\hat{x}_{i+k-1}) \right\|^2 + \left\| \hat{x}_{i+k} - x_{i+k} \right\|^2 + \left\| \hat{x}_{i+k+1} - L(\hat{x}_{i+k}) \right\|^2 \tag{11.69}$$

This procedure is repeated using the same window or overlapping windows by replacing the original time series x_i with the most recent $\hat{x}_i$, until there is no significant difference in $\hat{x}_i$ for successive iterations.

The noise level is then measured by the point-wise error defined as

$$\varepsilon_j = \left\| x_{j+1} - L(x_j) \right\| \tag{11.70}$$

By embedding the signal part of the dynamic equation into an m-dimensional phase space in the form $y_{n+m} = f(y_n, y_{n+1},..., y_{n+m-1})$, Grassberger et al. (1993) defined a corrected sequence as

$$x_{n+m}^{\text{corr}} = \frac{1}{a_m^{(n)}} \left[\sum a_i^{(n)} x_{n+1} + b^{(n)} \right] \text{ for all } n \tag{11.71}$$

By repeating the procedure, it is expected that it would converge to the correct y_n.

The noise reduction process requires an algorithm applicable to the entire noisy orbit to yield a new, less noisy orbit. It is an iterative process that yields a less noisy orbit each time. The method of Grassberger et al. (1993) modifies the ideas of Sauer (1992) and Cawley and Hsu (1992), resulting in a powerful noise reduction algorithm (Ott et al., 1994).

In another study (Jayawardena and Gurung, 2000), existing noise reduction techniques have been reviewed and their performances compared using synthetic time series artificially corrupted by varying levels of noise. The noise-reduced series have then been used to compute correlation dimensions and for making predictions by the dynamical systems approach. A brief description is given below.

For an artificially corrupted synthetic data, the correlation dimensions of raw series were estimated by the software RRCHAOS (Schouten and van den Bleek, 1994), which failed for the Hénon map with 20% noise. The data sets were then cleaned by the methods of Schreiber and Grassberger (1991), Grassberger et al. (1993), and Schreiber (1993b), after introducing slight variations to the respective original methods of the first two algorithms. In Schreiber's (1993b) original method, also known as the 0-th order noise reduction (Kantz, 1993), noise reduction has been done by embedding the time series as in Equation 11.6; however, the embedding vector includes equal numbers of past and future values and the present one. A set of nearest neighbours for each vector, x_n, inside the radius r, is then found and the central coordinates of the neighbourhoods, x_j, are simply averaged. In the cited study, a distance-based weighted average, as suggested by Cawley and Hsu (1992), has been used. In the original noise reduction method of Schreiber and Grassberger (1991), which assumes linearity in the neighbourhood, the fitting of the dynamics has to be done from the noisy data itself. This would, in certain circumstances, lead to singularity of the fitted equation, making the corrections very large. They did not make any corrections to the points if it is more than four times the average value. In the cited study, the SVD fitting (Press et al., 1992) has been used to avoid such a situation. Initially, the algorithm begins with the size of r equal to the noise level (which can be obtained from Equation 11.54) of the data, and if the prescribed number of neighbours is not found then its value is increased by a certain factor. If enough neighbours are not found even after increasing the neighbourhood size to a very large value, no corrections are made for such points. The results of the cited (Jayawardena and Gurung, 2000) study are summarized in Table 11.1.

A value of r_0 greater than 1.0 indicates that there is some noise reduction accompanied by an increase in the SNR taking place. From the values of r_0 for different noise reduction techniques, which vary between 1.72 and 4.3 (Table 11.1), it can be seen (Equation 11.58) that there is a corresponding improvement in the SNR from 4.71 to 12.67. The same conclusion

Table 11.1 Correlation dimensions of raw and noise-reduced synthetic data and the performances of the noise reduction algorithms

Time series	Correlation dimension				r_0			r_{dyn}		
	Raw	I	II	III	I	II	III	I	II	III
Hénon map[a]	1.2	1.23	1.44	1.33	2.88	2.55	4.3	6.96	3.65	12.56
Hénon map[b]	–	1.32	1.54	1.31	2.97	2.59	3.62	7.62	4.31	14.91
Lorenz map[a]	1.78	2.09	2.08	2.0	1.93	1.72	2.83	–	–	–
Lorenz map[b]	2.17	2.18	2.26	2.03	3.86	1.98	3.12	–	–	–

Note: Data set I: noise reduced by Schreiber's (1993b) method; data set II: noise reduced by the method of Schreiber and Grassberger (1991); data set III: noise reduced by the method of Grassberger et al. (1993); correlation dimensions for the data sets computed by Grassberger–Procaccia algorithm with efficient neighbour search.

[a] Raw data: 10% noise added; correlation dimension computed by the software RRCHAOS.
[b] Raw data: 20% noise added; correlation dimension computed by the software RRCHAOS.

can be made for r_{dyn}, which has ranged from 3.65 to 14.91, indicating that determinism of the data sets has also increased after the noise reduction. For real data sets, the same techniques can be used for noise reduction; however, there is no verification since the noise level is unknown. Schreiber (1993a) has suggested a method in which the correlation integral is divided into four different regions and a function is proposed in terms of the noise level. The noise level corresponding to the minimum value of the function is assumed to be the actual noise level of the data series.

For dynamical noise due to perturbation of the system, an error term γ_n is added as follows (Grassberger et al., 1993):

$$y_{n+1} = f(y_n) + \gamma_n \tag{11.72}$$

Dynamical noise is much more difficult to handle. Some details about dynamical noise, such as the 'shadowing problem',[17] can be found in Grebogi et al. (1990), Hammel (1990), and Farmer and Sidorowich (1991), among others.

Hammel (1990) introduced a way of noise reduction by shadowing for both additive and dynamical noise. As the noisy orbit $(p_k)_{k=0}^N$ is generated by the map f^k, errors are made at each iteration. The resulting points in the orbit satisfy $\left\| p_{k+1} - f(p)_k \right\| < \delta$ where δ is small. This type of noise is called dynamical noise. The noise reduction process then is to find a less noisy orbit near $(p_k)_{k=0}^N$. To make a less noisy orbit and to keep it nearby are conflicting issues that require some compromise. The original noisy orbit lies at the limit of nearness, whereas the noise-free true orbit $(x_k)_{k=0}^N$ lies somewhere near $(p_k)_{k=0}^N$.

11.9.3 Noise level

Noise reduction methods reduce the noise to some extent but there always remains some amount of noise in the data that may influence the application, prediction, and control of the dynamical system. It is therefore necessary to know how much noise remains. In other words, an estimate of the noise level is needed.

The noise level gives an idea about how badly a data set is contaminated. Most techniques of computing invariant measures fail if the data set contains as little as 2% noise (Schreiber, 1993a). Methods for determination of noise levels have been proposed by Cawley and Hsu (1992), Schouten et al. (1994), Schreiber (1993a), Diks (1999), and Jayawardena et al. (2008), among others. A summary of the contributions made in some selected studies is given below.

Using the correlation sum $C_m(r)$ defined by Grassberger and Procaccia (1983a,b) (see also Section 10.2.10), the correlation dimension D for noise-free data can be expressed as

$$C_m(r) \approx r^D \tag{11.73}$$

where

$$D = \underset{r \to 0}{\text{Limit}} \underset{N \to \infty}{\text{Limit}} D_m(r); \quad D_m(r) = \frac{d\ln(C_m(r))}{d(\ln(r))} \tag{11.74}$$

---

[17]The orbit of x_0 is said to be ϵ-shadow of the noisy orbit $(p_k)_{k=0}^N$ if $\left\| f^k(x_0) - p_k \right\| < \epsilon$ for $0 \le k \le N$.

In the variation of $D_m(r)$ as a function of r and m, four different regions can be identified. In region I, which is for small r, there exist large statistical fluctuations. Region II is dominated by noise and in this region $D_m(r) \approx m$. In region III, $D_m(r) \approx D$ and the scaling behaviour of the attractor can be observed. In region IV in which r is of the order of the size of the attractor, no scale invariance can be expected.

In terms of $D_m(r)$, Schreiber (1993a) used the following formula for calculating the noise level σ:

$$D_{m,m'}(r) = \frac{D_m(r) - D_{m'}(r)}{m - m'} = g\left[\frac{r}{2\sigma}\right] \tag{11.75}$$

where m' is the number of delay coordinates that should at least be equal to the box counting dimension, and

$$g(z) = \frac{2}{\sqrt{\pi}} \frac{ze^{-z^2}}{\mathrm{erf}(z)} \tag{11.76}$$

In regions II and III, all the curves of $\dfrac{D_m(r) - D_{m'}(r)}{m - m'}$ have the same functional form. In Equation 11.75, $D_m(r)$ and $D_{m'}(r)$ can be numerically evaluated from $C_m(r)$ and $C_{m'}(r)$, respectively. The noise level σ is estimated as that corresponding to value of the function $g(z)$ that best fits $D_{m,m'}(r)$, which can be obtained by least squares procedure. It is equivalent to minimizing

$$\sum_{m,i}\left[D_{m,m'}(r_i) - \frac{r_i \exp\left[-\dfrac{1}{4\sigma^2}r_i^2\right]}{\sigma\sqrt{\pi}\mathrm{erf}\left[\dfrac{r_i}{2\sigma}\right]} \right]^2 \tag{11.77}$$

By using a scalar time series with the real part of 100,000 iterates of Ikeda map with added Gaussian noise of amplitude 0.005, and embedding the data in $m = 2, 3, 4, ..., 10$ dimensions, Schreiber (1993a) computed the values of $C_m(r)$ for $r = 0$–0.025. The values of $C_{m,m'}(r)$ were then computed for $m' = 2$, and $m = 3, 4, ..., 10$. The computed value of the noise level has been 0.00516, which is in close agreement (about 3% error) with the added noise.

A more simple form for the correlation sum has been obtained by Schouten et al. (1994), based on the maximum norm. Using an upper bound of radius r as r_0 (different from r_0 in Equation 11.59), and δ as the maximum noise amplitude, they obtained the correlation sum as

$$C_m(r) = \left[\frac{r - \delta}{r_0 - \delta}\right]^D, \text{ for } \delta \leq r \leq r_0 \tag{11.78}$$

Diks (1999), based on the closed-form expression for the correlation sum in the presence of Gaussian noise obtained by Smith (1992) under the assumption that the clean attractor underlying the noisy data has an integer-valued dimension D, and a similar expression obtained by Oltmans and Verheijen (1997) under general conditions, derived the following expression for the correlation sum in the presence of noise when $C_m(r)$ is based on Euclidean norm:

$$C_m(r) = \frac{\phi e^{-m\pi K} m^{-D/2} 2^{-m} \sigma^{D-m} r^m}{\Gamma(m/2+1)} M\left(\frac{m-D}{2}, \frac{m}{2}+1, -\frac{r^2}{4\sigma^2}\right) \tag{11.79}$$

In Equation 11.79, ϕ is a constant and M is Kummer's confluent hypergeometric function, which has the following integral representation:

$$M(a,b,z) = \frac{\Gamma(b)}{\Gamma(a)\Gamma(b-a)} \int_0^1 e^{zt} t^{a-1} (1-t)^{b-a-1} dt \tag{11.80}$$

Using Equations 11.79 and 11.80, the correlation dimension and noise level for the time series can be estimated by a non-linear least squares method, at least in theory. However, because of the strong non-linearity in the equations, it is difficult in practice.

Recognizing the difficulties involved in non-linear least squares method, Jayawardena et al. (2008) have introduced a new method that employs a linear least squares estimation procedure to estimate the correlation dimension and the noise level. The method is based on the correlation sum form obtained by Diks (1999) coupled with the special property of the Kummer's confluent hypergeometric function. Starting from Equation 11.79 and 11.80, a relationship linking the correlation dimension D, the correlation sum $C_m(r)$ with respect to r, and the noise level σ has been shown to be (see Appendix 11.1 for detailed derivation)

$$D + 2\left[m\frac{d[\ln(C_m(r))]}{d[\ln(r)]} - \frac{d^2[\ln(C_m(r))]}{d[\ln(r)]^2} - \left(\frac{d[\ln(C_m(r))]}{d[\ln(r)]}\right)^2\right]\frac{\sigma^2}{r^2} = \frac{d[\ln(C_m(r))]}{d[\ln(r)]} \tag{11.81}$$

In this equation, the correlation dimension is linear with respect to σ^2. By substituting $\sigma = 0$ in Equation 11.81, the correlation dimension for noise-free chaotic data can be obtained as

$$D = \frac{d[\ln C_m(r)]}{d[\ln r]} \tag{11.82a}$$

and for noisy data, as $r \to 0$, as (see Appendix 11.2 for proof)

$$\lim_{r \to 0} \frac{d[\ln C_m(r)]}{d[\ln r]} = m \tag{11.82b}$$

Equation 11.81 can be rewritten in the form

$$y = D + 2\sigma^2 x \tag{11.83a}$$

where

$$y = \frac{d[\ln C_m(r)]}{d[\ln r]} \tag{11.83b}$$

$$x = \frac{1}{r^2}\left[m\frac{d[\ln C_m(r)]}{d[\ln r]} - \frac{d^2[\ln C_m(r)]}{(d[\ln r])^2} - \left(\frac{d[\ln C_m(r)]}{d[\ln r]}\right)^2 \right]$$

(11.83c)

which, in their finite difference approximations, take the form

$$y_n \approx r_n\frac{c_n - c_{n-1}}{\Delta r}$$

(11.83d)

$$x_n \approx \frac{(m-1)\Delta r(c_n - c_{n-1}) - r_n(c_{n-1} - 2c_n + c_{n+1}) - r_n(c_n - c_{n-1})^2}{r_n(\Delta r)^2}$$

(11.83e)

for $r_{n+1} - r_n = \Delta r$; $c_n = \ln C_m(r_n)$, r_n, $1 \le n \le L$, a given radius, and L, the number of test values of the radius r.

The least squares estimates of the noise level and the correlation dimension can then be shown to be

$$\bar{\sigma}^2 = \frac{\displaystyle\sum_{n=2}^{L-2}(y_{n+1} - y_n)(x_{n+1} - x_n)}{2\displaystyle\sum_{n=2}^{L-2}(x_{n+1} - x_n)^2}$$

(11.84)

and

$$\bar{D} = \frac{1}{L-2}\sum_{n=2}^{L-1}(y_n - 2\bar{\sigma}^2 x_n)$$

(11.85)

This method has been tested with five mathematical series, which are known to become chaotic under certain parameter conditions: the Hénon Map, the Lorenz equation, the Duffing equation, the Rössler equation, and Chua's circuit. The tests show consistent satisfactory results (Table 11.2). It is then applied to three real-world data series: the Southern Oscillation Index (SOI), eastern equatorial Pacific sea surface temperature anomaly index (SSTA), and the normalized Darwin–Tahiti mean sea level pressure differences. Although these results cannot be verified, it is reasonable to expect them to be acceptable since the method has been extensively verified using several examples.

Although the noise levels (σ) added to the artificial data sets generated by the above systems are known, the actual noise levels for the noisy data would be somewhat different. In the above study, the actual noise level has been calculated as follows:

$$\sigma_{actual} = \sqrt{\frac{1}{N-1}\sum_{i=1}^{N}(s_i - \bar{s}_i)^2}$$

(11.86)

where N is the sample number, and, $\{s_i, 1 \le i \le N\}$ and $\{\bar{s}_i, 1 \le i \le N\}$ are the noisy and clean data, respectively.

Table 11.2 Comparison of noise levels

Data set	Hénon			Lorenz			Duffing			Rössler			Chua		
σ	0.05	0.10	0.15	0.5	1.0	1.5	0.1	0.3	0.5	0.5	1.0	1.5	0.05	0.1	0.15
σ_{actual}	0.0708	0.1416	0.2123	0.7078	1.4156	2.1234	0.1416	0.4247	0.7078	0.708	1.4156	2.1234	0.0708	0.1416	0.2123
SNR	46.377	32.373	22.613	57.516	43.623	35.544	58.282	36.314	26.099	48.097	38.466	26.124	39.143	25.280	17.180
$\hat{\sigma}$	0.0771	0.1233	0.2203	0.8843	1.7152	2.7981	0.2251	0.2971	0.7026	0.5171	0.9295	1.3725	0.0649	0.1069	0.1786
$\hat{\sigma}_{Schreiber}$	0.1831	0.1998	0.2320	0.3521	0.4633	1.3830	0.5775	0.6386	0.8140	0.5211	0.6361	1.0732	0.0572	0.0854	0.1268
$\hat{\sigma} - \sigma$	0.0271	0.0233	0.0703	0.3843	0.7152	1.2981	0.1251	-0.0029	0.2026	0.0171	-0.0705	-0.1275	0.0149	0.0069	0.0286
$\hat{\sigma}_{Schreiber} - \sigma$	0.1331	0.0998	0.0820	-0.1479	-0.5367	-0.1170	0.4775	0.3386	0.3140	0.0211	-0.3639	-0.4268	0.0072	-0.0146	-0.0232
$\hat{\sigma} - \sigma_{actual}$	0.0063	-0.0183	0.0080	0.1765	0.2996	0.6747	0.0835	-0.1276	-0.0052	-0.1909	-0.4861	-0.7509	-0.0059	-0.0347	-0.0337
$\hat{\sigma}_{Schreiber} - \sigma_{actual}$	0.1123	0.0582	0.0197	-0.3557	-0.9523	-0.7404	0.4359	0.2139	0.1062	-0.1869	-0.7795	-1.0502	-0.0136	-0.0562	-0.0855

Note: σ, added noise level; σ_{actual}, actual noise level; $\hat{\sigma}$, estimated noise level by the proposed method; $\hat{\sigma}_{Schreiber}$, estimated noise level by Schreiber's method; SNR, signal-to-noise ratio.

Table 11.3 Standard errors between estimated and added (or actual) noise for the data sets

Data set	Added noise		Actual noise	
	Proposed method	*Schreiber's method*	*Proposed method*	*Schreiber's method*
Hénon	0.0455	0.1071	0.0121	0.0738
Lorenz	0.8840	0.3284	0.4382	0.7261
Duffing	0.1375	0.3835	0.0881	0.2870
Rössler	0.0847	0.3241	0.5281	0.7628
Chua	0.0190	0.0164	0.0293	0.0596

Table 11.2 gives the comparisons of the added, actual, and estimated noise levels by this method and by Schreiber's method (Schreiber, 1993a) together with the differences in the estimates of noise level by this method and the added noise level, the differences between the noise levels estimated by Schreiber's method and the added noise level, the differences between the noise levels estimated by this method and the actual noise level, and the differences in the noise levels estimated by Schreiber's method and the actual noise level. The corresponding standard errors between estimated and added (or actual) noise are given in Table 11.3, which shows that the standard errors for the actual noise are always lower by this method, whereas for the added noise they are lower than the corresponding values except for the Lorenz and Chua data. The standard errors should, however, be interpreted with caution because the sample size is only 3. In fact they are more close to the actual noise levels than to the added noise levels. This perhaps may be because pseudo-random numbers generated by the computer do not strictly follow a Gaussian distribution. It should also be noted that this method has the advantage that it is linear, whereas Schreiber's method is non-linear.

11.10 APPLICATION AREAS

The dynamical systems approach and chaos theory have been applied to many diverse fields, such as mathematics, meteorology, computer science, engineering, finance, physics, population dynamics, and biology. In the field of hydrology, the applications have been mainly in the prediction of rainfall and/or runoff, although a few applications in other water-related fields can also be found. Perhaps the earliest application of deterministic chaos is in weather prediction (Lorenz, 1963), which led to the well-known 'butterfly effect' (see also Chapter 10) and the conclusion that long-term prediction of chaotic systems is impossible. Subsequently, Rodriguez-Iturbe et al. (1989) analysed a storm event recorded at 15-s intervals in Boston in the United States, and a weekly rainfall series recorded in Genoa, Switzerland, and concluded that the short-time interval storm event exhibited chaotic behaviour, whereas the weekly time series did not. The former is reported to have a correlation dimension of 3.78, whereas the latter did not attain a saturation value. This claim has been supported with the accompanying positive Lyapunov exponent for the storm event. Chaotic behaviour of short-time rainfall has also been observed in the studies by Sharifi et al. (1990) and Tsonis and Elsner (1988, 1991), among others.

Jayawardena and Lai (1994) analysed daily rainfall data at three stations and daily stream-flow data at two stations in Hong Kong for possible chaos. They estimated the correlation dimensions, Lyapunov exponents, and KS entropies, and made predictions using different local approximation methods (see also Section 11.6). The applicability of the numerical estimation methods was investigated by using random processes with normal, log-normal, Gamma and Weibull distributions, ARMA processes with the same distributions as above, chaotic series generated by the logistic map with additive Gamma noise, and periodic data with additive noise generated by having the additive Gamma random series superimposed on a cosine wave. They also used the inverse approach for detecting chaos using the prediction error and the lead time of prediction as indicators. The results confirmed that an optimal embedding dimension can be found for chaotic series with additive noise. For a given embedding dimension, the prediction error increases with increasing lead time of prediction, which can be used for differentiating noise from deterministic chaos. Another observation they made was that the algorithm of Wolf et al. (1985) may sometimes give positive Lyapunov exponents for random and ARMA processes. A comparison of the dynamical systems approach with a stochastic approach has also been made by the authors, perhaps the first of such studies. In a separate study (Jayawardena and Gurung, 2000), several 'noisy' time series of river flow data have been 'cleaned' by a number of different methods.

Sivakumar et al. (1998, 1999a, 2002, 2007) and Islam and Sivakumar (2002) have reported the existence of chaos in daily rainfall in Singapore as evidenced by low correlation dimensions in six data series. They also addressed the issue of minimum data requirement for correlation dimension estimation, and the problem of noise reduction. Sivakumar and his co-workers have examined rainfall time series from different geographical locations and suggested the presence of chaos at different temporal scales (daily, 2-day, 4-day, and 8-day); analysed daily and monthly river discharge data also representing different geographic locations, climatic conditions, basin sizes, characteristics and land uses; addressed the issue of sediment dynamics in rivers; and carried out a comparative analysis of the phase space reconstruction and artificial neural networks methods applied to daily river flow prediction. A summary of their contributions is given in Sivakumar and Berndtsson (2010).

Ng et al. (2007) used streamflow data in Saugeen River in Canada from 1947 to 1992 and showed that the phase space plot gets compressed into a small attracting subspace. They analysed the phase space with varying delay times and correlation dimension estimates and concluded that the data exhibit weak chaos because of the high dimensionality obtained (correlation dimension of 10 for embedding dimension of 21 for original data and 8 and 17 for differenced data).

More recently, She and Yang (2010) used the generalized degrees of freedom (GDF) to determine the optimum embedding dimension and the number of nearest neighbours in their application to predict 40-day lead-time daily discharges in two Russian rivers. They used an adaptive local linear prediction method in which the embedding dimension and the number of nearest neighbours are combined as a parameter set (m, q) that changes adaptively with increasing forecast steps.

The summary above gives only a glimpse of the applications of the dynamical systems approach to hydrology and water resources fields. More applications to hydro-meteorological time series can be found in the works of Porporato and Ridolfi (1997), Regonda et al. (2004), and Sivakumar et al. (2007), among others.

11.11 CONCLUDING REMARKS

In this chapter, an attempt has been made to introduce the tools and techniques developed in the mathematics and physics literature for the analysis and prediction of non-linear time series with the hope that they will be useful to researchers looking for alternative methods of modelling and prediction of hydrological and environmental systems. It should be emphasized that these tools and techniques are not meant to replace existing ones, but rather to complement them. There are no general methods applicable to all non-linear systems. Techniques and methods of analysis are problem dependent, and care should be exercised in choosing the right approach for the problem in hand.

APPENDIX 11.1: DERIVATION OF EQUATION 11.81

With the substitution of

$$a = \frac{m-D}{2}, b = \frac{m}{2} + 1, z = -\frac{r^2}{4\sigma^2} \tag{A1-1}$$

in Kummer's confluent hypergeometric function (Equation 11.80), Equation 11.79 can be rewritten as

$$
\begin{aligned}
C_m(r) &= \frac{\phi e^{-m\tau K} m^{-D/2} 2^{-m} \sigma^{D-m} r^m}{\Gamma(m/2+1)} M\left(\frac{m-D}{2}, \frac{m}{2}+1; -\frac{r^2}{4\sigma^2}\right) \\
&= \frac{\phi e^{-m\tau K} m^{-D/2} 2^{-m} \sigma^{D-m} r^m}{\Gamma(m/2+1)} \frac{\Gamma(m/2+1)}{\Gamma((m-D)/2)\Gamma(D/2+1)} F\left(-\frac{r^2}{4\sigma^2}\right) \\
&= \frac{\phi e^{-m\tau K} m^{-D/2} 2^{-m} \sigma^{D-m} r^m}{\Gamma(D/2+1)\Gamma((m-D)/2)} F\left(-\frac{r^2}{4\sigma^2}\right)
\end{aligned}
\tag{A1-2}
$$

where

$$F(z) = \int_0^1 e^{zt} t^{a-1} (1-t)^{b-a-1} \, dt \tag{A1-3}$$

It can be proved that Equation A1-3 satisfies the condition

$$aF(z) + (z - b)F'(z) - zF''(z) = 0 \tag{A1-4}$$

(See Appendix 11.3 for the proof.)
Taking natural logarithms of Equation A1-2,

$$
\begin{aligned}
\ln[C_m(r)] &= \ln(\phi) - \frac{D}{2}\ln(m) - m\ln(2) + (D-m)\ln(\sigma) + m\ln(r) \\
&\quad - m\tau K + \ln[(F)] - \ln\left[\Gamma\left(\frac{D}{2}+1\right)\right] - \ln\left[\Gamma\left(\frac{m-D}{2}\right)\right]
\end{aligned}
\tag{A1-5}
$$

From Equation A1-1, we have

$$\frac{dz}{dr} = -\frac{r}{2\sigma^2} \tag{A1-6}$$

Since all the terms in Equation A1-5, except $\ln[C_m(r)]$, $m\ln(r)$, and $\ln[F(z)]$, are independent of r, their derivatives with respect to r are zero. Therefore,

$$
\begin{aligned}
\frac{d[\ln(C_m(r))]}{dr} &= \frac{d}{dr}[m\ln(r) + \ln(F(z))] \\
&= \frac{m}{r} + \frac{1}{F(z)}\frac{d}{dr}(F(z)) \\
&= \frac{m}{r} - \frac{r}{2\sigma^2}\frac{F'(z)}{F(z)}
\end{aligned}
\tag{A1-7}
$$

From which

$$\frac{F'(z)}{F(z)} = 2\sigma^2\left(\frac{m}{r^2} - \frac{d[\ln(C_m(r))]}{rdr}\right)$$

Denote

$$P(m,r) = \frac{m}{r^2} - \frac{d[\ln(C_m(r))]}{rdr}$$

Then

$$F'(z) = 2\sigma^2 P(m,\ r)F(z) \tag{A1-8}$$

Also

$$\frac{d}{dr}(F'(z)) = F''(z)\frac{dz}{dr} = -\frac{r}{2\sigma^2}F''(z)$$

and

$$\begin{aligned}
\frac{d}{dr}(F'(z)) &= \frac{d}{dr}[2\sigma^2 P(m,r)F(z)] \\
&= 2\sigma^2\left\{F(z)\frac{dP(m,r)}{dr} + P(m,r)\frac{dF(z)}{dr}\right\} \\
&= 2\sigma^2\left\{F(z)\frac{dP(m,r)}{dr} - \frac{r}{2\sigma^2}F'(z)P(m,r)\right\} \\
&= 2\sigma^2\left\{F(z)\frac{dP(m,r)}{dr} - \frac{r}{2\sigma^2}[2\sigma^2 P(m,r)F(z)]P(m,r)\right\} \\
&= 2\sigma^2\left\{F(z)\frac{dP(m,r)}{dr} - rP^2(m,r)F(z)\right\}
\end{aligned}$$

Therefore

$$\begin{aligned}
F''(z) &= -\frac{2\sigma^2}{r}\frac{d}{dr}(F'(z)) \\
&= -\frac{2\sigma^2}{r}\left\{2\sigma^2\left[\frac{dP(m,r)}{dr}F(z) - rP^2(m,r)F(z)\right]\right\} \tag{A1-9} \\
&= -4\sigma^4\left\{\frac{dP(m,r)}{rdr} - P^2(m,r)\right\}F(z)
\end{aligned}$$

Substituting Equations A1-8 and A1-9 into Equation A1-4, we get

$$aF(z) + (z-b)[2\sigma^2 P(m,r)F(z)] + 4\sigma^4 z\left\{\frac{dP(m,r)}{rdr} - P^2(m,r)\right\}F(z) = 0$$

which leads to

$$a + 2\sigma^2 (z - b) P(m, r) + 4z\sigma^4 \left\{ \frac{dP(m, r)}{rdr} - P^2(m, r) \right\} = 0$$

Then

$$a - 2\sigma^2 \left(\frac{r^2}{4\sigma^2} + b \right) P(m, r) - r^2\sigma^2 \left\{ \frac{dP(m, r)}{rdr} - P^2(m, r) \right\} = 0 \qquad \text{(A1-10)}$$

Since $a = (m - D)/2$, $b = m/2 + 1$, we have

$$\frac{m - D}{2} - 2\sigma^2 \left(\frac{r^2}{4\sigma^2} + \frac{m}{2} + 1 \right) P(m, r) - \sigma^2 \left[\frac{rdP(m, r)}{dr} - r^2 P^2(m, r) \right] = 0$$

Therefore

$$\frac{m - D}{2} - \left[\frac{r^2}{2} + \sigma^2 (m + 2) \right] P(m, r) - \sigma^2 \left[\frac{rdP(m, r)}{dr} - r^2 P^2(m, r) \right] = 0$$

and

$$m - D - r^2 P(m, r) - 2\sigma^2 (m + 2) P(m, r) - 2\sigma^2 \frac{rdP(m, r)}{dr} + 2\sigma^2 r^2 P^2(m, r) = 0$$

Thus, we have

$$m - r^2 P(m, r) = D + 2\sigma^2 \left\{ (m + 2) P(m, r) + \frac{rdP(m, r)}{dr} - r^2 P^2(m, r) \right\} \qquad \text{(A1-11)}$$

Because

$$P(m, r) = \frac{m}{r^2} - \frac{d[\ln(C_m(r))]}{rdr} \qquad \text{(A1-12)}$$

we have

$$\frac{dP(m, r)}{dr} = -\frac{2m}{r^3} + \frac{1}{r^2} \frac{d[\ln(C_m(r))]}{dr} - \frac{1}{r} \frac{d^2[\ln(C_m(r))]}{dr^2} \qquad \text{(A1-13)}$$

and

$$P^2(m,r) = \frac{m^2}{r^4} - \frac{2m}{r^3}\frac{d[\ln(C_m(r))]}{dr} + \frac{1}{r^2}\left[\frac{d[\ln(C_m(r))]}{dr}\right]^2 \tag{A1-14}$$

By substituting Equations A1-12 through A1-14 into Equation A1-11, we obtain

$$m - r^2 P(m,r) = D + 2\sigma^2 \left\{ \frac{m(m+2)}{r^2} - \frac{(m+2)}{r}\frac{d[\ln(C_m(r))]}{dr} - \frac{2m}{r^2} \right.$$

$$+ \frac{1}{r}\frac{d[\ln(C_m(r))]}{dr} - \frac{d^2[\ln(C_m(r))]}{dr^2} - \frac{m^2}{r^2}$$

$$\left. + \frac{2m}{r}\frac{d[\ln(C_m(r))]}{dr} - \left[\frac{d[\ln(C_m(r))]}{dr}\right]^2 \right\}$$

$$= D + 2\sigma^2 \left\{ \frac{(m-1)}{r}\frac{d[\ln(C_m(r))]}{dr} - \frac{d^2[\ln(C_m(r))]}{dr^2} - \left[\frac{d[\ln(C_m(r))]}{dr}\right]^2 \right\}$$

$$= D + \frac{2\sigma^2}{r^2} \left\{ (m-1)r\frac{d[\ln(C_m(r))]}{dr} - r^2\frac{d^2[\ln(C_m(r))]}{dr^2} - r^2\left[\frac{d[\ln(C_m(r))]}{dr}\right]^2 \right\}$$

But

$$m - r^2 P(m,r) = \frac{d[\ln(C_m(r))]}{d[\ln(r)]}$$

and

$$r\frac{d[\ln(C_m(r))]}{dr} + r^2\frac{d}{dr}\left[\frac{d[\ln(C_m(r))]}{dr}\right]$$

$$= r\frac{d}{dr}\left[r\frac{d[\ln(C_m(r))]}{dr}\right]$$

$$= \frac{d}{d[\ln(r)]}\left[\frac{d[\ln(C_m(r))]}{d[\ln(r)]}\right]$$

$$= \frac{d^2[\ln(C_m(r))]}{d[\ln(r)]^2}$$

Therefore,

$$D + 2\left[m\frac{d[\ln(C_m(r))]}{d[\ln(r)]} - \frac{d^2[\ln(C_m(r))]}{d[\ln(r)]^2} - \left(\frac{d[\ln(C_m(r))]}{d[\ln(r)]}\right)^2 \right]\frac{\sigma^2}{r^2} = \frac{d[\ln(C_m(r))]}{d[\ln(r)]} \tag{11.81}$$

APPENDIX 11.2: PROOF OF EQUATION 11.82B

For noisy data, from Equation A1-7,

$$\frac{d[\ln(C_m(r))]}{rdr} = \frac{m}{r^2} - \frac{1}{2\sigma^2} \frac{F'(z)}{F(z)}$$

$$\frac{d[\ln(C_m(r))]}{d[\ln(r)]} = m - \frac{r^2}{2\sigma^2} \frac{F'(z)}{F(z)}$$

Because $F(z), F'(z)$ are bounded for $z \to 0$,

$$\lim_{r \to 0} \frac{d[\ln C_m(r)]}{d[\ln r]} = m \tag{11.82}$$

APPENDIX 11.3: PROOF OF EQUATION A1-4

Kummer's confluent hypergeometric function has the integral representation

$$M(a,b,z) = \frac{\Gamma(b)}{\Gamma(a)\Gamma(b-a)} \int_0^1 e^{zt} t^{a-1} (1-t)^{b-a-1} \, dt \tag{A3-1}$$

First let

$$F(z) = \int_0^1 e^{zt} t^{a-1} (1-t)^{b-a-1} \, dt \tag{A3-2}$$

Then

$$F'(z) = \int_0^1 e^{zt} t^a (1-t)^{b-a-1} \, dt$$

$$F''(z) = \int_0^1 e^{zt} t^{a+1} (1-t)^{b-a-1} \, dt \tag{A3-3}$$

We have

$$\int_0^1 e^{zt} t^a (1-t)^{b-a-1} \, dt = -\frac{1}{b-a} \int_0^1 e^{zt} t^a d[(1-t)^{b-a}] \tag{A3-4}$$

By using integration by parts, Equation A3-4 is

$$\int_0^1 e^{zt} t^a (1-t)^{b-a-1} \, dt$$

$$= \frac{z}{b-a} \int_0^1 e^{zt} t^a (1-t)^{b-a} \, dt + \frac{a}{b-a} \int_0^1 e^{zt} t^{a-1} (1-t)^{b-a} \, dt$$

$$= \frac{z}{b-a} \int_0^1 e^{zt} t^a (1-t)(1-t)^{b-a-1} \, dt + \frac{a}{b-a} \int_0^1 e^{zt} t^{a-1} (1-t)(1-t)^{b-a-1} \, dt$$

$$= \frac{z}{b-a} \left[\int_0^1 e^{zt} t^a (1-t)^{b-a-1} \, dt - \int_0^1 e^{zt} t^{a+1} (1-t)^{b-a-1} \, dt \right]$$

$$+ \frac{a}{b-a} \left[\int_0^1 e^{zt} t^{a-1} (1-t)^{b-a-1} \, dt - \int_0^1 e^{zt} t^a (1-t)^{b-a-1} \, dt \right]$$

$$= \frac{z}{b-a} [F'(z) - F''(z)] + \frac{a}{b-a} [F'(z) - F(z)]$$

Then we have

$$aF(z) + (z-b)F'(z) - zF''(z) = 0 \tag{A1-4}$$

REFERENCES

Abarbanel, H.D.I. (1996): *Analysis of Observed Chaotic Data*, 2nd Edition. Springer, New York.

Abarbanel, H.D.I., Brown, R. and Kadtke, J.B. (1990): Prediction in chaotic nonlinear systems: Methods for time series with broad band Fourier spectra. *Physical Review A*, 41(4), 1782–1807.

Abarbanel, H.D.I., Brown, R., Sidorowich, J.J. and Tsimring, L.S. (1993): The analysis of observed chaotic data in physical systems. *Review of Modern Physics*, 65(4), 1331–1391.

Ataei, M., Khaki-Sedigh, A., Lohmann, B. and Lucas, C. (2003): Determination of embedding dimension using multiple time series based on singular value decomposition. In: Troch, I. and Breitenecker, F. (Eds.), *Proceedings 4th Mathmod*, Vienna, February 2003, pp. 190–196.

Bayly, B.J., Goldhirsch, I. and Orszag, S.A. (1987): Independent degrees of freedom of dynamical systems. *Journal of Scientific Computing*, (2)2, 111–121.

Brown, R. (1993): Orthogonal polynomials as prediction functions in arbitrary phase space dimensions. *Physical Review E*, 47, 3962–3969.

Casdagli, M. (1989): Nonlinear prediction of chaotic time series. *Physica D*, 35, 335–356. Reprinted in Ott et al. (1994).

Cawley, R. and Hsu, G.H. (1992): Local-geometric projection method for noise reduction in chaotic maps and flows. *Physical Review A*, 46(6), 3057–3082.

Diks, C. (1999): *Nonlinear Time Series Analysis, Methods and Applications*. World Scientific, Singapore.

Eckmann, J.P. and Ruelle, D. (1985): Ergodic theory of chaos and strange attractors. *Reviews of Modern Physics*, 57(3), 617–656.

Elsner, J.B. and Tsonis, A.A. (1997): Time series analysis and multi-season predictions of Atlantic basin hurricane activity. Abstract volume of the XXII General Assembly of the European Geophysical Society, Vienna, Austria, 21–25 April 1997.

Elsner, J.B. and Tsonis, J.B. (1992): Nonlinear prediction, chaos and noise. *Bulletin of the American Meteorological Society*, 73, 49–60.

Farmer, J.D. and Sidorowich, J.J. (1987): Predicting chaotic time series. *Physical Review Letters*, 59(8), 845–848. Reprinted in Ott et al. (1994).

Farmer, J.D. and Sidorowich, J.J. (1991): Optimal shadowing and noise reduction. *Physica D*, 47, 373–392.

Fraser, A.M. (1989a): Reconstructing attractors from scalar time series: A comparison of singular system and redundancy criteria. *Physica D*, 34D, 391–404.

Fraser, A.M. (1989b): Information and entropy in strange attractors. *IEEE Transactions on Information Theory*, 35(2), 245–262.

Fraser, A.M. and Swinney, H.L. (1986): Independent coordinates for strange attractors from mutual information. *Physical Review A*, 33, 1134–1140.

Gallager, R.G. (1968): *Information Theory and Reliable Communication*. Wiley, New York.

Grassberger, P., Hegger, R., Kantz, H., Schaffrath, C. and Schreiber, T. (1993): On noise reduction methods for chaotic data. *Chaos*, 3, 127–141. Reprinted in Ott et al. (1994).

Grassberger, P. and Procaccia, I. (1983a): Characterization of strange attractors. *Physical Review Letters*, 50(5), 346–349.

Grassberger, P. and Procaccia, I. (1983b): Measuring the strangeness of strange attractors. *Physica D*, 9, 189–208.

Grebogi, C., Hammel, S.M., Yorke, J.A. and Sauer, T. (1990): Shadowing of physical trajectories in chaotic dynamics: Containment and refinement. *Physical Review Letters*, 65(13), 1527–1530.

Grebogi, C., Ott, E., Pelikan, S. and Yorke, J.A. (1984): Strange attractors that are not chaotic. *Physica D*, 13(1–2), 261–268.

Hammel, S.M. (1990): A noise reduction method for chaotic systems. *Physics Letters A*, 148, 421–428. Reprinted in Ott et al. (1994).

Holzfuss, J. and Mayer-Kress, G. (1986): An approach to error-estimation in the application of dimension algorithms. In: Mayer-Kress, G. (Ed.), *Dimensions and Entropies in Chaotic Systems*. Springer-Verlag, New York, pp. 114–122.

Islam, S. and Sivakumar, B. (2002): Characterization and prediction of runoff dynamics: A nonlinear dynamical view. *Advances in Water Resources*, 25, 179–190.

Jayawardena, A.W. and Gurung, A.B. (2000): Noise reduction and prediction of hydrometeorological time series: Dynamical systems approach vs. stochastic approach. *Journal of Hydrology*, 228(3–4), 242–264.

Jayawardena, A.W. and Lai, F.Z. (1994): Analysis and prediction of chaos in rainfall and streamflow time series. *Journal of Hydrology*, 153(1–4), 23–52.

Jayawardena, A.W., Li, W.K. and Xu, P. (2002): Neighbourhood selection for local modelling and prediction of hydrological time series. *Journal of Hydrology*, 258(1–4), 40–57.

Jayawardena, A.W., Xu, P.C. and Li, W.K. (2008): A method of estimating the noise level in a chaotic time series. *Chaos*, 18(2), 023115-023115-11, doi:10.1063/1.2903757.

Judd, K. and Mees, A. (1995): On selecting models for non-linear time series. *Physica D*, 82, 426–444.

Kantz, H. (1993): Noise reduction by local reconstruction of the dynamics. In: Weigend, A.S. and Gershenfeld, N.A. (Eds.), *Time Series Prediction: Forecasting the Future and Understanding the Past, Studies in the Science of Complexity*, Proc., vol. XVII, Addison-Wesley, Reading, MA, pp. 475–490.

Kantz, H. (1994): A robust method to estimate the maximal Lyapunov exponent of a time series. *Physics Letters A*, 185, 77–87.

Kantz, H. and Schreiber, T. (1997): *Nonlinear Time Series Analysis*. Cambridge University Press, Cambridge, UK.

Kaplan, D. and Glass L. (1992): Direct test for determinism in a time series. *Physical Review Letters*, 68, 427–430. Reprinted in Ott et al. (1994).

Kennel, M.B., Brown, R. and Abarbanel, H.D.I. (1992): Determining embedding dimension for phase-space reconstruction using a geometrical construction. *Physical Review A*, 45, 3403–3414.

Kostelich, E.J. and Schreiber, T. (1993): Noise reduction in chaotic time series data: A survey of common methods. *Physical Review E*, 48(3), 1752–1763.

Kostelich, E.J. and Yorke, J.A. (1988): Noise reduction in dynamical systems. *Physical Review A*, 38, 1649–1652. Reprinted in Ott et al. (1994).

Lorenz, E.N. (1963): Deterministic non-periodic flow. *Journal of the Atmospheric Sciences*, 20, 130–141.

Mañé, R. (1981): On the dimension of the compact invariant sets of certain nonlinear maps. In: Rand, D.A. and Young, L.-S. (Eds.), *Dynamical Systems and Turbulence, Lecture Notes in Mathematics*, vol. 898. Springer-Verlag, Berlin, pp. 230–242.

Mees, A.I. and Judd, K. (1996): Parsimony in dynamical modeling. In: Kravtsov, Y. and Kadtke, J. (Eds.), *Predictability of Complex Dynamical Systems*, Springer, Berlin, pp. 123–142.

Mpitsos, G.J., Creech, H.C., Cohan, C.S. and Mendelson, M. (1987): Variability and chaos neurointegrative principles in self-organization of motor patterns. In: Hao, B. (Ed.), *Directions in Chaos*, World Scientific, Singapore, pp. 163–190.

Ng, W.W., Panu, U.S. and Lennox, W.C. (2007): Chaos based analytical techniques for daily extreme hydrological observations. *Journal of Hydrology*, 342(1–2), 17–41.

Oltmans, H. and Verheijen, P.J.T. (1997): Influence of noise on power-law scaling functions and an algorithm for dimension estimations. *Physical Review E*, 56, 1160–1170.

Ott, E. (1993): *Chaos in Dynamical Systems*, 1st Edition. Cambridge University Press, Cambridge, UK.

Ott, E., Sauer, T. and Yorke, J.A. (1994): *Coping with Chaos: Analysis of Chaotic Data and the Exploitation of Chaotic Systems*. Wiley, New York.

Packard, N.H., Crutchfield, J.P., Farmer, J.D. and Shaw, R.S. (1980): Geometry from a time series. *Physical Review Letters*, 45(9), 712–716. Reprinted in Ott et al. (1994).

Porporato, A. and Ridolfi L. (1997): *Nonlinear analysis of river flow time sequences. Water Resources Research*, 33(6), 1353–1367.

Press, W.H., Vetterling, W.T., Teukolsky, S.A. and Flannery, B.P. (1992): *Numerical Recipe in Fortran, The Art of Scientific Computation*. Cambridge University Press, Cambridge, MA.

Regonda, S., Sivakumar, B. and Jain, A. (2004): Temporal scaling in river flow: Can it be chaotic? *Hydrological Sciences Journal*, 49(3), 373–385.

Rodriguez-Iturbe, I., De Power, B.F., Sharifi, M.B. and Georgakakos, K.P. (1989): Chaos in rainfall. *Water Resources Research*, 25(7), 1667–1675.

Sauer, T. (1992): A noise reduction method for signals from nonlinear systems. *Physica D*, 58, 193–201.

Sauer, T., Yorke, J.A. and Casdagli, M. (1991): Embedology. *Journal of Statistical Physics*, 65(3–4), 579–616.

Schouten, J.C., Takens, F. and van den Bleek, C.M. (1994): Estimation of the dimension of a noisy attractor. *Physical Review E*, 50(3), 1851–1861.

Schouten, J.C. and van den Bleek, C.M. (1994): *RRCHAOS: A Menu-Driven Software Package for Chaotic Time Series Analysis*. Reactor Research Foundation, Delft, The Netherlands.

Schreiber, T. (1993a): Determination of the noise level of chaotic time series. *Physical Review E*, 48(1), R13–R16.

Schreiber, T. (1993b): Extremely simple noise reduction method. *Physical Review E*, 47(4), 2401–2404.

Schreiber, T. (1999): Interdisciplinary application of non-linear time series methods. *Physics Reports*, 308, 1–64.

Schreiber, T. and Grassberger, P. (1991): A simple noise-reduction method for real data. *Physics Letter A*, 160, 411–418.

Sharifi, M.B., Georgakakos, K.P. and Rodriguez-Iturbe, I. (1990): Evidence of deterministic chaos in the pulse of storm rainfall. *Journal of the Atmospheric Sciences*, 47(7), 888–893.

Shaw, R. (1984): *The Dripping Faucet as a Model Chaotic System*. Aerial Press, Santa Cruz, CA.

She, D. and Yang, X. (2010): A new adaptive local linear prediction method and its application in hydrological time series. *Mathematical Problems in Engineering*, 2010, 15 pp. Article ID 205438, doi:10.1155/2010/205438.

Sivakumar, B. and Berndtsson, R. (2010): Non-linear dynamics and chaos in hydrology. In: Sivakumar, B. and Berndtsson, R. (Eds.), *Advances in Data-Based Approaches for Hydrologic Modeling and Forecasting*. World Scientific, Singapore.

Sivakumar, B. and Jayawardena, A.W. (2002): An investigation of the presence of low-dimensional chaotic behaviour in the sediment transport phenomenon. *Hydrological Sciences Journal* 47(3), 405–416.

Sivakumar, B. and Jayawardena, A.W. (2003): Sediment transport phenomenon in rivers: An alternative perspective. *Modelling of Hydrologic Systems, Special Issue of Environmental Modelling and Software*, 18(8–9), 831–838.

Sivakumar, B., Jayawardena, A.W. and Fernando, T.M.K.G. (2002): River flow forecasting: Use of phase space reconstruction and artificial neural networks approaches. *Journal of Hydrology*, 265(1–4), 225–245.

Sivakumar, B., Jayawardena, A.W. and Li, W.K. (2007): Hydrologic complexity and classification: A simple data reconstruction approach. *Hydrological Processes*, 21(20), 2713–2728.

Sivakumar, B., Liong, S.Y. and Liaw, C.Y. (1998): Evidence of chaotic behavior in Singapore rainfall. *Journal of the American Water Resources Association*, 34(2), 301–310. Also reprinted in *Water Resources Journal*, (201), 40–50, 1999.

Sivakumar, B., Liong, S.Y., Liaw, C.Y. and Phoon, K.K. (1999a): Singapore rainfall behaviour: Chaotic? *Journal of Hydrologic Engineering, ASCE*, 4(1), 38–48.

Sivakumar, B., Phoon, K.K., Liong, S.Y. and Liaw, C.Y. (1999b): A systematic approach to noise reduction in chaotic hydrological time series. *Journal of Hydrology*, 219, 103–135.

Small, M. (2005): Applied nonlinear time series analysis: Applications in physics, physiology and finance. World Scientific Publishing Co., Inc., Hackensack, NJ. Available at http://www.worldscibooks.com/etextbook/5722/5722_chap1.pdf.

Smith, L.A. (1992): Identification and prediction of low dimensional dynamics. *Physica D*, 58, 50–76.

Sugihara, J. and May, R.M. (1990): Nonlinear forecasting as a way of distinguishing chaos from measurement error in time series. *Nature*, 344(19), 734–741. Reprinted in Ott et al. (1994).

Takens, F. (1981): Detecting strange attractors in turbulence. In: Rand, D.A. and Young, L.S. (Eds.), *Dynamical Systems and Turbulence, Lecture Notes in Mathematics, Proceeding of a Symposium held at the University of Warwick 1979/1980*, vol. 898, Springer, New York, pp. 366–381.

Theiler, J., Eubank, S., Longtin, A., Galdrikian, B. and Farmer, J.D. (1992a): Testing for nonlinearity in time series: The method of surrogate data. *Physica D*, 58, 77–94. Reprinted in Ott et al. (1994).

Theiler, J., Galdrikian, B., Longtin, A., Eubank, S. and Farmer, J.D. (1992b): Using surrogate data to detect nonlinearity in time series. In: Casdagli, M. and Eubank, S. (Eds.), *Nonlinear modeling and forecasting, Proceeding of the Workshop on Nonlinear Modeling and Forecasting, September 1990, Santa Fe, New Mexico*, vol. 12, Addison-Wesley, Reading, MA, pp. 163–188.

Tsonis, A.A. and Elsner, J.B. (1988): The weather attractor on very short time scales. *Nature*, 333, 545–547.

Tsonis, A.A. and Elsner, J.B. (1991): Fractal properties of rain in the time domain and Taylor's hypothesis. *Meteorology and Hydrology*, 2, 98–103.

Tsonis, A.A. and Elsner, J.B. (1992): Nonlinear prediction as a way of distinguishing chaos from random fractal sequences. *Nature*, 358, 217–220.

Whitney, H. (1936): Differentiable manifolds. *Annals of Mathematics*, 37, 645.

Wolf, A., Swift, J.B., Swinney, H. and Vastano, J.A. (1985): Determining Lyapunov exponents from a time series. *Physica D*, 16, 285–317.

World Scientific Books. Available at http://www.worldscibooks.com/etextbook/5722/5722_chap1.pdf.

Yule, G.U. (1927): On a method of investigating periodicities in disturbed series, with special reference to Wolfer's sunspot numbers. *Philosophical Transactions of the Royal Society of London, Series A*, 226, 267–298.

Chapter 12

Support vector machines

12.1 INTRODUCTION

Support vector machines (SVMs),[1] originally developed by Vladimir Vapnik and his co-workers (e.g., Boser et al., 1992; Guyon et al., 1993; Cortes and Vapnik, 1995; Schölkopf et al., 1995, 1996; Vapnik and Lerner, 1963; Vapnik and Chervonenkis, 1964, 1979; Vapnik et al., 1997), and belonging to the class of supervised learning (they go through statistical learning[2]) can be used for optical character recognition (OCR); pattern recognition such as handwriting, speech, images, etc.; classification and regression analysis; and time series prediction. The simplest form of SVMs is when a given set of input data has to be classified into two possible classes, thereby leading to binary classification. Mathematically, SVMs achieve the classification by constructing a hyperplane in a high-dimensional space. By aiming at a larger separation between the hyperplane and the nearest data point, a lower generalization error can be expected. SVMs have good generalization properties, but are rather slow in the testing phase. They are universal approximators just as multilayer perceptrons and radial basis functions (RBFs).

SVMs have several advantages over other learning methods. For example, they exhibit better generalization; they are interpretable as they have explicit dependence on data; the optimization will not lead to local minima since learning is by optimizing a convex function; they have fewer parameters compared with neural networks; and they give confidence measures. Optimization is done by minimizing the 'primal' form with respect to the primal variables and simultaneously maximizing the 'dual' with respect to the Lagrange multipliers (dual variables). SVM implementation needs a quadratic programming (QP) solver.

In preparing this chapter, the author has benefitted from material available in books by Vapnik (1982, 1998, 2000), SVM tutorials by Fletcher (2009), Burges (1998), Hsu et al. (2003), Smola and Schölkopf (1998), and Gunn (1998) and many others.

12.2 LINEARLY SEPARABLE BINARY CLASSIFICATION

In a linear classifier, a p-dimensional vector (a data point) can be separated by a $(p-1)$-dimensional hyperplane. There can be several such hyperplanes as illustrated in Figure 12.1 for a two-dimensional data set separated by different lines. The classifier with the maximum distance between the hyperplane and the nearest data point is known as the

---

[1] Machine is a name given to a family of functions $f(x, \alpha)$.
[2] See Appendix 12.1.

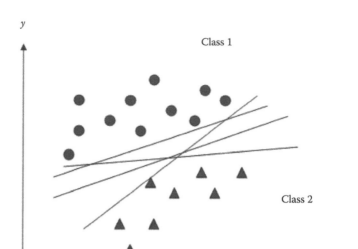

Figure 12.1 Different lines of separation.

maximum-margin hyperplane. It is also known as the perceptron of optimal stability. The samples on the margin are called support vectors.

In linearly separable binary classification, a p-dimensional training data set may be of the form

$$(x_i, y_i); \quad i = 1, 2, \ldots, L; \quad y_i \in \{-1, 1\}$$

The y_i's can be either fixed as above, or derived from a probability distribution. The assumption that the data are independently and identically distributed is more general than associating a fixed y with every x.

If $p = 2$, a line separating the data x_1, x_2 (x_1, x_2 are vectors) into two classes may be drawn. If $p > 2$, then a hyperplane is needed to separate $x_1, x_2, \ldots, x_p$ into two classes. The hyperplane can be of the form

$$w.x + b = 0 \tag{12.1}$$

where the weight w in the dot product[3] (or inner product) is orthogonal to the hyperplane, b is the bias, and $\dfrac{b}{\|w\|}$ is the perpendicular distance from the hyperplane to the origin (Figure 12.2). Support vectors are the training data closest to the separating hyperplane. They support the hyperplane, whereas the other vectors do not. Removal of non-support vectors does not affect the margin. The objective of SVMs is to maximize the distance between the hyperplanes and the closest member of both classes. The problem then is to select w and b such that the distance between the two parallel hyperplanes H_1 and H_2 (Figure 12.2) is maximized. The parallel hyperplanes H_1 and H_2 are respectively of the form

$$w.x + b = 1 \text{ for } H_2 \tag{12.2a}$$

---

[3] Dot product: $w.x = \displaystyle\sum_{i=1}^{L} w_i x_i.$

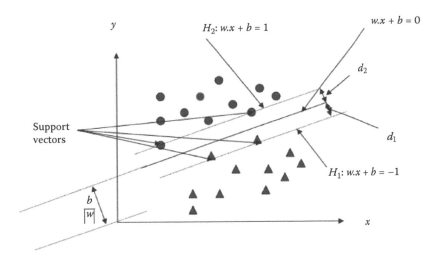

Figure 12.2 Hyperplane for linear binary classification.

and

$$w.x + b = -1 \text{ for } H_1 \qquad (12.2b)$$

It is also possible to have different values for y_i, for example, $y_i \in \{-5, 5\}$ as that would not affect the position of the separating hyperplane $w.x + b = 0$. In this case, H_1 and H_2 would be separated by a larger distance.

The data points that satisfy Equations 12.2a and 12.2b lie on the hyperplanes H_2 and H_1, respectively. In linearly separable data, the two hyperplanes can be chosen in such a way that there are no data points between them. To prevent the data falling onto the margins, the following constraints are imposed:

$$w.x_i + b \geq 1 \text{ for } x_i \text{ belonging to class 1} \quad (y_i = 1) \qquad (12.3a)$$

or

$$w.x_i + b \leq -1 \text{ for } x_i \text{ belonging to class 2} \quad (y_i = -1) \qquad (12.3b)$$

which can be combined and rewritten as

$$y_i(w.x_i + b) \geq 1 \text{ for all } 1 \leq i \leq L \qquad (12.3c)$$

If the distances are d_1 and d_2 (Figure 12.2), the margin, $|d_1| + |d_2|$, should be maximized. When $|d_1| = |d_2|$, it can be shown from the geometry that the margin is equal to $\frac{1}{\|w\|}$ where $\|w\|$ is the Euclidean norm of w. Maximizing $\frac{1}{\|w\|}$ is equivalent to minimizing $\|w\|$, and minimizing $\|w\|$ is equivalent to minimizing $\frac{1}{2}\|w\|^2$, a term that is useful in formulating a quadratic programming problem. The factor $\frac{1}{2}$ is added for mathematical convenience.

Minimization is subject to the constraints given in Equation 12.3, which can be reformulated into an optimization problem. The hyperplane that optimally separates the data is the one that minimizes

$$\frac{1}{2}\|w\|^2 \text{ subject to } y_i(w.x_i + b) \geq 1 \text{ for all } 1 \leq i \leq L \tag{12.4}$$

With the introduction of Lagrange multipliers[4] α_i, the problem, in its 'primal' form, L_P (w,b,α), can be formulated as follows:

$$\underset{w,b}{\text{Min}}\ \underset{\alpha}{\text{Max}}\quad L_P(w,b,\alpha) = \frac{1}{2}\|w\|^2 - \sum_{i=1}^{L}\alpha_i[y_i(w.x_i + b) - 1] \tag{12.5}$$

The primal form given in Equation 12.5 should be minimized to find optimum w and b and maximized to find α subject to $\alpha_i \geq 0$. This follows from the fact that the solution to minimizing $\frac{1}{2}\|w\|^2$ subject to constraints $y_i(w.x_i + b) \geq 1$ for $1 \leq i \leq L$ is given by the saddle point of the Lagrangian functional $L_P(w,b,\alpha)$ given in Equation 12.5 (Minoux, 1986). This can be done by setting the derivatives of L_P with respect to w and b to zero.

$$\frac{\partial L_P}{\partial w} = 0 \text{ gives } w = \sum_{i=1}^{L}\alpha_i y_i x_i \tag{12.6a}$$

$$\frac{\partial L_P}{\partial b} = 0 \text{ gives } \sum_{i=1}^{L}\alpha_i y_i = 0 \tag{12.6b}$$

In Equation 12.6b, only a few α_i's will be positive and the corresponding x_s's are called the support vectors.

The above formulation is known as the 'primal' form of optimization. The advantage in the Lagrangian formulation is that the constraints in Equation 12.3 will be replaced by constraints in the Lagrangian multipliers, which are easier to handle. Also, in the reformulated problem, the training data will only appear as dot products between vectors. The latter property enables the generalization of the method to non-linear problems.

The problem can also be formulated as its 'dual'. It is obtained by substituting Equations 12.6a and 12.6b into Equation 12.5, and takes the form

$$\text{Max}\quad L_D(\alpha) = \sum_{i=1}^{L}\alpha_i - \frac{1}{2}\sum_{i,j}\alpha_i\alpha_j y_i y_j x_i.x_j \tag{12.7}$$

---

[4] Lagrange multipliers are used for finding the optimum of a function subject to external constraints. For example, if the function $f(x, y)$ is to be maximized (or minimized) subject to the constraint $g(x, y) = C$, then finding the optimum of the Lagrangian function $L(x, y, \lambda) = f(x, y) - \lambda[g(x, y) - C]$ is equivalent to finding the optimum of the original function. λ is known as the Lagrangian multiplier.

subject to $\alpha_i \geq 0$, and $\sum_{i=1}^{L} \alpha_i y_i = 0$, which should be maximized with respect to α.

The SVM linear classifier can be considered as an extension of the perceptron. Equation 12.7 may also be written as

$$\text{Max} \quad \sum_{i=1}^{L} \alpha_i - \frac{1}{2}\alpha^T Q\alpha \tag{12.8a}$$

or

$$\text{Min} \quad \frac{1}{2}\alpha^T Q\alpha - \sum_{i=1}^{L} \alpha_i \tag{12.8b}$$

where Q is given by

$$Q_{ij} = y_i y_j x_i.x_j \tag{12.9}$$

The dual formulation ensures that the maximum of $L_D(\alpha)$ subject to the constraints $\alpha_i \geq 0$ occurs for the same values of w, b, and α as the minimum of $L_P(\alpha)$ subject to the constraints $\frac{\partial L_P}{\partial w} = 0$, $\frac{\partial L_P}{\partial b} = 0$, and $\frac{\partial L_P}{\partial \alpha} = 0$ (Fletcher, 1987).

This is a convex quadratic optimization problem[5] that requires a QP solver to solve. The solution of the quadratic programming problem will give α, and substitution in Equation 12.6a will give w. Owing to the high dimension of the vector w, it is easier to solve the 'dual' problem. The 'dual' requires only the dot product of each input vector. What remains to be determined is the threshold b.

Any support vector x_s satisfies

$$y_s(w.x_s + b) = 1 \tag{12.10}$$

which when substituted for w from Equation 12.6a gives

$$y_s \left(\sum_{m \in N_S} \alpha_m y_m x_m.x_s + b \right) = 1 \tag{12.11}$$

In Equation 12.11, N_S indicates the number of support vectors, which is determined by picking data points with $\alpha_i > 0$. Since $y_s^2 = 1$ (Equation 12.3), multiplying both sides of Equation 12.11 by y_s and rearranging terms gives

$$b = y_s - \sum_{m \in N_S} \alpha_m y_m x_m.x_s \tag{12.12}$$

---

[5] A convex optimization is one that satisfies $f_i(\alpha x + \beta y) \leq \alpha f_i(x) + \beta f_i(y)$ with $\alpha + \beta = 1$; $\alpha \geq 0$, $\beta \geq 0$. Any linear programming problem is a convex optimization problem. In fact, convex optimization can be considered as a generalization of linear programming.

A better estimate of b can be obtained by averaging over all the support vectors. Then,

$$b = \frac{1}{N_S} \sum_{s \in S} \left(y_s - \sum_{m \in S} \alpha_m y_m x_m . x_s \right).$$

(12.13)

The value of y' for any new value of x' is given as

$$y' = \text{sgn}(w.x' + b)$$

(12.14)

where 'sgn' is the signum[6] function This completes the determination of the SVM for linearly separable binary classification.

If the formulation is done with $b = 0$ in the primal, the hyperplane will pass through the origin, thereby reducing the number of degrees of freedom by one. Such a hyperplane is unbiased, whereas one that does not is biased. The corresponding dual is the same as before but without the equality constraint $\sum_{i=1}^{L} \alpha_i y_i = 0$. If the data are linearly separable, all the support vectors will lie on the margins and, hence, the set of support vectors is sparse.

The sequential steps of linear binary classification consist of constructing Q_{ij} using Equation 12.9, finding α by maximizing the dual (Equation 12.7) using a QP solver, calculating w using Equation 12.6a, determining the set of support vectors by finding the indices i for which $\alpha_i > 0$, calculating b using Equation 12.13, and classifying new x' using Equation 12.14. It is to be noted that every training data point has a Lagrangian multiplier, and the points corresponding to $\alpha_i > 0$ are called the support vectors. They lie on the hyperplanes H_1 or H_2, whereas all other training data points have $\alpha_i = 0$ and lie on either side of H_1 or H_2 (Figure 12.1). The equality in Equation 12.3 holds in the former case, while the inequality holds in the latter case. If all the non-support vectors are removed and the training repeated, the same separating hyperplanes would be found.

12.3 SOFT-MARGIN BINARY CLASSIFICATION

12.3.1 Linear soft margin

To handle data that are not exactly linearly separable, meaning that some data could be misclassified (Figure 12.3), the SVM for linearly separable data can be extended to allow a certain amount of misclassification by introducing a set of non-negative slack variables, $\xi_i \geq 0$. The objective then is to classify the data as cleanly as possible into two classes while still maximizing the distance to the nearest cleanly classified data. The optimization then becomes a trade-off between the large margin and a small error penalty. The constraints then would be

$$w.x_i + b \geq +1 - \xi_i \text{ for } y_i = 1$$

(12.15a)

$$w.x_i + b \leq -1 + \xi_i \text{ for } y_i = -1$$

(12.15b)

---

[6] Sgn (x) returns a value of 1 if $x > 0$, 0 if $x = 0$, and -1 if $x < 0$.

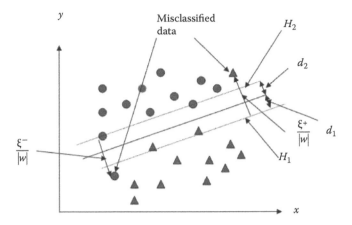

Figure 12.3 Soft-margin binary classification.

The constraints in Equations 12.15a and 12.15b can be combined into

$$y_i(w.x_i + b) - 1 + \xi_i \geq 0 \text{ for all } 1 \leq i \leq L \tag{12.15c}$$

The objective function aimed at reducing the number of misclassifications takes the form (the first term is the same as in Equation 12.4 and the extra term takes care of the additional errors introduced by the misclassifications)

$$\text{Min} \quad \left\{ \frac{1}{2}\|w\|^2 + C\left(\sum_{i=1}^{L} \xi_i\right)^k \right\} \tag{12.16}$$

subject to constraint given in Equation 12.15, with the additional constraint $0 \leq \alpha_i \leq C$, where C is a user-defined regularization parameter that controls the trade-off between the slack variable penalty, which increases with distance from the margin boundary and the size of the margin, and k is any positive integer. Equation 12.16 represents a convex programming problem, and $k = 1, 2$ correspond to a quadratic programming problem. When $k = 1$, the slack variables or their Lagrangian multipliers do not appear in the dual formulation (see Equation 12.21 below), which is an added advantage. Use of non-linear penalty functions makes it more difficult to find a global solution since the problem may become non-convex. Large values of C correspond to greater penalty to errors, implying that only the empirical risk[7] is minimized.

The solution to Equation 12.16 subject to constraints in Equation 12.15 is given by the saddle point of the Lagrangian primal form L_P given below (Minoux, 1986):

$$\text{Min} \quad L_P = \frac{1}{2}\|w\|^2 + C\sum_{i=1}^{L}\xi_i - \sum_{i=1}^{L}\alpha_i\left[y_i(x_i.w+b)-1+\xi_i\right] - \sum_{i=1}^{L}\mu_i\xi_i \tag{12.17}$$

---

[7] See Appendix 12.1.

subject to $\alpha_i \geq 0$, $\mu_i \geq 0$, which should be minimized with respect to w, b, and ξ_i, and maximized with respect to the Lagrangian multipliers α and μ. The μ_i's are introduced to satisfy the condition $\xi_i \geq 0$.

Differentiating Equation 12.17 with respect to w, b, and ξ_i and setting the derivatives to zero leads to the following Karush–Kuhn–Tucker (KKT)[8] conditions for the primal:

$$\frac{\partial L_P}{\partial w} = 0 \text{ gives } w = \sum_{i=1}^{L} \alpha_i y_i x_i \text{ (same as Equation 12.6a)} \tag{12.18a}$$

$$\frac{\partial L_P}{\partial b} = 0 \text{ gives } \sum_{i=1}^{L} \alpha_i y_i = 0 \text{ (same as Equation 12.6b)} \tag{12.18b}$$

$$\frac{\partial L_P}{\partial \xi_i} = 0 \text{ gives } C = \alpha_i + \mu_i \tag{12.18c}$$

Other KKT conditions include

$$y_i(x_i.w + b) - 1 + \xi_i \geq 0; \quad i = 1, 2, ..., L \tag{12.18d}$$

$$\xi_i \geq 0 \tag{12.18e}$$

$$\alpha_i \geq 0 \tag{12.18f}$$

$$\mu_i \geq 0 \tag{12.18g}$$

$$\alpha_i \{y_i(w.x_i) + b\} - 1 + \xi_i\} = 0 \tag{12.18h}$$

$$\mu_i \xi_i = 0 \tag{12.18i}$$

where the indices i and j correspond to the number and dimension of the training data, respectively. Equations 12.18h and 12.18i are the KKT complementarity conditions that enable the determination of b.

Substituting Equation 12.18 in Equation 12.17 gives the dual form, which is the same as Equation 12.7 for the linearly separable case except that Equation 12.18c with the condition $\mu_i \geq 0$ implies $\alpha_i \leq C$:

$$\text{Max} \quad \sum_{i=1}^{L} \alpha_i - \frac{1}{2} \alpha^T Q \alpha \text{ (same as Equation 12.8a)} \tag{12.19a}$$

or

$$\text{Min} \quad \frac{1}{2} \alpha^T Q \alpha - \sum_{i=1}^{L} \alpha_i \text{ (same as Equation 12.8b)} \tag{12.19b}$$

---

[8] See Appendix 12.2 for the definition of the KKT conditions.

subject to

$$\sum_{i=1}^{L} y_i \alpha_i = 0; \quad 0 \le \alpha_i \le C; \quad i = 1, 2, \dots, L$$

where Q is given by

$$Q_{ij} = y_i y_j x_i.x_j \text{ (same as Equation 12.9)} \tag{12.20}$$

The dual form of Equation 12.17 (for $k = 1$) may be written as

$$\text{Max} \quad L_D = \sum_{i=1}^{L} \alpha_i - \frac{1}{2} \sum_{i,j}^{L} \alpha_i \alpha_j y_i y_j x_i.x_j \tag{12.21}$$

subject to

$$0 \le \alpha_i \le C; \quad \sum_{i=1}^{L} \alpha_i y_i = 0$$

The solution to the primal form of the problem is given by

$$w = \sum_{i=1}^{N_S} \alpha_i y_i x_i \tag{12.22}$$

The threshold b is calculated using the KKT complementarity conditions in the same way (Equations 12.12 and 12.13) as for the linearly separable case with the support vectors corresponding to $0 < \alpha_1 \le C$. The value of y' corresponding to a given value of x' is calculated using Equation 12.14.

As before, the sequential steps of linear soft-margin binary classification consist of constructing Q_{ij} using Equation 12.9, choosing a suitable value for C (user defined), finding α by maximizing the dual (Equation 12.21) using a QP solver, calculating w using Equation 12.22, determining the set of support vectors by finding the indices i for which $0 \le \alpha_1 \le C$, calculating b using Equation 12.13, and classifying new x' using Equation 12.14.

12.3.2 Non-linear classification

In real systems, linear separability may not always be possible. The problem of separation could be made easier by mapping the original finite dimensional space into a higher-dimensional space. Linear classification in high-dimensional feature space is equivalent to non-linear classification in the low-dimensional input space. The objective is to map the given data into a 'feature space'[9] of high dimension (even infinite) with as wide a gap as possible between the two classes. It is done using a mapping function ϕ (Figure 12.4). Since

---

[9] For example, if the original finite dimensional space is $\{x_1, x_2, x_3\}$, then the feature space may be of the form $\{1, \sqrt{2}x_1, \sqrt{2}x_2, \sqrt{2}x_3, x_1^2, x_2^2, x_3^2, x_1 x_2, x_2 x_3, x_3 x_1, \dots\}$.

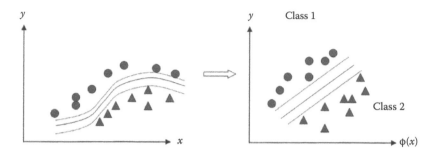

Figure 12.4 Mapping onto feature space.

the SVM algorithm depends on the data only through dot products in the feature space in the form $\phi(x_i).\phi(x_j)$, a kernel function that gives the dot products as $k(x_i, x_j) = \phi(x_i).\phi(x_j)$ is all what is needed. The price to be paid for mapping onto a high-dimensional feature space is the extremely high cost of computing resources both in memory and time. The explicit form of ϕ is not needed. The dot product is computed in the high-dimensional feature space. Admissible kernel functions that should satisfy Mercer's[10] condition include linear, polynomial, radial basis, and sigmoid-type functions, of which the RBF is a popular choice. If the kernel function k is symmetric and positive-definite, the data can be linearly separated. However, there are instances in which the RBF is not the best kernel to use, particularly when the feature space is very large. In such cases, a linear kernel may be chosen. Possible choices of kernel functions include

- Linear: $k(x_i, x_j) = x_i^T x_j$
- Polynomial: $k(x_i, x_j) = (\gamma x_i^T x_j + r)^d, \quad \gamma > 0$
- RBFs: $k(x_i, x_j) = e^{-\gamma \|x_i - x_j\|^2}$ where $\gamma = \dfrac{1}{2\sigma^2}$, σ the spread
- Sigmoid: $k(x_i, x_j) = \tanh(\gamma x_i^T x_j + r)$

where γ, r, and d are kernel parameters.

In the feature space of a higher dimension, x_i is mapped onto $\phi(x_i)$. Then, Equation 12.15c can be written as

$$y_i(w^T w \phi(x_i) + b) - 1 + \xi_i \geq 0$$

The term Q_{ij} (in Equation 12.9) would take the form

$$Q_{ij} = y_i y_j \phi(x_i).\phi(x_j) = y_i y_j k(x_i, x_j) \tag{12.23}$$

The rest of the formulation follows the same procedure as for the linear case and can be summarized as follows:

- Choose a suitable kernel function $k(x_i, x_j)$.
- Determine Q_{ij} from Equation 12.23.

---

[10] Mercer's condition arises out of Mercer's theorem, which guarantees the existence of a continuous symmetric function $\phi(u,v)$, which describes an inner product in some feature space, has an expansion of the form $\phi(u,v) = \displaystyle\sum_{k=1}^{\infty} a_k z_k(u) z_k(v)$ with $a_k > 0$.

- Choose a suitable value for the parameter C.
- Find α from

$$\text{Max} \quad \sum_{i=1}^{L} \alpha_i - \frac{1}{2}\alpha^T Q\alpha \tag{12.24a}$$

or

$$\text{Min} \quad \frac{1}{2}\alpha^T Q\alpha - \sum_{i=1}^{L} \alpha_i \tag{12.24b}$$

subject to

$$\sum_{i=1}^{L} y_i\alpha_i = 0; \quad 0 \le \alpha_i \le C; \quad i = 1, 2, \dots, L$$

This requires a QP solver.
- Find w from

$$w = \sum_{i=1}^{L} \alpha_i y_i \phi(x_i) \tag{12.25}$$

- Find the set of support vectors S by finding the indices i that satisfy the condition $0 \le \alpha_i \le C$.
- Determine b from

$$b = \frac{1}{N_S} \sum_{s \in S} \left[y_s - \sum_{m \in S} \alpha_m y_m \phi(x_m).\phi(x_s) \right] \tag{12.26}$$

- Classify any new point x' according to whether y' from Equation 12.27 below is +1 or –1.

$$y' = \text{sgn}(w.\phi(x') + b) \tag{12.27}$$

The description given thus far is for binary classification. The methods can be extended to multiclass classification by considering two classes at a time. Two of the popular methods used for multiclass classification are the 'one against many' and the 'one against one' approaches. In the former approach, each class is split out and all of the other classes are merged, whereas in the latter approach, $\frac{1}{2}k(k-1)$ models, where k is the number of classes, are constructed and trained with data from two classes. The latter approach has been followed in the SVM software package LIBSVM (Chang and Lin, 2011). Other methods are also available (see Hsu and Lin [2002] for a comparison).

12.4 SUPPORT VECTOR REGRESSION

12.4.1 Linear support vector regression

In support vector regression, the objective is to predict the output value y, for a given input value x, which is more than just classification. This requires an objective function, which usually is the sum of empirical risk[11] and a complexity term that makes the flatness[12] of the feature space. The training data has the same form as before $\{(x_i, y_i); i = 1, 2, ..., L\}$, but the dependent variable y takes the form

$$y_i = w.x_i + b \tag{12.28}$$

The penalty function used in regression is such that there is no penalty if $|z_i - y_i| < \varepsilon$, where z_i is the target (expected) value of the output y_i, and ε is an indicator of the error. The region $y_i \pm \varepsilon$ is called the ε-insensitive tube. Another variation to the penalty function is that the output variables lying outside the tube are given two slack variables, one from above (ξ^+), and the other from below (ξ^-) the tube. These are shown in the constraints below. The objective (error) function is of the form

$$\frac{1}{2}\|w\|^2 + C\sum_{i=1}^{L}(\xi_i^+ + \xi_i^-) \tag{12.29}$$

subject to

$$z_i \le y_i + \varepsilon + \xi^+$$

$$z_i \ge y_i - \varepsilon - \xi^-$$

$$\xi^+ \ge 0; \quad \xi^- \ge 0$$

By introducing the Lagrange multipliers $\alpha_i^+ \ge 0$; $\alpha_i^- \ge 0$; $\mu_i^+ \ge 0$; $\mu_i^- \ge 0$, the primal form of the problem can be stated as

$$\text{Min} \quad L_P = \frac{1}{2}\|w\|^2 + C\sum_{i=1}^{L}(\xi_i^+ + \xi_i^-) - \sum_{i=1}^{L}(\mu_i^+\xi_i^+ + \mu_i^-\xi_i^-) -$$

$$\sum_{i=1}^{L}\alpha_i^+(\varepsilon + \xi_i^+ + y_i - z_i) - \sum_{i=1}^{L}\alpha_i^-(\varepsilon + \xi_i^- - y_i + z_i) \tag{12.30}$$

---

[11] See Appendix 12.1.
[12] Flatness in this context implies small w.

Substituting for y_i and differentiating with respect to w, b, ξ^+, and ξ^-,

$$\frac{\partial L_P}{\partial w} = 0 \quad \Rightarrow \quad w = \sum_{i=1}^{L} (\alpha_i^+ - \alpha_i^-) x_i \tag{12.31a}$$

$$\frac{\partial L_P}{\partial b} = 0 \quad \Rightarrow \quad \sum_{i=1}^{L} (\alpha_i^+ - \alpha_i^-) = 0 \tag{12.31b}$$

$$\frac{\partial L_P}{\partial \xi^+} = 0 \quad \Rightarrow \quad C = (\alpha_i^+ + \mu_i^+) \tag{12.31c}$$

$$\frac{\partial L_P}{\partial \xi^-} = 0 \quad \Rightarrow \quad C = (\alpha_i^- + \mu_i^-) \tag{12.31d}$$

Substituting these, the dual form of the formulation that should be maximized with respect to α_i^+ and α_i^-, and which is easier to solve, can be written as

$$\text{Max} \quad L_D = \sum_{i=1}^{L} (\alpha_i^+ - \alpha_i^-) z_i - \varepsilon \sum_{i=1}^{L} (\alpha_i^+ - \alpha_i^-) - \frac{1}{2} \sum_{i,j}^{L} (\alpha_i^+ - \alpha_i^-)(\alpha_j^+ - \alpha_j^-) x_i . x_j \tag{12.32}$$

subject to $0 \le \alpha_i^+ \le C$; $0 \le \alpha_i^- \le C$; and $\sum_{i=1}^{L} (\alpha_i^+ - \alpha_i^-) = 0$. The dual formulation eliminates the dual variables μ_i^+ and μ_i^-.

The set of support vectors S corresponds to $\xi_i^+ = 0$, or $\xi_i^- = 0$ for $0 < \alpha < C$. The threshold b is obtained as before using the KKT conditions as

$$b = z_s - \varepsilon - \sum_{m \in S}^{L} (\alpha_m^+ - \alpha_m^-) x_m . x_s \tag{12.33}$$

and a better value is obtained as before by averaging

$$b = \frac{1}{N_s} \sum_{s \in S} \left\{ z_s - \varepsilon - \sum_{m \in S}^{L} (\alpha_m^+ - \alpha_m^-) x_m . x_s \right\} \tag{12.34}$$

For any given value of x', the corresponding y' is given by Equation 12.28, which when substituted for w from Equation 12.31a gives

$$y' = \sum_{i=1}^{L} (\alpha_i^+ - \alpha_i^-) x_i . x' + b \tag{12.35}$$

Thus, in the case of linear support vector regression, the problem is completely described in terms of the dot products of the data. This feature becomes useful when extending to non-linear support vector regression. In summary, the sequential steps of linear support vector regression consist of choosing suitable values for C and ε (user defined), finding α_i^+, α_i^- by maximizing the dual (Equation 12.32) using a QP solver, calculating w using Equation 12.31a, determining the set of support vectors by finding the indices i for which $0 \leq \alpha_i \leq C$ and $\xi_i = 0$, calculating b using Equation 12.34, and determining y' corresponding to a new x' using Equation 12.35.

12.4.2 Non-linear support vector regression

Similar to non-linear classification, there are many problems that are not linearly regressable. In the same way as in non-linear classification, such problems can be made easier to solve by mapping the original finite dimensional space into a higher-dimensional feature space using a suitable mapping function. Only the inner product of the mapped inputs in the feature space need be determined without the need to explicitly calculate the mapping function. In SVM terminology, this is called the 'kernel trick' approach. The steps involved are as follows:

- Choose a suitable kernel function $k(x_i, x_j)$.
- Choose suitable values for the parameters C and ε.
- Determine α_i^+ and α_i^- by maximizing the dual

$$\sum_{i=1}^{L} (\alpha_i^+ - \alpha_i^-) z_i - \varepsilon \sum_{i=1}^{L} (\alpha_i^+ - \alpha_i^-) - \frac{1}{2} \sum_{i,j}^{L} (\alpha_i^+ - \alpha_i^-)(\alpha_j^+ - \alpha_j^-) \phi(x_i).\phi(x_j) \tag{12.36}$$

  subject to the constraints $0 \leq \alpha_i^+ \leq C$; $0 \leq \alpha_i^- \leq C$; and $\sum_{i=1}^{L}(\alpha_i^+ - \alpha_i^-) = 0$. This is done using QP solver.
- Determine w from

$$w = \sum_{i=1}^{L} (\alpha_i^+ - \alpha_i^-) \phi(x_i) \tag{12.37}$$

- Determine the set of support vectors S by finding the indices i where $0 \leq \alpha \leq C$, and $\xi_i = 0$
- Determine b from

$$b = \frac{1}{N_s} \sum_{s \in S} \left\{ z_i - \varepsilon - \sum_{m=1}^{L} (\alpha_i^+ - \alpha_i^-) \phi(x_i).\phi(x_m) \right\} \tag{12.38}$$

- Determine y' for each new point x' from

$$y' = \sum_{i=1}^{L} (\alpha_i^+ - \alpha_i^-) \phi(x_i).\phi(x') + b \tag{12.39}$$

It is also to be noted that α_i^+ and α_i^- are non-zero only for $|z_i - y_i| \geq \varepsilon$. For $|z_i - y_i| < \varepsilon$, α_i^+ and α_i^- are each zero. Therefore, the expansion of w in terms of x_i is sparse. Only those x_i's with non-zero coefficients are sufficient to describe w. They are called support vectors. In the non-linear case, the optimization corresponds to finding the flattest function in the feature space, not in the input space.

12.5 PARAMETER SELECTION

The effectiveness of the SVM depends on the type of kernel used, the kernel parameters, and the soft-margin parameter C. If the Gaussian function is used, only one parameter, γ, needs to be specified. The best combinations of C and γ are selected by grid-type search with exponentially increasing values, such as $C \in \{2^{-5}, 2^{-3}, ..., 2^{13}, 2^{15}\}$, and $\gamma \in \{2^{-15}, 2^{-13}, ..., 2^{1}, 2^{3}\}$. Cross validation is carried out to choose the best combination. This approach is computationally expensive since the grid search must evaluate the model at many points for each parameter.

The development of an SVM regression model from training data requires three parameters, C, ε, and γ $\left(\gamma = \dfrac{1}{2\sigma^2}\right)$ if the RBF kernel is used. Proper choice of these parameters greatly affects the performance of the generalization capacity of the predictive model. Improper choice of the spread in the RBF, σ, can lead to underfitting or overfitting in the prediction stage. Small values of σ tend to map the input space close to a linear feature space, leading to a simplified model causing underfitting, whereas large values of σ tend to result in a complex non-linear mapping causing overfitting. Overfitting tends to lack generalization when presented with unseen data. An increase in the value of σ takes into account all support vector distances, whereas a smaller value of σ would ignore most of the support vectors. Although the best value of σ has to be chosen by trial and error, a reasonable guideline is $\dfrac{1}{k}$, where k is the number of input variables.

The regularization parameter C controls the trade-off between maximizing the margin and minimizing the training error, or in other words, the trade-off between the degree of complexity of the model and the fitting error. If it is too small, underfitting will result, and if it is too large, overfitting will occur. The optimal value of the parameter ε depends on the noise in the data, which is unknown.

Polynomial kernels have more parameters than RBFs that requires more tuning, which can be time consuming. In situations where the RBF-type kernels are not suitable, a linear kernel that requires only one parameter to be tuned appears to perform well.

12.6 KERNEL TRICKS

The non-linearity in support vector algorithm could be introduced by a number of ways. One of the ways is by data preprocessing. For example, the input data may be mapped onto a feature space by a mapping function and the standard SV algorithm applied in the feature space. Typically, a d-dimensional feature space of polynomial degree p has $^{d+p+1}C_p$ different features, making this approach computationally infeasible. An easier way is to look for a feature space in which the dot products only need to be found; that is, $k(x,x') = \phi(x).\phi(x')$, and not $\phi(.)$ explicitly, leading to Equations 12.35 through 12.38. The problem then is to find a kernel function $k(x,x')$ that corresponds to a dot product in some feature space. Some of the admissible kernel functions are given in Section 12.3.2.

'Kernel tricks' (Aizerman et al., 1964) are used to map an input space onto a higher-dimensional feature space without having to compute the mapping function (which normally is non-linear) explicitly. A linear algorithm operating in the feature space is equivalent to a non-linear algorithm in the original input space. The 'trick' only requires the dot products in the feature space. The mapping function is chosen so that the high-dimensional dot products can be computed in the original input space using the kernel function. The higher dimension can even be infinite.

12.7 QUADRATIC PROGRAMMING

For the solution of the SVM problem, a QP solver is necessary. QP belongs to a class of constrained optimization in which the objective function is a quadratic function of the variables and the constraints are linear functions of the variables, as opposed to linear programming (LP) where the objective function and the constraints (which may be equalities) are both linear functions of the variables. The objective function may be convex, in which case it is easier to solve, or non-convex in which case it is more difficult. Most QP solvers are designed for convex functions. LP problems are normally solved by the simplex method (Dantzig, 1962), whereas QP problems can be solved by an extensions of the simplex method, or the interior point method (or the Newton–Barrier method) (Karmarkar, 1984), or the Sequential Minimal Optimization (SMO) (Platt, 1999). In all constrained optimization problems, the first step is to determine whether a feasible solution exists.

QP is not trivial and the details are outside the scope of this chapter. There are several software packages available as QP solvers. A partial list of codes can be found at http://www.numerical.rl.ac.uk/qp/qp.html. MATLAB® (version 5.3/R11) has a QP solver that uses the interior point method.

12.8 LIMITATIONS AND PROBLEMS

One of the problems in any modelling exercise is the determination of relevant input variables. Too many will make the dimension of problem large thereby requiring more computing effort and resources, and too few will under-represent the real problem. Use of some data preprocessing can reduce the dimensionality thereby simplifying the problem. Principal component analysis (PCA) by which the correlated input variables are lumped together thereby reducing the dimensionality, Gamma test (Corcoran et al., 2003), and forward selection (Wang et al., 2006; Noori et al., 2010) are some of the techniques available and used in SVM, as well as other data-driven approaches of modelling.

12.9 APPLICATION AREAS

SVMs have been applied to pattern recognition (e.g., Cortes and Vapnik, 1995; Schölkopf et al., 1995, 1996; Burges and Schölkopf, 1997), image recognition (e.g., Osuna et al., 1997), voice recognition (e.g., Schmidt, 1996), and regression (e.g., Müller et al., 1997; Mukherjee et al., 1997). They have good generalization performance but rather slow in testing. In theory, SVMs can be used to make predictions of any variable that depends on one or more independent variables.

Past applications in the water sector include groundwater contamination (Asefa et al., 2005), ice break-up forecast in the reach of the Yellow River (Zhou et al., 2009), typhoon rainfall forecasting (Lin et al., 2009), monsoon rainfall prediction from general circulation model outputs using a statistical downscaling model (Ghosh, 2010), monthly flow predictions (Lin et al., 2006), daily river discharge prediction (Chan et al., 2002b; Jayawardena et al., 2000), soil moisture prediction (Gill et al., 2006), rainfall–runoff modelling (Dibike et al., 2001), runoff prediction using singular spectrum analysis and SVM (Sivapragasam et al., 2001), daily runoff forecasting by combining SVM with chaos theory (Yu et al., 2004), identification of appropriate model structure and parameters for rainfall–runoff modelling (Bray and Han, 2004), rainfall–runoff modelling and runoff classification (Ghimire, 2011), prediction of longitudinal dispersion coefficient in natural streams based on hydraulic and geometric data (Noori et al., 2009), assessment of the performance of an SVM model for monthly discharge prediction using different data preprocessing techniques (Noori et al., 2011), and prediction of monthly Southern Oscillation Index (Chan et al., 2002a), among several others.

SVMs can be a useful tool for early warning systems. For example, flood warnings can be issued on the basis of whether a certain water level or discharge is greater than a predefined threshold value, which can be formulated as a classification problem. The expected water level or discharge can be estimated by formulating the problem as an SVM regression. Similar formulations can be made for any type of process that has some input variables and corresponding output variables as long as there is a training data set.

12.10 CONCLUDING REMARKS

SVMs are a relatively new statistical learning method that can be used for classification, regression, pattern recognition, and several other types of data analysis. It has several advantages over other machine learning techniques. For example, every local solution of an SVM problem is also a global solution and that the solution is guaranteed to be unique if the objective function is convex. Through the use of 'kernel tricks' SVMs can avoid the problems posed by the 'curse of dimensionality'. This chapter gives the general formulations for both classification (binary, soft margin, and non-linear) and regression (linear and non-linear). For a deeper understanding of the mathematical details, a list of up-to-date references is also given.

APPENDIX 12.1: STATISTICAL LEARNING

In any real system, the complete input space (population) is not known. Only a sample of the input space is known. With this limited information, the objective is to understand the system and make predictions in a different part of the input space and at a future time via a model. Since the model is constructed using incomplete information, there is bound to be some error and the performance of the model for a certain sample is estimated by a loss function that is minimized to determine the model parameters. In statistical learning, the 'loss' is related to the 'risk'.

Statistical learning theory, originally developed for pattern recognition and classification and sometimes known as the Vapnik–Chervonenkis (VC) theory, deals with small samples but includes theories developed for large samples. It does not assume any causal relationship between the inputs and outputs and looks only for statistical dependency. The general problem of statistical learning is to minimize a risk functional $R(\alpha)$ of the form

$$R(\alpha) = \int L(y, f(x, \alpha)) \, dP(x, y)) \tag{A12.1}$$

in which the probability function $P(x,y)$ is unknown, but an independently and identically distributed sample $(x_1, y_1), (x_2, y_2), ..., (x_L, y_L)$ of space (x,y) is available, and that $L(y,f(x,\alpha))$ is a set of loss functions (α is a parameter), which may be of the form

$$L(y, f(x, \alpha)) = \begin{cases} 0 \text{ if } y = f(x, \alpha) \\ 1 \text{ if } y \neq f(x, \alpha) \end{cases} \tag{A12.2a}$$

$$L(y, f(x, \alpha)) = (y - f(x, \alpha))^2 \tag{A12.2b}$$

$$L(p(x, \alpha)) = -\log(p(x, \alpha)) \tag{A12.2c}$$

$$L(y, f(x, \alpha)) = \begin{cases} |f(x, \alpha) - y| - \varepsilon & \text{for } |f(x, \alpha) - y| \geq \varepsilon \\ 0 & \text{otherwise} \end{cases} \tag{A12.2d}$$

$$L(y, f(x, \alpha)) = \begin{cases} \dfrac{1}{2}(f(x, \alpha) - y)^2 & \text{for } |f(x, \alpha) - y| < \mu \\ \mu|f(x, \alpha) - y| - \dfrac{\mu^2}{2} & \text{otherwise} \end{cases} \tag{A12.2e}$$

Equations A12.2a, A12.2b, and A12.2c correspond to binary classification, regression estimation, and density estimation, respectively (Vapnik, 1998). The loss function given in Equation A12.2d is referred to as the ε-insensitive loss function (Vapnik, 1995) for SVM regression. This loss function is used to penalize errors that are greater than a threshold ε resulting in sparse representation of the decision rule, giving significant computational advantages. Other loss functions for regression include the Laplacian and Huber's (Equation 12.2e) (Huber, 1972). These three loss functions for regression do not produce sparseness in support vectors.

Empirical risk minimization (ERM)

In the absence of a knowledge about the probability function $P(x,y)$, the induction principle is used to replace the true risk functional $R(\alpha)$ of Equation A12.1 by the empirical risk functional as

$$R_{\text{emp}}(\alpha) = \frac{1}{L} \sum_{i=1}^{L} L(y_i, f(x_i, \alpha)) \tag{A12.3}$$

where L is the sample size. Expected (or true) risk is the probability of misclassifications (in the case of a classification problem) averaged over a large number of samples of unknown

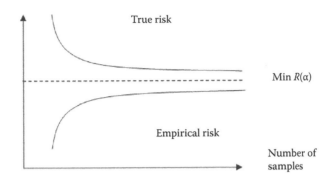

Figure 12.5 Variation of risk functional with number of samples.

probability distribution. This is unknown. Empirical risk is the probability of misclassifications computed from available data. The empirical risk is expected to converge to the true risk asymptotically as the number of training data samples increase (Figure 12.5).

If the loss function defined in Equation A12.2b is used, the empirical risk functional becomes

$$R_{emp}(\alpha) = \frac{1}{L} \sum_{i=1}^{L} (y - f(x, \alpha))^2 \qquad (A12.4)$$

The problem of minimizing the functional $R_{emp}(\alpha)$ is not trivial. It involves finding the value of the parameter α from a set of values that will define the specific function $f(x_i, \alpha^*)$ that makes the functional a minimum. The parameter may belong to a set of scalar quantities or a set of vectors.

Structural risk minimization (SRM)

SRM is an inductive principle in machine learning with finite training data sets that looks for a trade-off between the quality of fitting a model (function) to the training data set as measured by the empirical error and the complexity of the model (function) that approximates the training data as measured by the VC dimension[13] or the number of free parameters. Unlike in the empirical risk minimization, which attempts to minimize the empirical risk at any cost, the structural risk minimization looks for the optimal taking into account the amount of data and the quality of the approximation of the data by the chosen function. It is carried out by

- Using *a priori*[14] knowledge, choosing a class of functions such as polynomials, neural networks, RBFs, splines, etc., of increasing complexity

---

[13] The VC dimension h of a set of functions is defined as the maximum number of vectors that can be separated into two classes in all $2h$ combinations using functions of the set.

[14] Solving a learning problem with finite data requires selecting the 'structure' of a set of approximating (or loss) functions. This is normally assumed because without some prior assumed starting point, there is no way to generalize beyond observed data. Prediction of future is impossible without assuming some kind of relationship between the past and the future.

- Arranging the class of functions into a nested hierarchy in order of increasing complexity (degree of the polynomial, number of hidden layer neurons, etc., can be used as a measure of the complexity)
- Carrying out ERM on each subset (parameter estimation)
- Selecting the model that has the minimum value of the sum of empirical risk and VC confidence

The upper bound for the actual risk with probability $(1 - \eta)$ consists of two parts: empirical risk and VC confidence, as shown in the following equation (Vapnik, 1995):

$$R(\alpha) \leq R_{emp}(\alpha) + \sqrt{\frac{h\left(\ln\frac{2L}{h} + 1\right) - \ln\frac{\eta}{4}}{L}} \tag{A12.5}$$

where h is the VC dimension and L is the sample size. This is independent of the probability function $P(x,y)$.

If $\frac{L}{h}$ is large, the actual risk is determined by the empirical risk. Therefore, the minimization of actual risk can be achieved by minimizing the empirical risk. On the other hand, if $\frac{L}{h}$ is small, a small value of empirical risk does not guarantee a small actual risk. Achieving overall minimization thus requires the VC dimension as a controlling variable. With increasing complexity of the model, the training error will decrease but the VC confidence will increase, leading to overfitting and lacking generalization properties.

The SRM principle is implemented by dividing the class of models into nested subclasses on the basis of the VC dimension, training a series of models for each subclass with the objective of minimizing the empirical risk only, and choosing the trained model that has the sum of the empirical risk and the VC confidence a minimum. VC dimension is the most widely used measure of model complexity. The principle of SRM is similar to the principle of parsimony (sometimes referred to as the philosophical principle 'Occam's razor'), which is characterized by the Akaike information criterion (AIC), Bayes information criterion (BIC), and the minimum descriptive length (MDL), all of which penalize the model complexity (order of function or machine complexity) to achieve better generalization. A highly complex model that hugs the data (and the noise) or a too simplistic model would be far from reality because neither of them would possess good generalization capability when presented with unseen data. The objective should be to look for the model with optimal complexity. The VC dimension provides an estimate of the upper bound of Bayesian error. Determining the optimal VC dimension analytically for non-linear approximating functions is difficult, if not impossible, but could be carried out heuristically using early stopping criteria such as cross validation. The structure (nesting) has to be done before analysing the training data. In the context of neural networks, SRM is similar to cross validation (see Figure 8.9).

APPENDIX 12.2: KARUSH–KUHN–TUCKER (KKT) CONDITIONS

The Karush–Kuhn–Tucker (KKT) conditions state that the products of dual variables and constraints must vanish at the points of solution; that is,

$$\alpha_i^+(\varepsilon + \xi_i^+ - y_i + (w.x_i + b)) = 0 \tag{A12.6a}$$

$$\alpha_i^-(\varepsilon + \xi_i^- - y_i - (w.x_i + b)) = 0 \qquad\qquad (\text{A}12.6b)$$

$$(C - \alpha_i^+)\xi_i^+ = 0 \qquad\qquad (\text{A}12.6c)$$

$$(C - \alpha_i^-)\xi_i^- = 0 \qquad\qquad (\text{A}12.6d)$$

These lead to the conclusions that

i. Only samples (x_i, y_i) corresponding to $\alpha_i^+ = C$, $\alpha_i^- = C$ lie outside the ε-tube.
ii. $\alpha_i^+\alpha_i^- = 0$ implies α_i^+ and α_i^- cannot be non-zero simultaneously. Therefore,

$$\varepsilon - y_i + (w.x_i) + b \geq 0, \text{ and } \xi_i^+ = 0 \text{ if } \alpha_i^+ > 0$$

$$\varepsilon - y_i + (w.x_i) + b \leq 0, \text{ if } \alpha_i^+ > 0$$

The Lagrange multipliers are non-zero only for $|f(x_i) - y_i| \geq \varepsilon$; that is, for all samples inside the ε-tube, α_i^+, $\alpha_i^- = 0$.
The KKT condition for the primal problem may be stated as follows (Fletcher, 1987):

$$\frac{\partial L_P}{\partial w_k} = w_k - \sum_{i=1}^{L} \alpha_i y_i x_{ik} = 0; \quad k = 1, 2, ..., d \qquad\qquad (\text{A}12.7a)$$

$$\frac{\partial L_P}{\partial b} = -\sum_{i=1}^{L} \alpha_i y_i = 0 \qquad\qquad (\text{A}12.7b)$$

$$y_i(x_i.w + b) - 1 \geq 0; \quad i = 1, 2, ..., L \qquad\qquad (\text{A}12.7c)$$

$$\alpha_i \geq 0; \quad i = 1, 2, ..., L \qquad\qquad (\text{A}12.7d)$$

$$\alpha_i(y_i(w.x_i) + b) - 1 = 0; \quad i = 1, 2, ..., L \qquad\qquad (\text{A}12.7e)$$

They are satisfied at the solution of any constrained optimization problem including support vector machines. Since SVMs are a convex problem, the KKT conditions are necessary and sufficient for w, b, and α to be a solution (Fletcher, 1987). Therefore, solving the SVM problem is equivalent to finding a solution to the KKT conditions. The threshold (bias) b is found by the KKT 'complementarity' condition (Equation A12.7e) for any i for which $\alpha \neq 0$. An average value from a number of support vectors, however, provides a better estimate.

REFERENCES

Aizerman, M., Braverman, E. and Rozonoer, L. (1964): Theoretical foundations of the potential function method in pattern recognition learning. *Automation and Remote Control*, 25, 821–837.

Asefa, T., Kemblowski, M., Urroz, G. and McKee, M. (2005): Support vector machines (SVM's) for monitoring network design. *Groundwater*, 43(3), 413–422.

Boser, B.E., Guyon, I.M. and Vapnik, V.N. (1992): A training algorithm for optimal margin classifiers. In: Haussler, D. (Ed.), *Proceedings of the Annual Conference on Computational Learning Theory*, ACM Press, Pittsburgh, July 1992, pp. 144–152.

Bray, M. and Han, D. (2004): Identification of support vector machine for runoff modelling. *Journal of Hydroinformatics*, 6, 265–280.

Burges, C.J.C. (1998): A tutorial on support vector machines for pattern recognition. *Data Mining and Knowledge Discovery*, 2, 121–167. http://research.microsoft.com/pubs/67119/svmtutorial.pdf.

Burges, C.J.C. and Schölkopf, B. (1997): Improving the accuracy and speed of support vector learning machines. In: Mozer, M., Jordan, M. and Petsche, T. (Eds.), *Advances in Neural Information Processing Systems*, vol. 9. MIT Press, Cambridge, MA, pp. 375–381.

Chan, C.W., Chan, W.C., Jayawardena, A.W. and Harris, C.J. (2002a): Structure selection of neurofuzzy networks based on support vector regression. *International Journal of Systems Science*, 33(9), 715–722.

Chan, C.W., Jayawardena, A.W. and Choy, K.Y. (2002b): Modelling of river discharge using neural network with structure determined by the support vectors. *Proceedings of the 4th Asian Control Conference*, Singapore, 2002, pp. 618–623.

Chang, C.C. and Lin, C.J. (2011): *A Library for Support Vector Machines*. Department of Computer Science, National Taiwan University, Taipei, Taiwan. http://csie.ntu.edu.tw/~cjlin/papers/libsvm.pdf; http://csie.ntu.edu.tw/~cjlin/libsvm.

Corcoran, J., Wilson, I. and Ware, J. (2003): Predicting the geo-temporal variation of crime and disorder. *International Journal of Forecasting*, 19, 623–634.

Cortes, C. and Vapnik, V. (1995): Support vector networks. *Machine Learning*, 20, 273–297.

Dantzig, G.B. (1962): *Linear Programming and Extensions*. Princeton University Press, Princeton, NJ.

Dibike, Y.B., Velickov, S., Solomatine, D. and Abbott, M.B. (2001): Model induction with support vector machines, introduction and application. *ASCE Journal of Computing in Civil Engineering*, 15(3), 208–216.

Fletcher, R. (1987): *Practical Methods of Optimization*, 2nd Edition. John Wiley & Sons, Chichester, UK.

Fletcher, T. (2009): Support vector machines explained. http://www.tristanfletcher.co.uk/svm%20Explained.pdf.

Ghimire, P.R. (2011): Application of support vector machine for rainfall–runoff modeling in West Rapti River basin, Nepal. A Master Thesis submitted in partial fulfillment of the requirement for the Master's Degree in Disaster Management, National Graduate Institute for Policy Studies (GRIPS), Tokyo, Japan, International Centre for Water Hazard and Risk Management (ICHARM), Tsukuba, Japan.

Ghosh, S. (2010): SVM-PGSL Coupled approach for statistical downscaling to predict rainfall from GCM output. *Journal of Geophysical Research*, 115, D22102, doi:10.1029/2009JD013548.

Gill, M.K., Asefa, T., Kemblowski, M.W. and McKee, M. (2006): Soil moisture prediction using support vector machines. *Journal of the American Water Resources Association*, 42(4), 1033–1046.

Gunn, S.R. (1998): Support vector machines for classification and regression. Technical Report, Faculty of Engineering, Science and Mathematics, School of Electronics and Computer Science, University of Southampton. http://www.scribd.com/doc/48264898/SVM.

Guyon, I., Boser, B. and Vapnik, V. (1993): Automatic capacity tuning of very large VC-dimension classifiers. In: Hanson, S.J., Cowan, J.D. and Giles, C.L. (Eds.), *Advances in Neural Information Processing Systems*, vol. 5. Morgan Kaufmann Publishers, San Mateo, CA, pp. 147–155.

Hsu, C.W., Chang, C.C. and Lin, C.J. (2003): A practical guide to support vector classification. Technical Report Department of Computer Science and Information Engineering, National Taiwan University.

Hsu, C.W. and Lin, C.J. (2002): A comparison of methods for multi-class support vector machines. *IEEE Transactions on Neural Networks*, 13(2), 415–425.

Huber, P.J. (1972): Robust statistics: A review. *Annals of Statistics*, 43, 1041.

Jayawardena, A.W., Fernando, T.M.K.G., Chan, C.W. and Chan, W.C. (2000): Comparison of ANN, dynamical systems and support vector approaches for river discharge prediction. *Proceedings of the 19th Chinese Control Conference*, Hong Kong, China, December 6–8, pp. 504–508.

Karmarkar, N. (1984): A new polynomial-time algorithm for linear programming. *Combinatorics*, 4, 373–395.

Lin, G.F., Chen, G.R., Wu, M.C. and Chou, Y.C. (2009): Effective forecasting of hourly typhoon rainfall using support vector machines. *Water Resources Research*, 45, W08440, 11 pp., doi:10.1029/2009WR007911.

Lin, J.Y., Cheng, C.T. and Chau, K.W. (2006): Using support vector machines for long term discharge prediction. *Hydrological Sciences Journal*, 51(4), 599–612.

Minoux, M. (1986): *Mathematical Programming: Theory and Algorithms*. John Wiley & Sons, Chichester, UK.

Mukherjee, S., Osuna, E. and Girosi, F. (1997): Nonlinear prediction of chaotic time series using a support vector machine. *Proceedings of the IEEE Workshop on Neural Networks for Signal Processing*, vol. 7, Amelia Island, FL, pp. 511–519.

Müller, K.R., Smola, A.J., Rätsch, G., Schölkopf, B., Kohlmorgen, J. and Vapnik, V. (1997): Predicting time series with support vector machines. *Proceedings of ICANN'97*, Springer Lecture Notes in Computer Science, vol. 1327, pp. 999–1004.

Noori, R., Hoshyaripour, G., Ashrafi, K. and Nadjar-Araabi, B. (2010): Uncertainty analysis of developed ANN and ANFIS models in predicting of carbon monoxide daily concentration. *Atmospheric Environment*, 44, 476–482.

Noori, R., Karbassi, A., Farokhnia, A. and Dehghani, M. (2009): Predicting the longitudinal dispersion coefficient using support vector machine and adaptive neuro-fuzzy inference system techniques. *Environmental Engineering Science*, 26, 1503–1510.

Noori, R., Karbassi, A.R., Moghaddamnia, A., Han, D., Zokaei-Ashtiani, M.H., Farokhnia, A. and Ghafari Gousheh, M. (2011): Assessment of input variables determination on the SVM model performance using PCA, Gamma test and forward selection techniques for monthly streamflow prediction. *Journal of Hydrology*, 401(3–4), 177–189, doi:10.1016/j.hydrol.2011.02.021.

Osuna, E., Freund, R. and Girosi, F. (1997): Training support vector machines: An application to face detection. *IEEE Conference on Computer Vision and Pattern Recognition*, pp. 130–136.

Platt, J. (1999): Fast training of support vector machines using sequential minimal optimization. In: Schölkopf, B., Burges, C.J.C. and Smola, A.J. (Eds.), *Advances in Kernel Methods – Support Vector Learning*. MIT Press, Cambridge, MA, pp. 185–208.

Schmidt, M.S. (1996): Identifying speaker with support vector networks. *Proceedings of the 28th Symposium on the Interface (Interface-96)*, Sydney, Australia.

Schölkopf, B., Burges, C. and Vapnik, V. (1995): Extracting support data for a given task. In: Fayyad, U.M. and Uthurusamy, R. (Eds.), *Proceedings, First International Conference on Knowledge Discovery & Data Mining*, 1995. AAAI Press, Menlo Park, CA.

Schölkopf, B., Burges, C. and Vapnik, V. (1996): Incorporating invariances in support vector learning machines. In: von der Malsburg, C., von Seelen, W., Vorbrüggen, J.C. and Sendhoff, B. (Eds.), *Artificial Neural Networks ICANN'96*, Springer Lecture Notes in Computer Science, vol. 1112. Berlin, 1996, pp. 47–52.

Sivapragasam, C., Liong, S.Y. and Pasha M.F.K. (2001): Rainfall–runoff forecasting with SSA-SVM approach. *Journal of Hydroinformatics*, 3, 141–152.

Smola, A.J. and Schölkopf, B. (1998): A tutorial on support vector regression. NeuroCOLT2 Technical Report Series, NC2-TR-1998-030, October 1998. http://www.neurocolt.com.

Vapnik, V. (1998): *Statistical Learning Theory*. John Wiley & Sons, New York.

Vapnik, V. (2000): *The Nature of Statistical Learning Theory*. Springer, New York.

Vapnik, V. and Chervonenkis, A. (1964): A note on one class of perceptrons. *Automation and Remote Control*, 25(1), 112–120.

Vapnik, V. and Chervonenkis, A. (1979): *Theory of Pattern Recognition* [in Russian]. Nauka, Moscow, 1974. German Translation: W. Wapnik and A. Tscherwonenkis, *Theorie der Zeichenerkennung*. Akademie-Verlag, Berlin.

Vapnik, V., Golowich, S. and Smola, A. (1997): Support vector method for function approximation, regression estimation, and signal processing. In: Mozer, M.C., Jordan, M.I. and Petsche, T. (Eds.), *Advances in Neural Information Processing Systems*, vol. 9. MIT Press, Cambridge, MA, pp. 281–287.

Vapnik, V. and Lerner, A. (1963): Pattern recognition using generalized portrait method. *Automation and Remote Control*, 24, 774–780.

Vapnik, V.N. (1982): *Estimation of Dependences Based on Empirical Data*. Springer, Berlin.

Wang, X.X., Chen, S., Lowe, D. and Harris, C.J. (2006): Sparse support vector regression based on orthogonal forward selection for generalized kernel model. *Neurocomputing*, 70, 462–474.

Yu, X.Y., Liong, S.Y. and Babovic, V. (2004): EC-SVM approach for real-time hydrological forecasting. *Journal of Hydroinformatics*, 6, 209–223.

Zhou, H., Li, W., Zhang, C. and Liu, J. (2009): Ice break-up forecast in the reach of Yellow River: The support vector machines approach. *Hydrology and Earth Systems Science Discussions*, 6, 3175–3198.

Chapter 13

Fuzzy logic systems

13.1 INTRODUCTION

Traditional logic theory involves reasoning based on binary sets, which have two valued logic, true or false, yes or no, zero or one. In real life, much of the information that we come across and process is not so crispy but involves some degree of fuzziness. The truth value may range between the completely true value and the completely false value, leading to a partial truth. The key idea in fuzzy systems is to allow a partial truth to prevail, which can be numerically described by a specific function, referred to as a membership function that takes values between 0 and 1. For example, the discharge in a river may be perceived as high or low without the precise knowledge of the quantitative rate of flow. In other words, a quantitative description is translated into a qualitative linguistic description, and vice versa at a later stage. In this case, the concept of 'high' or 'low' is subjective and context dependent. The mathematics of the fuzzy set[1] theory was introduced by Zadeh (1965) in which the values of the membership functions (values of true and false) operate over the range (0,1). His idea was to replace the binary logic 'yes/no' by a five-level classification of the form 'definitely yes', 'probably yes', 'may be', 'probably no', and 'definitely no'. Negnevitsky (2005) defines fuzzy logic as a set of mathematical principles for knowledge representation based on degrees of membership rather than on the crisp membership of the classical binary logic. Fuzzy logic enables embedding uncertain or imprecise reasoning in everyday life to computers that operate in exact deterministic ways. Fuzzy logic models are conceptually easy to understand, flexible, tolerant to imprecise data, and can handle non-linear functions of arbitrary complexity and built on the experience of experts. They translate imprecise linguistic information sets into computer-usable numerical language. However, they cannot learn well from the data. In general, since knowledge acquisition is difficult and the universe of discourse[2] of each input variable needs to be divided into several intervals, fuzzy logic systems are restricted to fields where expert knowledge is available and the number of input variables is small.

The general structure of a fuzzy logic system is illustrated in Figure 13.1. It consists of a knowledge base that includes a database and a rule base, and three layers of information processing between the external input and output data. The main problems in building fuzzy systems include the selection of the relevant input and output variables, choice of the

---

[1] A fuzzy set is a set without a crisp, clearly defined boundary. It can contain elements with only a partial degree of membership. In fuzzy logic, the truth of any statement becomes a matter of degree. Reasoning in fuzzy logic is generalizing the familiar binary Boolean logic (e.g., yes or no, 0 or 1, open or closed, true or false). Fuzzy logic also permits in-between values (e.g., may be, probably, 0.2 and 0.7, half open, half closed, partially true, not totally false, etc.). The function that maps the degree of truth (e.g., high or low) within the limits 0 and 1 as a function of the value of the variable (e.g., runoff) is called a membership function.

[2] The 'universe of discourse' is the range of all possible values for an input to a fuzzy system.

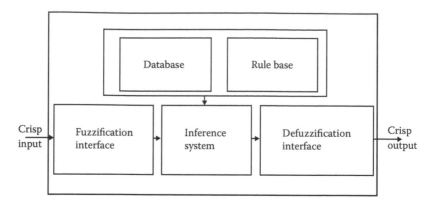

Figure 13.1 Structure of a fuzzy logic system.

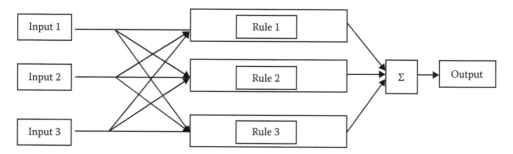

Figure 13.2 Information flow in a fuzzy system.

possible term sets for each linguistic variable, choice of the type of membership functions, fuzzification of the crisp input and output variables, derivation of the rule set, and aggregation of the outcomes of the rules and defuzzification. It should be noted that the choice of membership functions is rather subjective, but is not due to randomness. The output of the aggregation process is a single fuzzy set for each output variable. The flow of information in a fuzzy system with three inputs and three rules is illustrated in Figure 13.2.

13.2 FUZZY SETS AND FUZZY OPERATIONS

13.2.1 Fuzzy sets

In a classical set A, an object can belong to it or not belong, which implies that the sets A and NOT[3] A ($1 -$ set A) contain the entire universe because everything that is not contained in set A is contained in set NOT A. It has one ordered pair $\{x, (0,1)\}$ present for each element x of the universe of discourse X. Zero in the second element in the ordered pair indicates non-membership and 1 indicates membership, which may be written as

$$f_A(x) = \begin{cases} 1 & x \in X \\ 0 & x \notin X \end{cases} \tag{13.1}$$

---

[3] The definitions of NOT and other logical operators are given in Section 13.2.2.

where $f_A(x)$ is the membership function, which has a value of 0 or 1. For example, consider a rainfall of 100 mm/h. In a crisp system, the question 'Is the rainfall heavy?' has two possible answers: yes or no. If 100 mm/h is the mathematical boundary between 'heavy' and 'not heavy', the answer to any given rainfall intensity is unique, yes or no.

The human brain recognizes information in a slightly different way; there is no crisp boundary between 'heavy' and 'not heavy', although, mathematically it could be defined in a crisp way. In a fuzzy system that is based on natural human reasoning and perception, the boundary between 'heavy' and 'not heavy' is not sharp; it varies continuously. In fuzzy sets, the smooth transition from 'belonging to a set' and 'not belonging to a set' is gradual and is characterized by membership functions that have values ranging from 0 to 1. A rainfall greater than 100 mm/h is certainly perceived as heavy with a membership function value of 1; 90 mm/h can also be perceived as heavy with a membership function value of 0.9; 30 mm/h can be perceived as 'not heavy', with a membership function of 0.3. This means that the range of rainfall intensities between 0 and 150 mm/h, which is also known as the universe of discourse, can be identified by varying degrees of 'heaviness' as reflected by their membership functions (Figure 13.3).

A fuzzy set A in X can be defined as a set of ordered pairs A expressed as

$$A = \{(x, f_A(x)) | x \in X\} \tag{13.2}$$

and is characterized by a membership function $f_A(x)$, which has values in the interval $(0,1)$ with the value of $f_A(x)$ representing the degree of membership of x in A. A value close to unity means a higher degree of membership of x in A and vice versa.

A classical (crisp) set can be thought of as a special case of a fuzzy set where the membership function has only two values, 0 and 1. If $X = \{x_1, x_2, x_3, ..., x_n\}$ is a finite set, then a fuzzy set A is denoted (notation) as

$$A = \left\{ \frac{f_A(x_1)}{x_1}, \frac{f_A(x_2)}{x_2}, ..., \frac{f_A(x_n)}{x_n} \right\} \tag{13.3}$$

where $f_A(x_i)$ is the membership function of x_i in A. For example, a set of fuzzy numbers close to 2 may be written as

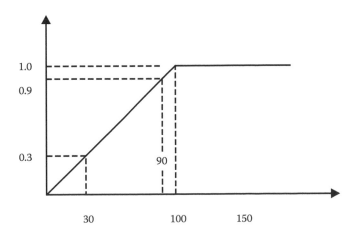

Figure 13.3 Linear membership function for 'heaviness'.

$$A = \left\{ \frac{0}{0} + \frac{0.5}{1} + \frac{1.0}{2} + \frac{0.5}{3} + \frac{0}{4} \right\} \tag{13.4}$$

which can be represented by a triangular membership function and therefore called a triangular fuzzy number of approximately 2 (Figure 13.4). A fuzzy number can also be represented by a continuous membership function as shown in Figure 13.5. Similarly, Figure 13.6 shows a trapezoidal fuzzy number x approximately in the interval (b,c).

A fuzzy number is a special case of fuzzy sets. It has an increasing part, a decreasing part, and sometimes a flat part. The simplest fuzzy number is a triangle, which has zero values outside the universe of discourse and piece-wise linear parts in between. The piece-wise parts need not necessarily be symmetrical. A triangular fuzzy number can be specified by three values: the most likely value with a membership function value of unity, the lowest

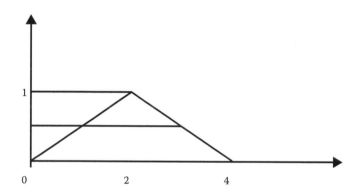

Figure 13.4 A triangular fuzzy number (x is approximately equal to 2).

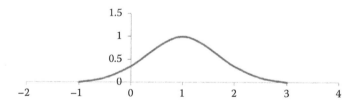

Figure 13.5 A continuous fuzzy number (x is approximately equal to 1).

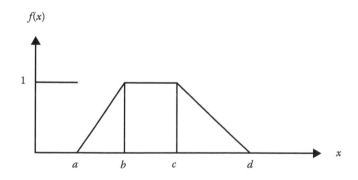

Figure 13.6 A trapezoidal fuzzy number (x is approximately in the interval (b,c)).

possible value with a membership function value of zero, and the highest possible value with a membership function value of zero. The membership function will have values of zero outside the latter two values. The interval between the lowest and highest possible values is called the support of the fuzzy number.

13.2.2 Logical operators AND, OR, and NOT

For two crisp binary sets A and B, the truth values of the Boolean logical operations 'AND', 'OR', and 'NOT' are shown in Table 13.1. In the context of fuzzy sets, these are known as fuzzy intersection (AND), fuzzy union (OR), and fuzzy complement (NOT), respectively. Sometimes, the 'AND' operation is referred to as conjunction, the 'OR' operation disjunction, and the 'NOT' operation negation.

The 'AND' and 'OR' operations and their equivalents in the fuzzy context can be performed by using a number of functions. The two most popular ones for the 'AND' operation are the minimum (*min*) function and the product (*prod*) function, whereas for the 'OR' operation they are the maximum (*max*) function and the probability (*probor*) function. It is reported that there is no great difference in the performance of the rule system with respect to the choice of the operator (Bárdossy and Duckstein, 1995; p. 48). Some useful definitions are given below.

13.2.2.1 Intersection

The intersection of two fuzzy sets with membership functions $f_A(x)$ and $f_B(x)$ is a fuzzy set D in X, written as $D = A \cap B$, and with a membership function defined as

$$f_D(x) = \min[f_A(x), f_B(x)] \; x \in X \tag{13.5}$$

It is the largest fuzzy set contained in both A and B. The intersection is equivalent to the Boolean operator 'AND'.

13.2.2.2 Union

Union of two fuzzy sets with membership functions $f_A(x)$ and $f_B(x)$ is a fuzzy set C in X, written as $C = A \cup B$, and with a membership function defined as

$$f_C(x) = \max[f_A(x), f_B(x)] \; x \in X \tag{13.6}$$

It is the smallest fuzzy set containing both A and B. Union is equivalent to the Boolean operator 'OR'. Figure 13.7 illustrates the union and intersection operators.

Table 13.1 Logical operators 'AND', 'OR', and 'NOT'

A	B	A 'AND' B	A 'OR' B	A 'NOT' A	B 'NOT' B
0	0	0	0	1	1
0	1	0	1	1	0
1	0	0	1	0	1
1	1	1	1	0	0

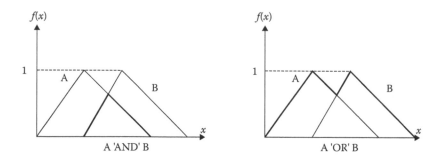

Figure 13.7 Union and intersection of two fuzzy membership functions.

13.2.2.3 Other useful definitions

Other useful operations include the complement, containment, algebraic product, algebraic sum, and algebraic difference, which are defined as follows:

a. The complement A' of A, which is a negation of the specified membership function is defined as

$$f_{A'}(x) = 1 - f_A(x) \tag{13.7}$$

It has the same meaning as the Boolean operator 'NOT'.
b. The containment of a fuzzy set A is in the fuzzy set B if and only if $f_A(x) \le f_B(x)$.
c. The algebraic product of A and B, denoted by AB, can be defined in terms of the membership functions of A and B as follows:

$$f_{AB}(x) = f_A(x)f_B(x) \tag{13.8}$$

The 'dual' of algebraic product is defined as $(A'B')' = A + B - AB$.
d. The algebraic sum of A and B, denoted by $A + B$, can be defined in terms of the membership functions of A and B as follows:

$$f_{A+B}(x) = f_A(x) + f_B(x) \tag{13.9}$$

A necessary condition for the above definition to be valid is $f_A(x) + f_B(x) \le 1$ for all x.
e. The algebraic difference of A and B, denoted by $|A - B|$, can be defined in terms of the membership functions of A and B as follows:

$$f_{|A-B|}(x) = |f_A(x) - f_B(x)| \tag{13.10}$$

f. The above definitions of the maximum and minimum values depend on one of the two values in the argument even when the range of the argument is large; that is, one argument having a large value. To overcome this problem, other definitions of intersection and union such as T-norms and T-conorms can be used. They are respectively defined as

$$f_{AB}^{\text{T-norm}}(x) = T\{f_A(x), f_B(x)\} \tag{13.11}$$

where the binary operator T represents the product of $f_A(x)$ and $f_B(x)$, and is referred to as the triangular norm, or T-norm for short. It has the following properties:

$$T(0,0) = 0; \quad T(a,1) = T(1,a) = a \tag{13.12a}$$

$$T(a,b) \leq T(c,d) \text{ if } a \leq c \text{ and } b \leq d \text{ (Monotonicity)} \tag{13.12b}$$

$$T(a,b) = T(b,a) \text{ (Symmetry)} \tag{13.12c}$$

$$T(a,T(b,c)) = T(T(a,b),c) \text{ (Associativity)} \tag{13.12d}$$

Similarly, the fuzzy union ('OR') can be specified by the T-conorm (sometimes referred to as S-norm), which is defined as follows:

$$f_{AB}^{T\text{-conorm}}(x) = S\{f_A(x), f_B(x)\} \tag{13.13}$$

where the binary operator S represents the addition of $f_A(x)$ and $f_B(x)$, and is referred to as the triangular conorm, or T-conorm for short. It has the following corresponding properties:

$$S(1,1) = 1; \quad S(a,0) = S(0,a) = a \tag{13.14a}$$

$$S(a,b) \leq S(c,d) \text{ if } a \leq c \text{ and } b \leq d \text{ (Monotonicity)} \tag{13.14b}$$

$$S(a,b) = S(b,a) \text{ (Symmetry)} \tag{13.14c}$$

$$S(a,S(b,c)) = S(S(a,b),c) \text{ (Associativity)} \tag{13.14d}$$

The widely used T-norms are the minimum T-norm defined as

$$T_{min}(a, b) = min(a, b) \tag{13.15}$$

and the product T-norm, defined as

$$T_{prod}(a, b) = ab \tag{13.16}$$

The widely used T-conorm, which is dual to the minimum T-norms, is defined as

$$S_{max}(a, b) = max\{a, b\} \tag{13.17}$$

and the probabilistic sum T-conorm, which is dual to the product T-norms, defined as

$$S_{prob}(a, b) = a + b - a.b \tag{13.18}$$

A T-norm represents a fuzzy intersection, whereas a T-conorm (S-norm) represents a fuzzy union. They can be used to combine criteria in multicriteria problems.
g. Alpha (α)-cut is a set of elements with degree of membership $\geq \alpha$.

13.2.3 Linguistic variables

In everyday life, an object can be described in many ways using modifiers (adjectives and/ or adverbs) to describe a characteristic of the object or the object itself. Such descriptions that use natural languages convey imprecise meanings. For example, a person may be described as 'very tall', 'tall', 'somewhat tall', or 'not tall' depending on one's perception of 'tallness'. The modifiers used describe the degree of tallness in a linguistic way. A set of linguistic variables for speed, for example, may be defined as very slow, slow, moderately fast, fast, and very fast. Similarly, a set of linguistic variables for temperature may be defined as very cold, cold, warm, hot, and very hot. For other variables, similar kind of linguistic variables can be assigned depending on the perception of its effect. Each linguistic variable can be assigned a membership value that will indicate its degree of effectiveness. For example, for speed, anything less than about 50 km/h may be perceived as slow, about 60 km/h as normal, about 70 km/h as moderately fast, 80 km/h as fast, and more than about 100 km/h as very fast. The universe of discourse in this case is [0,150], 150 being a set value for the maximum possible speed. For temperature, anything less than about 0°C may be perceived as very cold, about 10°C as cold, about 20°C as warm, about 30°C as hot, and more than about 40°C as very hot. The universe of discourse is [–10,50], the two values being set to represent the possible range of temperatures. Linking numerical values with the linguistic variables is subjective and problem specific. Linguistic variables are used to describe vague concepts in order to make them more precise (or imprecise) by assigning increasing (or decreasing) values of membership functions.

13.3 MEMBERSHIP FUNCTIONS

Membership functions map each point in the input (and output) space to a membership value (degree of membership) which ranges from 0 to 1. The complete input space is sometimes referred to as the universe of discourse. Many types of membership functions can be used in fuzzy logic systems, some discrete and some continuous. Hardlimiter (see also Figure 8.6a) can be considered as a membership function that has a crisp boundary, such as the value of a variable identified as either 'high' or 'not high'. Commonly used membership functions in fuzzy logic systems include triangular, trapezoidal, Gaussian, bell-shaped, and sigmoidal. The former two are discrete, while the latter three are continuous. Membership functions need not be symmetrical.

13.3.1 Triangular

The triangular membership function (Figure 13.8) is the simplest and has been used in many applications. It has three parameters a, b, and c with $a \leq b \leq c$ and is defined as

$$f(x:a,b,c) = \begin{cases} 0 & \text{for } x \leq a \\ \dfrac{x-a}{b-a} & \text{for } a \leq x \leq b \\ \dfrac{c-x}{c-b} & \text{for } b \leq x \leq c \\ 0 & \text{for } x > c \end{cases} \tag{13.19}$$

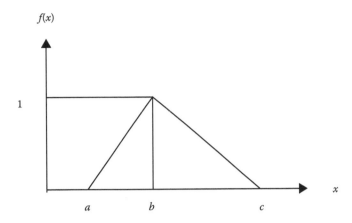

Figure 13.8 Triangular membership function.

13.3.2 Trapezoidal

The trapezoidal membership function has four parameters a, b, c, and d with $a \le b \le c \le d$, identified as the left foot point, left shoulder point, right shoulder point, and right foot point (Figure 13.9), and defined as (Jang et al., 1997)

$$f(x : a,b,c,d) = \begin{cases} 0 & \text{for } x < a \\ \dfrac{x-a}{b-a} & \text{for } a \le x \le b \\ 1 & \text{for } b \le x \le c \\ \dfrac{d-x}{d-c} & \text{for } c \le x \le d \\ 0 & \text{for } x > d \end{cases} \tag{13.20}$$

The triangular function can be considered as a special case of the trapezoid in which the two shoulders b and c merge together. Triangular and trapezoidal membership functions have discontinuities at the intersections of the straight line segments. Smooth trapezoids can be generated by replacing the discrete parts ab, bc, cd by continuously differentiable

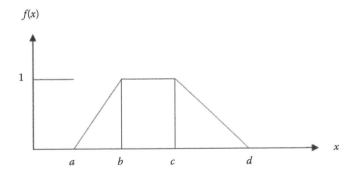

Figure 13.9 Trapezoidal membership function.

non-linear functions such as polynomials, sine and cosine functions, etc. For example, a smooth continuous trapezoid can be defined as (Jang et al., 1997)

$$f(x:a,b,c,d) = \begin{cases} 0 & \text{for } x < a \\ \dfrac{1}{2} + \dfrac{1}{2}\cos\left(\dfrac{x-b}{b-a}\pi\right) & \text{for } a \leq x \leq b \\ 1 & \text{for } b \leq x \leq c \\ \dfrac{1}{2} + \dfrac{1}{2}\cos\left(\dfrac{x-c}{d-c}\pi\right) & \text{for } c \leq x \leq d \\ 0 & \text{for } x > d \end{cases}$$

(13.21)

Similar continuous functions can also be obtained for the triangles.

13.3.3 Gaussian

Gaussian is a widely used function in many different types of transformations. It is symmetric and has two parameters, the mean μ and the standard deviation σ, and is defined as (Figure 13.10)

$$f(x:\mu,\sigma) = e^{-\frac{(x-\mu)^2}{2\sigma^2}}$$

(13.22)

13.3.4 Asymmetric Gaussian

Asymmetric Gaussian is a two-sided composite of two different Gaussians (Figure 13.11).

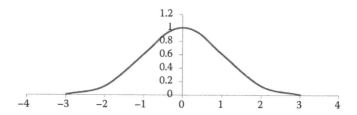

Figure 13.10 Gaussian membership function.

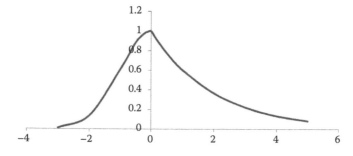

Figure 13.11 Asymmetric Gaussian ($\mu = 0$, $\sigma = 1$ and $\mu = 0$, $\sigma = 2$).

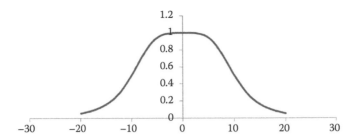

Figure 13.12 Bell-shaped Gaussian with *a* = −10, *b* = 2, and *c* = 0 (Equation 13.23).

13.3.5 Generalized bell-shaped Gaussian

The bell-shaped Gaussian (Figure 13.12) is symmetric and has three parameters with a flat top. It is defined as

$$f(x, a, b, c) = \frac{1}{1 + \left| \dfrac{x - c}{a} \right|^{2b}} \tag{13.23}$$

where c is the centre, a the beginning of the rising part, and b is a positive parameter.

13.3.6 Sigmoidal

The logistic sigmoid is bounded, continuous and continuously differentiable. It attains its maximum (1) and minimum (0) values asymptotically and is defined as

$$f(x) = \frac{1}{1 + e^{-rx}} \tag{13.24}$$

where r is the steepness parameter (see also Figure 8.6h).

In addition to these, polynomial-based membership functions referred to as *S*-curve, *Z*-curve, and pi-curve have also been used in the MATLAB® toolbox. The *S*-curve starts from zero and reaches the maximum value of unity asymptotically, whereas the *Z*-curve starts from the maximum value unity and reaches the zero value asymptotically. The *S*-curve is the mirror image of the *Z*-curve. The pi-curve has the start and end at zero.

13.3.7 Singleton

A fuzzy singleton is a fuzzy set whose support is a single point in the universe of discourse with a membership function of unity (Figure 13.13). It is used for Takagi–Sugeno–Kang (TSK) (see Section 13.5.3) zero-order models only.

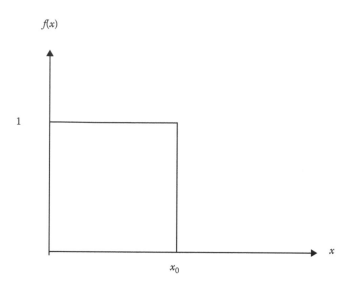

Figure 13.13 Singleton fuzzy set.

13.4 FUZZY RULES

A fuzzy control system is obtained by a set of rules, which in many instances is based on common sense. The fuzzy rule base consists of rules that include all possible fuzzy relations between inputs and outputs. However, it is to be noted that the number of rules increases exponentially with increasing number of inputs, leading to what is known as the 'curse of dimensionality'. The rules relate the combined linguistic subsets of input variables to a convenient linguistic output subset based on the expert knowledge and/or on the available theory.

Fuzzy rules may be specified using the knowledge of experts directly and/or supplemented by available data, or may be not known explicitly but the variables are specified by experts, or may have to be constructed purely from data. The truth value of a fuzzy rule is known as the degree of fulfilment (also known as the firing level or degree of confidence), which takes values in the interval (0,1).

A fuzzy rule includes statement as 'IF–THEN' with two parts. The first part that starts with 'IF' and ends before 'THEN' is referred to as the predicate (or premise, or antecedent), which combines in a harmonious manner the subsets of input variables. After the 'THEN' comes the consequent part, which includes the convenient fuzzy subset of the output based on the antecedent part. They are expressed in the 'IF–THEN' format, such as

'IF' the rainfall is high, 'THEN' runoff is high.
'IF' the degree of saturation of the soil is high, 'THEN' runoff is high.
'IF' the storage in the upstream reservoir is low, 'THEN' downstream runoff is low, etc.

Sometimes, the input subsets within the antecedent part are combined, most often with the logical operator 'AND', whereas the rules are usually combined with the logical operator 'OR'. When the antecedent of a given rule has more than one part (e.g., IF *rainfall is high* AND *soil moisture is high* THEN *runoff is high*), the fuzzy operator is applied to obtain one number that represents the result of the antecedent for that rule (MathWorks). This number is then applied to the output function. For the 'AND' operation, *min* (minimum) and *prod* (product) are supported by MATLAB (MATLAB Fuzzy Logic Toolbox). For the 'OR' operation, *max* (maximum) and the *probor* (probability) are supported. The activation of a rule is the deduction of the conclusion. The *prod* activation (multiplication) scales the membership curves, thus preserving the initial shape, rather than clipping them as the *min* activation does.

In fuzzy systems, there are no mathematical equations and model parameters. All the uncertainties, non-linearities, and model complexities are included in the descriptive fuzzy inference[4] procedure in the form of IF–THEN statements. There are basically two types of rule systems, namely the Mamdani (Mamdani and Assilian, 1975) type and the TSK type (Takagi and Sugeno, 1974, 1985). In the Mamdani type, the fuzzy rule is also expressed in linguistic form. According to the TKS rule system, the fuzzy rule is expressed as a mathematical function of the input variables, which is more appropriate for neuro-fuzzy systems (Sen, 1998; 2009).

Interpretation of the IF–THEN rule consists of evaluating the antecedent after fuzzifying the input and applying fuzzy operators (when the antecedent consists of two or more parts), and then applying the result of the antecedent, which should be a single number. The latter process is known as implication. In binary logic, the interpretation is that if the antecedent is true, the consequent is also true. In fuzzy logic, if the antecedent is true to some degree, then the consequent is also true to the same degree. A two-part rule consisting of fuzzy numbers A, B, and C is of the form

IF x is A AND y is B THEN z is C

and can be extended to multiparts. All parts of the rule should be calculated simultaneously and converted into a single number. Rules should overlap to avoid a single rule determining the outcome. It is also possible to have multipart consequents, such as

IF x is A AND y is B THEN z_1 is C AND z_2 is D

In such a case with multipart consequents, the output set is modified by truncating using the *min* function or scaling using the *prod* function. All consequents in a multipart case are affected equally by the results of the antecedent. The shape of the multivariate fuzzy membership function depends on the shapes of the individual univariate membership functions and on the fuzzy operators used to represent the T-norm.

The above description is for a single rule. In a real-life problem, there could be several rules leading to several fuzzy outputs. Such outputs are then aggregated to a single fuzzy output set, which should be defuzzified to obtain a single number. In contrast to crisp rules, fuzzy rules accommodate partial and simultaneous fulfilment of the truth.

A multipart rule set may look like

IF x is A_1 and y is B_1, THEN z is C_1 and the firing level may be denoted as α_1
IF x is A_2 and y is B_2, THEN z is C_2 and the firing level may be denoted as α_2
IF x is A_3 and y is B_3, THEN z is C_3 and the firing level may be denoted as α_3
...
...
IF x is A_n and y is B_n, THEN z is C_n and the firing level may be denoted as α_n

---

[4] Fuzzy inference is a method that interprets the values in the input vector and, based on some set of rules, assigns values to the output vector. All rules are applied in parallel, and the order of presentation of the rules is unimportant.

Here, the fuzzy rule maps the intersection of n variables.

Every rule has a *weight* (a number between 0 and 1), which is applied to the number given by the antecedent. Generally this weight is unity and so it has no effect at all on the implication process. However, it may sometimes be necessary to have a higher weight for one rule relative to others. Once the rule is applied, the consequent (output) is a fuzzy set represented by a membership function, which has a value representing its linguistic characteristics.

Each rule leads to a fuzzy conclusion. These component conclusions then need some kind of summation.

13.5 FUZZY INFERENCE

The fuzzy inference system (FIS) maps a given input to a corresponding output using fuzzy logic. It combines the components such as membership functions, fuzzy logic operators, and rules. It can also be thought of as the rule evaluation of a fuzzy system. A FIS can be thought of as consisting of three stages. At the input stage, the input variables are mapped to appropriate membership functions. At the processing stage, the rules are invoked to generate outputs for each rule, which are then combined in some manner to obtain an overall result for all the rules. At the output stage, the combined result is converted to output values, which become the end product. There are four well-known inference mechanisms in fuzzy logic systems: Mamdani (Mamdani and Assilian, 1975), TSK (Takagi and Sugeno, 1985), Tsukamoto (1979, 1994), and Larsen (Larsen and Yager, 2000). Of these, the widely used ones are the Mamdani type and the TSK type, both of which are supported by MATLAB Fuzzy Logic Toolbox. The Mamdani type, originally proposed in 1975 to control a steam engine and boiler combination, has undergone modifications over the years; however, the basic concepts remain the same. In this type, the output membership functions will also be fuzzy sets, which need defuzzification after aggregation. The defuzzification requires finding the centroid of a two-dimensional function, which has to be done by some kind of integration. The TSK and Tsukamoto types do not require defuzzification at the output because the outputs are expressed as mathematical functions of the inputs.

The results of individual rules can be combined into an overall fuzzy consequent in different ways. The maximum algorithm ($= \max\{f_A(x), f_B(x)\}$) is widely used.

13.5.1 Fuzzy or approximate reasoning

The rationale behind fuzzy reasoning is the compositional rule of inference, which for two-valued logic is the 'modus ponens' that has its roots in propositional logic. It states that if the antecedent of a conditional claim is true, then the consequent must also be true. For example,

- If there is heavy rain, then there is flooding (conditional claim).
- There is flooding (fact).
- Therefore, there must be heavy rain (consequent).

Symbolically, this logic may be expressed as

- If x is A, then y is B (conditional claim; x is the antecedent and y is the consequent).
- x is A (fact).
- Therefore, y must be B (consequent).

In approximate reasoning, the logic can be stated as

- If x is A, then y is B.
- x is A'.
- y must be B'.

where A' is close to A and B' is close to B. This is called 'generalized modus ponens' with modus ponens as a special case.

For multipart antecedents, the max–min composition uses the minimum value of the membership functions of each part as the degree of fulfilment, whereas the max–product composition uses the product of the membership functions of each part as the degree of fulfilment.

13.5.2 Mamdani fuzzy inference system

13.5.2.1 Fuzzification of inputs

The inputs and outputs are in most cases crisp numbers within the applicable range (universe of discourse), which can be fuzzified using chosen membership functions. Before fuzzification, linguistic terms should be assigned to the crisp values of the variables within the universe of discourse. This process is called domain partition, or discretization or quantization. The number of linguistic terms to use is problem specific and rather subjective and not necessarily unique. The discretization should be sufficiently fine to describe the process variation adequately, keeping the computer memory storage requirement as the limiting condition. Overlapping of membership functions is essential for smooth mapping. For example, to describe flow ranging from 0 to 50 m³/s, linguistic terms such as dry weather flow, low flow, normal flow, high flow, and flood flow may be used and assigned to equivalent crisp flow as follows:

$$f_D(Q) = \begin{cases} 1 - \dfrac{Q}{10} & 0 \leq Q \leq 10 \\ 0 & Q \geq 10 \end{cases} \quad \text{Dry weather flow (DWF)} \tag{13.25a}$$

$$f_L(Q) = \begin{cases} \dfrac{Q}{10} & 0 \leq Q \leq 10 \\ 2 - \dfrac{Q}{10} & 10 \leq Q \leq 20 \end{cases} \quad \text{Low flow (LF)} \tag{13.25b}$$

$$f_N(Q) = \begin{cases} \dfrac{Q}{10} - 1 & 10 \leq Q \leq 20 \\ 3 - \dfrac{Q}{10} & 20 \leq Q \leq 30 \end{cases} \quad \text{Normal flow (NF)} \tag{13.25c}$$

$$f_H(Q) = \begin{cases} \dfrac{Q}{10} - 2 & 20 \leq Q \leq 30 \\[2mm] 4 - \dfrac{Q}{10} & 30 \leq Q \leq 40 \end{cases} \qquad \text{High flow (HF)} \hspace{2cm} (13.25\text{d})$$

$$f_F(Q) = \begin{cases} \dfrac{Q}{10} - 3 & 30 \leq Q \leq 40 \\[2mm] 1 & Q > 40 \end{cases} \qquad \text{Flood flow (FF)} \hspace{2cm} (13.25\text{e})$$

The domain partition described in Equations 13.25a through 13.25e can be represented by the five membership functions defined above as illustrated in Figure 13.14, and the corresponding membership function values for some specific values of Q are given in Table 13.2.

The output from the fuzzification process is a fuzzy number in the interval (0,1) that represents the degree of membership of the input variable. It is then applied to the output function, which should be a single value.

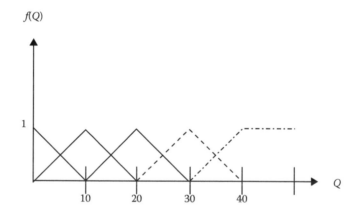

Figure 13.14 Membership functions for different linguistic variables (Equations 13.25a through 13.25e).

Table 13.2 Membership function values for some specific values of Q

| Q (m³/s) | Membership function value | | | | |
	DWF	LF	NF	HF	FF
0	1	0	0	0	0
5	0.5	0.5	0	0	0
10	0	1	0	0	0
15	0	0.5	0.5	0	0
20	0	0	1	0	0
25	0	0	0.5	0.5	0
30	0	0	0	1	0
35	0	0	0	0.5	0.5
40	0	0	0	0	1
>40	0	0	0	0	1

13.5.2.2 Application of fuzzy operators 'AND' or 'OR'

After the fuzzification, the degree of membership function (the degree to which the rule is satisfied) is known. For multipart antecedents, the fuzzy operator (usually 'AND') is applied to obtain a single value that represents the combined result of all parts of the rule. This value is then applied to the consequent membership function. The output is a single truth value. There are several methods of applying the 'AND' and 'OR' operations in a fuzzy logic systems of which the *min* (minimum) and *prod* (product) for 'AND', and *max* (maximum) and *probor* (probability) for 'OR' are the widely used ones. The probabilistic 'OR' is also known as the algebraic sum.

13.5.2.3 Implication from antecedent to consequent

In some situations, certain rules may have a greater impact than others. In such cases, the rules have to be multiplied by weighting factors that reflect their relative importance. If all the rules are equally important, there is no need to multiply by a weighting factor. This step should precede implementing the implication. A single number given out by the antecedent forms the input to the implication process from which an output fuzzy set comes out as the output. Implication is applied to each rule using the same operators 'AND' and 'OR', which respectively truncates the output fuzzy set with the *min* (minimum) function and scales the output fuzzy set with the *prod* (product) function.

Truncating (or clipping), sometimes known as α-cut, is the most common method of correlating the rule consequent with the truth value of the rule antecedent. It cuts the consequent membership function at the level of the antecedent truth. By this method, the truncated fuzzy set loses some information. In the case of a triangular membership function, the top part of the triangle is clipped off, leaving a trapezium below. This trapezium is considered as the fuzzy consequent part for the first rule. Because of the clipping, this trapezium does not have a membership value of unity and is therefore not normal[5] but convex.[6] For the second and subsequent rules, there will be similar trapeziums but with different bases, thus giving consequent fuzzy sets for each individual rule. They are then combined together with the 'OR' operator, or maximum function, by stacking the trapeziums on top of each other one by one with some parts overlapping. This may lead to an irregular shape, which is neither normal nor convex. This process is called the *min–max* operation. It depends only on the minimum value of the degree of fulfilment of each rule and ignores all other remaining degrees of fulfilment of the antecedent fuzzy sets. The method does not take into consideration the contributions from other parts of the antecedent for that rule. Yet this method is often preferred because it is less complex, and generates an aggregated output surface that is easier to defuzzify. Figure 13.15 gives a schematic illustration of this procedure.

For two rules R_1 and R_2 given below, the Mamdani implication uses the minimum operator

R_1: If x is A_1 and y is B_1, then z is C_1.
R_2: If x is A_2 and y is B_2, then z is C_2.

The firing levels (or the degrees of fulfilment) α_1 and α_2 for x_0, y_0 are given as

$$\alpha_1 = \min(A_1(x_0), B_1(y_0)) \tag{13.26a}$$

---

[5] Not normal in this context means the clipped membership function does not have its maximum value of unity.
[6] Convexity implies that the membership function has an increasing part and decreasing part.

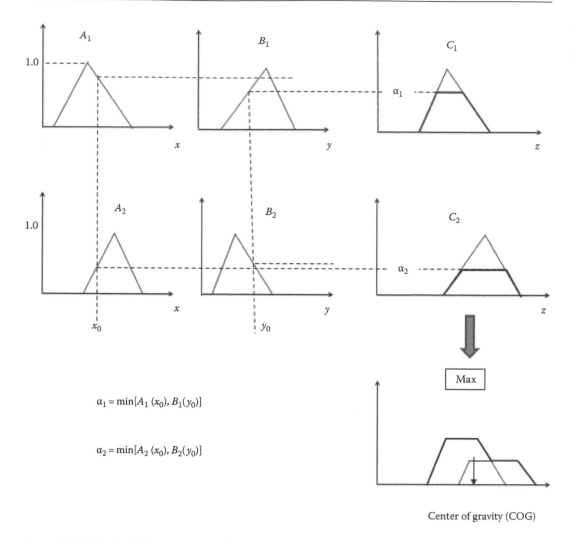

$$\alpha_1 = \min[A_1(x_0), B_1(y_0)]$$

$$\alpha_2 = \min[A_2(x_0), B_2(y_0)]$$

Center of gravity (COG)

Figure 13.15 Mamdani inference system for a two-rule two-input fuzzy inference system with triangular membership functions using the min–max operation (x, y are input variables, z is output variable; vertical axis gives the membership function value; A_1, A_2, B_1, B_2 are the antecedent membership functions, and C_1 and C_2 are the consequent membership functions).

$$\alpha_2 = \min(A_2(x_0), B_2(y_0)) \tag{13.26b}$$

The rule outputs respectively are

$$C_1'(z) = \min(\alpha_1, C_1(z)) \tag{13.27a}$$

$$C_2'(z) = \min(\alpha_2, C_2(z)) \tag{13.27b}$$

The overall output is obtained as

$$C(z) = \max(C_1'(z), C_2'(z)) = \max\{\min(\alpha_1, C_1(z)), \min(\alpha_2, C_2(z))\} \tag{13.28}$$

Alternatively, the product of the degree of fulfilment in the antecedent part can be adopted as the combined degree of fulfilment for the consequent part. In this method, all parts of the antecedent are taken into consideration but the multiplication may lead to very small values because each degree of fulfilment is less than unity. This is particularly so in the case of multipart antecedents. This procedure is known as scaling or the *prod–max* operation. It preserves the original shape of the fuzzy set and loses less information. On the negative side is the fact that the combined degree of fulfilment may approach zero and therefore may not represent the real situation. Figure 13.16 gives a schematic illustration of the product operation procedure.

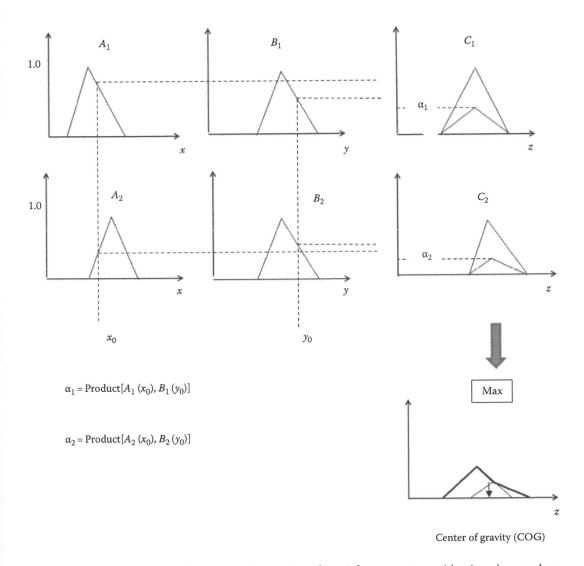

$$\alpha_1 = \text{Product}[A_1(x_0), B_1(y_0)]$$

$$\alpha_2 = \text{Product}[A_2(x_0), B_2(y_0)]$$

Center of gravity (COG)

Figure 13.16 Product operation for a two-rule two-input fuzzy inference system with triangular membership functions using the prod–max operation (x, y are input variables, z is output variable; vertical axis gives the membership function value; A_1, A_2, B_1, B_2 are the antecedent membership functions, and C_1 and C_2 are the consequent membership functions).

13.5.2.4 Aggregation of consequents across the rules

Aggregation is the process of combining the consequents of all rules (previously truncated or scaled) into a single fuzzy set. The inputs to the aggregation process are the truncated output functions returned by the implication process for each rule. The output of the aggregation process is a single fuzzy set for each output variable. It can be done via the *max* (maximum), *prob* (probabilistic), and *sum* (sum of the outputs of each rule) functions. The order in which the rules are applied is unimportant.

13.5.2.5 Defuzzification

Defuzzification is the last step in the fuzzy inference process since the final output of a fuzzy system has to be a crisp number. The input to the defuzzification process is the output fuzzy set from the aggregation process, and the output is a single number. Defuzzification is done according to the membership function of the output variable. There are several methods of which the centroid method is perhaps the most widely used. It returns the centre of area under the curve. Other methods such as the bisector of area, middle of maximum (average of the maximum value of the output set), largest maximum, and smallest of the maximum can also be used. Defuzzification causes some loss of information.

a. **Centroid method**

The centroid method (centre of gravity [COG] method) is defined as

$$z_0 = \frac{\sum_{i=1}^{k} z_i C(z_i)}{\sum_{i=1}^{k} C(z_i)} \tag{13.29}$$

where z_0 is the defuzzified output value, $C(z)$ is the membership function of the output variable, and k is the number of rules. For continuous membership functions, it can be expressed as

$$z_0 = \frac{\int_w z C(z)\,dz}{\int_w C(z)\,dz} \tag{13.30}$$

Equation 13.29 can be numerically calculated.

b. **Bisector of area**

In this method, the area under the composite consequent membership function is divided into two equal halves. The bisector satisfies the relationship

$$\int_{z_1}^{z_0} f(z)\,dz = \int_{z_0}^{z_2} f(z)\,dz \tag{13.31}$$

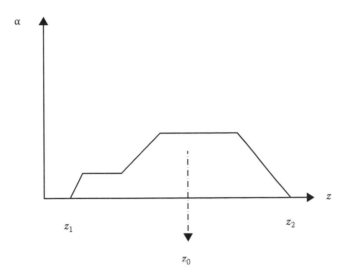

Figure 13.17 Bisector of area method of defuzzification.

where z_1 and z_2 respectively are the minimum and maximum values of the outputs, and z_0 corresponds to the location of the bisector (Figure 13.17).

c. Arithmetic average method

This method gives equal weights to each fuzzy output. The aggregated fuzzy output is sampled at a number of equidistant points, and the arithmetic average is calculated in a way similar to Simpson's rule of integration, as

$$z = \frac{1}{n+2}\left\{\sum_{i=1}^{n} z_i + z_{max} + z_{min}\right\}$$ (13.32)

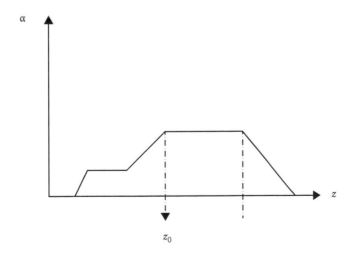

Figure 13.18 Smallest of maximum method of defuzzification.

where n is the number of intervals and z_{max} and z_{min} are the maximum and minimum values of the output.

d. Smallest of maxima

In this method, the smallest of the membership value range that has the maximum is chosen (Figure 13.18).

e. Largest of maxima

This is similar to the previous method except that the largest value is taken instead of the smallest value (Figure 13.19).

f. Middle of maxima

This method gives the output value corresponding to the mean of the range for the maximum membership function value (Figure 13.20).

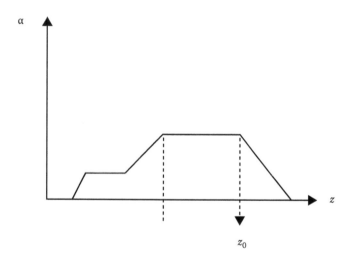

Figure 13.19 Largest of maximum method of defuzzification.

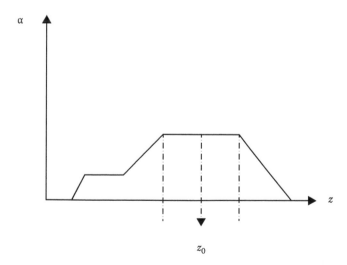

Figure 13.20 Mean of maximum method of defuzzification.

13.5.3 Takagi–Sugeno–Kang (TSK) fuzzy inference system

Fuzzification of inputs, application of fuzzy operators, and implication are the same as for the Mamdani type. The main difference is that there is no defuzzification since the outputs from the rules are crisp. The TSK implication uses local linear models to model a global non-linear system heuristically but making it look like a transparent system. It uses a single spike, known as a singleton, or a linear function of the antecedent variables as the membership function of the consequent. A singleton is a fuzzy set with a membership function that is unity at a single point on the universe of discourse and zero everywhere else. The simplest and the most commonly used TSK implication uses a constant and is known as the zero-order TSK fuzzy model. It is of the form

Rule 1: If x is A AND y is B, THEN z is a constant.

For the zero-order TSK model to be a smooth function of the inputs, there should be sufficient overlap of the antecedent membership functions. The overlapping of membership functions in the consequent does not have a significant effect on the smoothness of the input–output behaviour. For multirule systems, all consequent membership functions are represented by singleton spikes.

Using local linear functions for two rules R_1 and R_2, the antecedent–consequent relationships are of the form

R_1: If x is A_1 and y is B_1, then z_1 is $a_1x + b_1y + c_1$.
R_2: If x is A_2 and y is B_2, then z_2 is $a_2x + b_2y + c_2$.

The firing levels α_1 and α_2, as before for x_0, y_0, using the minimum operator, are given as

$$\alpha_1 = \min(A_1(x_0), B_1(y_0)) \tag{13.33a}$$

$$\alpha_2 = \min(A_2(x_0), B_2(y_0)) \tag{13.33b}$$

The rule outputs (linear in this case, but a constant or a higher-order function may also be used) respectively are

$$z_1 = a_1x_0 + b_1y_0 + c_1 \tag{13.34a}$$

$$z_2 = a_2x_0 + b_2y_0 + c_2 \tag{13.34b}$$

This means that for a two rule linear system, there are six (a_1, b_1, a_2, b_2, c_1, c_2, according to Equation 13.34) parameters; that is, $2p$ where p is the number of parameters for each rule. With k rules, the number of parameters increases to kp. They are estimated by minimizing the sum of squares of the errors. Since the number of parameters to be estimated increases with increasing number of rules, a compromise is often needed between the complexity of the system and the efficiency of verification.

The crisp overall output is a weighted average of individual outputs from each rule. For the two rules, it is

$$z_0 = \frac{\alpha_1 z_1 + \alpha_2 z_2}{\alpha_1 + \alpha_2} \tag{13.35}$$

If there are 'k' rules, the crisp overall output for all the rules is given as a weighted average of outputs for each rule in the form

$$z_0 = \frac{\displaystyle\sum_{i=1}^{k} \alpha_i z_i}{\displaystyle\sum_{i=1}^{k} \alpha_i} \tag{13.36}$$

Figure 13.21 illustrates the TSK fuzzy inference system.

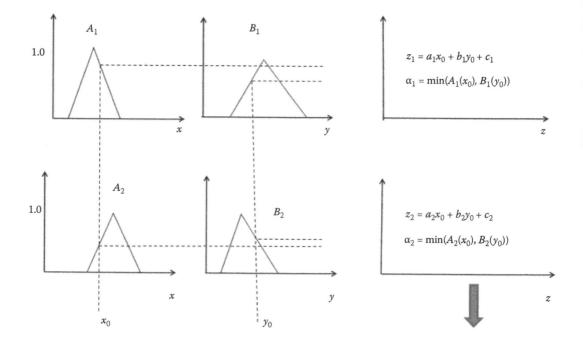

Figure 13.21 TSK inference system for a two-rule two-input fuzzy inference system with triangular membership functions (x, y are input variables, z is output variable; vertical axis gives the membership function value; A_1, A_2, B_1, B_2 are the antecedent membership functions; consequent is a crisp output). The antecedent membership functions of a TSK fuzzy inference system can be determined by a process called clustering.

13.5.3.1 Clustering

Clustering is a mathematical tool that attempts to discover structures or certain patterns in a data set, where the objects inside each cluster show a certain degree of similarity. Binary clustering allows an object to belong to only one of several clusters with crisp boundaries, whereas fuzzy clustering allows each object to belong to more than one cluster with different degrees of membership functions (between 0 and 1) and with vague or fuzzy boundaries between clusters.

Clustering belongs to the class of unsupervised learning. An important criterion used for clustering is the distance between different data points. In the case of data with same units, a frequently used measure of distance is the Euclidean distance, although other measures such as the Hamming distance, Chebyshev distance (or Tchebychev distance), Minkowski distance, Canberra distance, and angular separation (Pedrycz, 2005) can also be used. The Euclidean distance between two points $\{(x_i, y_i), (x_j, y_j)\}$ in a two-dimensional space is defined as

$$d_{ij} = \sqrt{(x_i - x_j)^2 + (y_i - y_j)^2} \tag{13.37}$$

The Euclidean distance between two vectors x_i and x_j

$$d(x_i, x_j) = \|x_i - x_j\| = \left[\sum_{k=1}^{m} (x_{ik} - x_{jk})^2 \right]^{1/2} \tag{13.38}$$

which can also be extended to multidimensional space. In fuzzy logic systems, clustering is extensively used to obtain antecedent membership functions.

Clustering can be done using different algorithms that lead to different cluster shapes. The commonly used ones in the TSK-type adaptive neuro-fuzzy inference system (ANFIS) include the Gustafson–Kessel algorithm (Gustafson and Kessel, 1979), the Gath and Geva algorithm (Gath and Geva, 1989), and algorithms based on expectation maximization (EM) identification of Gaussian mixture models (Abonyi, 2002). More details of cluster analysis can be found in a book by Höppner et al. (1999). Two of the popular algorithms are briefly described below.

a. **k-means algorithm**

This is one of the widely used methods of unsupervised learning. For example, it is used in artificial neural networks, particularly in radial basis function–type networks (see also Section 9.6.4). It is based on assigning k-number of clusters *a priori* and finding the centres of the k clusters using the distance measure as a criterion. The clusters should be as far apart as possible. Initially, the centres of the k clusters are assumed usually randomly. Each data point is then assigned to the nearest cluster depending on the distance between that point and the centre of the cluster. A new set of cluster centres is then calculated using the centroid as the cluster centre, and the reassignment is repeated until the differences between two successive cluster centres become insignificant as measured by some threshold criterion. Weaknesses in the k-means algorithm include arriving at suboptimal classification as a result of the choice of the value of k, which is unknown *a priori*, as well as the initial choice of the centres. These may be avoided by trying different values of k and different initial values of centres.

b. **c-means algorithm**

The c-means algorithm (Dunn, 1973; Bezdek, 1981) is similar to the k-means algorithm but uses fuzzy numbers (degrees of membership functions) instead of crisp numbers used in the latter. Unlike in k-means clustering where a data point can belong only to one cluster, in c-means clustering a data point can belong to two or more clusters. The method partitions the data into c-fuzzy subsets.

The algorithm starts with an initial guess for the cluster centres that are intended to be the mean locations of each cluster. It then assigns every data point a membership grade for each cluster (Equations 13.39 and 13.40). By repeated updating of the cluster centres and the membership grades for each data point, the algorithm optimizes the locations of cluster centres using an objective function J_m (Equation 13.41) that represents the distance from any given data point to a cluster centre weighted by that data point's membership grade. Considering a finite set of elements $\{x_1, x_2, \ldots x_N\}$, partitioned into C clusters, the steps involved are

1. Initialize the fuzzy membership matrix $U = [u_{ij}]$, $U^{(0)}$, $i = 1, n$; $j = 1, C$, subject to $\sum_{j=1}^{C} u_{ij} = 1$ for all $i = 2,3,\ldots,n$, and that every cluster is non-empty and different from the entire set.
2. For $k = 1$, calculate the centre vectors $C^{(k)} = [c_j]$ with $U^{(k)}$ using Equation 13.40.
3. For $k = k + 1$, update $U^{(k)}$ to $U^{(k+1)}$ using Equation 13.39.
4. STOP if $\|U^{(k+1)} - U^{(k)}\| < \varepsilon$. Otherwise, return to step 2.

The applicable equations are

$$u_{ij} = \left\{ \sum_{k=1}^{C} \left(\frac{\|x_i - c_j\|}{\|x_i - c_k\|} \right)^{\frac{2}{m-1}} \right\}^{-1} \tag{13.39}$$

$$c_j = \frac{\sum_{i=1}^{N} u_{ij}^{m} x_i}{\sum_{i=1}^{N} u_{ij}^{m}} \tag{13.40}$$

$$J_m = \sum_{i=1}^{N} \sum_{j=1}^{C} u_{ij}^{m} \|x_i - c_j\|^2 \tag{13.41}$$

where u_{ij} is the degree of membership of x_i $[0,1]$ in the cluster j, x_i is the i-th value of d-dimensional measured data, c_j is the d-dimension centre of the cluster, J_m is the objective function to be minimized, $\|*\|$ is any norm expressing the distance between any measured data and the centre, ε is some threshold lying between 0 and 1, k is an iteration counter, and m is the fuzziness index, which can be any real number greater than 1 but usually set to 2 (Bezdek, 1981). When the value of m is closer to unity, it resembles binary classification, whereas when it is approaching infinity, the degree of fuzziness also increases. The original assumed value of $m = 2$ has been subsequently optimized using Genetic Algorithm (Alata et al., 2008).

In Equation 13.39, u_{ij} is an $N \times C$ matrix with N rows and C columns. For a data set with two clusters and five data points, $N = 5$ and $C = 2$, the matrix would look like the following (left for k-means and right for c-means):

$$
u_{N \times C} = \begin{bmatrix} 1 & 0 \\ 0 & 1 \\ 1 & 0 \\ 0 & 1 \\ 1 & 0 \\ 0 & 1 \end{bmatrix} \quad u_{N \times C} = \begin{bmatrix} 0.9 & 0.1 \\ 0.8 & 0.2 \\ 0.7 & 0.3 \\ 0.6 & 0.4 \\ 0.5 & 0.5 \\ 0.4 & 0.6 \end{bmatrix} \tag{13.42}
$$

In the case of k-means algorithm, u_{ij} is either zero or 1; in the case of c-means algorithm, $\sum_{j=1}^{C} u_{ij} = 1$, for all i, and $0 < \sum_{i=1}^{N} u_{ij} < N$ for all N. MATLAB Fuzzy Logic Toolbox supports both k-means and c-means algorithms. The information returned from MATLAB c-means clustering algorithm can be used to generate TSK-type fuzzy inference systems that best model a given set of input–output data using a minimum number of rules.

13.5.4 Tsukamoto inference system

The antecedent of the Tsukamoto system is the same as that of the Mamdani or TSK systems. The consequent of each fuzzy rule, however, is represented by a fuzzy set with a monotonic membership function (Figure 13.22). The output of each rule is therefore a crisp value corresponding to the firing strength of the rule. The weighted average of each rule gives the overall output. There is no defuzzification needed. The firing levels of each rule is given as

$$
\alpha_i = \min(A_i(x_0), B_i(x_0)) \quad i = 1,2,\ldots,k \tag{13.43}
$$

where k is the number of rules. The consequent of each rule is a crisp output z_i given as

$$
\alpha_i = C_i(z_i) \tag{13.44}
$$

Since $C_i(z_i)$ is monotonic, there is only one value of z corresponding to a given value α_i of α, and the overall crisp output, computed by the COG method, is given as

$$
z_0 = \frac{\sum_{i=1}^{k} \alpha_i z_i}{\sum_{i=1}^{k} \alpha_i} \quad \text{(same as Equation 13.36)}
$$

13.5.5 Larsen inference system

In the Larsen inference system, the fuzzy implication is done using the product operator. The firing level of the i-th rule is given as

$$
\alpha_i = \mathrm{prod}(A_i(x_0), B_i(x_0)) \quad i = 1,2,\ldots,k \tag{13.45}
$$

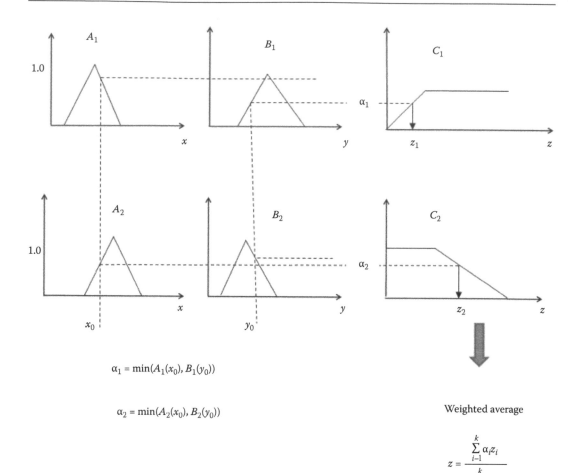

$$\alpha_1 = \min(A_1(x_0), B_1(y_0))$$

$$\alpha_2 = \min(A_2(x_0), B_2(y_0))$$

Weighted average

$$z = \frac{\sum\limits_{i-1}^{k} \alpha_i z_i}{\sum\limits_{i-1}^{k} \alpha_i}$$

Figure 13.22 Tsukamoto inference system for a two-rule two-input fuzzy inference system with triangular antecedent membership functions and monotonic piece-wise linear consequent membership functions (x, y are input variables, z is output variable; vertical axis gives the membership function value; A_1, A_2, B_1, B_2 are the antecedent membership functions, and C_1 and C_2 are the consequent membership functions that should be monotonic but not necessarily piece-wise linear).

where k indicates the number of rules. The membership function value of the consequent is given as

$$C(w) = \max(\alpha_i C_i(w)) \quad i = 1,2,\dots,k \tag{13.46}$$

The crisp output can be obtained by using any defuzzification method.

Sometimes it is more efficient computationally to use a single output membership function, referred to as a singleton, rather than a distributed one. With a singleton, a weighted average can be used. This is the approach adopted by the TSK type where the output membership functions are either constants or some mathematical function that, in most cases, is linear.

For a Gaussian membership function, the degree of fulfilment can be obtained using a T-norm such as the product operator in the form

$$\alpha_i = f_{A_{1,i}}(x_1).f_{A_{2,i}}(x_2)...f_{A_{n,i}}(x_n) \qquad (13.47a)$$

$$= \exp\left[-\left(\frac{(x_1 - \mu_{1,i})^2}{2\sigma_{1,i}^2} + \frac{(x_2 - \mu_{2,i})^2}{2\sigma_{2,i}^2} + ... + \frac{(x_n - \mu_{n,i})^2}{2\sigma_{n,i}^2}\right)\right] \qquad (13.47b)$$

where μ's and σ's are the means and the standard deviations of each Gaussian membership function. If the standard deviations are the same for all membership functions, Equation 13.47 simplifies to

$$\alpha_i = \exp\left[-\left(\frac{(x_1 - \mu_{1,i})^2 + (x_2 - \mu_{2,i})^2 + ... + (x_n - \mu_{n,i})^2}{2\sigma_i^2}\right)\right] \qquad (13.48)$$

This assumption simplifies the computational procedure and also reduces the number of parameters in each antecedent from $2n$ to $n + 1$, thereby making it more parsimonious. The total number of parameters would then be $k(n + 1)$ where k is the number of rules.

13.6 NEURO-FUZZY SYSTEMS

Neuro-fuzzy (or fuzzy-neuro) refers to the combination of artificial neural networks and fuzzy logic. Hybrid artificial intelligence, first proposed by Jang (1993), synergizes the human-like reasoning style of fuzzy systems with the connectionist structure of artificial neural networks. Neuro-fuzzy systems are universal approximators with the ability to incorporate 'IF–THEN' rules. Fuzzy models cannot learn from data but are interpretable, whereas neural networks can learn from data and are accurate but not easily interpretable. The neuro-fuzzy modelling research field can be divided into linguistic fuzzy modelling that focuses on interpretability and precise fuzzy modelling that focuses on accuracy. These two types are respectively identified by the Mamdani model (Mamdani and Assilian, 1975) and the TSK model (Takagi and Sugeno, 1974, 1985).

This approach is becoming one of the major areas of interest because it gets the benefits of neural networks as well as of fuzzy logic systems, and removes their individual disadvantages by combining them on the common features. Neural networks and fuzzy logic have some common features, such as distributed representation of knowledge, model-free estimation, ability to handle data with uncertainty and imprecision, etc. Fuzzy logic has tolerance for imprecision of data, while neural networks have tolerance for noisy data. A neural network's learning capability provides a good way to adjust expert's knowledge, and it automatically generates additional fuzzy rules and membership functions to meet certain specifications. This reduces the design time and cost. On the other hand, the fuzzy logic approach possibly enhances the generalization capability of a neural network by providing more reliable output when extrapolation is needed beyond the limits of the training data.

The basic unit of a fuzzy neural system is called a 'fuzzy neuron'. It has its roots in the biological neuron, and together with the learning mechanism constitutes the computational process of a fuzzy neural system. Unlike in a simple neural network (Figure 13.23a

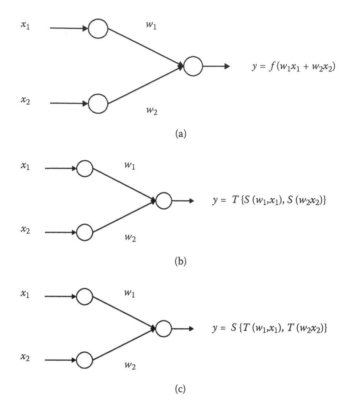

$$y = f(w_1x_1 + w_2x_2)$$

(a)

$$y = T\{S(w_1,x_1), S(w_2x_2)\}$$

(b)

$$y = S\{T(w_1,x_1), T(w_2x_2)\}$$

(c)

Figure 13.23 (a) Simple neural net, (b) fuzzy 'AND' neural net, and (c) fuzzy 'OR' neural net.

through 13.23c) where the inputs and weights are real numbers, and where the input–weight (multiplication) combinations are summed (addition) and the output is obtained after applying an activation function such as the sigmoid to give

$$y = f\left(\sum_{i=1}^{n} x_i w_i\right) \tag{13.49}$$

a hybrid neural network receives crisp inputs, weights and activation functions, all real numbers in the interval (0,1), and the output is taken as the T-norm or T-conorm. The weights and inputs are combined and aggregated using T-norm, or T-conorm, or some other operation. The two common fuzzy neurons are the 'AND fuzzy neuron', which gives the minimum, and the 'OR fuzzy neuron', which gives the maximum, which respectively can be expressed as

$$y = T(S(w_i, x_i)) = \min\{\max(w_i, x_i)\} \quad i = 1, n \tag{13.50a}$$

$$y = S(T(w_i, x_i)) = \max\{\min(w_i, x_i)\} \quad i = 1, n \tag{13.50b}$$

They are also known as the min–max composition and the max–min composition, respectively. Other types of fuzzy neurons include the Implication–OR fuzzy neuron (Eklund et al., 1991, 1992) and the Kwan and Cai fuzzy neuron (Kwan and Cai, 1994).

13.6.1 Types of neuro-fuzzy systems

There can be several types of neuro-fuzzy systems, which can be broadly classified as follows:

- Regular (standard) neural networks that use multiplication, addition, and sigmoidal activation functions. They are universal approximators.
- Regular fuzzy neural networks that use fuzzy signals and/or weights and sigmoidal activation functions. They are not universal approximators.
- Hybrid neural networks that use crisp inputs, outputs, and activation functions. They use operators such as *T*-norms and *T*-conorms (Equations 13.12 through 13.18) to combine the inputs and weights. They may not use multiplication, addition, or sigmoidal functions because the results of these operations are not necessarily in the unit interval. Furthermore, the inputs, weights, and outputs of a hybrid neural network all lie within the unit interval [0,1]. The processing element of a hybrid neural network is called a fuzzy neuron and there are several types of them.
- Hybrid fuzzy neural networks that use fuzzy signals and or weights. They use operators such as *T*-norms, *T*-conorms, or other operators to combine the inputs and weights. They are universal approximators (Buckley and Hayashi, 1994a,b).

A typical fuzzy neural network, or FuNN (Figure 13.24), that uses the multilayer percep-tron with a modified form of back-propagation consists of five layers identified as

- Layer 1 – Input layer (crisp). The output of the node is the degree to which the input satisfies the linguistic label associated with the node. Triangular, trapezoidal, and Gaussian-type membership functions can be used as the membership functions.
- Layer 2 – Fuzzification layer.
- Layer 3 – Rule layer.
- Layer 4 – Action element layer.
- Layer 5 – Defuzzification layer (crisp).

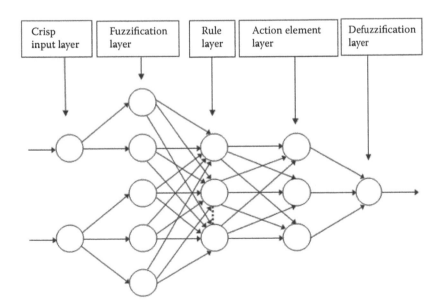

Figure 13.24 Structure of a fuzzy neural network.

According to the mode of learning, there can be different versions, such as

- Fixed version – Rules change; membership function do not change.
- Adaptive version with an extended back-propagation – All connection weights change during training.
- Either fixed or adaptive versions – Structural learning with forgetting algorithm.
- Adaptive version with Genetic Algorithm.

13.6.1.1 Umano and Ezawa (1991) fuzzy-neural model

A typical FuNN model works as follows (Fullér, 1995):
Consider a set of fuzzy rules such as

IF x is A_i THEN y is B_i, where A_i and B_i are fuzzy numbers, $i = 1,..., n$.

Each rule in the above can be interpreted as a training pattern for a multilayer neural network, where the antecedent part of the rule is the input and the consequent part of the rule is the desired output of the neural network. The training set derived from above can be written in the form

$$\{(A_1, B_1),...,(A_n, B_n)\}$$

This concept can be extended to multiple input–single output fuzzy systems as well as multiple input–multiple output fuzzy systems.

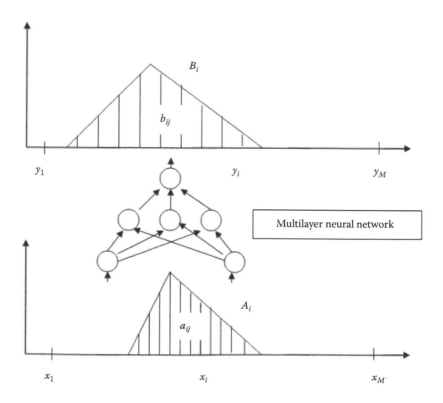

Figure 13.25 Network with membership values as inputs and outputs.

The fuzzy IF–THEN rules can be implemented by the standard error back-propagation algorithm by representing a fuzzy set by a finite number of its membership values. A discrete version of the input–output pairs can be obtained as follows:

$$\{(A_i(x_1),\ldots,A_i(x_N)),\ (B_i(y_1),\ldots,B_i(y_M))\} \text{ for } i = 1,\ n$$

where

$$x_j = \alpha_1 + \frac{(j-1)(\alpha_2 - \alpha_1)}{(N-1)} \text{ for } 1 \le j \le N \tag{13.51}$$

$$y_i = \beta_1 + \frac{(i-1)(\beta_2 - \beta_1)}{(M-1)} \text{ for } 1 \le i \le M \tag{13.52}$$

and the intervals (α_1, α_2) contain all the A_i's, and (β_1, β_2) contain all the B_i's; $M, N \ge 2$ are positive integers. Using the notation $a_{ij} = A_i(x_j)$, and $b_{ij} = B_i(y_j)$, the fuzzy neural network can be converted into a crisp network with n inputs and m outputs, which can be trained by the back-propagation algorithm (Figure 13.25). This is the method proposed by Umano and Ezawa (1991) where a fuzzy set is represented by a finite number of its membership values.

13.7 ADAPTIVE NEURO-FUZZY INFERENCE SYSTEMS (ANFIS)

In any fuzzy inference system, the membership functions and rules are somewhat subjective because human-determined membership functions vary from person to person and from time to time, and are rarely optimal. This is particularly so in systems with large data sets. The ANFIS can serve as a basis for constructing a set of rules and appropriate membership functions for a set of input–output pairs that minimize the output error. In an adaptive network, each node performs a certain function on the incoming signals and parameters relevant to that node and produces an output signal. The parameters are updated using training data according to some learning procedure based on the gradient descent method.

ANFIS is more complex than ordinary fuzzy inference systems and has several limitations. It can only support the TSK type with zero- or first-order consequent membership functions; should also have only one output that is obtained using weighted defuzzification; all output membership functions must be the same type; the number of output membership functions should be equal to the number of rules; and has unit weight for each rule.

The antecedent part of an ANFIS consists of IF–THEN rules and mathematical functions (restricted to linear) in the consequent part. The input space is partitioned into a number of fuzzy regions, and the mathematical functions in the consequent part describe the fuzzy system's behaviour in these regions. Construction of a TSK-type ANFIS consists of determining the antecedent membership functions in the rule set, which is usually done using either prior knowledge of the system, or by some other method. The parameters of the consequent are determined by the least squares method. The determination of the antecedent parameters (membership function) is a non-linear optimization problem, which is normally carried out using the back-propagation algorithm. However, the back-propagation algorithm, which uses gradient descent methods, has certain inherent drawbacks. Other methods of gradient-free non-linear optimization algorithms can be used but the computational complexities of such methods limit their application. Therefore, resort is often made to heuristic approaches such as fuzzy clustering to carry out the non-linear optimization. Clustering determines the fuzzy partition of the data into a matrix $U = [f_{i,k}]$, a $c \times N$ matrix where $f_{i,k}$ represent the degree of membership in clusters c. In TSK-type ANFIS, each cluster is represented by one rule.

ANFIS begins with an initially assumed fuzzy inference system and fine tunes its membership functions using back-propagation. In conventional fuzzy systems, the rules are decided by experts on the basis of prior knowledge. When such expert knowledge is not available, the membership functions for each input variable are decided empirically or by trial and error. The ANFIS model, first proposed by Jang (1993), uses a mixture of back-propagation and least mean square procedure to train the network. The architecture of an ANFIS consists of five layers as described below (Figure 13.26).

> **Layer 1:** This layer represents the membership function of A_i and its specific degree of membership. Typically, a bell-shaped Gaussian function with a maximum of unity and a minimum of zero is used. It may be of the form

$$O_i^1 = f_{A_i}(x) = \exp\left(-\left(\frac{x - c_i}{a_i}\right)^2\right) \tag{13.53}$$

> where O_i^1 is a node function or membership function of A_i, which specifies the degree to which x satisfies A_i, x is the input to node i, A_i is the linguistic label (high, medium, low, etc.) associated with this node, $f_{A_i}(x)$ is the membership function, and a_i and c_i are the premise (or antecedent) parameters. The shape of the function changes with changing values of the parameters. Triangular and trapezoidal functions can also be used in this layer. The output of this layer gives the degree to which a given input x_i satisfies the fuzzy descriptor A_i. In ANFIS architectural notation, this layer is represented by square nodes and they have parameters.

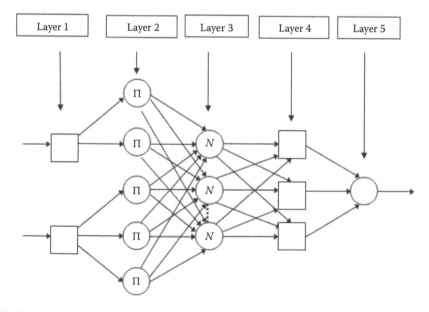

Figure 13.26 Structure of an ANFIS.

Layer 2: Nodes in this layer multiply the incoming signals by the membership function values and send the products to the next layer. For example, for a two-input ANFIS

$$w_i = f_{A_i}(x).f_{B_i}(y); \quad i = 1, 2 \tag{13.54}$$

where w_i represents the firing strength of a rule, and $f_{A_i}(x)$ and $f_{B_i}(y)$ are the membership functions of the inputs x and y, respectively. Other functions, such as the minimum function that perform the 'AND' fuzzy operation, can also be used. Nodes in this layer are denoted by circles inscribed with Π. Nodes denoted by circles have no parameters.

Layer 3: In this normalizing layer, the ratio of the i-th rule firing strength to the sum of all rules firing strengths is calculated

$$\bar{w}_i = \frac{w_i}{w_1 + w_2}; \quad i = 1, 2 \tag{13.55}$$

The output from this layer is the normalized firing strength of a rule. Nodes in this layer are denoted by circles inscribed with N.

Layer 4: Nodes in this layer have the function

$$O_i^4 = \bar{w}_i g_i = \bar{w}_i (p_i x + q_i y + r_i) \tag{13.56}$$

where p_i, q_i, and r_i are the consequent parameters. It calculates the weighted consequent. Nodes in this layer are denoted by squares.

Layer 5: This node gives the summation of outputs from incoming signals as

$$O_i^5 = \sum \bar{w}_i g_i = \frac{\sum w_i g_i}{\sum w_i} \tag{13.57}$$

This adaptive network corresponds to a TSK-type fuzzy inference system. A similar ANFIS architecture can be constructed for a Tsukamoto-type fuzzy inference system by replacing the monotonic non-linear consequent membership function by a piece-wise linear approximation with two consequent parameters corresponding to maximum and minimum values of the membership function. The method can also be extended to Mamdani-type fuzzy inference systems where a defuzzification is required, but the procedure is rather complicated.

ANFIS has two sets of parameters, the antecedent parameters such as the parameters of the membership function (centre and spread in the case of Gaussian membership function), or the parameters of any continuously and piece-wise differentiable functions such as trapezoidal and triangular, and the consequent parameters such as the parameters of the TSK or Tsukamoto type fitted to the consequent. In adaptive mode, these parameters need updating by exposing to training data sets. Several types of updating such as using the back-propagation with gradient descent only for all parameters, gradient descent with one-pass least squares estimate, sequential least squares estimate, and hybrid learning consisting of back-propagation for the antecedent parameters and least squares estimate method for the consequent membership function can be used. Of these, the hybrid approach is widely used.

13.7.1 Hybrid learning

Hybrid learning combines back-propagation with the gradient descent method and least squares estimate method. The optimization trains the membership functions and parameters to match the target output data. It can be carried out in batch mode or sequential (online) mode. For a two-input TSK-type ANFIS, the overall output from layer 5 is of the form

$$O_i^5 = \frac{w_1}{w_1 + w_2} g_1 + \frac{w_2}{w_1 + w_2} g_2$$

$$= (\bar{w}_1 x) p_1 + (\bar{w}_1 y) q_1 + (\bar{w}_1) r_1 + (\bar{w}_2 x) p_2 + (\bar{w}_2 y) q_2 + (\bar{w}_2) r_2$$

(13.58)

which is linear in the consequent parameters. Hybrid learning consists of two passes – a forward pass and a backward pass. In the forward pass, antecedent parameters are fixed and the functional signals (node outputs) move forward up to layer 4, and the consequent parameters are identified by the least squares estimate method. In the backward pass, error rates (signals) propagate backwards and the antecedent parameters are updated by the gradient descent method, while keeping the consequent parameters fixed. For fixed antecedent parameters, the least squares estimate gives optimal consequent parameters.

Training is terminated using a fixed number of epochs or a preassigned error measure as the stopping criterion. The trained ANFIS should then be validated using a set of unseen data to test whether it can generalize the input–output pattern. Care should be taken to avoid overfitting.

Other hybrid fuzzy neural network models include the approximate reasoning-based intelligent control, or ARIC (Berenji, 1992), and an improvement to it called the generalized ARIC, or GARIC (Berenji and Khedkar, 1992). Both ANFIS and GARIC models are not easy to interpret compared with Mamdani-type fuzzy systems. More recently, models such as NEFCON (Nauck, 1994), NEFCLASS (Nauck and Kruse, 1996), and NEFPROX (Nauck and Kruse, 1997) have been developed for neuro-fuzzy control, classification, and regression, respectively. They all implement Mamdani-type fuzzy systems and thus use special learning algorithms.

13.8 APPLICATION AREAS

Fuzzy logic systems have found a large number of applications in industry. Examples in Japan include the development of braking and stopping system for railways, and washing cycle in washing machines, both by Hitachi; vacuum cleaners by Matsushita; autofocussing system in cameras by Canon; and air conditioner operation by Mitsubishi. In the hydrological field, applications include water level forecasting (See and Openshaw, 1999, 2000; Alvisi et al., 2005); flood forecasting (e.g. Tareghian and Kashefipour, 2007; Nayak et al., 2005); infiltration modelling (Bárdossy and Disse, 1993); rainfall–runoff modelling (Hundecha et al., 2001; Xiong et al., 2001); and hydrological time series modelling (Nayak et al., 2004).

Example 13.1

An example of the application of the Mamdani fuzzy inference system is given here to illustrate the fuzzy logic approach using hydrological data from a river basin in Sri

Lanka.[7] The Gin River basin in Sri Lanka has a catchment area of 947 km^2 and the river has a length of 113 km. Daily discharge measurements at an upstream gauging station at Thawalama (6°20′33.3″ N; 80°19′50″ E), a downstream gauging station at Agaliya (6°10′24.33″ N; 80°10′31.82″ E), and four rainfall measurements made at Annilkanda, Deniyaya, Hiniduma, and Thawalama for the period 1997–1999 were used for model calibration while the corresponding data for the period 2000–2003 were used for validation with 1- and 2-day lead times. The rainfall data of the four stations were averaged by the Thiessen polygon method. For the fuzzy logic model, the two input variables were initially partitioned into three ranges, low, medium, and high, while the output variable was partitioned into five ranges, very low, low, medium, high, and very high. Triangular and trapezoidal functions were used as membership functions. The membership functions were as follows:

For rainfall (RF):

$$M_{low}(RF) = \begin{cases} 1 & \text{for } 0 < RF < 40 \\ 2 - \dfrac{RF}{40} & \text{for } 40 < RF < 80 \end{cases} \tag{13.59a}$$

$$M_{medium}(RF) = \begin{cases} \dfrac{RF}{40} - 1 & \text{for } 40 < RF < 80 \\ 3 - \dfrac{RF}{40} & \text{for } 80 < RF < 120 \end{cases} \tag{13.59b}$$

$$M_{high}(RF) = \begin{cases} \dfrac{RF}{40} - 2 & \text{for } 80 < RF < 120 \\ 1 & \text{for } 120 < RF \end{cases} \tag{13.59c}$$

These membership functions are illustrated in Figure 13.27a.
For upstream discharge ($Q_{upstream}$):

$$M_{low}(Q_{upstream}) = \begin{cases} \dfrac{Q_{upstream}}{50} & \text{for } 0 < Q_{upstream} < 50 \\ 2 - \dfrac{Q_{upstream}}{50} & \text{for } 50 < Q_{upstream} < 100 \end{cases} \tag{13.60a}$$

$$M_{medium}(Q_{upstream}) = \begin{cases} \dfrac{Q_{upstream}}{100} - 0.5 & \text{for } 50 < Q_{upstream} < 150 \\ 2.5 - \dfrac{Q_{upstream}}{100} & \text{for } 150 < Q_{upstream} < 250 \end{cases} \tag{13.60b}$$

---

[7] This study was carried out in partial fulfilment of the requirement for the Master's Degree in Disaster Management by J.D. Amarasekara (2012) under the supervision of the author.

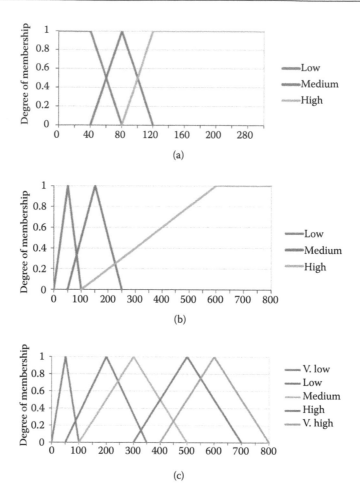

Figure 13.27 (a) Membership functions of upstream rainfall (horizontal axis gives the rainfall values in mm/day), (b) membership functions of upstream discharge (horizontal axis gives the discharge values in m³/s), and (c) membership functions for downstream discharge (horizontal axis gives the discharge values in m³/s).

$$M_{\text{high}}(Q_{\text{upstream}}) = \begin{cases} \dfrac{Q_{\text{upstream}}}{500} - 0.2 & \text{for } 100 < Q_{\text{upstream}} < 600 \\ 1 & \text{for } 600 < Q_{\text{upstream}} \end{cases} \tag{13.60c}$$

These membership functions are illustrated in Figure 13.27b.
For downstream discharges ($Q_{downstream}$):

$$M_{\text{very low}}(Q_{\text{downstream}}) = \begin{cases} \dfrac{Q_{\text{downstream}}}{50} & \text{for } 0 < Q_{\text{downstream}} < 50 \\ 2 - \dfrac{Q_{\text{downstream}}}{50} & \text{for } 50 < Q_{\text{downstream}} < 100 \end{cases} \tag{13.61a}$$

$$
M_{\text{low}}(Q_{\text{downstream}}) = \begin{cases} \dfrac{Q_{\text{downstream}}}{150} - 0.33 & \text{for } 50 < Q_{\text{downstream}} < 200 \\[2ex] 2.33 - \dfrac{Q_{\text{downstream}}}{150} & \text{for } 200 < Q_{\text{downstream}} < 350 \end{cases} \tag{13.61b}
$$

$$
M_{\text{medium}}(Q_{\text{downstream}}) = \begin{cases} \dfrac{Q_{\text{downstream}}}{200} - 0.5 & \text{for } 100 < Q_{\text{downstream}} < 300 \\[2ex] 2.5 - \dfrac{Q_{\text{downstream}}}{200} & \text{for } 300 < Q_{\text{downstream}} < 500 \end{cases} \tag{13.61c}
$$

$$
M_{\text{high}}(Q_{\text{downstream}}) = \begin{cases} \dfrac{Q_{\text{downstream}}}{200} - 1.5 & \text{for } 300 < Q_{\text{downstream}} < 500 \\[2ex] 3.5 - \dfrac{Q_{\text{downstream}}}{200} & \text{for } 500 < Q_{\text{downstream}} < 700 \end{cases} \tag{13.61d}
$$

$$
M_{\text{very high}}(Q_{\text{downstream}}) = \begin{cases} \dfrac{Q_{\text{downstream}}}{200} - 2 & \text{for } 400 < Q_{\text{downstream}} < 600 \\[2ex] 4 - \dfrac{Q_{\text{downstream}}}{200} & \text{for } 600 < Q_{\text{downstream}} < 800 \end{cases} \tag{13.61e}
$$

These membership functions are illustrated in Figure 13.27c.

The maximum number of rules depends on the number of inputs and the number of ranges in each input. However, some rules may be superfluous. Therefore, by trial and error, the rule set, totalling nine rules, for this initial attempt is fixed as follows:

1. IF upstream rainfall is *low* AND upstream discharge is *low* THEN downstream discharge is *very low*.
2. IF upstream rainfall is *low* AND upstream discharge is *medium* THEN downstream discharge is *medium*.
3. IF upstream rainfall is *low* AND upstream discharge is *high* THEN downstream discharge is *high*.
4. IF upstream rainfall is *medium* AND upstream discharge is *low* THEN downstream discharge is *low*.
5. IF upstream rainfall is *medium* AND upstream discharge is *medium* THEN downstream discharge is *medium*.
6. IF upstream rainfall is *medium* AND upstream discharge is *high* THEN downstream discharge is *high*.
7. IF upstream rainfall is *high* AND upstream discharge is *low* THEN downstream discharge is *medium*.
8. IF upstream rainfall is *high* AND upstream discharge is *medium* THEN downstream discharge is *high*.
9. IF upstream rainfall is *high* AND upstream discharge is *high* THEN downstream discharge is *very high*.

The functional representation of the input–output relationships for 1- and 2-day lead times were assumed to be of the form

Downstream discharge at time $t = f$ (upstream discharge at time $[t - 1]$, rainfall at time $[t - 1]$)

Downstream discharge at time $t = f$ (upstream discharge at time $[t - 2]$, rainfall at time $[t - 2]$)

The membership functions were then gradually increased by partitioning the output discharge data into seven segments while keeping the partitioning of the inputs unchanged. The membership functions were triangular and trapezoidal as before, and the coordinates of the memberships functions (Figure 13.28) are given in Table 13.3. For this configuration, 15 rules were chosen by trial and error as shown below:

1. IF upstream rainfall is *low* AND upstream discharge is *very low* THEN downstream discharge is *very low*.
2. IF upstream rainfall is *low* AND upstream discharge is *low* THEN downstream discharge is *low*.
3. IF upstream rainfall is *low* AND upstream discharge is *medium* THEN downstream discharge is *medium*.
4. IF upstream rainfall is *low* AND upstream discharge is *high* THEN downstream discharge is *medium–high*.
5. IF upstream rainfall is *low* AND upstream discharge is *very high* THEN downstream discharge is *high*.
6. IF upstream rainfall is *medium* AND upstream discharge is *very low* THEN downstream discharge is *low*.
7. IF upstream rainfall is *medium* AND upstream discharge is *low* THEN downstream discharge is *low–medium*.
8. IF upstream rainfall is *medium* AND upstream discharge is *medium* THEN downstream discharge is *medium*.

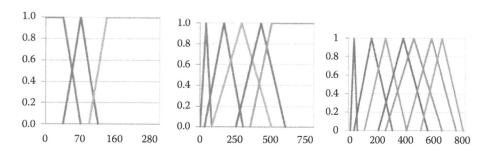

Figure 13.28 Membership functions of upstream rainfall (left), upstream discharge (centre), and downstream discharge (right) for the 15-rule case.

Table 13.3 Coordinates of the membership functions for the downstream discharge for the 15-rule case

Coordinates	Very low	Low	Low–medium	Medium	Medium–high	High	Very high
a	0	25	100	200	250	400	500
b	25	150	250	375	450	575	650
c	50	300	400	550	650	750	800

Note: See Figures 13.8 and 13.9 for the definition of a, b, and c.

9. IF upstream rainfall is *medium* AND upstream discharge is *high* THEN downstream discharge is *high*.

10. IF upstream rainfall is *medium* AND upstream discharge is *very high* THEN downstream discharge is *very high*.

11. IF upstream rainfall is *high* AND upstream discharge is *very low* THEN downstream discharge is *low–medium*.

12. IF upstream rainfall is *high* AND upstream discharge is *low* THEN downstream discharge is *medium*.

13. IF upstream rainfall is *high* AND upstream discharge is *medium* THEN downstream discharge is *medium–high*.

14. IF upstream rainfall is *high* AND upstream discharge is *high* THEN downstream discharge is *high*.

15. IF upstream rainfall is *high* AND upstream discharge is *very high* THEN downstream discharge is *very high*.

This process could be continued by fine tuning the partitioning and the rule set. Figure 13.29 shows such a partitioning in which the upstream rainfall and upstream discharge data are partitioned into five segments each and the downstream discharge data into nine segments. The membership functions were triangular and trapezoidal as before, and the coordinates of the membership functions are given in Tables 13.4a through 13.4c. Because of the complexity of this case, the best combination of membership functions and rules were obtained using the Fuzzy Logic Toolbox of MATLAB.

The rule set consisting of 25 rules for this case, obtained by trial and error is as follows:

1. IF upstream rainfall is *very low* AND upstream discharge is *very low* THEN downstream discharge is *extremely low*.

2. IF upstream rainfall is *very low* AND upstream discharge is *low* THEN downstream discharge is *very low*.

3. IF upstream rainfall is *very low* AND upstream discharge is *medium* THEN downstream discharge is *low*.

4. IF upstream rainfall is *very low* AND upstream discharge is *high* THEN downstream discharge is *medium*.

5. IF upstream rainfall is *very low* AND upstream discharge is *very high* THEN downstream discharge is *high*.

6. IF upstream rainfall is *low* AND upstream discharge is *very low* THEN downstream discharge is *very low*.

7. IF upstream rainfall is *low* AND upstream discharge is *low* THEN downstream discharge is *low*.

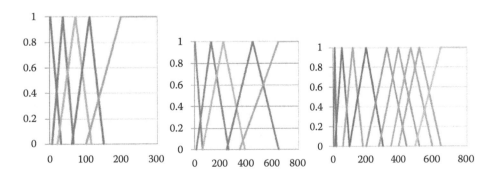

Figure 13.29 Membership functions of upstream rainfall (left), upstream discharge (centre), and downstream discharge (right) for the 25-rule case.

Table 13.4a Coordinates of the upstream rainfall membership functions for the 25-rule case

Coordinates	Very low	Low	Medium	High	Very high
a	−30	5	20	60	100
b	0	35	70	110	200
c	30	65	115	150	300
d					340

Note: See Figures 13.8 and 13.9 for the definitions of a, b, c, and d.

Table 13.4b Coordinates of the upstream discharge membership functions for the 25-rule case

Coordinates	Very low	Low	Medium	High	Very high
a	−35	10	60	250	350
b	0	130	225	450	650
c	60	265	390	650	800
d					1050

Note: See Figures 13.8 and 13.9 for the definitions of a, b, c, and d.

Table 13.4c Coordinates of the downstream discharge membership functions for the 25-rule case

Coordinates	Extremely low	Very low	Low	Low medium	Medium	Medium high	High	Very high	Extremely high
a	0	10	60	100	200	275	350	400	500
b	10	55	120	200	325	400	475	525	650
c	20	100	180	300	450	525	600	650	800

Note: See Figures 13.8 and 13.9 for the definitions of a, b, and c.

8. IF upstream rainfall is *low* AND upstream discharge is medium THEN downstream discharge is *medium.*
9. IF upstream rainfall is *low* AND upstream discharge is *high* THEN downstream discharge is *medium–high.*
10. IF upstream rainfall is *low* AND upstream discharge is *very high* THEN downstream discharge is *high.*
11. IF upstream rainfall is *medium* AND upstream discharge is *very low* THEN downstream discharge is *low.*
12. IF upstream rainfall is *medium* AND upstream discharge is *low* THEN downstream discharge is *low–medium.*
13. IF upstream rainfall is *medium* AND upstream discharge is *medium* THEN downstream discharge is *medium.*
14. IF upstream rainfall is *medium* AND upstream discharge is *high* THEN downstream discharge is *high.*
15. IF upstream rainfall is *medium* AND upstream discharge is *very high* THEN downstream discharge is *very high.*
16. IF upstream rainfall is *high* AND upstream discharge is *very low* THEN downstream discharge is *low–medium.*
17. IF upstream rainfall is *high* AND upstream discharge is *low* THEN downstream discharge is *medium.*
18. IF upstream rainfall is *high* AND upstream discharge is *medium* THEN downstream discharge is *medium–high.*

19. IF upstream rainfall is *high* AND upstream discharge is *high* THEN downstream discharge is *high*.

20. IF upstream rainfall is *high* AND upstream discharge is *very high* THEN downstream discharge is *very high*.

21. IF upstream rainfall is *very high* AND upstream discharge is *very low* THEN downstream discharge is *medium*

22. IF upstream rainfall is *very high* AND upstream discharge is *low* THEN downstream discharge is *medium–high*.

23. IF upstream rainfall is *very high* AND upstream discharge is *medium* THEN downstream discharge is *high*.

24. IF upstream rainfall is *very high* AND upstream discharge is *high* THEN downstream discharge is *very high*.

25. IF upstream rainfall is *very high* AND upstream discharge is *very high* THEN downstream discharge is *extremely high*.

The relative performance of the three cases with 9, 15, and 25 rules, and the accompanying partitioning as measured by four indicators, is shown in Table 13.5. It is clear that as the complexity of the fuzzy logic system increases, the agreement with the actual values becomes better and better. A typical time series plot of predicted and actual downstream discharges and a scatter plot are shown in Figures 13.30 and 13.31.

Example 13.2

A similar example using the TSK fuzzy inference system using hydrological data from the Fu River basin (26°30′–28°37′ N; 115°30′–119°10′ E) in China[8] is presented next. The Fu River has a catchment area of 16,493 km^2 and a length of 348 km. Daily discharge measurements made at the upstream station Liaojiawan (27°58′ N; 116°24′ E) across Fu River, and at Loujiachun (27°59′ N; 116°18′ E) across Linshui River, a tributary of Fu River, were used as inputs while the corresponding daily discharge measurements made at the downstream station Lijiadu (28°13′ N; 116°10′ E) across Fu River were used as outputs. The period of record for model calibration was for 1960–1975 while that for validation was for 1977–1979.

For a TSK fuzzy system with m inputs each of which has n partitions, the total number of fuzzy rules is n^m. Therefore, the number of rules increases with increasing number of inputs and partitions. The number of inputs in this case is two, thus requiring two cases with different partitions. With three partitions (low, medium, and high), there will be $3^2 = 9$ rules. They can be written as follows:

1. IF upstream discharge q_1 is *low* AND upstream discharge q_2 is *low* THEN downstream discharge $Q_1 = a_1 q_1 + b_1 q_2 + c_1$.

2. IF upstream discharge q_1 is *low* AND upstream discharge q_2 is *medium* THEN downstream discharge $Q_2 = a_2 q_1 + b_2 q_2 + c_2$.

3. IF upstream discharge q_1 is *low* AND upstream discharge q_2 is *high* THEN downstream discharge $Q_3 = a_3 q_1 + b_3 q_2 + c_3$.

4. IF upstream discharge q_1 is *medium* AND upstream discharge q_2 is *low* THEN downstream discharge $Q_4 = a_4 q_1 + b_4 q_2 + c_4$.

5. IF upstream discharge q_1 is *medium* AND upstream discharge q_2 is *medium* THEN downstream discharge $Q_5 = a_5 q_1 + b_5 q_2 + c_5$.

6. IF upstream discharge q_1 is *medium* AND upstream discharge q_2 is *high* THEN downstream discharge $Q_6 = a_6 q_1 + b_6 q_2 + c_6$.

7. IF upstream discharge q_1 is *high* AND upstream discharge q_2 is *low* THEN downstream discharge $Q_7 = a_7 q_1 + b_7 q_2 + c_7$.

[8] This study was carried out in partial fulfilment of the requirement for the Master's Degree in Disaster Management by Bing Zhu (2012) under the supervision of the author.

Table 13.5 Comparison of the performance indicator values for 9, 15, and 25 rules for the year 2008

Performance indicator	9 Rules		15 Rules		25 Rules	
	1-day lead time	*2-day lead time*	*1-day lead time*	*2-day lead time*	*1-day lead time*	*2-day lead time*
Mean absolute error (MAE)	35.344	34.005	27.010	29.452	24.326	25.600
Relative root mean square error (RRMSE)	0.688	0.656	0.616	0.572	0.544	0.564
Coefficient of efficiency (EF)	0.566	0.606	0.652	0.700	0.729	0.708
Coefficient of determination (CD)	0.808	0.806	0.904	0.903	1.211	1.122

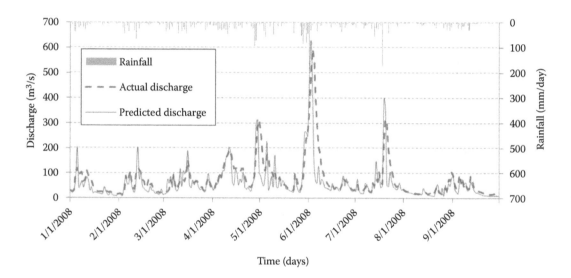

Figure 13.30 Predicted and observed discharges at the downstream station for Gin River (2-day lead time).

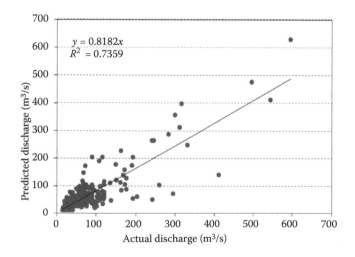

Figure 13.31 Scatter plot of predicted and actual discharges for the year 2008 (2-day lead time).

8. IF upstream discharge q_1 is *high* AND upstream discharge q_2 is *medium* THEN downstream discharge $Q_8 = a_8q_1 + b_8q_2 + c_8$.
9. IF upstream discharge q_1 is *high* AND upstream discharge q_2 is *high* THEN downstream discharge $Q_9 = a_9q_1 + b_9q_2 + c_9$.

In this rule set, q_1 and q_2 are the observed upstream discharge value as the two inputs; Q is the downstream discharge value as output; a_i, b_i, c_i are the coefficients of the piecewise linear regression relationships of the consequent connecting the inputs to the output. Membership functions for Loujiachun (Figure 13.32) upstream discharge (q_1)

$$w_L = \begin{cases} 1 & \text{for } 0 < q_1 < 74 \\ \dfrac{225}{188} - \dfrac{q_1}{376} & \text{for } 74 < q_1 < 450 \end{cases} \tag{13.62a}$$

$$w_M = \begin{cases} \dfrac{q_1}{376} - \dfrac{37}{188} & \text{for } 74 < q_1 < 450 \\ \dfrac{718}{493} - \dfrac{q_1}{986} & \text{for } 450 < q_1 < 1436 \end{cases} \tag{13.62b}$$

$$w_H = \begin{cases} \dfrac{q_1}{986} - \dfrac{225}{493} & \text{for } 450 < q_1 < 1436 \\ 1 & \text{for } 1436 < q_1 \end{cases} \tag{13.62c}$$

Membership functions for Liaojiawan (Figure 13.33) upstream discharge (q_2) Membership function:

$$w_L = \begin{cases} 1 & \text{for } 0 < q_2 < 131 \\ \dfrac{5}{4} - \dfrac{q_2}{524} & \text{for } 131 < q_2 < 655 \end{cases} \tag{13.63a}$$

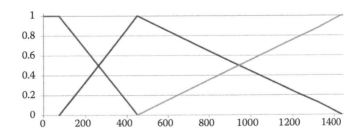

Figure 13.32 Membership functions for Loujiachun (q_1).

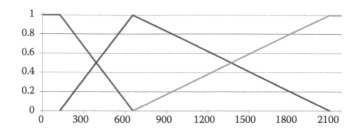

Figure 13.33 Membership functions for Liaojiawan (q_2).

$$w_M = \begin{cases} \dfrac{q_2}{524} - \dfrac{1}{4} & \text{for } 131 < q_2 < 655 \\[2ex] \dfrac{2083}{1428} - \dfrac{q_2}{1428} & \text{for } 655 < q_2 < 2083 \end{cases}$$ (13.63b)

$$w_H = \begin{cases} \dfrac{q_2}{1428} - \dfrac{655}{1428} & \text{for } 655 < q_2 < 2083 \\[2ex] 1 & \text{for } 2083 < q_2 \end{cases}$$ (13.63c)

In the TSK model, there is no defuzzification. Crisp outputs are given by the consequent relationship assumed, which in this case is linear. The model output can be expressed as

$$Q_{\text{com}} = \sum_{i=1}^{N} \frac{w_i}{\sum\limits_{i=1}^{N} w_i} (a_i q_1 + b_i q_2 + c_i)$$ (13.64a)

which may be expanded as

$$Q_{\text{com}} = \frac{w_1}{\sum\limits_{i=1}^{N} w_i} (a_1 q_1 + b_1 q_2 + c_1) + \frac{w_2}{\sum\limits_{i=1}^{N} w_i} (a_2 q_1 + b_2 q_2 + c_2)$$

$$+ \ldots + \frac{w_9}{\sum\limits_{i=1}^{N} w_i} (a_9 q_1 + b_9 q_2 + c_9)$$ (13.64b)

where the a_i, b_i, c_i have to be estimated by the least squares method by comparing the computed output with the expected output and the w_i's represent the degrees of membership for each rule. Both min ('AND') and product ('OR') operators have been used.

After the estimation of the a_i, b_i, c_i values by the least squares method, the rule set can be written as

1. IF upstream discharge q_1 is *low* AND upstream discharge q_2 is *low* THEN downstream discharge $Q_1 = 1.38q_1 + 0.47q_2 + 7.77$.
2. IF upstream discharge q_1 is *low* AND upstream discharge q_2 is *medium* THEN downstream discharge $Q_2 = 2.34q_1 + 0.70q_2 - 137.23$.
3. IF upstream discharge q_1 is *low* AND upstream discharge q_2 is *high* THEN downstream discharge $Q_3 = 7.20q_1 - 0.70q_2 - 715.27$.
4. IF upstream discharge q_1 is *medium* AND upstream discharge q_2 is *low* THEN downstream discharge $Q_4 = -0.40q_1 + 2.35q_2 + 318.07$.
5. IF upstream discharge q_1 is *medium* AND upstream discharge q_2 is *medium* THEN downstream discharge $Q_5 = 1.02q_1 + 1.26q_2 + 117.67$.
6. IF upstream discharge q_1 is *medium* AND upstream discharge q_2 is *high* THEN downstream discharge $Q_6 = 2.20q_1 + 1.02q_2 + 1021.6$.
7. IF upstream discharge q_1 is *high* AND upstream discharge q_2 is *low* THEN downstream discharge $Q_7 = 0.92q_1 + 0.76q_2 + 356.45$.
8. IF upstream discharge q_1 is *high* AND upstream discharge q_2 is *medium* THEN downstream discharge $Q_8 = 1.05q_1 + 0.83q_2 - 1061.31$.
9. IF upstream discharge q_1 is *high* AND upstream discharge q_2 is *high* THEN downstream discharge $Q_9 = 1.18q_1 + 0.88q_2 + 196.34$.

For the product operator, 'OR', a different set of parameters is obtained. They are as follows:

1. IF upstream discharge q_1 is *low* AND upstream discharge q_2 is *low* THEN downstream discharge $Q_1 = 1.42q_1 + 0.46q_2 + 6.82$.
2. IF upstream discharge q_1 is *low* AND upstream discharge q_2 is *medium* THEN downstream discharge $Q_2 = 2.06q_1 + 0.86q_2 - 41.09$.
3. IF upstream discharge q_1 is *low* AND upstream discharge q_2 is *high* THEN downstream discharge $Q_3 = 17.68q_1 - 3.29q_2 + 3473.61$.
4. IF upstream discharge q_1 is *medium* AND upstream discharge q_2 is *low* THEN downstream discharge $Q_4 = -0.43q_1 + 2.47q_2 + 286.25$.
5. IF upstream discharge q_1 is *medium* AND upstream discharge q_2 is *medium* THEN downstream discharge $Q_5 = 1.14q_1 + 1.16q_2 + 66.4$.
6. IF upstream discharge q_1 is *medium* AND upstream discharge q_2 is *high* THEN downstream discharge $Q_6 = 3.02q_1 + 0.72q_2 + 1109.63$.
7. IF upstream discharge q_1 is *high* AND upstream discharge q_2 is *low* THEN downstream discharge $Q_7 = 2.81q_1 - 4.07q_2 + 1497.64$.
8. IF upstream discharge q_1 is *high* AND upstream discharge q_2 is *medium* THEN downstream discharge $Q_8 = 1.77q_1 + 0.01q_2 + 940.28$.
9. IF upstream discharge q_1 is *high* AND upstream discharge q_2 is *high* THEN downstream discharge $Q_9 = 1.22q_1 + 0.84q_2 + 149.33$.

Table 13.6 Performance indicators with three clustering centres

Performance indicator	Minimum		Product		
	Calibration	Verification	Calibration	Verification	Optimum value
MAE	84.176	84.495	84.882	84.793	0
RMSE	172.711	135.945	176.292	136.288	0
RRMSE	0.436	0.4222	0.445	0.423	0
EF	0.931	0.9468	0.928	0.947	1.0
CD	0.964	0.9602	0.957	0.968	1.0

Table 13.7 Performance indicators with five clustering centres

Performance indicator	Min and max		Product and probability		
	Calibration	Verification	Calibration	Verification	Optimum value
MAE	75.113	74.579	74.961	73.664	0
RMSE	148.517	133.085	148.028	128.143	0
RRMSE	0.375	0.413	0.374	0.398	0
EF	0.949	0.949	0.949	0.953	1.0
CD	1.054	1.032	1.053	1.031	1.0

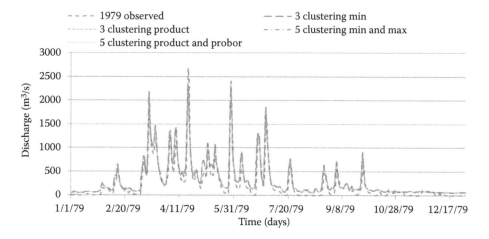

Figure 13.34 Time series plot of observed and predicted discharges for the year 1979.

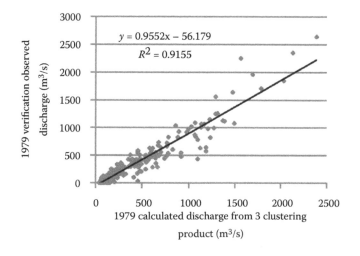

Figure 13.35 Scatter plot of observed and predicted discharges for the year 1979 using three cluster centres.

In this example, the partitioning was done by clustering the data. The peak points in the triangular membership functions correspond to the centres of the clusters. It can be seen that the accuracy of prediction increases as the number of clusters increase (Tables 13.6 and 13.7; Figures 13.34 and 13.35).

Neural networks have been used in the design of membership functions of fuzzy systems. The normal practice of designing membership functions in fuzzy systems is by using expert knowledge and/or *a priori* information about the system to design and tune the membership functions. This approach is time consuming. By using neural network learning capabilities, the process can be automated, thereby reducing time and cost and improving performance. The idea of using neural networks to design the membership functions and the use of the gradient descent method for tuning the parameters that define the shape of the membership functions is equivalent to learning in a feed-forward neural network (Takagi and Hayashi, 1991). Many Japanese and Korean companies use fuzzy neural systems in the operation of equipment such as photocopying machines (e.g., Sanyo, Ricoh) (Morita et al., 1992), electric fans with remote controllers (Sanyo, 1991; Nikkei Electronics, 1991), washing machines (e.g., Hitachi, Toshiba, Sanyo) (Narita et al., 1991), and many other household electrical appliances.

13.9 CONCLUDING REMARKS

In this chapter, an attempt has been made to introduce the basic concepts of fuzzy logic systems in the context of systems approach of modelling. Although by no means comprehensive, it is hoped that the material presented would be helpful for using the fuzzy logic approach for hydrological and environmental systems modelling.

REFERENCES

Abonyi, J. (2002): Modified Gath–Geva fuzzy clustering for identification of Takagi–Sugeno fuzzy models, systems, man and cybernetics, Part B: Cybernetics. *IEEE Transactions*, 32(5), 612–621.

Alata, M., Molhim, M. and Ramini, A. (2008): Optimizing of fuzzy c-means clustering algorithm using GA. *World Academy of Science, Engineering and Technology*, 39, 224–229.

Alvisi, S., Mascellani, G., Franchini, M. and Bárdossy, A. (2005): *Water Level Forecasting Through Fuzzy Logic and Artificial Neural Network Approaches.* Published in Hydrology and Earth System Science Discussions, EGU.

Amarasekara, J.D. (2012): Development of a flood forecasting model for the Gin River and Kelani River basins in Sri Lanka using fuzzy logic approach. Thesis submitted in partial fulfilment for the Master's degree in disaster management, National Graduate Institute for Policy Studies (GRIPS) and International Centre for Water Hazard and Risk Management (ICHARM) under the auspices of UNESCO (unpublished).

Bárdossy, A. and Disse, M. (1993): Fuzzy rule-base models for infiltration. *Water Resources Research*, 29(2), 373–382.

Bárdossy, A. and Duckstein, L. (1995): *Fuzzy Rule-Based Modelling with Applications to Geophysical, Biological and Engineering Systems.* CRC Press, Boca Raton, FL.

Berenji, H.R. (1992): A reinforcement learning-based architecture for fuzzy logic control. *International Journal of Approximate Reasoning*, 6, 267–292.

Berenji, H.R. and Khedkar, P. (1992): Learning and tuning fuzzy logic controllers through reinforcements. *IEEE Transactions on Neural Networks*, 3, 724–740.

Bezdek, J.C. (1981): *Pattern Recognition with Fuzzy Objective Function Algorithms.* Plenum Press, New York.

Buckley, J.J. and Hayashi, Y. (1994a): Fuzzy neural networks: A survey. *Fuzzy Sets and Systems*, 66(1), 1–13.

Buckley, J.J. and Hayashi, Y. (1994b): Fuzzy neural networks. In: Zadeh, L.A. and Yager, R.R. (Eds.), *Fuzzy Sets, Neural Networks and Soft Computing*. Van Nostrand Reinhold, New York, pp. 233–249.

Dunn, J.C. (1973): A fuzzy relative of the ISODATA process and its use in detecting compact well-separated clusters. *Journal of Cybernetics*, 3, 32–57.

Eklund, P., Fogström, M. and Forsström, J. (1992): A generic neuro-fuzzy tool for developing medical decision support. In: Eklund, P. (Ed.), *Proceedings EPP92, International Seminar on Fuzzy Control through Neural Interpretations of Fuzzy Set*s Åbo Akademis tryckeri, Åbo, pp. 1–27.

Eklund, P., Virtanen, H. and Riisssanen, T. (1991): On the fuzzy logic nature of neural nets. *Proceedings of Neuro-Nimes*, pp. 293–300.

Fullér, R. (1995): Neural Fuzzy Systems, Åbo Akademis tryckeri, Åbo, 249 pp. http://users.abo.fi/rfuller/nfs.html.

Gath, I. and Geva, A.B. (1989): Unsupervised optimal fuzzy clustering. *IEEE Transactions on Pattern Analysis and Machine Intelligence*, 7, 773–781.

Gustafson, E.E. and Kessel, W.C. (1979): Fuzzy clustering with a fuzzy covariance matrix. *Proc. IEEE CDC*, San Diego, CA, 1979, pp. 761–766.

Höppner, F., Klawonn, R., Kruse, T. and Runkler, T. (1999): *Fuzzy Cluster Analysis*. John Wiley, Chichester, UK.

Hundecha, Y., Bárdossy, A. and Theisen, H. (2001): Development of a fuzzy logic-based rainfall–runoff model. *Hydrological Sciences Journal*, 46(3), 363–376.

Ishikawa, M. (1996): Structural learning with forgetting. *Neural Networks*, 9, 501–521.

Jang, J.S.R. (1993): ANFIS: Adaptive-network-based fuzzy inference systems. *IEEE Transactions on Systems, Man and Cybernetics*, 23(1993), 665–685.

Jang, J.S.R., Sun, C.T. and Mizutani, E. (1997): Neuro-fuzzy and soft computing. In: *MATLAB Curriculum Series*. Prentice Hall, Upper Saddle River, NJ. ISBN No. 0-13-261066-3.

Kasabov, N. (1998): ECOS: A framework for evolving connectionist systems and the ECO learning paradigm. *Proc. of ICONIP'98*, Kitakyushu, Japan, IOS Press, vol. 3, pp. 1232–1235.

Kasabov, N., Kim, J.S., Watts, M. and Gray, A. (1997): FuNN/2 – A fuzzy neural network architecture for adaptive learning and knowledge acquisition, information sciences applications.

Kwan, H.K. and Cai, Y. (1994): A fuzzy neural network and its application to pattern recognition. *IEEE Transactions on Fuzzy Systems*, 2(3), 185–193.

Larsen, H.L. and Yager, R.R. (2000): A framework for fuzzy recognition technology, systems, man and cybernetics, Part C: Applications and reviews. *IEEE Transactions*, 30(1), 65–76.

Mamdani, E.H. and Assilian, S. (1975): An experiment in linguistic synthesis with a fuzzy logic controller. *International Journal of Man Machine Studies*, 7(1), 1–13.

MathWorks Inc.: MATLAB User manual for MATLAB Fuzzy Logic Tool Box.

Morita, T., Kanaya, M. and Inagaki, T. (1992): Photo-copier image density control using neural network and fuzzy theory. *Proceedings of the Second International Workshop on Industrial Fuzzy Control and Intelligent Systems*, pp. 10–16.

Narita, R., Tatsumi, H. and Kanou, H. (1991): Application of neural networks to household applications. *Toshiba Review*, 46, 935–938 (in Japanese).

Nauck, D. (1994): A fuzzy perceptron as a generic model for neuro-fuzzy approaches. In: *Proceedings 2nd German GI-Workshop Fuzzy-Systems '94*, München.

Nauck, D. and Kruse, R. (1996): Neuro-fuzzy classification with NEFCLASS. In: Kleinschmidt, P., Bachem, A., Derigs, U., Fischer, D., Leopold-Wildburger, U. and Möhring, R. (Eds.), *Operations Research Proceedings*, Berlin, 1995, pp. 294–299.

Nauck, D. and Kruse, R. (1997): Function approximation by NEFPROX. *Proc. Second European Workshop on Fuzzy Decision Analysis and Neural Networks for Management, Planning, and Optimization (EFDAN'97)*, Dortmund, pp. 160–169.

Nayak, P.C., Sudheer, K.P., Rangan, D.M. and Ramasastri, K.S. (2004): A neuro-fuzzy computing technique for modelling hydrological time series. *Journal of Hydrology*, 291, 52–66.

Nayak, P.C., Sudheer, K.P., Rangan, D.M. and Ramasastri, K.S. (2005): Short-term flood forecasting with a neurofuzzy model. *Water Resources Research*, 41, W04004.

Negnevitsky, M. (2005): *Artificial Intelligence: A Guide to Intelligent Systems*, Revision/New Edition. Addison-Wesley, Harlow, England, 407 pp. ISBN 0-321-20466-2.

Nikkei Electronics (1991): New trend in consumer electronics: Combining neural networks and fuzzy logic. *Nikkei Electronics*, 528, 165–169 (in Japanese).

Pedrycz, W. (2005): *Knowledge-Based Clustering: From Data to Information Granules*. John Wiley & Sons, Chichester, UK, 336 pp.

Sanyo (1991): Electric fan series in 1991, Sanyo News Release (March 14, 1991). Sanyo (in Japanese).

See, L. and Openshaw, S. (1999): Applying soft computing approaches to river level forecasting. *Hydrological Sciences Journal*, 44(5), 763–779.

See, L. and Openshaw, S. (2000): A hybrid multi-model approach to river level forecasting. *Hydrological Sciences Journal*, 45(4), 523–535.

Sen, Z. (1998): Fuzzy algorithm for estimation of solar irrigation from sunshine duration. *Solar Energy*, 63, 39–49.

Sen, Z. (2009): *Fuzzy Logic and Hydrological Modelling*. CRC Press, Boca Raton, FL.

Takagi, H. and Hayashi, I. (1991): NN-driven fuzzy reasoning. *International Journal of Approximate Reasoning*, 3, 191–212.

Takagi, T. and Sugeno, M. (1974): Application of fuzzy algorithms for control of simple dynamic plant. *IEE Proceedings*, 12, 1585–1588.

Takagi, T. and Sugeno, M. (1985): Fuzzy identification of systems and its applications to modelling and control. *IEEE Transactions on Systems, Man and Cybernetics*, 15, 116–132.

Tareghian, R. and Kashefipour, S.M. (2007): Application of fuzzy systems and artificial neural networks for flood forecasting. *Journal of Applied Sciences*, 7, 3451–3459.

Tsukamoto, Y. (1979): *An Approach to Fuzzy Reasoning Method, Advances in Fuzzy Set Theory and Applications*. North Holland, Amsterdam, pp. 137–149.

Tsukamoto, Y. (1994): Some issues of reasoning in fuzzy control: Principle, practice and perspective. *International Conference on Tools with Artificial Intelligence*, pp. 192–196.

Umano, M. and Ezawa, Y. (1991): Execution of approximate reasoning by neural network. *Proceedings of FAN Symposium*, pp. 267–273 (in Japanese).

Xiong, L., Shamseldin, A.Y. and O'Connor, K.M. (2001): A non-linear combination of the forecasts of rainfall-runoff models by the first-order Takagi–Sugeno fuzzy system. *Journal of Hydrology*, 245, 196–217.

Zadeh, L.A. (1965): Fuzzy sets. *Information and Control*, 8, 338–353.

Zhu, B. (2012): Hydrological forecasting based on TSK fuzzy logic system in Fu River basin. Thesis submitted in partial fulfilment for the Master's degree in disaster management, National Graduate Institute for Policy Studies (GRIPS) and International Centre for Water Hazard and Risk Management (ICHARM) under the auspices of UNESCO (unpublished).

Chapter 14

Genetic algorithms (GAs) and genetic programming (GP)

Genetic algorithms are search algorithms based on the mechanics of natural selection and natural genetics. They combine survival of the fittest among string structures with a structured yet randomized information exchange to form a search algorithm.

David E. Goldberg
Genetic Algorithms in Search, Optimization and Machine Learning, 1989

14.1 INTRODUCTION

Since the original works by Holland (1975, 1993), genetic algorithms (GAs), being a form of evolutionary computing, have found applications in many areas of science and engineering. They are inspired by Darwin's theory of evolution and imitate nature's selection of fitter or stronger genes according to a mechanism dictating the survival of the fittest. This adaptive nature lends GAs to be applied to problems that require progressive modifications such as in parameter optimization. The main components of GA consists of

- Genetic representation for potential solutions
- A way to create an initial population
- An evaluation function that plays the role of the environment rating solutions in terms of their fitness
- A selection method to choose two reproductive solutions
- Genetic operators that alter the composition of the offsprings
- Estimation of GA parameters (population size, probability of applying genetic operators, etc.)

GAs operate on a population (a set of possible solutions) on the basis of biological genetics and natural selection, and are designed to produce successive populations having an increasing number of individuals (solutions) with desirable characteristics. They are designed in such a way that best individuals proliferate over generations. The objective function corresponding to each individual determines its 'fitness'. The selection procedure possesses a guided randomness rather than being entirely random, and thus leads the populations of the subsequent generations increasingly towards the optimum. The 'best so far' individual or the best set of parameters that correspond to the minimum/maximum objective function is recorded over the generations.

GAs operate on a coding of the parameters, rather than on the parameters themselves. Each parameter is encoded into a string of finite length made up of binary numbers. These strings are then concatenated to form one string that is regarded as one individual (structure). Several such individuals constitute a population.

All living organisms consist of cells that have the same set of chromosomes. They are strings of DNA that identify the organism. A chromosome consists of genes, which are blocks of DNA. A complete set of genetic material (all chromosomes) is called a genome. The fitness of an organism is measured by the success of the organism in its life.

In GA terminology, a set of solutions (represented by chromosomes) is called a population. Successive populations are generated in such a way that the new population is better than the old one. The new solutions, selected according to their fitness, are referred to as the 'offsprings'. This procedure is repeated until some stopping criterion is satisfied.

In solving a particular problem, the first attempt is to look for some particular solution from a set of possible solutions. The space containing all feasible solutions is known as the search space. Each point (multidimensional) in the search space is a feasible solution and the aim is to look for the best solution from the search space that may not necessarily be the optimal. Methods of finding the optimal solution include 'hill climbing', 'simulated annealing', and 'genetic algorithms'.

GAs consist of the following steps:

1. Generation of a random population of 'n' chromosomes (feasible solutions for the problem).
2. Evaluation of the fitness function $f(x)$ of each chromosome x in the population.
3. Generation of a new population by repeating the following:
   - Select two parent chromosomes from a population according to their fitness.
   - With a crossover probability, cross over the parents to form a new offspring. If no crossover is performed, the offspring will be an exact copy of the parents.
   - With a mutation probability, mutate new offspring at each locus (position in chromosome).
   - Place a new offspring in a new population.
4. Use new generation for repeating the algorithm.
5. If the stopping criterion is satisfied, stop.
6. Go to step 2.

The questions that need clarification include how to create chromosomes, what type of encoding to use, and how to select parents for crossover. The concept of elitism is used to ensure that at least one best solution is copied without changes to a new population so that the best solution can survive to the end of the run.

14.2 CODING

The equivalent of chromosomes in an optimization problem are the parameters of the model or function. Any possible set of parameters is represented by a set of genetic structures that collectively correspond to a point (multidimensional) in the search space. Each structure can be represented by a string of binary numbers. A string of length L has L bit positions each of which is occupied by a 0 or 1. For an 'n' parameter optimization problem, there will be 'n' binary structures, one for each parameter. For example, let

$$S_i^L = \{B_L, B_{L-1}, ..., B_3, B_2, B_1\}_i \text{ give the bit value corresponding to bit positions}$$

$$\{L, L-1, ..., 3, 2, 1\}$$

where S_i^L is the i-th structure of the string of length L and B_k is a binary number at the k-th bit position. The string attains its maximum value of $2^L - 1$ when all the bits are occupied by 1's and its minimum value is zero. The length of the string depends on the required precision. If the i-th parameter x_i that corresponds to the i-th structure is known to be in the range $\{a_i \leq x_i \leq b_i\}$, then the precision of the representation, or the smallest possible division of the parameter range is $\frac{(a_i - b_i)}{(2^L - 1)}$. As can be seen, the precision increases with increasing string length. The mapping from the binary string S_i^L to the real value x_i is accomplished first by converting the binary string S_i^L from base 2 to base 10 as

$$\left[S_i^L = \{B_L, B_{L-1}, ..., B_3, B_2, B_1\}_i \right]_2 = \left[\sum_{k=1}^{L} 2^{k-1} B_k \right]_{10} = x_i \tag{14.1}$$

and then finding the corresponding real number as

$$x_i = b_i + x_i \frac{(a_i - b_i)}{(2^L - 1)} \tag{14.2}$$

The initial population is randomly generated by L binary numbers to fill up the positions in one structure and repeating it for all n structures. At this stage, some prior information about the feasible range of values of the parameters could help accelerate the search process.

The selection process for reproduction is based on an objective function that may be a measure of error. The structures can be arranged according to the objective function and those with better (higher or lower depending on the objective function) values have a higher chance of propagating the desired characteristics to the next generation. The basis for this is the schema theory (Holland, 1993), which asserts that structures having a fitness above that of the average for the whole population tend to occur more frequently in the next generation (Holland, 1993). More details of the schema theory can be found in Goldberg (1989).

The roulette wheel method in which each set of structures in the ranking is assigned a pie-shaped slice in proportion to its fitness, and selecting the structures at which the pointer stops after an arbitrary rotation, is a widely adopted method.

14.3 GENETIC OPERATORS

The basic genetic operators are copying, crossover, masking, and mutation. These operations are illustrated in Figure 14.1 for an example in which the string length is 10. For clarity, the binary numbers in the two structures are shown in upper and lower cases.

Copying is the complete replacement of all the bit positions of the resulting structure by those of the original. Single-point crossover refers to the exchange of corresponding bit values beyond a certain point on the string. The crossover point is randomly decided. In Figure 14.1, the crossover points are indicated by vertical lines. Multiple-point crossovers refer to the cases when there are more than one crossover points, which are again decided randomly. For masking, a masking string that has the same length as the original structures has to be randomly generated. The masking string determines which of the original structures dominate a certain bit position of the resulting structure. If the bit position of the masking string is 1, then the corresponding bit positions of the two resulting structures will be those of the first and second original structures, respectively. If the bit position of masking string is 0, then they will be those of the second and first, respectively. Mutation corresponds to the rare perturbation of a population's biological characteristics to prevent evolutionary dead ends. Mutation means that the elements of DNA are a bit changed. Such

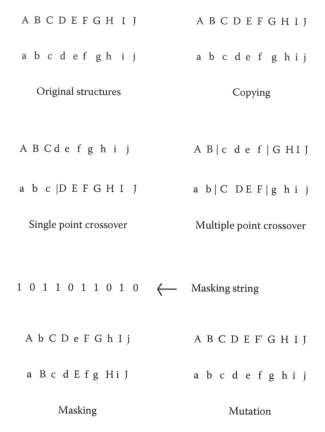

A B C D E F G H I J   A B C D E F G H I J

a b c d e f g h i j   a b c d e f g h i j

Original structures   Copying

A B C d e f g h i j   A B | c d e f | G H I J

a b c | D E F G H I J   a b | C D E F | g h i j

Single point crossover   Multiple point crossover

1 0 1 1 0 1 1 0 1 0   ⟵   Masking string

A b C D e F G h I j   A B C D E F' G H I J

a B c d E f g H i J   a b c d e f g h i j

Masking   Mutation

Figure 14.1 Genetic operators.

changes are mainly caused by errors in copying genes from parents. In GAs, it is usually adopted to prevent stagnation of a suboptimal population. This operator is used less frequently compared with the others. In Figure 14.1 above, one of the bits (*F*) becomes *F'* after mutation. It is to be noted that 1 becomes 0 and 0 becomes 1 when mutation takes place.

14.4 PARAMETERS OF GA

The parameters of GAs (population size, probability of applying genetic operators, probability distributions to be applied during ranking, etc.) are problem dependent as with many other optimization techniques. They are usually not known *a priori* and very often a trial-and-error procedure is needed to determine them.

14.5 GENETIC PROGRAMMING (GP)

In the recent past, genetic programming (GP), an evolutionary algorithm (EA)-based model, has been used to emulate the rainfall–runoff process (Liong et al., 2002; Whigham and Crapper 2001; Savic et al., 1999; Jayawardena et al., 2005, 2006) as a viable alternative to traditional rainfall–runoff models. GP has the advantage of providing inherent functional input–output relationships as compared with traditional black box models, and therefore can offer some possible interpretations of the underlying processes.

GP is an automatic programming technique for evolving computer programs to solve, or approximately solve, problems (Koza, 1992). In engineering applications, GP is frequently applied to model structure identification problems. In such applications, GP is used to infer the underlying structure of either a natural or experimental process in order to model the process numerically.

GP is a member of the EA family. EAs are based on Darwin's natural selection theory of evolution where a population is progressively improved by selectively discarding the not-so-fit populations and breeding new offsprings from better populations. EAs work by defining a goal in the form of a quality criterion and then use this goal to measure and compare solution candidates in a stepwise refinement of a set of data structures and return an optimal or near-optimal solution after a number of generations. Evolutionary strategies (ES), GAs, and evolutionary programs (EP) are three early variations of evolutionary algorithms, whereas GP is perhaps the most recent variant of EAs. These techniques have become extremely popular owing to their success at searching complex non-linear spaces and their robustness in practical applications.

The basic search strategy in GP is a genetic algorithm (Goldberg, 1989). GP differs from the traditional GA in that it typically operates on parse trees instead of bit strings. A parse tree is built up from a terminal set (the variables in the problem) and a function set (the basic operators used to form the function). Figure 14.2 shows such an example for the expression $\frac{1}{2a}\left[-b + \sqrt{b^2 - 4ac}\right]$ represented as a parse tree. The functions are mathematical or logical operators, and terminals are constants and variables. As in any standard evolutionary algorithm, in order to optimize its fitness value during the evolving process, the trees in GP are dynamically modified by genetic operators.

The 'tree size' of this expression is 7, where tree size is the maximum 'node depth' of a tree and node depth is the minimum number of nodes that must be traversed to get from the 'root node' of the tree (Figure 14.2) to the selected node.

The problem of finding a function in symbolic form that fits a given set of data is called symbolic regression. Genetic symbolic regression is a special application of GP in the field of symbolic regression. The aim of symbolic regression is to determine a functional relationship between input and output data sets. Symbolic regression is error-driven evolution and it may be linear, quadratic, or higher-order polynomial. Details of GP can be found in Liong et al. (2002). The function set may contain operators such as +, −, ×, /, *, power(x,y), exp(x), etc. The ends of a parse tree are called 'Terminals', which are actually the input variables. The nodes are functions taken from the 'Function Set'.

As a GA, GP proceeds by initially generating a population of random parse trees, calculates their fitness – a measure of how well they solve the given problem, and subsequently

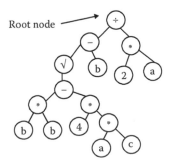

Figure 14.2 GP parse tree for the quadratic equation $\frac{1}{2a}\left[-b + \sqrt{b^2 - 4ac}\right]$.

selects the better parse trees for reproduction and variation to form a new population. Crossover in GP is similar to that in GA. Genetic material is exchanged between two parents to create two offsprings. In GP, subtrees are exchanged between the two parents. This process of selection, reproduction, and variation is iterated until some stopping criterion is satisfied.

GP has the unique feature that it does not assume any functional form of the solution, and that it can optimize both the structure of the model and its parameters. Since GP evolves an equation relating the output and input variables, it has the advantage of providing inherent functional relationship explicitly over techniques such as ANNs. This gives the GP approach the automatic ability to select input variables that contribute beneficially to the model and to disregard those that do not.

For rainfall–runoff modelling, the mathematical relationship may be expressed as

$$Q_{t+\delta\Delta t} = f(R_t, R_{t-\Delta t}, \ldots, R_{t-\omega\Delta t}, Q_t, Q_{t-\Delta t}, \ldots, Q_{t-\omega\Delta t}) \tag{14.3}$$

where Q is the runoff, R is the rainfall intensity, δ (with $\delta = 1, 2, \ldots$) refers to how far into the future the runoff prediction is desired, ω (with $\omega = 1, 2, \ldots$) implies how far back the recorded data in the time series are affecting the runoff prediction, while Δt stands for time step.

14.6 APPLICATION AREAS

GAs have found applications in several areas of science and engineering such as pattern recognition, signal processing, machine learning, time series prediction, and optimization. In the hydrological field, the applications have been mainly for parameter optimization. Examples include their use for the calibration of Xinanjiang model (Wang et al., 1991), storm water management model (SWMM) (Liong et al., 1995), ARNO model (Franchini, 1996), moisture and contaminant transport model (Jayawardena et al., 1997), and tank model (Fernando, 1997).

14.7 CONCLUDING REMARKS

This short and last chapter gives a basic introduction to GAs and GP. GAs aim to successively generate superior solutions starting from an initial solution. The approach can be used in many areas of science and engineering. One of the main application areas is for parameter optimization in complex models. GP, which has many common features as in GAs, has the added advantage of providing inherent functional input–output relationships as compared with traditional black box models, and therefore can offer some possible interpretations of the underlying processes.

REFERENCES

Fernando, D.A.K. (1997): On the application of artificial neural networks and genetic algorithms in hydro-meteorological modelling. PhD thesis, The University of Hong Kong.
Franchini, M. (1996): Use of genetic algorithms combined with a local search method for automatic calibration of conceptual rainfall–runoff models. *Hydrological Sciences Journal*, 41(1), 21–39.

Goldberg, D. (1989): *Genetic Algorithms in Search, Optimization and Machine Learning*. Addison Wesley, Reading, MA.

Holland, J.H. (1975): *Adaptation in Natural and Artificial Systems*. The University of Michigan Press, Ann Arbor, MI.

Holland, J.H. (1993): *Adaptation in Natural and Artificial Systems*. MIT Press, Cambridge, MA.

Jayawardena, A.W., Fernando, D.A.K. and Dissanayake, P.B.G. (1997): Genetic algorithm approach of parameter optimisation for the moisture and contaminant transport problem in partially saturated porous media. *Proceedings of the 27th Congress of the International Association for Hydraulic Research*, vol. 1, ASCE, San Francisco, August 10–15, pp. 761–766.

Jayawardena, A.W., Muttil, N. and Fernando, T.M.K.G. (2005): Rainfall–runoff modelling using genetic programming. In Zerger, A. and Argent, R.M. (Eds.), *MODSIM 2005 International Congress on Modelling and Simulation: Advances and Applications for Management and Decision Making*, Modelling and Simulation Society of Australia and New Zealand, Melbourne, Australia, December 12–15, pp. 1841–1847. ISBN 0-9758400-2-9.

Jayawardena, A.W., Muttil, N. and Lee, J.H.W. (2006): Comparative analysis of a data-driven and GIS-based conceptual rainfall–runoff model. *Journal of Hydrologic Engineering, ASCE*, 11(1), 1–11.

Koza, J. (1992): *Genetic Programming: On the Programming of Computers by Natural Selection*. MIT Press, Cambridge, MA.

Liong, S.Y., Chan, W.T. and ShreeRam, J. (1995): Peak flow forecasting with genetic algorithms and SWMM. *Journal of Hydraulic Engineering, ASCE*, 121(8), 613–617.

Liong, S.Y., Gautam, T.R., Khu, S.T., Babovic, V. and Muttil, N. (2002): Genetic programming: A new paradigm in rainfall–runoff modelling. *Journal of American Water Resources Association*, 38(3), 705–718.

Savic, D.A., Walters, G.A. and Davidson, G.W. (1999): A genetic programming approach to rainfall–runoff modeling. *Water Resources Management*, 13, 219–231.

Wang, Q.J. (1991): The genetic algorithm and its application to calibrating conceptual rainfall–runoff models. *Water Resources Research*, 27(9), 2467–2471.

Whigham, P.A. and Crapper, P.F. (2001): Modelling rainfall–runoff relationships using genetic programming. *Mathematical and Computer Modelling*, 33, 707–721.

Index

Page numbers followed by f, t and n indicate figures, tables and notes, respectively.